T4-AJS-572

Economics and Ecological Risk Assessment

Applications to Watershed Management

Environmental and Ecological Risk Assessment

Series Editor

Michael C. Newman
College of William and Mary
Virginia Institute of Marine Science
Gloucester Point, Virginia

Published Titles

Coastal and Estuarine Risk Assessment
Edited by
Michael C. Newman, Morris H. Roberts, Jr., and Robert C. Hale

Risk Assessment with Time to Event Models
Edited by
Mark Crane, Michael C. Newman, Peter F. Chapman, and John Fenlon

Species Sensitivity Distributions in Ecotoxicology
Edited by
Leo Posthuma, Glenn W. Suter II, and Theo P. Traas

Regional Scale Ecological Risk Assessment: Using the Relative Risk Method
Edited by
Wayne G. Landis

Economics and Ecological Risk Assessment

Applications to Watershed Management

Edited by

Randall J. F. Bruins

Matthew T. Heberling

CRC PRESS

Boca Raton London New York Washington, D.C.

Library of Congress Cataloging-in-Publication Data

Economics and ecological risk assessment : applications to watershed management / edited by Randall J.F. Bruins and Matthew T. Heberling.
p. cm. — (Environmental and ecological risk assessment ; 5)
Includes bibliographical references and index.
ISBN 1-56670-639-4 (hardcover : alk. paper)
1. Watershed management. 2. Water quality management. 3. Ecological risk assessment. 4. Environmental economics. 5. Watershed management–Case studies. I. Bruins, Randall J.F. II. Heberling, Matthew T. III. Environmental and ecological risk assessment series ; 5. IV. Title. V. Series.

TC409.E29 2004
333.73—dc22 Cau 2004018535

Direct all inquiries to CRC Press, 2000 N.W. Corporate Blvd., Boca Raton, Florida 33431.

Visit the CRC Press Web site at www.crcpress.com

International Standard Book Number 1-56670-639-4
Library of Congress Card Number 2004018535
Printed in the United States of America 1 2 3 4 5 6 7 8 9 0
Printed on acid-free paper

Forewords

AN ECOLOGICAL RISK ASSESSOR'S VIEW OF ENVIRONMENTAL ECONOMICS[a]

From the perspective of ecological risk assessors and other applied ecologists, the increasing influence of economics in environmental decision making is problematical. That is because it is a challenge to their world view and because it creates a challenge to their scientific practice. However, it is an unavoidable challenge that also provides opportunities.

Competing World Views

Nearly all ecologists chose their career because they are attracted to nature and wish to do something to help preserve it. It is no accident that the largest environmental conservation organization in the United States, the Nature Conservancy, was founded by the largest professional organization of ecologists, the Ecological Society of America. Most ecological risk assessors are ecologists who saw an opportunity to help control threats to the natural environment from pollution and other anthropogenic agents. Hence, their view tends to be that, when an economic activity posing a significant hazard to a population or community has been identified, it should be precluded or remediated. Ecological risks are sufficient justification for action.

In contrast, economists tend to be attracted to the manmade world of getting and spending. They are still taught that free markets are agents of Adam Smith's invisible hand, which works for the common good through the pursuit of individual self-interest.[1] In that view, environmental regulations are likely to be bad because they distort markets, thereby crippling the invisible hand.

These generalizations are caricatures, but they capture an important point that has inhibited interactions between the fields. Putting a dollar value on nature makes most ecologists very uneasy if not hostile. They recognize that not every stream and meadow can be protected, but they would rather be subjective advocates for streams and meadows than balance the dollar value of those ecosystems against the dollar value of a development. They tend to believe that economics is an inappropriate approach to valuing ecosystems because ecosystems are not fungible. That is, the economic gain from development of a natural area is not qualitatively equivalent to the environmental loss. In contrast, neoclassical economists believe that their science is an appropriate way to assign value and balance competing values. Economists who accept the need for regulation in public policy use contingent valuation and other tools for monetizing nonmarket values, but unlike ecologists, they still view the dollar as a measure of all things.[1] Some individuals, such as Robert Costanza, have tried to bridge the gap by creating an ecological economics, but their efforts have not been well received by the economic community. Economists argue that the services of nature could not be worth $33 trillion, as estimated by Costanza et al. because the world's economies do not generate that much wealth.[2] An ecologist can agree by saying, of course, that we

[a] The views expressed in this foreword are those of the author and do not necessarily reflect the view or policies of the U.S. Environmental Protection Agency.

cannot afford to buy another Earth. This dichotomy of worldviews is reflected in the observation that, while cost-benefit analysis has increasingly influenced environmental regulatory decisions, ecological issues have had little influence on economics. Ecological risk assessors are frustrated that their issues are not more influential, but they are often reluctant to embrace economics as a means of increasing their influence.

Technical Challenges

Participation in cost-benefit analysis presents two technical challenges to ecological risk assessors. First, they must be able to predict, at least probabilistically, the levels of effects on particular attributes of real ecological entities. Second, they must be able to predict those effects for every attribute of every ecological entity that may have value.

Quantifying the ecological benefits of a regulatory action requires that any effects that would occur in the absence of regulation be specified. The specification must have sufficient clarity and detail to allow an economist to assign a dollar value. Often the effects have been vaguely defined (e.g., population impairment). In some cases, the units of effect have been deliberately designed to be indicative of a broad legal goal (e.g., the Index of Biotic Integrity) rather than a real property. Further, the effects metric is often a criterion value or some other point estimate of a safe level. Hence, the ecological risk assessment determines whether the safe level will be exceeded but not what will happen if it is. To provide useful and appropriate input to a cost-benefit analysis, ecological risk assessors must be able to estimate the effect (e.g., 40% reduction in fish species) corresponding to a prescribed exposure. This in turn requires pushing the limits of predictive ecology to generate exposure-response models in the absence of complete and reliable data on the properties of the agent, its effects on organisms in the receiving system, the interactions among organisms, and the influence of varying abiotic properties of the receiving system. This implies a need to devote much more time and resources to ecological risk assessment and to the toxicology, chemistry, ecology, and other sciences that support it. This need is illustrated by the many years and millions of dollars that have been devoted to trying to predict the effects of management actions on the abundance of salmon in the Columbia River.

If ecological risk assessment is to properly contribute to the balancing of costs and benefits, it must be complete. Currently, ecological risk assessments address a sensitive species, a representative species or ecosystem, or a small set of such species or ecosystems. It is assumed that if those few endpoints are protected, all important ecological attributes will be protected. That strategy will not suffice in a cost-benefit decision structure. Since the industry or developer can readily estimate all of the costs of a regulatory action, the ecological risk assessor should develop an equally complete estimate of the ecological benefits. However, this is an impossible task. Every species has value and different agents have different effects on them that change their value in different ways. The value of a species is different in different ecological contexts. For example, a grass that is normally a minor constituent of a prairie may be critically important to herbivores in a drought year. And species are not the only ecological entities with value. Communities, ecosystems, and regions have their own valued properties. These range from the prevention of floods by watershed vegetation to the aesthetic experience provided by a lichen community on a rock. Hence, an assessment of the benefits of dam removal to restore salmon in the Columbia River should quantify the innumerable benefits of free-flowing rivers, not just the market economic benefits of harvesting restored wild salmon stocks.

Lessons for Ecological Risk Assessors

Ecological risk assessors whose values are offended by placing a dollar value on whooping cranes or tupelo swamps can support environmental advocacy groups and fight for enforcement of the environmental laws that require environmental protection, regardless of cost. However, to be effective in their professional practice, they must recognize that the influence of cost-benefit analysis on environmental regulation is increasing. This need not be a bad thing. It is an opportunity to reveal to decision makers the extent of benefits derived from the environment. Many decision makers have little appreciation for the sensitivity of ecological receptors or their various benefits to society. However, the case studies presented in this book reveal that ecological risk assessors have not done a good job of meeting this demand. In many cases, they could not even estimate the risks to the few assessment endpoints selected for their analysis. The economic tools presented in the case studies may be weak in terms of their ability to unambiguously quantify the full range of environmental benefits, but their ability to quantify economic outcomes related to ecological changes exceeds the capacity of the ecological tools to predict those changes. We need to do better.

In sum, this book presents a challenge to ecological risk assessors that we cannot ignore. It may not be possible to quantify all the benefits of not destroying an ecosystem or not extinguishing a species. However, we must be able to at least quantify the conspicuous ecological benefits. For example, we should be able to quantify the expected effects on abundances of rare mussel species of riparian vegetation or strip mine restoration in the Clinch River. That will require more resources for ecological risk assessment and the supporting environmental sciences. Ecologists would do well to read this book, examine its case studies, and seriously consider their implications for ecological research and assessment.

Glenn W. Suter II
Science Advisor
National Center for Environmental Assessment
U.S. Environmental Protection Agency
Cincinnati, Ohio

REFERENCES

1. Fullerton, D., and Stavins, R., How economists see the environment, in *Economics of the Environment*, Stavins, R., Ed., W. W. Norton and Company, New York, 2000.
2. Bockstael, N.E., Freeman III, A.M., Kopp, R.J., Portney, P.R., and Smith, V.K., On measuring economic values for nature, *Environ. Sci. Technol.*, 34, 1384, 2000.

RISK AND ACTION — FIGURING OUT THE RIGHT THING TO DO

Economic valuation and ecological risk assessment both are proposed as ways of helping society figure out the right thing to do in particular kinds of problem situations involving ecological risk. The problem of right action has engaged philosophers for millennia with the result that we know some things, but not everything, about its solution. One thing we know is that action cannot be justified by the facts alone; it takes factual and normative premises. That is, to derive conclusions about action, we need some statements about how the world works and some statements about what we believe are good or right.

Ecological risk presents a difficult challenge to justified action; all too often the facts are incomplete and disputed, while the normative grounds for action are bitterly debated. Ecologists as go-it-alone advocates for the ecosystem encounter two serious problems: (1) their training provides little background in dealing with normative or value questions, and there is a continuing temptation to fall back on crude precautionary principles (ecological risks should be precluded or remediated, whatever the cost); and (2) advocates with only facts in their intellectual arsenals are tempted to escalate the fact-claims when the urgency of their agenda is challenged. Advocates who fall into these traps risk being seen as crying wolf and marginalizing themselves in a body politic that seeks to resolve most of its workaday conflicts by finding a balance.

Economists bring a value system to the table, along with some technical tools for implementing it. Furthermore the value system is one that considers the consequences of a proposed action, beneficial and adverse, and seeks to find a balance. There are reasons why collaboration among economists and ecological risk assessors holds promise, but there are also reasons why collaboration is not always welcomed.

Economic theories of value define individual good as the satisfaction of individual preferences. The resulting measures of value accurately reflect individual preferences, as intended, but have also the more controversial property that the preferences of the well-off count for more. Measures of benefit and cost for individuals are summed to calculate social-welfare changes in a modern and sophisticated attempt to implement Bentham's venerable idea of "the greatest good for the greatest number." Two points must be made about this value system. First, it is a fairly coherent instantiation of a legitimate philosophy — utilitarianism. Second, from utilitarians of other stripes and from adherents of competing moral philosophies, it encounters a suite of criticisms including the following:

- *Preferences may be ill-considered and ephemeral (as may costs, which are subjective, too).* Decisions about ecological risk may commit society to long-lived outcomes (at worst, some natural entity may be lost forever), yet some critics worry that human preferences might be whimsical, ill-considered, and shaped primarily by the past so that they travel poorly into the future. The sense that preferences are impermanent may lead to a quest for some more enduring foundation for value.
- *More generally, economists refuse any obligation to justify preferences.* The sense that preferences may be ill-informed, ill-considered, and unjustified may motivate arguments that conservation decisions should be based on natural science, without much consideration of what it is that humans value.

- *Value may involve more than preference.* Utilitarianism's greatest 19th-century critic, Immanuel Kant, insisted that aesthetic judgments, while subjective, involve much more than preference ("I know what I like"); such judgments can make a claim to interpersonal agreement because they can be based on good reasons and shared experiences. The Kantian aesthetic leads to arguments that certain natural entities have intrinsic value — a good of their own, independent of human caring.
- *The great moral questions should be addressed by principles, not values.* Kant argued that universal moral principles could be found to address the truly important decisions that human and social life requires, an argument that effectively relegates preferences to a set of issues that are morally less important. This perspective leads to a search for moral principles that imply human duties toward natural entities.
- *The great moral questions are best addressed in terms of rights that must be respected.* Rights-based theories of the good offer an array of positions, some of which have implications for the protection of nature. For example, libertarians might argue that people's rights to enjoy nature oblige other people not to befoul it.
- *It is not just about humans.* Standard economics assumes, along with many other strands of philosophy, that humans are the only entities whose concerns matter. However, this position has been attacked from many quarters. There are utilitarians who argue that animal welfare matters, Kantian aesthetes who argue that natural entities may have intrinsic value, and rights-based deontologists who argue that rights should be extended to natural entities.
- *Perhaps it is not about humans at all.* The basic program of deep ecology is to take any or all of the basic moral-philosophy approaches and expand the set of entities that matter — that is, entities whose welfare counts, that have rights, and that have a good of their own — independently of human concern or patronage.

One idea that many of these viewpoints have in common, and in contrast with the broad sweep of economic thinking, is that natural entities are not always (in the extreme, one might say never) fungible with money. Of course, money is not really the issue; it is, after all, just a convenient token of value in exchange. The real point of contention is substitutability; that is, whether an ideal utility-generating bundle of goods and services is (or should be) substitutable for natural entities. Nonfungibility arguments posit that trade-offs involving natural entities and a bundle of ordinary goods and services are inappropriate in general, or in particular circumstances that can be defined.

Having recognized a broad array of normative arguments about valuation, where do I come out? For most workaday decisions involving ecological risk, I believe the nonfungibility argument is overstated and unhelpful, trade-offs are appropriate, and decision processes should seek to balance benefits and costs. In this context, the economic approach to valuation does decently well at evaluating trade-offs, and the solutions it endorses have the great virtue of increasing aggregate welfare (the size of the game), even if it makes little serious attempt to resolve the distributional questions. Nevertheless, when unique environments are threatened, the higher stakes justify departures from business-as-usual decision rules. In such cases, precautionary rules such as the safe minimum standard of conservation should be taken seriously.

The argument to this point has been that economic valuation offers a sensible way of *thinking about* value in the context of workaday decisions about ecological risk. However, economic valuation claims also to offer empirical methods and procedures that serve to quantify benefits, costs, and trade-offs. To get this right, the economist would need detailed and reliable factual information about how the world works (at least, that part of the world relevant to a particular ecological risk) and how people value the changes in ecological goods, services, and amenities that would result from risk-mitigation strategies. A perfect account of economic value, in the context of ecological risk, is scarcely conceivable; the demands for valid information are too great. Economic valuation studies in complex problem contexts seem always to involve some data deficiencies, and often these data deficiencies motivate some compromises with the theoretical requirements for ideal valuation. This does not support any claim that ecological risk assessment does better, with respect to data. First, when economic valuation is built on a foundation of ecological risk assessment, many of the data problems that limit the validity of economic estimates occur in the natural science input, that is, in the assessment of the ecological risk. Second, many standalone ecological risk assessments are simply less ambitious than would be required to support credible economic valuation. This sort of risk assessment asks less of itself and the data, which seems hardly a virtue.

This volume, uniquely in my experience, brings together a conceptual framework for combining economic valuation and ecological risk assessment with six case studies that explore the possibilities of applying it in specific applied contexts, and it provides some context for the whole effort within the institutional realities of American environmental management agencies. In each of the first three case studies, integrated ecological risk assessment (ERA) and economic valuation (EV) is an ideal not quite accomplished. In the Platte River case, it does no injustice to an economic analysis that has some interesting standalone features to point out that it came later than the ERA, exhibits only modest linkage to it, and was designed with relatively little stakeholder input. In the Big Darby case, the ecological risk analysts were forced to make the best of some data deficiencies, and integrated ERA-EV was never really on the cards; the EV came later and was performed by a different team of researchers. Similarly, in the Clinch River case the ERA preceded the EV, and the EV was based only loosely upon it. In all three cases, some good science and some good economics were applied to a practical environmental management issue, but the result fell short of integrated ERA-EV. Notably, the resulting ERAs and EVs displayed one difference that was thoroughly predictable: the ERAs focused more on ecosystem baselines and changes, while the EVs focused more on management alternatives.

In a later chapter, Lazo et al. succeed in integrating ERA and EV by reversing the order in which they are applied. A particular economic valuation method is used to discover public preferences among an array of broadly defined restoration alternatives, and then ERA techniques are used to develop more detailed management plans to implement the top-ranked restoration alternatives. Without endorsing their approach in its entirety, we can concede that they might be on to something.

Given this volume's grand objectives, it is unsurprising that the end result is both a monument to the hard work and ingenuity of its contributors and the genuine progress they have made, and a cautionary tale about how far we collectively are from plausibly declaring victory. Yet, the dream of integrated ecological risk assessment and economic valuation remains valid. Ecological risk matters, and plans for action must be based on some sense of how the world works *and* some sense of what the public values.

Alan Randall
Department of Agricultural, Environmental, and Development Economics
The Ohio State University
Columbus, Ohio

About the Editors

Randall J. F. Bruins has been an environmental scientist with the U.S. Environmental Protection Agency since 1980 and has a longstanding interest in interdisciplinary research. Randy holds an M.A. in zoology from Miami University (Ohio) and a Ph.D. in environmental science from The Ohio State University. In 23 years with the USEPA, he has assessed human health risks associated with the disposal of wastewater treatment residuals and municipal solid wastes and ecological risks from chemicals and other stressors in freshwater ecosystems. Dr. Bruins has also conducted research on the relationship between wetland agricultural practices and flooding risk in central China. In his current position with USEPA's National Center for Environmental Assessment in Cincinnati, Ohio, his focus is on improving methods for estimating the societal benefits of ecosystem protection through better integration of ecological risk assessment and economic analysis. He is a member of the International and U.S. Societies for Ecological Economics and the American Society for Ecological Engineering and is referee for the journal *Ecological Engineering*.

Matthew T. Heberling is currently an economist in the U.S. Environmental Protection Agency's National Center for Environmental Assessment (NCEA). He holds a Ph.D. in agricultural economics from The Pennsylvania State University, where he specialized in environmental and natural resource economics. Dr. Heberling joined NCEA in 2001 to work in a program of research integrating ecological risk assessment and economic analyses and now is focused on understanding the ecological risks, costs, and benefits associated with water quality management. The purpose of this research is to show how water quality standard setting and watershed restoration may be linked for analysis, concentrating on the ecological-economic trade-offs. His research experience also includes using economic valuation methods to examine individuals' preferences for recreational fishing and to prioritize stream restoration. He is a member of the American Agricultural Economics Association and the Northeastern Agricultural and Resource Economics Association.

Contributors

John C. Allen
Department of Agricultural Economics
University of Nebraska
Lincoln, Nebraska

P. David Allen II
Stratus Consulting Inc.
Marquette, Michigan

Douglas Beltman
Stratus Consulting Inc.
Boulder, Colorado

Richard C. Bishop
Department of Agricultural and Applied Economics
University of Wisconsin
Madison, Wisconsin

Randall J.F. Bruins
National Center for Environmental Assessment
U.S. Environmental Protection Agency
Cincinnati, Ohio

David J. Chapman
Stratus Consulting Inc.
Boulder, Colorado

Richard A. Cole
U.S. Army Corps of Engineers
Institute for Water Resources
Alexandria, Virginia

Heida L. Diefenderfer
Battelle Marine Sciences Laboratory
Pacific Northwest National Laboratory
Sequim, Washington

Steven R. Elliott
Department of Economics
Miami University
Oxford, Ohio

O. Homer Erekson
Henry W. Bloch School of Business and Public Administration
University of Missouri
Kansas City, Missouri

Jon D. Erickson
Rubenstein School of Environment and Natural Resources
University of Vermont
Burlington, Vermont

Timothy D. Feather
Planning and Management Consultants, Ltd.
Carbondale, Illinois

John Gowdy
Department of Economics
Rensselaer Polytechnic Institute
Troy, New York

Matthew T. Heberling
National Center for Environmental Assessment
U.S. Environmental Protection Agency
Cincinnati, Ohio

Caroline Hermans
Rubenstein School of Environment and Natural Resources
University of Vermont
Burlington, Vermont

Keith Hofseth
U.S. Army Corps of Engineers
Institute for Water Resources
Alexandria, Virginia

Dennis E. Jelinski
Departments of Biology and Geography
Queens University
Kingston, Ontario, Canada

James A. Kahn
Williams School of Commerce
Washington and Lee University
Lexington, Virginia

Bettina Klaus
Department of Agricultural Economics
University of Nebraska
Lincoln, Nebraska

Diana Lane
Stratus Consulting Inc.
Boulder, Colorado

Jeffery K. Lazo
National Center for Atmospheric Research
Research Applications Group
Boulder, Colorado

Karin Limburg
School of Environmental Science and Forestry
State University of New York
Syracuse, New York

Orie L. Loucks
Department of Zoology
Miami University
Oxford, Ohio

Tara A. Maddock
National Center for Environmental Assessment
U.S. Environmental Protection Agency
Cincinnati, Ohio

Donna S. McCollum
Department of Zoology
Miami University
Oxford, Ohio

David Mills
Stratus Consulting Inc.
Boulder, Colorado

Joy D. Muncy
U.S. Army Corps of Engineers
Institute for Water Resources
Alexandria, Virginia

Audra Nowosielski
Department of Economics
Rensselaer Polytechnic Institute
Troy, New York

Robert V. O'Neill
Environmental Sciences Division
Oak Ridge National Laboratory
Oak Ridge, Tennessee

John Polimeni
Department of Economics
Rensselaer Polytechnic Institute
Troy, New York

Robert Raucher
Stratus Consulting Inc.
Boulder, Colorado

Robert D. Rowe
Stratus Consulting Inc.
Boulder, Colorado

Victor B. Serveiss
National Center for Environmental Assessment
U.S. Environmental Protection Agency
Washington, D.C.

Leonard A. Shabman
Resources for the Future
Washington, D.C.

Marc Smith
Ohio Environmental Protection Agency
Columbus, Ohio

Karen Stainbrook
College of Environmental Science and Forestry
State University of New York
Syracuse, New York

Steve Stewart
Department of Hydrology and
Water Resources
University of Arizona
Tucson, Arizona

Elizabeth Strange
Stratus Consulting Inc.
Boulder, Colorado

Raymond J. Supalla
Department of Agricultural Economics
University of Nebraska
Lincoln, Nebraska

Ronald M. Thom
Battelle Marine Sciences Laboratory
Pacific Northwest National Laboratory
Sequim, Washington

Amy Wolfe
Environmental Sciences Division
Oak Ridge National Laboratory
Oak Ridge, Tennessee

Osei Yeboah
Department of Agricultural Economics
Auburn University
Auburn, Georgia

Acknowledgments

The editors wish to acknowledge Bette Zwayer, Pat Daunt, Patricia L. Wilder, Dan Heing, Teresa Shannon, and Lana Wood for their assistance in manuscript preparation; Luella Kessler and Linda Ketchum for research assistance; and Ruth Durham, Donna Tucker, and David Bottimore for management of reviews. Many individuals provided peer review comments on a U.S. Environmental Protection Agency (USEPA) report from which more than half of the material for this book was derived. Reviewers external to USEPA were Darrell Bosch of the Department of Agricultural and Applied Economics, Virginia Tech University; Robert Costanza, Gund Institute for Ecological Economics, University of Vermont; and Peter deFur of Environmental Stewardship Concepts. USEPA reviewers of one or more chapters included Anne Grambsch, Anne Sergeant, Brian Heninger, Catriona Rogers, Christopher Miller, Dan Petersen, Glenn Suter II, John Powers, Keith Sargent, Lester Yuan, Mark L. Morris, Matt Massey, Michael Slimak, Sabrina Ise-Lovell, Stephen Newbold, Susan Herrod-Julius, Tara Maddock, Victor Serveiss, Wayne Munns, William O'Neil, and William Wheeler. Reviews of contributed chapters were accomplished using an anonymous round-robin process among the contributors, and we extend special thanks to Rebecca Ephroymson of Oak Ridge National Laboratory for her participation in this process. Any errors, however, remain the responsibility of the editors and contributors. We also acknowledge Glenn Suter II, Chris Cubbison, Mike Troyer, and Haynes Goddard for participation in the review of grant proposals that formed the core of this research effort, and Barbara Cook for invaluable assistance in grant management. Mike Troyer prepared maps appearing in several chapters. We acknowledge the important work of Suzanne Marcy, and members of the USEPA Risk Assessment Forum, who initiated the watershed ecological risk assessments that provided a basis for this research, and Victor Serveiss, who later assumed leadership of the watershed ecological risk assessment effort. We acknowledge Jackie Little and Nancy Keene of TN and Associates for their assistance in the organization of a workshop held in 2001 in Cincinnati, Ohio, and the attendees of that workshop. We thank Steve Lutkenhoff for his longstanding support of our work. Finally, we are indebted to our wives, Connie King Bruins and Jacqueline Heberling, for their understanding and encouragement throughout this project.

List of Acronyms

AF Acre-foot
AFS American Fisheries Society
AI Adaptive implementation
ASCA Alternative-specific constants — Option A
ASCB Alternative-specific constants — Option B
AWQC Ambient water quality criteria
BLM Bureau of Land Management
BMPs Best management practices
BOD Biological oxygen demand
BOR Bureau of Reclamation
CA Conjoint analysis
CAFOs Confined animal feeding operations
CBA Cost-benefit analysis
CEA Cost-effectiveness analysis
CENR Committee on Environment and Natural Resources
CERCLA Comprehensive Environmental Response, Compensation and Liability Act
CFR Code of Federal Regulations
COD Chemical oxygen demand
CPP Continuing planning process
CS Compensating surplus
CSO Combined sewer overflow
CVM Contingent valuation method
CWA Clean Water Act
DDT Dichloro-diphenyl-trichloroethane
DEM Digital elevation models
DO Dissolved oxygen
DOI Department of Interior
DPSIR Driving forces, pressures, state, impacts, response
DWR Department of Water Resources
EBI Effective buying income
EIA Economic impact analysis
EIS Environmental impact statement
EMAP Environmental Monitoring and Assessment Program
EMC Environmental Management Council (Dutchess County, NY)
EPT Ephemeroptera, Plecoptera, and Trichoptera index
ERA Ecological risk assessment
ESA Endangered Species Act
FERC Federal Energy Regulatory Commission
GAIA Graphic analysis for interactive assistance
GDSS Group decision-support system
GIS Geographic information systems
GR-SAM Geographically referenced social accounting matrix
HEA Habitat equivalency analysis
HRC Habitat-based replacement cost
I&E Impingement and entrainment
IBI Index of biotic integrity
IBM International Business Machine Corporation
ICI Invertebrate community index
IMPLAN Impact analysis for planning
IO Input-output model
IWA Integrated watershed assessment
KAF Knowledge adjustment factor
KL Knowledge level
MCDA Multi-criteria decision aid
MIwb Modified index of well-being
MOA Memorandum of agreement
MOS Margin of safety
MPC Municipality preliminary screener
MRS Marginal rate of substitution
NCR National Cash Register Company
NEPA National Environmental Policy Act
NGO Non-governmental organization
NMFS National Marine Fisheries Service
NOAA National Oceanic and Atmospheric Administration

NPDES National Pollution Discharge Elimination System
NPS Nonpoint source
NRC National Research Council
NRCS Natural Resource Conservation Service
NRDA Natural Resource Damage Assessment
NYSCD New York State Conservation Department
OECD Organization for Economic Cooperation and Development
OEPA Ohio Environmental Protection Agency
OPA Oil Pollution Act
PCB Polychlorinated biphenyl
POTWs Publicly-owned treatment works
PROMETHEE Preference Ranking Organization METHod of Enrichment Evaluation
PRWCMT Platte River Whooping Crane Maintenance Trust
QHEI Qualitative habitat evaluation index
RAP Remedial action plan
RCDP Restoration and compensation determination plan
REA Resource equivalency analysis
ROD Record of decision
RUM Random utility model
SAB Science Advisory Board
SAM Social accounting matrix
SCS Soil Conservation Service
SDSS Spatial decision-support system
TMDL Total maximum daily load
TN Total nitrogen
TNC The Nature Conservancy
TP Total phosphorus
TSS Total suspended solids
TVA Tennessee Valley Authority
TVE Total value equivalency
UAA Use attainability analysis
UN-L University of Nebraska-Lincoln
USACE U.S. Army Corps of Engineers
USC United States Code
USEPA U.S. Environmental Protection Agency
USFS U.S. Forest Service
USFWS U.S. Fish and Wildlife Service
USGS U.S. Geological Survey
UT-K University of Tennessee-Knoxville
VEA Value equivalency analysis
W-ERA Watershed ecological risk assessment
WIC Wappingers Creek Intermunicipal Council
WQS Water quality standards
WRC Water Resources Council
WRDA Water Resources Development Act
WRPA Water Resources Planning Act
WTA Willingness to accept
WTP Willingness to pay

Table of Contents

CHAPTER 1

Introduction

Randall J.F. Bruins and Matthew T. Heberling

CONTENTS

THE IMPORTANCE OF INTEGRATED, WATERSHED-LEVEL ANALYSIS

Aquatic ecosystems — such as coasts, estuaries, wetlands, lakes, rivers, and streams — provide many services to human society. They supply water and food, they assimilate wastes, they offer means of transportation and energy generation, they provide habitat for many species that humans value, and they offer recreation, aesthetics, and inspiration. In taking advantage of these services, humans have stressed these ecosystems. Alteration of stream corridors, changes in patterns of flow, introduction of nonindigenous species, and pollution by toxicants, nutrients, sediments, heat, and oxygen-demanding substances have diminished aquatic ecosystems' ability to continue providing the services that society values.

As social awareness has increased, efforts have been made to better manage these ecosystems and reduce human impacts upon them. In the United States, these efforts have included increased regulation and mitigation of pollution; increased attention to the ecological impacts of water resource projects; modification of agricultural practices and subsidies; and efforts by urban, suburban and rural communities to better steward their aquatic ecological resources through monitoring,

1-56670-639-4/05/$0.00+$1.50

planning and collective action. Most of these efforts have been accompanied by a recognition that aquatic ecosystems have complex interactions with their surrounding landscapes. As a result, the watershed increasingly is seen as a basic unit for aquatic ecosystem analysis and management.

This book is concerned with two types of analysis that are both important for aquatic ecosystem management: ecological risk assessment (ERA) and economic analysis. Both have been recognized as necessary, and their use is provided for in law and regulation, yet because they arise from very different philosophical traditions, they have tended to remain separate in both theory and practice.[1,2] This separation hampers environmental management. Analysts from these respective traditions often fail to coordinate their efforts, frequently misunderstand one another's terminology and approaches, or disagree as to what is important, thus providing decision-makers with incomplete or confusing information. Decision-makers may also assume that these analyses ought to be separate and thus fail to recognize the wealth of insight that their effective integration could produce.

ERA has been defined as "a process for collecting, organizing and analyzing information to estimate the likelihood of undesired effects on nonhuman organisms, populations or ecosystems."[3] Recommended procedures for carrying out ERA have been published by the U.S. Environmental Protection Agency (USEPA),[4] and the practice has been employed for a wide variety of ecological problems and settings. For example, a 1999 report by the Committee on Environment and Natural Resources (CENR) documented the use of ERA by five U.S. federal agencies to regulate the uses of toxic substances and pesticides, for the control of nonindigenous species, and to remediate and determine compensation for damage caused by chemical releases.[5] The general principles of ERA also underlie many important regulatory protections for aquatic ecosystems in the United States, such as state-issued water quality standards (WQS), but watersheds themselves are not usually the subject of ERA. Routine management approaches, however, including the monitoring and enforcement of WQS, cannot address certain kinds of aquatic ecosystem impairment. Some undesired effects are caused by human-caused insults (hereafter termed "stressors") for which there are no standards; these include, for example, introduced organisms and altered habitat. Some are a complex result of multiple kinds of stressors, and in some cases the causes remain unclear without further study. Moreover, some aquatic ecosystems host unique resources (such as rare species or habitats) having special requirements that are not adequately understood. In addition, it is often unclear, without focused analysis, whether a given set of proposed actions to correct these problems will be effective. In these cases, ERA that is carried out at the spatial scale of the watershed, here termed watershed ERA (W-ERA), may be useful.

As is further described in Chapter 3, W-ERA focuses on the key ecological resources and management goals for the watershed, rather than on regulatory standards alone. The approach directly engages stakeholders in the determination of assessment goals and scope, identifies all relevant threats, and applies scientific methods to the identification of causes, risks, and uncertainties of adverse effects. The resulting information is intended to be useful for the design of approaches for ecosystem protection or restoration, whether these measures are physical or institutional, regulatory or driven by incentives, or governmental or community-based — or some combination of these.

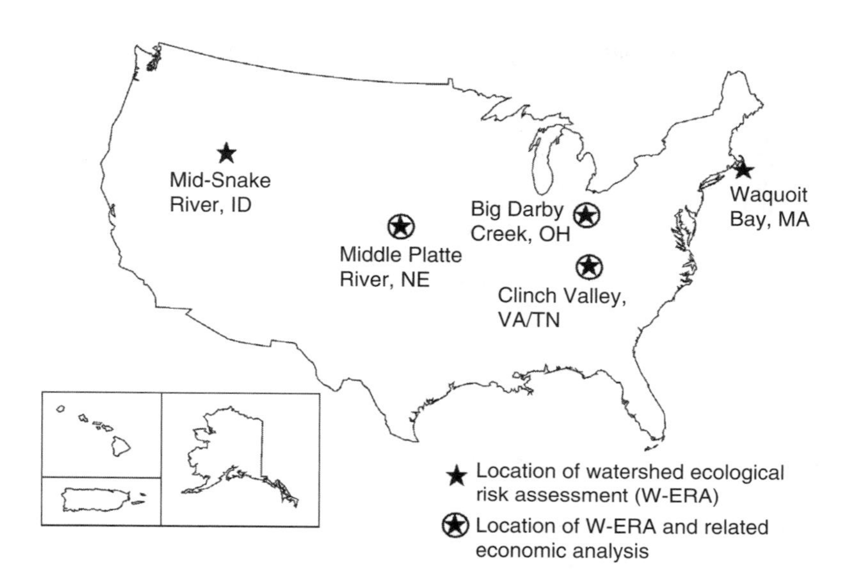

Figure 1.1 Locations in the USA of five watershed ecological risk assessment studies undertaken by USEPA and other partners. Comparison economic analyses were undertaken at three of the five locations as indicated.

In 1993, USEPA initiated W-ERA in five watersheds to evaluate the feasibility and usefulness of this approach (Figure 1.1).[5,6] The outcomes from some of these assessments, and their usefulness for management, have been described in the literature,[7–12] and W-ERA guidance has been made available as a web-based training unit.[13] Prior to this book, however, little information has been available on approaches for integrating economic analysis with ERA in a watershed management context (see CENR[5] and Appendix 9-A).

Economists study choices made by individuals or other entities relating to the allocation of scarce resources across competing uses (see Chapter 5). Watershed management choices involve complex and uncertain trade-offs of current and future financial and ecological resources. Economics offers an analytic framework for determining whether a given choice appears to provide a net benefit to society. Depending on the approach used, economic analysis can also address impacts on affected parties, illuminate negotiation processes, and help evaluate the long-term sustainability of various outcomes. However, the integration of W-ERA and economic analysis that is needed to realize these insights entails theoretical, technical, and procedural challenges.

GOAL AND GENESIS OF THIS BOOK

The goal of this book is to enhance the management of aquatic ecosystems by improving the integration of ERA and economic analysis. This book is intended for technically educated readers with an interest in improving environmental management, including researchers, analysts, advocates, and decision-makers working at

Table 1.1 Case studies of the integration of watershed ecological risk assessment and economic analysis, funded by USEPA in 1999

Study Area	Project Title	Principal Investigators and Grantee Institution
Big Darby Creek watershed, Ohio	"Determining biodiversity values in a place-based ecological risk assessment"	O. Homer Erekson[a] and Orie L. Loucks Miami University, Oxford, Ohio
Upper Clinch Valley, Virginia and Tennessee	"A trade-off weighted index approach to integrating economics and ecological risk assessment"	James Kahn[a] and Steven Stewart[a] University of Tennessee-Knoxville
Central Platte River floodplain, Nebraska	"A strategic decision modeling approach to management of the middle Platte ecosystem"	Raymond Supalla University of Nebraska-Lincoln

[a] No longer at grantee institution; see list of contributors.

local, state, regional, or national levels. It is based on experience in the United States, but many of the principles discussed are broadly applicable.

This book originated with a program of USEPA-funded research to investigate the integration of ERA and economics, with an emphasis on the watershed as the scale for analysis. In 1998, the National Center for Environmental Assessment of USEPA's Office of Research and Development solicited applications for assistance to conduct case studies of the integration of ERA and economic analysis. Research was required to include original economic analysis conducted in collaboration with an ongoing ERA, to reflect the state of the science of ERA and economics, and to be relevant to decision-making with respect to the problem being assessed. In 1999, following peer review of proposals, economic case studies were funded in conjunction with three of the five aforementioned W-ERAs (Figure 1.1, Table 1.1).

The ecological settings and resources of concern differed among the three locations. The degree of progress made by each W-ERA team prior to initiation of the economic study varied as well, and diverse methodological lenses were brought to these problems by the respective economic teams. But the commonalities between these three studies were also considerable in that each involved the watershed scale, each introduced economists to the ERA process, and each confronted the challenging task of interpreting ecological risks in economic terms and in a manner that would be meaningful to decision-makers.

Building on those commonalities, a workshop was held at USEPA in Cincinnati, Ohio in 2001 to review progress on the studies, to discuss environmental problems involving other watershed settings, and to discuss the ideal characteristics of a generalized approach for conducting studies of this type. Based on the workshop results, a conceptual approach for the integration of ERA and economic analysis in watersheds was developed. Reports of the three case studies and a description of the conceptual approach for integration were described in a USEPA report (*Integrating Ecological Risk Assessment And Economic Analysis In Watersheds: A Conceptual Approach And Three Case Studies*[14]) and also form the nucleus for the present volume (Chapters 9–12). To this nucleus, chapters on other methods and perspectives and three additional case studies have been added, broadening this exploration beyond the USEPA research context.

ORGANIZATION

Because ERA and economic analysis stem from different intellectual traditions, most readers will not be familiar with the methods and terminology of both. Therefore, an effort is made to limit jargon and to carefully define and cross-reference terms and concepts. Since an abundance of acronyms in such a work is practically unavoidable, each acronym is defined at its first use in every chapter, and a list of acronyms is compiled in the front matter for the convenience of the bewildered reader.

Part I of this book, *Background, Concepts, and Methods*, introduces some basic concepts and terminology of ERA and economics, especially as applied to watershed management. **Chapter 2** provides an historical overview of the federal role in watershed planning and management in the United States, which traces the changes in governmental approach that have resulted from changes in both the understanding of environmental problems and popular notions of governance. The chapter attributes many past and present shortcomings of federal management to an inadequate appreciation of the ecological connectedness of land and water, as well as to the difficulty inherent in valuing ecological resources in a manner consistent with the Constitutional objective to "promote the general welfare."

Chapter 3 summarizes the USEPA's *Guidelines for Ecological Risk Assessment*.[4] These *Guidelines*, which were published in 1998 following a 10-year period of consensus-building involving scientists both within and outside the USEPA, describe the principles underlying ERA and provide a procedural approach for conducting assessments that is designed to be broadly applicable. The chapter also presents some critiques of these methods, and it discusses their application by the USEPA to the watershed scale.

As is discussed in Chapter 2, the USEPA has important programmatic authority under the Clean Water Act (CWA), but it lacks the project authority, often critical to watershed management efforts, that is vested in the land and water resource management agencies. Therefore, **Chapter 4** illustrates how ERA can be applied within the six-step water resources planning process established for federal agencies.[15] It describes the use of ERA in U.S. Army Corps of Engineers ecological restoration projects, and in Appendix 4-A it presents an example of the use of ERA to evaluate project alternatives for a hypothetical salt marsh rehabilitation effort.

Chapter 5 then introduces basic concepts and methods for the economic analysis of environmental problems — the discipline commonly referred to as *environmental economics*. It explains what is meant by the term *economic value*, how economists measure the value of environmental changes, and how those measurements are incorporated in analytic approaches such as cost–benefit analysis. It discusses game theory, a field of economics that is concerned with the study of interacting decision-makers and has applications to environmental management, and it introduces the emerging body of practice and critique loosely referred to as *ecological economics*. Appendix 5-A introduces two techniques used to elicit, by means of questionnaires, the economic value that individuals would place on hypothetical environmental changes; these "stated preference" techniques are the contingent valuation method and conjoint analysis (or choice modeling).

Chapters 6 and 7 each comment, from differing perspectives, on the limited degree of interaction between ERA and economics that is currently seen within the CWA-mandated water quality standards (WQS) program, a program with important influence on watershed management practices. Chapter 6 explains current uses of ERA and economic concepts in WQS development; it presents conceptual arguments for improved integration, yet stops short of procedural recommendations. Chapter 7, on the other hand, presents a vision for integration accomplished through an adaptive implementation process, whereby stakeholders and regulators periodically reexamine the risks, benefits, and uncertainties associated with standard setting in a degraded aquatic system and revise restoration goals based on public preferences. Since both of these chapters, and later chapters as well, make frequent reference to the CWA concept of *biological* (or *biotic*) *integrity*, and to the ecological indices sometimes used to measure it, Appendix 6-A describes four such indices of biotic integrity used by the State of Ohio.

Economic value, as Chapter 5 explains, is determined based on trade-offs that individuals would be willing to make. Environmental law sometimes requires that ecological damages be compensated by proportionate restoration. While a legally required exchange does not have the same economic characteristics as one that is freely chosen, it nonetheless requires the establishment of a "currency" as a basis for equivalence. **Chapter 8** introduces the reader to various restoration "currencies," derived from ecological and economic concepts, that are used in the determination of natural resource damage compensation. It focuses especially on habitat equivalency analysis, the most commonly used approach.

Chapter 9 culminates Part I by proposing a conceptual approach for the integration of ERA, economics, and other disciplinary methodologies in the context of watershed management. The chapter begins by referring to several procedural approaches, which are compiled in Appendix 9-A, and criteria that have been applied to environmental management. It then outlines a new approach that draws from their common elements but is more explicit as to how ecological and economic analyses should interact. (A second Appendix, 9-B, briefly introduces sociocultural assessment methods that may serve to complement ecological and economic analyses.[a]) This new conceptual approach for integration serves as a point of reference for critical discussion of the case studies presented next.

Part II, *Applications*, presents six case studies, each demonstrating the use of a different economic method. The first three (described in Chapters 10–12) were part of a USEPA-sponsored program of research and demonstration, as described in the previous section. They sought to apply the USEPA's *Guidelines for Ecological*

[a] Health risk assessments may also be required, but these methods are familiar to many readers and are not discussed in this text.

Risk Assessment[4] in stepwise fashion; economic analysis was not a part of that process but was begun as a separate initiative several years later. The other three (described in Chapters 13–15) are more varied in purpose and approach.

The case studies presented in **Chapters 10 and 11** both address the protection of streams that have unusually high diversity of fish and mussels, including many rare species. Big Darby Creek in central Ohio is located in the broad plains of the eastern corn belt, where agriculture predominates but suburban development is expanding. In contrast, the Clinch and Powell Rivers ("Clinch Valley") are located in the mountainous terrain of southwest Virginia and northeast Tennessee, where agriculture is confined to narrow floodplains and coal mining, though declining, remains an important influence. In these studies, economists applied stated preference techniques — the contingent valuation method in the case of Big Darby Creek and conjoint analysis in the Clinch Valley — to put the value of protecting these unique species, and the high quality environments that support them, into an economic context.

Chapter 12 focuses not on estimating value but on resolving conflict. The Platte River watershed encompasses portions of Wyoming, Colorado, and Nebraska; to date these states have been unable to reach agreement on the provision of sufficient water and restored habitat to meet the needs of threatened and endangered species, including several migratory bird species, in a critical reach of river located in central Nebraska. In the case study, economists used game theory to search for solutions most likely to satisfy the preferences of the interested factions.

Chapter 13 describes an effort to predict the likely impacts of economic development on stream ecology in two Dutchess County, New York watersheds within the Hudson River catchment. This ongoing study seeks to combine economic simulation (input–output modeling), spatiotemporal analysis of land use changes, and spatial analysis of stream biological integrity to help Dutchess County residents better understand the potential long-term consequences of their immediate choices.

Chapters 14 and 15 demonstrate the application of methods for scaling restoration to balance losses. In Chapter 14, the economic value of a proposed set of remedial actions (i.e., clean-up) and other ecological enhancements in a polluted watershed is calibrated to equal the value of natural resource damage using a method termed *total value equivalency*. The method is applied in the assessment of damages resulting from polychlorinated biphenyl (PCB) discharges into the Lower Fox River and Green Bay in Wisconsin and Michigan. Chapter 15 presents a determination of *habitat replacement cost*, a method that balances an amount of harm inflicted on populations of particular species with the amount of habitat restoration that would restore those populations. The method is used to determine the monetary amount needed to restore habitat sufficient to compensate for damages to fish and shellfish populations caused by the intake of power-plant cooling water from Plymouth Bay in Massachusetts.

Part III, *Conclusions*, consists of one final chapter, **Chapter 16**, which examines the commonalities of these studies and draws general conclusions. The chapter describes the barriers to ecological–economic integration that still remain, and it makes recommendations for further research.

UNIQUE CONTRIBUTIONS

This book makes several unique contributions to environmental management. First, it places economic analysis into a context that is familiar to risk assessors. Because it uses the specific procedures and terminology of ERA, it will help ERA practitioners better understand how those procedures can be integrated with economic analysis. The conceptual approach presented in Chapter 9 borrows heavily from the USEPA's *ERA Framework*. The case studies demonstrate how risk assessment outcomes — that is, probabilities of adverse changes in ecological assessment endpoints — figure into economic analysis, and they sensitize the reader to the difficulties that economists face in using those results. They also illustrate for risk assessors the importance of the "with–without" context that is familiar to economists. Whereas risk assessors sometimes focus mainly on identifying risks associated with current situations and trends, or on identifying exposure targets for reducing those risks, economists most often focus on choices between alternative actions. Therefore, economists demand a comparison of current and future risks "with and without" a given action. The economist's perspective, evident both in the conceptual approach and the case studies, prods the risk assessor to use ERA in a way that maximizes its value to decision-makers. The case studies allow comparison of six different economic approaches.

Second, the risk assessment perspective employed in this book poses interesting challenges for the economist. Economists sometimes use relatively vague statements about the ecological improvements expected under a given policy to elicit the monetary amounts individuals would pay to obtain the policy, either because they lack more specific information on ecological changes or as a way to match the individuals' understanding of the ecosystems. ERA, on the other hand, uses the best-available data and methods to quantify the linkages between human activities, the stressors they produce, and the ensuing effects on particular ecological endpoints. The resulting statements about risk are as specific as possible about the nature and magnitude of effects expected, but they may also include descriptions of uncertainties. Translating these statements into terms amenable to economic analysis is difficult, as these case studies illustrate, but the challenge must be accepted if these sciences are to be integrated.[15]

Finally, this book introduces a conceptual approach for integrating ERA and economic analysis in the context of watershed management (see Chapter 9, especially Figure 9.1). The approach draws its elements from existing USEPA guidance, as well as from other environmental management frameworks developed by various agencies and advisory bodies. By synthesizing these elements in a way that emulates yet expands the *ERA Framework*, which is a familiar tool in the field of environmental management, it communicates the essential principles of integration to an important audience.

REFERENCES

1. Norgaard, R., The case for methodological pluralism, *Ecol. Econ.*, 1, 37, 1989.
2. Shogren, J.F. and Nowell, C., Economics and ecology: A comparison of experimental methodologies and philosophies, *Ecol. Econ.*, 5, 101, 1992.

3. Suter, G.W., Efroymson, R.A., Sample, B.E., and Jones, D.S., *Ecological Risk Assessment for Contaminated Sites*, Lewis Publishers, Boca Raton, FL, 2000.
4. USEPA, Guidelines for ecological risk assessment, EPA/630/R-95/002F, Risk Assessment Forum, U.S. Environmental Protection Agency, Washington, D.C., 1998.
5. CENR, Ecological risk assessment in the federal government, CENR/5-99/001, Committee on Environment and Natural Resources of the National Science and Technology Council, Washington, D.C., 1999.
6. Butcher, J.B., Creager, C.S., Clements, J.T., et al. Watershed level aquatic ecosystem protection: Value added of ecological risk assessment approach, Project No. 93-IRM-4(a), Water Environment Research Foundation, Alexandria, VA., 1997, 342 pp.
7. Diamond, J.M. and Serveiss, V.B., Identifying sources of stress to native aquatic fauna using a watershed ecological risk assessment framework, *Environ. Sci. Technol.*, 35, 4711, 2001.
8. USEPA, Waquoit Bay Watershed ecological risk assessment: The effect of land derived nitrogen loads on estuarine eutrophication, EPA/600/R-02/079, U.S. Environmental Protection Agency, Office of Research and Development, National Center for Environmental Assessment, Washington, D.C., 2002.
9. USEPA, Clinch and Powell Valley Watershed ecological risk assessment, EPA/600/R-01/050, U.S. Environmental Protection Agency, Office of Research and Development, National Center for Environmental Assessment, Washington, D.C., 2002.
10. Serveiss, V.B., Applying ecological risk principles to watershed assessment and management, *Environ. Manage.*, 29, 145, 2002.
11. USEPA, Ecological risk assessment for the Middle Snake River, Idaho, EPA/600/R-01/017, U.S. Environmental Protection Agency, Office of Research and Development, National Center for Environmental Assessment, Washington, D.C., 2002.
12. Valiela, I., Tomasky, G., Hauxwell, J., et al. Producing sustainability: Management and risk assessment of land-derived nitrogen loads to shallow estuaries, *Ecol. Appl.*, 10, 1006, 2000.
13. Serveiss, V., Norton, S., and Norton, D., Watershed ecological risk assessment, The Watershed Academy, U.S. Environmental Protection Agency, 2002, on-line training module at http://www.epa.gov/owow/watershed/wacademy/ acad2000/ecorisk. Accessed Sept. 29, 2004.
14. USEPA, Integrating ecological risk assessment and economic analysis in watersheds: A conceptual approach and three case studies, EPA/600/R-03/140R; NTIS PB2004-101634, National Center for Environmental Assessment, U.S. Environmental Protection Agency, Cincinnati, OH, 2003.
15. Suter, G.W., Adapting ecological risk assessment for ecosystem valuation, *Ecol. Econ.*, 14, 137, 1995.

CHAPTER 2

Watershed Planning and Management in the United States

Richard A. Cole, Timothy D. Feather, and Joy D. Muncy

CONTENTS

1-56670-639-4/05/$0.00+$1.50

INTRODUCTION

This chapter reviews the history of government-led planning and management of resources in a watershed context, leading up to the contemporary approaches now dominant in the United States. The recurrent institutional themes that contribute fundamentally to the uncertainty and risk associated with the present approaches are highlighted. This history chronicles the growing intensity and diversity of demands placed on public management of natural resources and the environment. It recounts the diversification of legislated objectives and agency authorities that now compete for attention in pursuit of sustained improvement of human welfare. While the promotion of the general welfare, as set forth in the preamble to the U.S. Constitution, has remained a consistent national goal for public planning and management throughout U.S. history, much less consistency is evident in the approaches taken to achieve national objectives in service of that goal.

One of the most fundamental impediments to better watershed planning is a chronic inability to confidently measure and predict change in human welfare in some universally accepted manner and expression. Despite many improvements, a profound gap remains between accepted economic valuation and the total sum of all of the watershed-based values contributing to human welfare improvement from management. Profound differences continue to exist between those managers and stakeholders who accept economic valuation (value expressed in monetary units) as an inclusive indicator of value and those who do not. These differences originated more than a century ago, pitting "conservationist" against "preservationist" philosophies, and continue to impede the most fundamental conflict resolution. This divisiveness has been and continues to be abetted by fragmented management authorities, the sometimes corruptive influence of special-interest groups on federal agencies, the absence of a common watershed planning framework, the difficulties inherent in fully engaging stakeholders in the planning process, insufficient scientific knowledge of natural and social processes, irreducible uncertainty in forecasts of management outcomes, and insufficient motivation to do better.

The history of watershed planning and management is characterized more by the disjointed fits and starts associated with these impediments than by smooth transition to enlightened integration of agency efforts. Even so, many advances have been made, and government agencies have accrued most, if not all, of the authorities needed to more effectively promote the general welfare. Many specific impediments remain in the way agencies communicate, cooperate, coordinate, and otherwise behave, however, and some gaps in authority remain to be filled. Yet this history also shows promise for substantial future improvements through watershed-framed planning and management, including evidence of an evolving interagency framework that could be used to greater advantage.

BRIEF HISTORY

Awakening of Concerns and Early Strategies (1864–1900)

The history of watershed planning and management in the United States traces back to a few highly influential figures that initiated public awareness of potential watershed-based problems in the nineteenth century. George Perkins Marsh, a widely traveled statesman from Vermont with first-hand knowledge of watershed degradation and its apparent effects, expressed early concerns about degraded watershed conditions in the 1840s and published a seminal book on the subject in 1864.[1] Influenced by Marsh, the first U.S. forestry agent, F.B. Hough, spearheaded a campaign of leading scientists to urge legislatures in state and federal government to protect and manage watersheds. That call for legislative action was followed in 1878 by John Wesley Powell's report to Congress on the potential for agricultural development in the arid territorial West. Powell recognized that climate and land condition determined irrigation-water supply and advocated the use of watershed boundaries for administering arid-land development.[2,3]

Powell's report informed a coalition of influential irrigation-development and timber-reserve interests who lobbied for passage of the Forest Preserve Act of 1891, which authorized the President to set aside forest preserves from western public domain lands to protect water supplies.[4] The Act was vague about what other uses were permitted, however, and promoters of multiple-use and preservation philosophies ardently disagreed from the start. Preservationists, led by John Muir, pressed to limit use to recreation, similar to national parks.[2] A preferred preservationist model was exemplified by the New York State legislature, which had, in 1885, set aside the Adirondack and Catskill forest preserves to remain "forever wild" for watershed protection and recreation. The preservationists were countered by Western timber, grazing, and water development interests, who successfully advocated a broader array of uses. Following study by the National Science Academy, Congress passed the Forest Management Act of 1897, which permitted regulated timber harvest, mining, grazing, and water use consistent with maintaining sustainable water and timber supplies.

These policies were pursued even though scientific understanding of watershed processes was rudimentary and anecdotal. Hard data were scarce; river discharge was not monitored consistently until a few years after the U.S. Geological Survey (USGS) was created in 1879. "Authoritative opinion" characterized the debates over watershed effects.[5] Those opinions often differed sharply over the degree to which watershed condition and management influenced water supply and variability. For example, the leading water management authority of the time, the U.S. Army Corps of Engineers (USACE), claimed there was little effect,[6] despite Marsh, Powell, Hough, and others. Knowledge about the process connections between watershed condition, climate, and water runoff quantity, quality, and dynamics would remain primarily qualitative until well into the twentieth century. This propensity to establish policy in advance of well-researched understanding of watershed process has persisted to this day.[7]

The Federal Government Takes the Lead (1901–1931)

Once installed as President in 1901, Theodore Roosevelt aggressively established a policy to manage public natural resources for "the greatest good for the greatest number for the longest time,"[8] which he believed was consistent with the Constitutional imperative "to promote the general welfare." The administration promoted a stronger scientific basis for agency management decisions, a strong moral character to resist short-sighted exploitation and corruption, and agency self-motivation and self-regulation to achieve the greater-good objective. The management philosophy was called *conservation*. This philosophy emphasized sustained yield of timber, water, and other natural resources for beneficial use, and it set the stage for gauging management success by measures of economic efficiency. Gifford Pinchot, head of the Bureau of Forestry, led the administration's promotion of this "gospel of efficiency."[4,8] In contrast, preservation philosophies claimed values beyond the economic concepts of the time. Conflicts between conservation and preservation philosophies were mollified by restricting most use of the national parks and monuments to non-consumptive use (except some range use) and by following multiple-use principles in the forest preserves and waterways. This vision for natural resource management was to dominate federal and state government until the environmental movement of the 1960s and 1970s.

Influenced by Pinchot and Frederic Newell, an engineer in the USGS, Roosevelt pushed quickly for the Reclamation Act of 1902, including authorization for the Reclamation Service (to become the Bureau of Reclamation in 1923). The Reclamation Service soon set about planning irrigation supply reservoirs in the Gila River, the Rio Grande, and other Western basins using data from recently established USGS gauges. In 1907, Roosevelt appointed the Inland Waters Commission, which was charged with preparing a comprehensive plan for water resources development, emphasizing commercial navigation. The Commission reported "that every river system is a unit from its source to its mouth and should be managed as such."[2] It was the first of three early commissions that emphasized a need for integrated resource management in a drainage-basin context.[9]

Responding to land management needs in 1905, separate forestry programs in the departments of Agriculture and the Interior were integrated to form the U.S. Forest Service (USFS) and the National Forest System. Pinchot became the first USFS director. At that time, the national forests were all west of the Mississippi. A movement to create more public land from private holdings, primarily for recreation in the eastern United States, fueled legislation to create national forests from private-land purchase. Congress resisted the recreational basis for acquisition but, in the Weeks Act of 1911, justified federal purchases for national forests to protect water supplies for navigation.[4] The law was passed despite the fact that the federal agency responsible for navigable waters, the USACE, continued to doubt the importance of land condition in maintaining water supply.[2,6]

As director of the USGS, Newell advocated scientific quantification of watershed process relationships including interactions between land condition and water supply.[10] Other than USGS monitoring, however, federal research was incidental until a collaborative effort including the USFS and the U.S. Weather Bureau was initiated in 1910.[5] It remained notably unique and low profile for nearly two decades, in part because other concerns diverted attention. A long divisive fight over whether to consolidate the national parks in the USFS was finally settled in separate authorization of the National Park Service and the national park system in 1916.[4] This was the first of many major conflicts in natural resource and environmental administration to derive from competition for funding and authority and to result in further fragmentation of management authority.

The agencies tended to make what they could of limited data and analyses to improve their competitive positions for funding and new authority.[2,4] To gain advantage in addressing problems that came before the Administration or Congress, they tended to promote solutions most congruent with their existing authorities and expertise. The response of the USACE to Congressional inquiry about flood control was typical. For several decades, the USACE had integrated flood control into levee structures designed to serve its navigation mission on the lower Mississippi, with the tacit approval of Congress.[11] Floods in 1912 and 1913 overwhelmed the levees, but the USACE still dismissed a more integrated approach — including levees, land treatment, and reservoirs — for more and bigger levees alone.[4,11] Some states disagreed with the USACE approach. The disastrous 1913 flood in Ohio stimulated passage of the Ohio Conservancy Act in 1914, the first state or federal legislation in the United States resulting in management units and plans based on watershed boundaries.[12] The Ohio approach, which relied on a combination of reservoirs and levees to reduce flood damage, was to serve as a model for other state and federal agencies, including the USACE. In 1917, Congress gave the USACE explicit authority to conduct flood control projects and watershed studies, establishing the basis for its present watershed planning authority. Extensive failure of flood control levees along the Mississippi in 1927 finally convinced the USACE that reservoirs had a place in flood control.

Following World War I, Congress passed laws emphasizing expansion of commerce through hydropower, navigable waterways, and highway development. Poor drainage, erosion, and flooding were chronic problems for transportation engineers, who learned from project complications as they implemented them. One chronic source of degraded watershed conditions was intense wildfires, which were typically followed by increased flooding and erosion of roads. The USFS was authorized to lead wildfire prevention nationally, justified in part by the need to protect watersheds, by applying what turned out to be a misguided policy that incorrectly anticipated scientific understanding of fire ecology and repression effects.

Congress grew more interested in the greater efficiency purported for multipurpose projects as it broadened federal water resource authority from navigation and water supply to flood control and hydropower. In 1927, it authorized the USACE to conduct 180 comprehensive river basin surveys nationwide to determine how hydroelectric power might be coordinated with other water resource development purposes.[11] Upon becoming President in 1928, Herbert Hoover restored executive interest in conservation and multipurpose watershed planning to more efficiently promote public welfare improvement. He directed the USACE to reorganize its management boundaries according to river basins. The Bureau of Reclamation (BOR) was authorized to build the first federal multipurpose project (water supply, hydropower, and flood control) at Boulder (Hoover) Dam.[12] Congress also expanded USFS research, which soon assumed a lead in experimental research of small-watershed processes and management at Coweeta Experimental Forest in North Carolina and other regional research centers.

Consolidating Federal Control (1932–1960)

The Great Depression dominated the agenda of Franklin Roosevelt's New Deal administration and justified greater consolidation of executive power to jump-start the economy. The New Deal adopted the USACE comprehensive river-basin studies to quickly put in place massive water development projects and programs, consistent with early twentieth-century conservation philosophy. However, the first clear example of multipurpose land and water resource management integration in a watershed context emerged under state sponsorship in Ohio in 1933.[12] The Muskingum Watershed Conservancy District added water and soil conservation purposes to flood control in response to severe drought and soil erosion. That same year, the USACE basin study for the Tennessee River was set aside for development under a separate Tennessee Valley Authority (TVA). The TVA was exceptional for its breadth of authority, spanning flood control, erosion control, recreation, improved public health, rural housing, and other services. Such inclusiveness allowed the TVA to circumvent the complications and costs normally associated with integrating diverse agencies in pursuit of improved public welfare.

On the landscape, the dustbowl drought of the early 1930s revealed more clearly than ever the cumulative effect of poor land-use practices in middle America. Along with erosion, flooding, and other problems, reservoirs were filling with sediment at unanticipated rates. On public domain lands, livestock overgrazing was identified as the main cause for badly deteriorated range and watershed condition. The Taylor Grazing Act was passed in 1934 to regulate grazing on the remaining public domain land under the administration of a new agency, the U.S. Grazing Service, which was later joined with the General Land Office to form the Bureau of Land Management (BLM) in 1946.

On private lands, poor crop-culture and grazing practices were most linked to erosion, sedimentation, and flood problems. To facilitate private improvements, the Soil Conservation Service (SCS) was established and authorized in 1935 to develop land-use classification approaches for diagnosing and treating erosion and stream-sediment problems in assistance to private landowners.[13] Congress identified land treatment as a necessary addition to levee and reservoir measures in the Flood Control Act of 1936. It gave the SCS authority to study watershed processes to facilitate retardation of runoff and erosion on private lands through subsidy of and cooperation with the landowners. The Act also first established the need for the water resource agencies to show that project benefits exceeded costs. In 1944, the SCS was authorized to consider reservoir and land treatments for small-watershed flood problems, complementing the main-stem activities of the USACE and the BOR. In 1954, its authority was extended to provide watershed-planning support to the states, tribes, and other federal agencies.

Many improvements were made in the 1930s and 1940s to develop complementary authorities in the land and water resource management agencies, which resulted in a more comprehensive watershed planning authority but required substantial integration of activities across agencies. The alternative was a separate single-agency authority over each river basin's planning and management as characterized by the TVA. The TVA model generated congressional interest in establishing other river-basin management authorities of similar comprehensiveness, but none materialized before World War II delayed all water resource projects. Natural resource planning had been centralized by the New Deal in the National Resources Planning Board, which quite successfully integrated planning among federal, state, and local institutions. The Board also dampened the influence of Congress on federal water resource agencies, however, and diminished Congressional recognition by project beneficiaries.[9,14] The justification for consolidated executive power in natural resources planning waned as the economy recovered during the war, and the Board was eliminated in 1943. Federal water resource project construction soon made up for the development hiatus during the war. Unlike the TVA model, project authorization proceeded incrementally and, for the most part, independently of larger river-basin considerations, but with returned constituent recognition of congressional influence.[9]

As the economy boomed and the cold war settled in following World War II, government-sponsored research began to generate environmental concerns, especially in the U.S. Fish and Wildlife Service (USFWS). New investigations, funded largely by the Atomic Energy Commission and the Department of Defense, rapidly advanced ecosystem understanding,[15] as reflected in Eugene Odum's hugely influential systems-oriented textbook.[16] His text showed a new generation of natural resource scientists and managers how contaminants might be transferred through ecosystems to wildlife, fish, and humans. Environmental concern spread from radioactive materials to toxic metals and pesticides (e.g., Rudd[17]), and to widespread oxygen depletion and petrochemical pollution in waters. Earlier investments in watershed research began to pay off during the 1940s, and funding expanded into areas of water quality effects. Results from USFS, SCS, USFWS, and academic field studies began to unravel pathways for contaminant movements through watersheds.

During this period, federal environmental authorities were dispersed among various agencies. Filling a gap in most agencies, the USFWS and USFS became the leading sources of ecological expertise among the natural resource agencies. The USFWS grew more concerned about environmental threats to fish and wildlife and led the call from within the federal government to enact corrective legislation.[18] Despite the growing evidence of local and state ineffectiveness, federal legislative sentiments continued to favor state regulation of environmental quality.[9] These concerns, however, had yet to capture national press interest, and public investment had yet to grow beyond local issues.

Perceived threats to fish and wildlife were to become important motivators for the environmental movement of the 1960s and 1970s.[18] A secondary motivation was increasing wilderness recreation on multiple-use public lands. Deterioration of recreational land had become of increasing concern in the USFS as road development for timber cutting and other uses advanced rapidly through the national forests. USFS research showed that fish and wildlife and wilderness destruction were connected through the watershed, especially through the sometimes-devastating effects of forest roads on riparian and aquatic habitat condition, a controversial issue to this day.

Democratization and Devolution (1960–1990)

After World War II, rapid expansions of the U.S. population, family income, leisure time, mass communication, and the national highway system contributed to an unprecedented demand for outdoor recreation on public land and waters. The public became more aware of the natural environment and more involved in its management.

Old tensions returned between those wanting preservation for recreation and those dedicated to multiple-use conservation on the national forests. But instead of the privileged few that argued in the era of Muir and Pinchot, the new debate extended well into the grass roots of a large and well-informed middle class.[18]

By the early 1960s, government-sponsored research had accumulated enough evidence of worsening environmental conditions to provoke massive media reaction, led by the release of Rachel Carson's *Silent Spring* in 1962.[19] Support escalated for nongovernmental organizations (NGOs) concerned about the environment. Suspected collusion between industrial special interests and government agencies, epitomized by Eisenhower's warnings about "the military–industrial complex," extended to agricultural, timber, mining, power generation, navigation, highway transportation, real estate development, and other special-interest relationships that were believed to threaten the general welfare by degrading the environment and natural resources. The pesticide issue was especially divisive. Private agricultural interests and Department of Agriculture scientists reacted negatively to what they perceived to be widespread ignorance and threats to U.S. food production. Coupled with growing disillusionment over the war in Vietnam, public trust in government protections of the environment eroded, and a voting majority was mobilized to support the "green decade" of federal environmental legislation.

Federal legislation responded in four general ways that were to alter natural resource planning and management. It opened government process more to public participation, including intervention through suits brought by public interest groups. It devolved more authority to the states and local agencies. Simultaneously, it promoted greater integration of federal, state, and local agency functions. Perhaps most revolutionary, it introduced tough environmental regulation of the federal agencies themselves, enforced mostly through the authority of the new U.S. Environmental Protection Agency (USEPA), formed in 1970. The paternalistic concept of an efficiently integrated, self-regulated, and publicly closed conservation planning and management process was no longer trusted, even if it was intended for the public good. Instead, a higher order of integration was demanded, one that may have been less efficient but included more public oversight of government actions to ensure grassroots protection of the general welfare. The stage had been set for this democratization by the rise of universal suffrage, universal education, the universal reach of mass-medium journalism, and the nearly universal improvement of economic welfare, accompanied by a newfound wariness honed by the chronic threat of World War III.

Once again, the national forests were central to controversy, mostly between recreational and timber interests. The Multiple Use Sustained Yield Act of 1960 moved the USFS into a more inclusive and integrated planning and management process, explicitly including watersheds, recreation and fish and wildlife, and encouraged more open public participation. Later, the National Forest Management Act of 1976 required plans to be developed for each of the national forests with extensive provision for public involvement. In that same year, the BLM first received holistic authority much like that of the USFS, broadening its rangeland focus and raising watershed consciousness. With passage of the Wilderness Act in 1964, the preservationists had an additional means for preserving wilderness recreation as well.

The last federal legislation to substantially shape agency concepts of large-scale watershed-based planning was the Water Resources Planning Act (WRPA) of 1965. The WRPA provided a framework for a more open and integrated, multipurpose, water-resources planning process through comprehensive river-basin initiatives. The U.S. Water Resources Council (WRC) was created to oversee the process and to improve the decision framework for water-resources development. Level B planning done by regional commissions linked national programmatic planning for water resource development priorities at Level A to specific project planning at Level C through the physical and social conditions existing in each river basin. While quite inclusive and integrative, the objectives of the WRPA became marginalized as environmental agendas became separately established in subsequent U.S. environmental and natural resource law, and as federal water resource development responsibility devolved more to the states.

Water resource agencies remained loyal to structural measures while an environmentally sensitized public turned toward more use of nonstructural solutions, such as in watersheds and floodplains.[9,20,21] For example, the restoration of natural floodplain and watershed conditions for flood- and soil-erosion control was increasingly favored over engineered levees, dams, and weirs, in part because those solutions also restored lost recreational opportunities and biodiversity at risk of extinction. The WRPA also did not do enough to defuse long-standing public criticism of construction agency planning, which appeared to favor approaches fitting the agency's construction emphasis and traditional special interests.[20,21] An increasing emphasis on nonfederal cost sharing further undermined the close networks that had developed between special interest groups, federal construction agencies, and Congress.[9] Less than two decades after its creation, President Reagan eliminated the WRC and its comprehensive approach to river basin planning.

Water-resource and environmental planning went separate ways in part because water resource development agencies had failed to plan carefully enough to prevent environmental damage. The decline of species had been worrisome enough by 1966 for Congress to authorize the listing of severely threatened species in the United States and to purchase habitat for their support. Federal and NGO interests in wild-river recreation and water quality improvement were separately involved in their own national-priority planning process. In 1968, the Wild and Scenic Rivers Act first set aside eight river segments and authorized procedures through which many more segments have since been set aside.[9] Study of interstate water quality issues pointed to the need for more comprehensive water quality legislation.

Despite the potential for more effective integration of environmental and developmental objectives in the National Environmental Policy Act (NEPA) of 1969 and a few other environmental laws (e.g., the Coastal Zone Management Act), the net effect of environmental law on public resource management was both disintegrative and polarizing. While the USEPA became the single most powerful agent in the middle of this contentious planning environment, other federal and state agencies contributed a share, many influenced by special interest NGOs. The USEPA consolidated regulatory power and environmental oversight for all other federal, state, and local agencies. Once the strong regulatory laws of 1972–1973 were passed, eclipsing the more integrative NEPA, the USEPA policy emphasized elimination of environmental threats through USEPA-determined environmental standards and best practices, and strong regulation enforcement. Environmental NGOs increasingly consolidated general public interest into special-interest influence on natural resource and environmental agencies, but especially on the USEPA. Partly in response, prodevelopment NGOs strengthened their traditional agency relationships and applied influence to administrative and congressional oversight of the USEPA, USFWS, and other regulatory agencies. Often forcing planners to work with incomplete information and allowing little tolerance for "adaptive management," this strategy set environmental objectives against development objectives and hobbled chances for more integrated resource management in pursuit of a more generalized public welfare improvement.

Water resource policy also contributed to this polarization. The WRC replaced the old standards used for water resource development with new guidelines,[22] which instituted the present approach to water resource project planning and implementation. Unlike the former approach, which treated environmental objectives as equal to other objectives, the new approach treated them as constraints dictated by environmental law.

One of the two most powerful pieces of regulatory legislation influencing the present course of watershed-based planning in the United States was the 1972 amendment of the Federal Water Pollution Control Act (commonly known as the Clean Water Act or CWA). The CWA authorized the USEPA, in cooperation with other federal agencies, states, and local agencies and industries, to develop comprehensive programs for water quality protection and improvement with regard for fish and wildlife needs, recreational purposes, and public water supply. The early regulatory approach was much more successful at eliminating point sources than nonpoint sources of pollutants,[23] which required more comprehensive and integrative planning and management measures. An amendment in 1987 refocused the Act on controlling intractable nonpoint sources of pollution through a more integrative and adaptive watershed management approach. The states were to develop and implement management programs on a watershed basis for USEPA approval, administer federal grants in aid, and conduct periodic program evaluation. Twenty states now have fully developed programs. The states in turn often encourage a local watershed approach to correcting water quality problems. Other local watershed groups organize independently or for somewhat different purposes. For example, under Section 404 of the CWA, the USACE can establish general permits for filling the Nation's waters, including wetlands, based on watershed or wetland management plans (e.g., Special Area Management Plans or SAMPs) to regulate loss of aquatic resources.

The Endangered Species Act (ESA) of 1973 also is a strong influence. Its purpose is to provide a program for conserving threatened and endangered species, including their protection and recovery, and a means for conserving supporting ecosystems through protection of critical habitat. All federal agencies were to use their authorities to further the Act's purposes, but only the USFWS and the National Marine Fisheries Service (NMFS) were assigned species listing, critical habitat determination, and enforcement authority. The ESA is considered an action of last resort[24] that does little to prevent the need for listing in the first place. As a consequence, the presence of listed species severely constrains negotiations for any achievement of objectives that do not protect the listed species.

Since the Water Resources Development Act of 1986, most federal water resource development has required a nonfederal cost-share and full assumption of operation and maintenance upon project implementation. The USACE also gained explicit authority to do environmental improvement projects (see Chapter 4), thereby elevating environmental needs from reactive constraints to proactive objectives. However, this legislation also limits the ecosystem restoration and watershed planning authority of the USACE to projects cost-shared by nonfederal authorities. Nonfederal sponsors gained more control over the prioritization and placement of each project, including how each will operate within a watershed context.

Searching for Unifying Approaches (1990–Present)

Impediments to Objective Achievement

Two to three decades after their enactment, many of the objectives identified in the environmental and natural resources legislation of the 1960s and 1970s had proved difficult to fully achieve. The list of protected species continued to grow steadily and few were recovered. Water quality continued to be degraded by diffuse, nonpoint sources in watersheds, which were growing in influence. Controversy over environmental and resource development objectives continued to slow or even paralyze planning and management as NGOs increasingly used the courts to influence agency actions. The reasons were complex, but the most chronic impediments seemed to be embedded in the fragmented authorities and policies of a government bureaucracy influenced by the divisive forces of competitive special interests.

In response, agency analysts and academics widened the temporal and spatial scopes of planning and management. Several related approaches emerged as promising ways to more comprehensively address management needs, the most notable being integrated resource management, adaptive management, and ecosystem management. Watershed management was not viewed as a distinct approach, but rather as one good way to geographically frame any of those approaches.

Integrated Resource Management

Broadly defined, integrated resource management is the coordinated management of two or more interrelated resources.[9] Some definitions emphasize process integration alone while others stress optimal integration of process for the most desirable outcome. Heathcote's[25] recent work on integrated watershed management is a good example of emphasis on process integration. The most common examples of process and outcome integration involve multi-objective management for maximum benefit from a single natural or engineered structure, or assemblage of such structures, such as management of forest stands for timber, watershed, range, wildlife, and recreation use (e.g., Hoekstra and Capp[26]), or management of a dam for water supply, hydropower, and recreation.[9] These forms of integrated management often were practiced within agencies. Specific perspectives might enter through the resources of forests,[27] coasts,[28] or lakes.[29] More recent definitions are more inclusive of agencies and publics. In Europe, the Technical Advisory Committee[30] defines integrated water resource management very comprehensively, including integration of all land and water processes in a river basin for maximization of human welfare while sustaining vital ecosystems. The USACE appears to be moving in that direction via its most recent strategic plan.[31] Most examples, at least in part, incorporate planning framed by watershed boundaries.

Adaptive Management

Adaptive management integrates learning from management processes and outcomes into the next round of planning. It is most used in planning environments where management outcomes are uncertain and is compatible with either integrated resource management or ecosystem management. While adaptive management has been compared to learning the hard way or trial and error, Holling[32] and Walters[33] have brought theoretic rigor to the concept. Its central element is an institutional means by which new learning can be systematically and continuously integrated with existing knowledge into a constantly improving management framework. A favored approach is development of a process simulation model or other organizational means that can be modified with new knowledge developed through monitoring and analysis. Such models can determine the effects of statistical uncertainty using sensitivity analysis or by building in stochastic elements.

Ecosystem Management

Ecosystem management goes back conceptually to the 1960s[34] and has been described through various points of view (e.g., Grumbine,[35] Boyce and Haney,[36] Vogt et al.[37]). Wording in the NEPA and the ESA implied that an ecosystem perspective was a useful approach to management resulting in environmental and species sustainability. Franklin[38] found that ecosystem sustainability, including all of the useful and potentially useful resources, is a common goal. Another common element is the definition of management boundaries based on natural form and process, such as lakeshore, watershed, or vegetation structure and composition. Watersheds are now widely recognized as a practical means for integrating objectives of both terrestrial and aquatic resource management in many locations. Planning is often facilitated by a model of the targeted ecosystem. In addition, ecosystem management often includes both natural and social processes and features linked through material and energy pathways, many of which are manageable.

Strongly influenced by the ESA, the USFS now emphasizes management for a sustainability that includes native biodiversity and endangered species, using an ecosystem management philosophy that embraces all of its management objectives. Some of the more controversial aspects of recent forest management decisions are traditional watershed management issues[39] addressed within watershed boundaries. During the past decade, the USFS also has led practical application of adaptive management principles (e.g., Rausher et al.[40] and the incorporation of human uses and values in watershed analyses[41]). It had already adopted integrated resource management principles to improve internal coordination, cooperation, and communication (e.g., Hoekstra and Capp[26]). The BLM, National Park Service, and USFWS have also adopted ecosystem management philosophies more or less modeled after that of the USFS, yet less fully developed in agency policy.

Water resource agencies continue to emphasize integrated water resource management over ecosystem management. However, the USACE increasingly includes ecosystem restoration objectives with other water resource objectives at the local project, watershed, and river basin levels. The BOR has used the NEPA authority to integrate ecosystem restoration objectives and adaptive management into operation of existing projects, such as Glen Canyon Dam.[42] In 1991, the SCS became the Natural Resources Conservation Service (NRCS) to reflect a broadening policy perspective. The Clean Water Action Plan of the Clinton Administration placed more emphasis on integrating federal, state, and local watershed planning to meet the water quality standards of the CWA.

Institutional Mechanisms

One practical impediment to interagency coordination and communication until recently was the varied approaches taken to strategic planning. Programmatic intentions often were unclear within as well as among federal agencies, adding a layer of uncertainty on top of project-level planning. In 1994, the Government Performance and Results Act was passed, requiring that all federal agencies publish strategic plans that set objectives and timetables for achievement. Agency response has varied, and most strategic plans can be improved as instruments for improved integration, but the process has the potential for more clearly establishing programmatic priorities that take into consideration interagency integration needs.

Interagency memoranda are another means for institutionalizing agency coordination. In 1995, for example, 17 federal agencies signed an administrative Memorandum of Understanding committing them to a coordinated ecosystem approach to management.[43] Another example is the Memorandum of Agreement (MOA) drafted by the USEPA, USFWS, and NMFS in 1999 to formalize earlier integration of ESA considerations into watershed management planning for water quality standards developed under the CWA.[44]

WATERSHED PLANNING AND MANAGEMENT TODAY

Frameworks and Geography

Much of watershed planning in the United States has evolved within the context of three general federal frameworks:[45] *water resource project planning* is the most thoroughly developed framework after years of formalization through the WRC and agency application. It increasingly facilitates multi-objective project planning including both environmental and economic development objectives. The USACE and the NRCS do most of this type of planning (e.g., Yoe and Orth[46]). The USFS has led development of programmatic *comprehensive adaptive management planning*, which combines elements of integrated resource management, ecosystem management, and adaptive management for land and water resources (e.g., USFS[47,48]). Other land management agencies tend to follow the USFS lead. The most rapidly evolving framework is the USEPA-facilitated programmatic *CWA planning* carried out by both federal and state agencies. The primary intent of this framework is to develop consistency in water-quality management plans by better coordinating assessments, protection, and restoration of watersheds under the auspices of different federal, state, tribal, and local agencies. Responding to criticism about its single-minded objective, more recent facilitation has recognized the need for integrating water quality objectives with other watershed planning objectives. Several key facilitative documents have been developed by the USEPA.[49–53]

The CWA planning must be considered in all federal watershed planning, regardless of the primary objective, and it is the most geographically inclusive watershed planning in the United States. It requires all federal agencies with significant land holdings to adopt a watershed management and planning process compatible with state and other agency planning for water quality considerations authorized by the CWA. So far 20 states have adopted watershed planning and management for the CWA purposes. About one third of the nation's lands are managed by federal agencies including a large majority under the USFS and BLM authority. Both agencies are mandated to include the CWA purposes in their planning process. Local communities have initiated thousands of watershed-based planning and management processes in small watersheds, many in response to water quality improvement needs as scripted by the CWA, but others focusing on flood damage reduction, recreation, or other purposes. Other planning is authorized at a river-basin scale; recent examples include the Chesapeake Bay Program, the Comprehensive Everglades Restoration Plan, the CalFed Bay-Delta Program, and the Interior Columbia Basin Ecosystem Management Project.

Resource Development and Regulatory Objectives

Increasing emphasis is being placed on environmental outcomes regardless of the watershed planning origin or framework. All public resource-development projects must meet the environmental standards developed and administered by the USEPA, USFWS, and NMFS (Figure 2.1). Most resource management agencies have incorporated an environmental sustainability goal into their missions, implying the intent to more fully embrace desired environmental conditions as management objectives rather than constraints, but continuing to derive direction from environmental standards. To the extent that environmental sustainability goals and environmental standards are compatible, this trend indicates more thorough acceptance of the general need to meet standards and often to exceed them, even as some specific requirements are questioned. Promotion of economic welfare objectives by federal resource-development projects has been, in effect, constrained by environmental protection needs and sustainability goals.

Resource development objectives will continue to be important in both land and water resource planning and management, but the need to meet the objectives of the CWA, ESA, and other environmental legislation is now widely accepted by resource planners and increasingly dominates the watershed planning process (Figure 2.1). Most resource planners assume that as long as environmental standards are carefully considered, the constraints they place on land and water development objectives

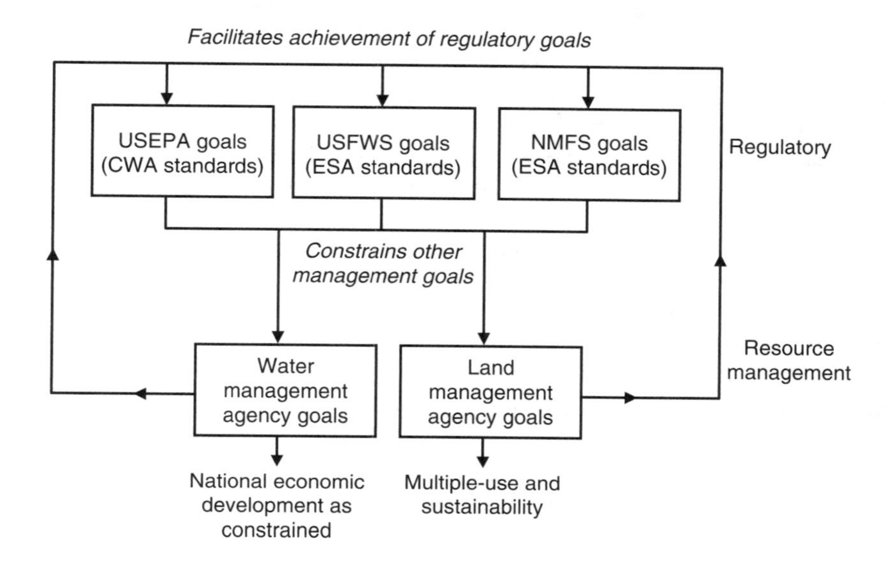

Figure 2.1 Federal agencies interact to form a natural resource management system larger than the individual parts through laws requiring that environmental standards be set and met.

should result in greater net public welfare, even if exceptions to the general rule should occur at specific projects. Serving this principle places great responsibility on the regulatory agencies to develop standards consistent with programmatically improving and sustaining the national interests, including adequate opportunity for resource agency review. It also obligates the resource agencies to thoroughly articulate their concerns when opportunities arise.

The USEPA must rely on state, local, and private capabilities for planning and implementing projects to achieve environmental standards because it has virtually no separate project development authority to meet its strategic goals and objectives. It needs other agency project development authority or private investment to facilitate that end. State and local authorities often seek out partnerships with authorized federal water resource-development agencies to leverage funds and technical know-how or to help facilitate a more effective watershed plan. The USACE ecosystem restoration authority is especially relevant when water quality improvements result in habitat improvements for vulnerable species. In contrast with USEPA's authority, the federal water resources agencies are limited to project planning and implementation, and since 1986 are excluded from assuming operation and maintenance responsibility, which must be turned over to the nonfederal sponsors. Those projects built before 1986 and now operated under federal authority can be programmatically integrated, such as through water-quality compliance with the CWA, an ESA recovery plan, or a river-basin commission plan.

The source of motivation and scale of watershed planning and management authorities are often related. Large-scale planning and management typically is motivated from within the region by perception of threat to regional economies such as decline of fisheries, loss of tourism, insufficient water quality and supply, or increased vulnerability to flood and storm damage. In such situations, environmental degradation often is perceived to be integral to the threats and appropriately addressed with other objectives. This perception also motivates planning and management of some smaller watersheds. But other small-scale watershed planning is strongly influenced by the need to meet environmental standards, most often for water quality, independent of or counter to perceived economic development needs. The ESA objective can be a motivator for watershed planning for private lands, but most often is a constraint on development. On public lands, however, recovery and sustainability of native biodiversity have become high-profile objectives of resource management.

Project and Program Planning Interactions

The need to meet national environmental objectives with fragmented authorities in both federal and state agencies forces agency interdependencies and awkwardly integrates their actions into a programmatic watershed planning cycle (Figure 2.2). In step one of the cycle, the federal, state, and local agencies with programmatic

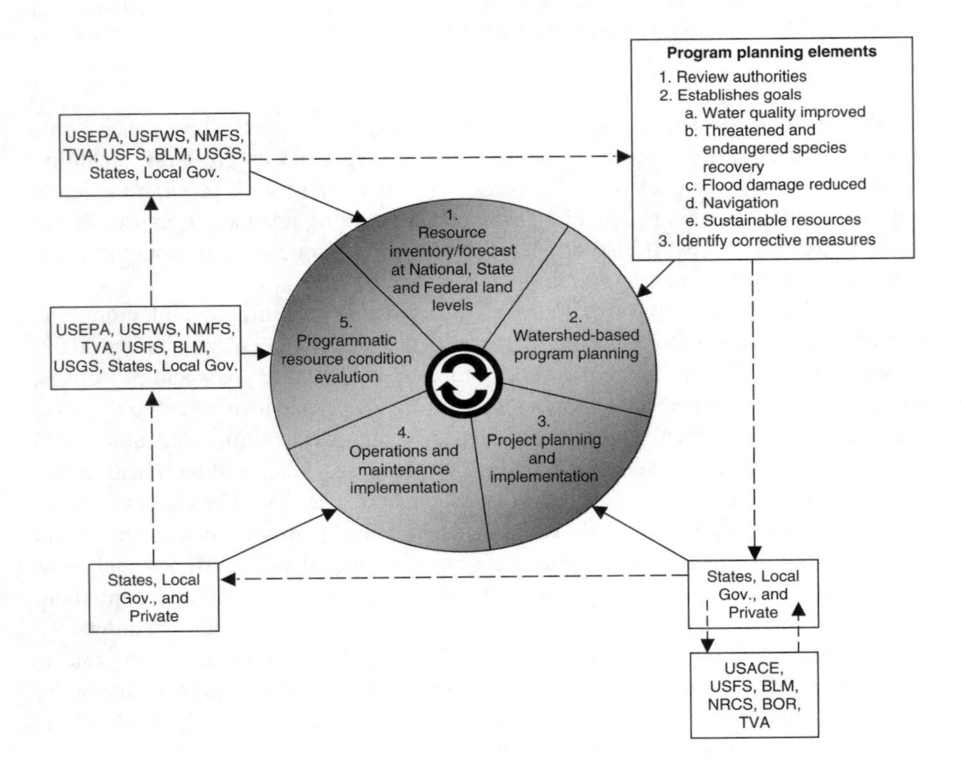

Figure 2.2 The *de facto* programmatic planning cycle, including a project-planning loop, that results from federal, state, and local agency watershed planning interactions with each other and with private institutions.

authorities complete resource inventories and forecasts of resource and environmental conditions but stop short of plan implementation. Progressing to step two in Figure 2.2, the federal and state agencies with programmatic authorities review the issues and relevant authorities, establish programmatic goals and objectives, and evaluate alternative measures for achievement. The degree of integration is determined by the extent to which all objectives and possible measures are included in the recommended plan.

With its broad programmatic authority but limited project development authority the USEPA depends on the state agencies and private firms to provide the project planning and management needed to complete USEPA program goals (the third step in Figure 2.2). The USFWS and NMFS have broad ESA programmatic authorities, but they work with private landholders to develop habitat conservation plans and with other agencies to develop species recovery plans and projects. Because habitat degradation is often linked to water quality, the USFWS and NMFS also work with the USEPA and the states to set priorities. The land-management agencies have relatively broad programmatic and project management authorities on public lands, but they also coordinate with one another, and with private land owners, to develop watershed plans for meeting the CWA and ESA objectives.

Implementation of plans usually requires project development and modification of some kind by the nonfederal agencies, which often solicit project planning and development assistance from the USACE, NRCS, or other federal agencies at step three. Once implemented, the projects are transferred to state or other nonfederal institutional operation and maintenance programs (step four). Many of the federal water-resource projects authorized before 1986 continue to be operated and maintained by federal agencies. These may be brought into the programmatic planning process to achieve new project-operation objectives, such as water quality or ESA recovery plan objectives.

Federal agencies with regulatory or land management authority usually remain in an oversight position to monitor and evaluate management results, including the results of project implementation, and to enter those results into subsequent inventories, forecasts, and planning (fifth step in Figure 2.2). Adaptive management links the fifth step with the next cycle of forecasting and planning.

The trend toward more emphasis on federal, state, and local partnerships extends into most management of public resources. The states and other local interests are increasingly in the planning "driver's seat," sometimes reluctantly so, looking to the federal government to facilitate with money and technical help, including facilitation of watershed planning when appropriate. The resulting interactions are often complicated by confusion over who might be in control. Many of the contemporary issues challenging watershed planning and management have roots in the changing and sometimes uncertain roles of local, state, and federal agencies in the planning process.

Common Elements in Watershed Planning Frameworks

Despite apparent differences in agency decision frameworks, many steps occur in common when watershed planning is effective. These steps typically are required for making smart decisions in any complex situation.[54,55] Four steps dominate the decision process and follow a relatively rigid sequence. These include (1) identifying the problem needing a solution, (2) identifying the objectives to be achieved by the solution, (3) identifying alternative solutions to the problem (including the do-nothing option), and (4) identifying the anticipated consequences of alternative solutions (evaluating objective achievement). The evaluation of objective achievement usually requires identification of the uncertainties affecting the decision and of the decision-maker's willingness to accept risk. When conflicts arise, objectives typically require adjustment for trade-offs among interested stakeholders during the evaluation step. Finally, present and future decisions need to be coordinated such as through adaptive management and strategic planning. These latter steps are not necessarily followed in strict sequence since planning environments change, requiring flexibility.

Effective watershed planning is a continuous process within the context of some continuous programmatic authority. Most watershed planning is incompletely informed by the state of existing ecological, social, and economic science and is an imperfect work in progress. Implementation of management measures usually entails uncertainty and risks. Therefore, effective watershed planning sooner or later requires an adaptive process at project and program levels to more fully meet efficiency, effectiveness, completeness, and acceptability criteria.[a] An agency's role in project planning and implementation may be temporary, such as when watershed project planning is done by the USACE, but this does not dispel the need for adaptive management under the programmatic authority of some other responsible agency or NGO.

Effective watershed planning is an inclusive, integrative process that considers all objectives and measures. It includes all stakeholder groups affected by planning and implementation. It provides for the financial and other resources necessary to complete the process. It typically involves more than one state and federal agency to bring the necessary complement of authorities to planning and implementation. However, the planning processes of all agencies follow a similar basic sequence, varying mostly in the details and with respect to mission focus.

Effective watershed planning is based in watershed science, including understanding of the interactions between hydrologic process and topographic features. The most effective programmatic planning takes advantage of independently researched science. This often materializes in a physical model that links proposed-measure performance to physical indicators of objective achievement, such as indicators of water quality, hydroregime behavior, species population performance, and the integrity of aquatic and riparian ecosystem structure and function. Such models are useful means for facilitating the "institutional memory" needed for adaptive management.

WATERSHED PLANNING PROBLEMS

Many agency documents play up the advantages of watershed planning and minimize the difficulties. Some useful, contemporary sources of insights about problems and constraints include the NRC,[7] The Meridian Institute,[56] and the USEPA.[51,57] The

[a] These four criteria, established by the Water Resources Council[22] for use in plan evaluation, are defined in Chapter 4.

large majority of watershed-planning problems involve chronic and pervasive impediments to communication, coordination, and cooperation. However, these are symptomatic of more fundamental problems. The watershed planning issues, constraints, and common mitigation strategies are described in this section and are summarized in Table 2.1. Issues and constraints fall into six broad categories including fragmented authorities and missions, decision frameworks, stakeholder inclusion, technical, analytical, and motivational.

Fragmented Authority and Mission

The proliferation and fragmentation of government authority has been the most fundamental of the chronic constraints operating on the efficiency, effectiveness, completeness, and acceptability of watershed-planning and management designed to achieve resource development and environmental objectives. Attempts to translate authorized national objectives for resource management and environmental protection into decisions that "promote the general welfare" are complicated by the absence of universally comparable and widely accepted measures of welfare. Whereas some managers and stakeholders accept an economic metric as sufficiently inclusive of all important welfare improvement, others insist that important environmental values cannot as yet, if ever, be measured in monetary units, and yet must be sustained.

Both history and contemporary examples of watershed planning reveal that incomplete and incremental approaches have not sustained public satisfaction for long. This remains as true today as when the USACE stubbornly rejected interagency integration of land and water measures for water supply and flood damage reduction. Growth in the complexity of government at federal, state, and local levels has made interagency cooperation, coordination, and mission integration among its most fundamental operational challenges. Increasing devolution of responsibility to the states and decreasing federal agency funding have reinforced an historic proclivity of federal agencies to align with special interests for competitive advantage against other such associations. While this challenge requires many avenues of effort, an essential element is a comprehensive management-planning framework for identifying the objectives and roles of government at each level, and the integrative measures needed for management success.

Planners, of course, are not the decision-makers who have to make the hard choices that may harm a few to serve the greater good. But the best planning provides reliable information necessary to develop and compare alternative approaches for choosing the most cost-effective ways to realize desirable standards and other objectives in pursuit of improved general welfare, including redistributive compensation for those unjustly harmed by decisions. It also provides information to decision-makers about who benefits and who pays, thereby separating special interest benefits from the more general national interests served by plans.

An important source of local suspicion of state and federal agency information brought to the watershed planning process is perception of bias towards the predilections of certain stakeholders. Since the earliest years of new authorization, the water resource agencies have been pressed by the commercial navigation and

Table 2.1 Summary of the more prevalent watershed planning problems and constraints, and the typical remedies used or recommended from federal agency experience

Watershed Planning Problem/Constraint	Mitigation Strategies
Fragmented Authorities and Missions	
Fragmented agency authorities	Develop more integrative frameworks, better training; MOAs
Fragmented agency sense of mission	Review all missions; more comprehensive training
Unclear agency roles in planning process	Clarification of agency roles early in planning process
Agency planners are too narrowly focused	Improve planner selection, increase training investment
Agency competition and reluctance to share power	Clarify mission statements and integrative frameworks, MOAs
Decision Frameworks	
Incomplete planning–decision framework used	Expand on existing frameworks; develop explicit guidance
Inadequate linking of decision-support system elements	Improve guidance; invest in better decision-support systems
Inflexible, top-down planning process	Clarify planning guidance, facilitate an inclusive public process
Delayed, ambiguous, and misrepresentative communication	Provide fast, clear, accurate, and complete information
Stakeholder Inclusion	
Incomplete stakeholder representation	Develop guidance for stakeholder inclusion; monitor for gaps
Limited stakeholder instinct for decision process	Develop guidance for stakeholder inclusion and education
Stakeholder distrust of planning process	Develop an explicit process and sensitive facilitation
Stakeholder ignorance of watershed process	Develop guidance for stakeholder inclusion and education
Stakeholder biases and inflexibility	Develop guidance for stakeholder inclusion and education
Stakeholder impatience with complex planning process	Describe framework early and give group chance to improve it
Stakeholder inability to understand risk and uncertainty	Develop guidance for stakeholder inclusion and education
Lack of strong, local leadership among stakeholders	Develop stakeholder selection guidance
Lack of objective, trustworthy information	Involve watershed specialists and invest more in research
Unclear definition of trade-offs for stakeholders	Identify all services/costs, improve models/methods
Technical	
Less-than-adequate science at watershed-planning scales	Reexamine approach; proceed adaptively; invest in research

Incomplete data for basic-process understanding	Reexamine approach; proceed adaptively: invest in research
Inadequate data for characterizing watershed attributes	Invest in data; proceed adaptively
Fragmented data management	Integrate fragmented research authorities into national database
Data gathering and management expense	Share data; invest in models/methods that use data efficiently
Analytical	
Watershed boundaries sometimes are ill-defined	Examine relevancy of uncertainty; reexamine approach
Risk and uncertainty are not clearly defined	Invest in representative data and models
Models of natural process are not comprehensive enough	Invest in basic and adaptive-management model development
Models do not link natural and social process	Invest in basic and adaptive-management model development
Ecological models are rudimentary	Invest in basic and adaptive-management model development
Motivational	
Inadequate funding for local watershed planning	Develop guidance for fund-raising
Fragmented sources of funding	Develop guidance for fund-raising and clarify agency roles
Complicated procedures for obtaining funding	Develop guidance for fund-raising and reducing red-tape
Buying local ownership in planning process	Clearly define and fairly distribute benefits and costs

irrigation interests, the land management agencies by the timber and range interests, and the environmental and recreation agencies by the environmental NGOs. Serving the interests of the whole public is not necessarily in the best interest of each agency if their special-interest "customers" feel too compromised by the result. The more visceral motivation to grow, or at least sustain, agency funding and power often competes with thorough integration of planning and management processes through agency cooperation, coordination, and communication.

This is not to say that agencies should avoid missions that incidentally benefit special interests as they serve the more general interests. Special interests pay attention to agency efforts, and often were the prime movers for the agency's legislated authority. The USEPA, for example, is charged with protecting the environment, not with deciding whether it is worthwhile doing so in all cases (in typical fragmented fashion, specific authorities do require benefits assessment); thus, one should expect a "bias" toward environmental interests as long as it serves all Americans well. Similarly, the USDA has a bias in favor of the needs of agricultural interests and appropriately so as long as the general welfare is not diminished. The USACE has a bias toward navigation and flood control interests, but it also has a clear mandate, along with other water management agencies, to develop water resources for the more general welfare through a measure of national economic development while simultaneously protecting the environment according to law. The USFWS also responds favorably to NGO interest in fish and wildlife recreation and endangered species recovery consistent with its mission, which is: "working with others to conserve, protect, and enhance fish, wildlife, and plants and their habitats for the continuing benefit of the American people."[58]

Most agency mission statements recognize that the general welfare should take priority over that of special interest groups, which are rarely even mentioned in contemporary statements. The mission of the USEPA, for example, is to protect the health and life support systems *of all Americans*, but not at unnecessary cost to those same Americans, even if environmental interests would be willing to have all Americans pay a premium to achieve their particular view of protection needs. To that end the USEPA Strategic plan[59] claims the "EPA has been promoting innovation to enhance existing programs and develop new approaches with the potential for achieving better and more cost-effective environmental results." It seems clear that the federal agencies quite consistently hold that the general interest is to be promoted over the special interest, in word if not always in deed. The technocratic approach pursued in the early twentieth century, which often resulted in a close alignment of agencies and special interests, has been firmly rejected by a public that demands more openness and equal access to agency decisions.

An agency's emphasis on serving the needs of a select few can place blinders on agency planners, who are rewarded for facilitating "customer satisfaction." This can translate into diverting benefits for the special interests away from an optimum distribution of benefits across all missions and objectives established to promote the general welfare. Public land and water resource managers underestimated preferences behind the environmental movement and contributed to a distrust that favors strong environmental regulation of resource development agencies. The larger concept of government purpose — the overall improvement of public welfare — continues to be overwhelmed in the immediacy of achieving explicitly authorized objectives.

Despite historic lessons, agencies continue to fragment their approaches to watershed planning, including planner training. Many agency planners have not been schooled in planning theory independent of the processes, and attendant biases, of their own agencies. Even when planning is more enlightened, the pressures to perform in the agency's best interest are intense. Only explicit guidance and performance rewards appropriate for broadening the planning perspective are likely to open up the field of planning vision to achieve results that transcend the limited objectives of each participating agency.

The federal government has slowly adapted to fragmented authorities and missions through various patch-up policies in laws, presidential orders, interagency agreements and understandings, and a more transparent strategic planning process. Natural resource and environmental agencies now have the statutory authority needed to integrate policies into more cohesive and comprehensive planning and management. The impediments now seem to rest less with inadequate authorities than with redirecting budget priorities within agencies toward development of more inclusive and integrative planning frameworks and methodologies. By definition, no one agency can succeed independently of the others.

Inadequate Decision Frameworks

Agency planning frameworks for the most part remain focused on their own missions and exist in various states of completion. This impediment has grown in complexity with each new authority assigned since the early years of the conservation movement. Frustrations over inadequate planning frameworks emerge in broad analyses such as the one by Golden and Rogers[60] entitled *Moving the Watershed Planning Process from Quagmire to Success*. Closer examination of the common agency watershed planning frameworks reveals that differences exist primarily in the details. Much could be accomplished to resolve differences by use of common process terminology, by developing a more coordinated sequence of planning processes, and by developing greater commonality in procedural mechanics. Natural resource and environmental authorities have stabilized in recent years, providing a unique opportunity for agencies to achieve framework integration needs.

Leaving room for flexibility is critical, however. Watershed-based planning often is perceived as inflexible, prolonged, and top-down, too often resulting in a restrictive master plan reminiscent of the paternalistic conservation philosophy that dominated natural resource management before the environmental movement. This model is strongly resisted by local interests desiring more flexibility and local control.[7] Experience shows that sticklers for framework protocol risk failure in watershed planning and that there is no certain way, no one-size-fits-all, that guarantees success. An inflexible, top-down approach to planning is likely to fail to meet the unique needs that typically arise in most complex planning environments. On the other hand, local confusion can be sewn by unnecessary inconsistency in planning terminology and procedures brought to the planning process from the unique perspectives of different agencies.

Decisions are made complex by circumstances that cause numerous decision consequences, affecting numerous interests in uncertain ways. While the basic elements of a decision process might be well defined in principle (e.g., Hammond et al.[54]), the democratization of the decision-making process, with the intent of getting meaningful public input, has yet to be consistently realized. Sorting decision-relevant information from noise is crucial to keeping the process on track. Effective decision-support systems of staff, equipment, consultants, and facilitative data management and analytic tools need to be established and managed to identify, gather, and analyze information needed in formats most appropriate for effective communication among planners, stakeholders, and decision-makers.

The most mundane failures driven by lack of clear, accurate, and complete communication often muddle decision frameworks. This can result from unwise staff budgeting, cost cutting, and administrative attention to detail. A complex planning process often is prolonged by real time requirements for organizing all of the necessary stakeholders, assembling all of the necessary information, waiting on necessary analyses, and allowing plenty of time for stakeholder negotiations. Avoiding delays and confusion associated with poor communication practice are generally worth any additional expense for attention to detail, staff help, equipment, and supplies.

Stakeholder Inclusion

The first rule of effective watershed planning, like other resource management planning, is to include all stakeholders representing all of the relevant objectives and issues. One of the most common reasons for planning dissatisfaction is the failure to account for all negative outcomes and missed opportunities. This lesson was hard learned, and too often forgotten, by the land and water conservation agencies, which focused intently on one or two development purposes. The most basic challenge is assurance that all watershed services get recognized and all public interests are fairly considered in the vision of management success.

Identifying and recruiting the appropriate local leadership is a critical detail that can make or break the planning process if not done carefully. It typically proves effective to invite local political leaders to select a steering committee. This is just as true at the local level as it was at the federal level, when Congress rejected the tightly centralized river-basin management model of the New Deal administration, typified by the TVA, in favor of greater congressional involvement. Because watershed boundaries cross political and administrative boundaries, the need for a coordinating watershed authority complicates the process. At least as important is recruitment of local opinion leaders — often not political leaders — especially if they have a challenging point of view. In addition, the NRC[7] recommended building in scientific review of watershed-management plans to ensure technical feasibility.

Because few stakeholders are well informed about watershed planning, the process needs to be clearly conveyed to all participants. Appropriate guidance indicating the common issues that occur in most watershed planning should provide insight into establishing local planning control, stakeholder representation needs, funding opportunities, and selection of planning facilitators. Many stakeholders enter complex planning processes insufficiently prepared for the time and hard work that good decisions require. Untrained stakeholders are rarely prepared for a rigorous planning process, including clarification of objectives, accounting for uncertainty, or balancing the effort they need to expend against the value of well-informed judgment.[54,55] Many have little understanding of the physical basis for watershed planning. Watershed education is a common need, but experience indicates that learning and understanding via "hands-on" field visits, and in the context of the actual planning process, often prove more useful than more formal approaches.

Most stakeholders, agencies included, are self selected and come to a planning process naturally focused on their own needs. Trade-off negotiations are complicated by stakeholders with long histories of antagonistic interaction and distrust of information brought by their counterparts, often because previous information proved to be inaccurate or deceptive. The different levels of skill among individual facilitators are especially evident in situations where stakeholder inflexibility is anticipated. Nothing helps resolve suspicions of dishonesty in stakeholder trade-off analyses and negotiation more than having trusted representation and information made available from an objective source. Even circumstantial evidence of sidebar alignments of special interests and agencies adds to the suspicion that the "books are cooked" on all sides of the negotiation process.

Technical

Too often, more faith is placed in the science underlying the planning process than is warranted,[7] as in the watershed planning framework first offered by the USEPA.[49] A fair and open analysis of trade-offs is often essential and yet is constrained by the quality of information and planning facilitation. In those situations where the technical difficulties have been eclipsed by the sales pitch for watershed planning, trust can quickly erode as the hard realities of information deficiencies surface. The NRC[7] summarized some of the major challenges that have emerged out of the history of watershed management. One of the major impediments in the past has been technical; simply defining and tracking watershed processes exceeded the science and the tools available for data management and analysis. As a consequence, planning models/protocols have been based on weak databases with marginal credibility. While the information age has done much to alleviate basic data collection and analytical impediments, many technical and analytical problems remain.

The willingness of political leaders to accept scientific generalities in place of the more necessary but expensive technical and analytical specifics is almost as prevalent today as it was when the national forests were first set aside over a century ago. Naiman et al.[61] have reiterated the point that too often the political process "cart" gets ahead of the scientific "horse." Scientists are increasingly unable to respond to the scale of the issues presented before them because policy development and responsive management frequently proceed without "an adequate empirical foundation." The NRC[7] concluded that while many watershed science issues are well understood in principle, many complex and uncertain aspects remain at the scales that now are being increasingly incorporated into watershed-planning processes. "When faced with complexity and uncertainty, watershed planning and management must make provisions for ongoing monitoring and basic science research."[62]

The science of watershed management is no better than the data that go into it. At least two types of data are needed. First, long-term records are needed for watershed processes in experimental watersheds exposed to cultural modification of the kind faced in complex watershed planning. Second, more data are needed to achieve objective multipurpose analyses in a particular watershed setting, often to validate the prediction of watershed models and to monitor the effects of watershed plan implementation. The concern for shortfalls in data quality and quantity shows up repeatedly, such as in the NRC reports on the Total Maximum Daily Load (TMDL) approach[23] and ecological indicators.[63] Social and economic data are more likely to be widely available because U.S. Bureau of the Census data are accessible with appropriate computer software.[7] However, very specific watershed distribution data often need to be gathered first-hand.

The NRC[7] recommends that existing data collection be reviewed carefully by the National Oceanic and Atmospheric Administration, the USGS and other agencies with major data collection responsibilities to ensure the placement and frequency of collection are optimal for needs. If watershed management and planning are to expand to the level anticipated by agencies such as the USEPA, much must be done to set up the most cost-effective data-collection network. The USEPA has recently set as a priority the cooperative development with the USGS of a national geographical information system for water-quality management.[51] The NRC[7] recommends that the Federal Geographic Data Committee, which has the primary responsibility for establishing the National Spatial Data Infrastructure, take the lead in developing national data standards, a central clearinghouse, and maintaining a single national watershed database. Within that context, agencies should be encouraged to coordinate and link their databases.

Analytical

Better decision-support systems are needed to facilitate the development of alternatives and analysis of benefits and costs, including their distribution across stakeholders.[7] Estimating all of the significant outputs from a watershed-management plan in ways that are meaningful for trade-off analysis and plan selection is one of the most problematic technical challenges. Cost–benefit analysis is useful for those values that can be translated into monetary terms but falls short for those increasingly important environmental features that are difficult to represent as dollar values.

Decision-support systems typically include the computer-programmed databases, simulation models, decision models, and user interfaces that aid decision-making. Simulation models in a decision-support system are especially useful for analysis of the benefits and trade-offs associated with each alternative plan. Predictive tools are most useful when uncertainty is somehow expressed or can be characterized by sensitivity analysis. Except for water runoff records and a limited amount of water-quality data, the record for watershed and biological processes is poorly developed for widespread use in uncertainty and risk analysis. Few habitat or ecosystem models address uncertainty and risks at all, let alone well.

Numerous analytical tools are available that meet limited watershed planning needs. Donigian et al.[64] and NRC[7] provided summary reviews and entries to the literature for many of the models in use. The NRC,[7] however, summarized the criteria for simulation in a contemporary watershed model and concluded that no existing model comes close to meeting all of the criteria. Unsophisticated users, such as local commissions faced with meeting state water quality standards, may not be aware of the limitations of such tools and can become easily disappointed with the results. Model assumptions commonly are understated, dismissed, or ignored. Models often are difficult for managers and decision-makers to use.

Motivation

Devolution of authority to states and local governments in recent decades has added a level of complexity not so evident in the early years of planning when the federal government bore much of the cost. Motivating the local community to buy ownership in a watershed-planning process organized at the state and national levels is one key to success, but local motivation often is frustrated by inadequate financing. Experience with small-watershed coalitions suggests that adequate commitment to funding from state and federal sources is a crucial source of motivation. In a recent meeting of the National Watershed Forum,[56] funding difficulties overshadowed other issues discussed by 480 community leaders involved in watershed planning and management oriented toward water-quality outputs. For many local watershed-planning groups, the requirement to conduct planning and management often seems to be driven by obscure benefits that will be derived outside the watershed. An oft-repeated local refrain is the need for state and federal funding aid consistent with the mix of benefits derived within and beyond the managed watershed.

It is important for self-motivated, local planning groups to line up money and in-kind services in advance of starting projects or risk discovering later that stakeholders cannot afford it, creating a sense of failure and negativism toward any subsequent planning process that might be motivated from outside. Progress often will need to be slow and may have to start with inexpensive and easily accomplished measures. Early success needs to be recognized, even celebrated. Local watershed planning organizations encounter many practical impediments to watershed planning stemming from limited local resources and insufficient federal and state facilitative services.[56] Local communities are not always aware of their options across agency services. The USEPA has invested heavily in providing Internet-based information for CWA watershed planning, but it has yet to fully develop alternative options or links to other Internet sites.

SUMMARY

The use of watershed boundaries to frame natural resource and environmental planning and management in the United States can be traced back to the nineteenth century. However, watershed-framed planning and management has had, and continues to have, mixed success. The primary impediments have changed little. As practiced, watershed planning often has been too limited in the scope and clarity of its framework, too frustrated by conflicting objectives and fragmented authorities, too limited by insufficient scientific data and analysis, too ignorant of stakeholder needs, too short on an objective agency process, and too naive about all relevant outcomes. Similar impediments have frustrated other comprehensive approaches to regional planning. Improvements have been slow to materialize despite many theoretical, technical, and political advances, and a national goal that has consistently focused on sustaining and improving human welfare.

Basic to improved planning is a transformation of the tension existing between resource development interests and environmental preservation interests into an inclusive decision-making process that is focused on the general improvement of human welfare. History suggests to some analysts that this is too difficult, if not impossible, and others would argue that it is not desirable. Some environmentalists (and perhaps others) would argue that a focus on human welfare is inappropriate. But the promotion of human welfare, both present and future, is likely to guide government decisions until public preferences and the Constitution are changed. For a more integrated approach to work, all agencies need to be clearly focused on national objectives and a common goal, which history has consistently identified as improvement of the general public welfare. Numerous authorities and administrative communications confirm that neither Congress nor the various presidential administrations believe any single division of government can do this well enough alone because each lacks something necessary to ensure a completely effective result. Only government at large — across federal, state, and local jurisdictions — can hope to aspire to this ideal.

The contentious history of natural resource development and environmental protection is aggravated by the difficulty of integrating resource development, often expressed in dollars, and nonmonetary environmental values into a rational framework for evaluating improvement in public welfare. Legislation born of the environmental movement, while justified by the need to improve checks and balances over resource development agencies and private business via a more complete planning process, further fragmented planning and management authority. It is important in future planning to improve the integration of national economic development objectives with other national objectives that cannot be expressed in monetary terms, and to search for improved ways to overcome this basic impediment to the watershed planning process in pursuit of public welfare. Natural resources and environmental objectives such as protections for threatened and endangered species, which cannot be justified in economic terms, require inclusion among national objectives, as shown by several decades of public support for the principles of the Endangered Species Act. In many instances, however, more can be done to avoid the last-resort constraints of environmental objectives on economic development by more thoughtful agency interaction at the programmatic and strategic planning level, including the development of a more facilitative watershed-based planning framework.

The need to transcend fragmentation and special-interest divisiveness has forced the responsible federal, state, and local agencies into a *de facto* programmatic planning and management cycle that is increasingly watershed based and driven by the need to achieve environmental standards. However, each agency brings its particular experience and concepts to the process, impeding communication and coordination, if that experience is not leavened by respect for other agency missions and the greater public good. While the theory of watershed-based planning and management is strong, the practice remains less certain, not because of political realities — Congress and administrations have their Constitutionally granted prerogatives — but because of the many historical and present-day impediments to integration around common purpose.

Despite the history that has frustrated a more inclusive watershed planning process, review of the existing practice of watershed planning and management in federal agencies shows many commonalties in the different approaches and the likelihood that a single, more inclusive framework can be revealed and refined. Most of the existing impediments are bureaucratic and imposed by a limited sense of agency mission and public interest. The first step in transcending those impediments is to clarify the commonalities among the frameworks in use and to delineate the links among them. Accomplishing this will require compensating for the political realities of decision-making within the watershed approach. In cases where the agencies and political jurisdictions involved have little in common, substantial compensation will be required, but in other cases an examination of commonalities may reveal ample opportunity for the use of an integrated watershed approach.

REFERENCES

1. Marsh, G.P., *Man and Nature or Physical Geography As Modified by Human Nature*, Charles Scribner, New York, 1864.
2. Dana, S.T. and Fairfax, S.K., *Forest and Range Policy: Its Development in the United States*, McGraw-Hill Publishing Company, New York, 1980.
3. deBuys, W., Ed., *Seeing Things Whole: The Essential John Wesley Powell*, Island Press/Shearwater Books, Washington, D.C., 2001.
4. Hays, S.P., *Conservation and the Gospel of Efficiency: The Progressive Conservation Movement, 1890–1920*, Harvard University Press, Cambridge, MA, 1959.
5. Satterlund, D.R. and Adams, P.W., *Wildland Watershed Management*, John Wiley & Sons, Inc., New York, NY, 1992.
6. Douglass, J.E., Annotated bibliography of publications on watershed management by the Southeast Forest Experiment Station, 1928–1970, Research Paper SE-93, U.S. Department of Agriculture Forest Service, Southeastern Forest Experiment Station, Ashville, NC, 1972.
7. NRC, *New Strategies for America's Watersheds*, Committee on Watershed Management, National Research Council, National Academy Press, Washington, D.C., 1999.
8. Pinchot, G., *Breaking New Ground*, Harcourt Brace Jovanovich, New York, 1947.
9. Muckleston, K.W., Integrated water management in the United States, in *Integrated Water Management: International Experiences and Perspectives*, Mitchell, B., Ed., Belhaven Press, New York, 1990.
10. Newell, F.H., What may be accomplished by reclamation, *Ann. Am. Acad. Polit. Social Sci.*, 33, 1909.
11. Arnold, J.L., The Evolution of the 1936 Flood Control Act, Office of History, U.S. Army Corps of Engineers, Fort Belvoir, Virginia, 1988.
12. Mitchell, B., The evolution of integrated resource management, in *Integrated Approaches to Resource Planning and Management*, Lang, R., Ed., University of Calgary Press, Calgary, Alberta, 1990.
13. Helms, D., Natural Resources Conservation Service, in *A Historical Guide to the U.S. Government*, Kurian, G.T., Harahan, J.P., and Kettl, K.F., Eds., Oxford University Press, New York, 1998.
14. White, G.F., *Strategies of American Water Management*, University of Michigan Press, Ann Arbor, MI, 1969.
15. Golley, F.B., *A History of the Ecosystem Concept in Ecology: More Than the Sum of the Parts*, Yale University Press, New Haven, CT, 1993.
16. Odum, E.P., *Fundamentals of Ecology*, W.B. Saunders Company, Philadelphia, PA, 1953.
17. Rudd, R.L., *Pesticides and the Living Landscape*, University of Wisconsin Press, Madison, WI, 1964.
18. Hays, S.P., *Explorations in Environmental History: Essays by Samuel P. Hays*, University of Pittsburgh Press, Pittsburgh, PA, 1998.
19. Carson, R., *Silent Spring*, Houghton Mifflin Company, New York, 1962.
20. NRC, *Restoration of Aquatic Ecosystems: Science, Technology and Public Policy*, National Research Council, Commission on Geosciences, Environment and Resources, Washington, D.C., 1992.
21. Shabman, L., Environmental activities in corps of engineers water resources programs: Charting a new direction, IWR Report-93-PS-1, Institute for Water Resources, U.S. Army Corps of Engineers, Alexandria, VA, 1993.
22. WRC, Economic and environmental principles and guidelines for water and related land resources implementation studies, U.S. Water Resources Council, Washington, D.C., 1983.

23. NRC, *Assessing the TMDL Approach to Water Quality Management*, National Research Council, National Academy Press, Washington, D.C., 2001.
24. NRC, Science and the Endangered Species Act, Prepublication copy of Executive Summary, National Academy of Sciences, Washington, D.C., 1995.
25. Heathcote, I.W., *Integrated Watershed Management: Principles and Practices*, John Wiley & Sons, Inc., New York, 1998.
26. Hoekstra, T.W. and Capp, J., Integrating Forest Management for Wildlife and Fish, General Technical Report NC-122, North Central Forest Experiment Station, Forest Service, U.S. Department of Agriculture, St. Paul, MN, 1988.
27. Kidd, C.V. and Pimentel, D., *Integrated Resource Management: Agroforestry for Development*, Harcourt Brace Jovanovich, New York, 1992.
28. Cicin-Sain, B. and Knect, R.W., *Integrated Coastal and Ocean Management: Concepts and Practices*, Island Press, Washington, D.C., 1998.
29. MacKenzie, S.H., *Integrated Resource Planning and Management: The Ecosystem Approach in the Great Lakes Basin*, Island Press, Washington, D.C., 1996.
30. TAC (Technical Advisory Committee), *Integrated Water Resources Management*, Global Water Partnership, Stockholm, Sweden, 2000.
31. Army Corps of Engineers, Civil Works Program Strategic Plan, Draft Report, U.S. Army Corps of Engineers, Washington, D.C., 2002.
32. Holling, C.S., Bazykin, A., Bunnell, P. et al. *Adaptive Environmental Assessment and Management*, Wiley-Interscience, New York, 1978.
33. Walters, C.J., *Adaptive Management of Renewable Resources*, Macmillan Publishing Company, New York, 1986.
34. Van Dyne, G.M., Ed., *The Ecosystem Concept in Natural Resource Management*, Academic Press, New York, 1969.
35. Grumbine, R.E., What is ecosystem management?, *Conser. Bio.*, 8, 1994.
36. Boyce, M.S. and Haney, A.W., Eds., *Ecosystem Management: Applications for Sustainable Forest and Wildlife Resources*, Yale University Press, New Haven, CT, 1997.
37. Vogt, K.A., Gordon, J., Wargo, J. et al. *Ecosystems: Balancing Science With Management*, Springer-Verlag, New York, 1997.
38. Franklin, J.F., Ecosystem management: An overview, in *Ecosystem Management: Applications for Sustainable Forest and Wildlife Resources*, Boyce, M.S. and Haney, A.W., Eds., Yale University Press, New Haven, CT, 1997.
39. Sedell, J., Sharpe, M., Apple, D.D., Copenhagen, M., and Furniss, M., Water and the forest service, USFS-660, U.S. Department of Agriculture, Forest Service, Washington, D.C., 2000.
40. Rausher, H.M., Lloyd, F.T., Loftis, D.L., and Twery, M.J., A practical decision-analysis process for forest ecosystem management, *Comp. Electron. Agri.*, 27, 2000.
41. Fight, R.D., Kruger, L.E., Hansen-Murray, C., Holden, A., and Bays, D., Understanding human uses and values in watershed analysis, PNW-GTR-489, Pacific Northwest Research Station, U.S. Department of Agriculture, Forest Station, Portland, OR, 2000.
42. NRC *Downstream: Adaptive Management of Glen Canyon Dam and the Colorado River Ecosystem*, National Research Council, National Academy Press, Washington, D.C., 2000.
43. Federal Highway Administration, Memorandum of understanding to foster the ecosystem approach, Federal Highway Administration, 1995. Accessed Oct. 1, 2002 at www.fhwa.dot.gov/legsregs/directives/policy/memoofun.htm.
44. USEPA, Memorandum of agreement between the Environmental Protection Agency, Fish and Wildlife Service and National Marine Fisheries Service regarding enhanced coordination under the Clean Water Act and Endangered Species Act; notice, *Fed. Reg.*, 66, 2001.

45. Cole, R.A., Feather, T.D., and Letting, P.K., Improving watershed planning and management through integration: A critical review of Federal opportunities, IWR Report 02-R-6, U.S. Army Corps of Engineers, Institute for Water Resources, Alexandria, VA, 2002.
46. Yoe, C.E. and Orth, K.D., Planning manual, IWR Report 96-R-21, Institute for Water Resources, U.S. Army Corps of Engineers, Alexandria, VA, 1996.
47. USFS, National forest system land resource management planning; final rule, *Fed. Reg.*, 65, 67513, 2000.
48. USFS, USDA Forest Service strategic plan, 2000 revisions, U.S. Forest Service, Department of Agriculture, Washington, D.C., 2000.
49. USEPA, The watershed protection approach: Annual report 1992, EPA/840/S-93/001, Office of Water, U.S. Environmental Protection Agency, Washington, D.C., 1993.
50. USEPA, Biological criteria: Technical guidance for streams and small rivers. Revised edition, EPA/822/B-096/001, U.S. Environmental Protection Agency, Office of Water, Washington, D.C., 1996.
51. USEPA, Designing an information management system for watersheds, EPA/841/R-97/005, Office of Water, U.S. Environmental Protection Agency, Washington, D.C., 1997.
52. USEPA, Catalog of federal funding sources for watershed protection, EPA/841/B-97/008, Office of Water, U.S. Environmental Protection Agency, Washington, D.C., 1997.
53. USEPA, Protecting and restoring America's watersheds: Status, trends and initiatives in watershed management, EPA/840/R-00/001, Office of Water, U.S. Environmental Protection Agency, Washington, D.C., 2001.
54. Hammond, J.S., Keeny, R.L., and Raiffa, H., *Smart Choices: A Practical Guide to Making Better Decisions*, Harvard Business School Press, Boston, MA, 1999.
55. Gregory, R., Using stakeholder values to make smarter environmental decisions, *Environment*, *42*, 34, 2000.
56. TMI, Final Report of the National Watershed Forum: Building Partnerships for Healthy Watersheds, The Meridian Institute, Arlington, VA, 2001.
57. USEPA, A review of statewide watershed management approaches, Office of Water, U.S. Environmental Protection Agency, Washington, D.C., 2002.
58. USFWS, Mission, Accessed September 30, 2004 at www.fws.gov.
59. USEPA, Strategic plan, EPA/190/R-00/002, Office of the Chief Financial Officer, U.S. Environmental Protection Agency, Washington, D.C., 2000.
60. Golden, B.F. and Rogers, J.W., Moving the watershed planning process from quagmire to success, in Proceedings Watershed '96, A National Conference on Watershed Management and Protection, Washington, D.C., U.S. Environmental Protection Agency.
61. Naiman, R.J., Magnuson, J.J., McKnight, D.M., Stanford, J.A., and Karr, J.R., Freshwater ecosystems and their management: A national initiative, *Science*, 270, 1995.
62. Stanford, J.A. and Poole, G.C., A protocol for ecosystem management, *Perspect. Ecosys. Manage.*, 6, 1996.
63. NRC, *Ecological Indicators for the Nation*, National Research Council, National Academy Press, Washington, D.C., 2000.
64. Donigian, A.S. Jr., Huber, W.C., and Barnwell, T.O. Jr., Models of nonpoint source water quality for watershed assessment and management, in Proceedings Watershed '96, A National Conference on Watershed Management and Protection, Baltimore, Maryland, Washington, D.C., U.S. Environmental Protection Agency.

CHAPTER 3

Introduction to Ecological Risk Assessment in Watersheds

Randall J.F. Bruins and Matthew T. Heberling[a]

CONTENTS

The previous chapter has explained that the emergence, in the 1960s, of scientific and public concern over visible environmental degradation resulted in strong environmental regulations administered by the newly formed U.S. Environmental Protection Agency (USEPA). These regulations subsequently constrained the economic objectives of federal watershed management. A more recent outcome of that scientific concern, and one that has played an important role in shaping many regulatory actions, was the development by the USEPA of methods for ecological risk assessment (ERA). As this book is concerned with the integration of ERA and economic analysis in the management of watersheds, this chapter lays necessary groundwork for this integration by explaining the basic elements of ERA and its uses in watershed management. The goal is to provide a sufficient background to make the succeeding chapters understandable to non-practitioners of ERA; for a comprehensive introduction

[a] The views expressed in this chapter are those of the authors and do not necessarily reflect the views or policies of the U.S. Environmental Protection Agency.

1-56670-639-4/05/$0.00+$1.50

to the topic, readers should refer to the USEPA's *Guidelines for Ecological Risk Assessment*.[1] Readers already familiar with ERA can safely skip the first section of this chapter, titled "Framework and Methods for Ecological Risk Assessment," which summarizes the steps of ERA presented in those *Guidelines*. The last two sections of this chapter present critiques and watershed applications of ERA, respectively, and Chapter 4 complements this brief introduction with a more extensive discussion of the uses of ERA in a project planning framework.

FRAMEWORK AND METHODS FOR ECOLOGICAL RISK ASSESSMENT

The U.S. Council on Environmental Quality has defined risk as "the possibility of suffering harm from a hazard," where a hazard is "a substance or action that can cause harm." Risk assessment is defined as "the technical assessment of the nature and magnitude of risk."[2] The Presidential/Congressional Commission on Risk Assessment and Risk Management defined risk as "the probability of a specific outcome, generally adverse, given a particular set of conditions" and risk assessment as "an organized process … to describe and estimate the likelihood of adverse health outcomes…."[3] Risk assessment thus includes both qualitative description (i.e., the "nature" of a possible "harm") and quantitation (i.e., its "magnitude"). Magnitude can apply both to the harmful effect itself (e.g., how many individuals or populations will be harmed, and to what degree) and to the possibility that the harm will occur. Possibility encompasses the concepts of probability (or likelihood) and uncertainty. In common usage the term *risk* often equates to likelihood, but in risk assessment a naked probability has little meaning apart from a qualitative and quantitative description of the probable harm and of the uncertainty associated with both the harm and its probability. In this text we use the term *adverse effects* rather than *harm*, and we use *risk* to encompass the nature, probability, and uncertainty of adverse effects.

The terms *probability* and *uncertainty* are closely related. "Uncertainty with respect to natural phenomena means that an outcome is unknown or not established and is therefore in question."[4] Uncertainty that is attributable to natural variability ("inherent uncertainty") is considered irreducible and often is described using probability distributions. Uncertainty that is due to incomplete knowledge ("knowledge uncertainty") is considered reducible given additional information.[a,4,5]

ERA is a scientifically based process for framing and analyzing human-caused risks to ecological resources.[1,6–8] In some of its elements it follows a framework defined earlier for human health risk assessment,[9] but it differs because of special problems presented in the assessment of ecological risks. The definition of *human health* is not especially problematic for health risk assessors, and the general public places a high value on human health protection measures (even if there is sometimes debate about what those measures should be).[b] Assessment of risks vis-à-vis human health is therefore

[a] By some definitions, inherent uncertainty is termed *variability*, and the term *uncertainty* is reserved for knowledge uncertainty.[52]
[b] While the World Health Organization has defined human health broadly as "a state of complete physical, mental and social well-being and not merely the absence of disease or infirmity," health risk assessment as practiced by environmental agencies is concerned only with hazards causing damage, injury, or harm.[2,3] Human health, for risk assessors, is thus the absence of these adverse conditions.

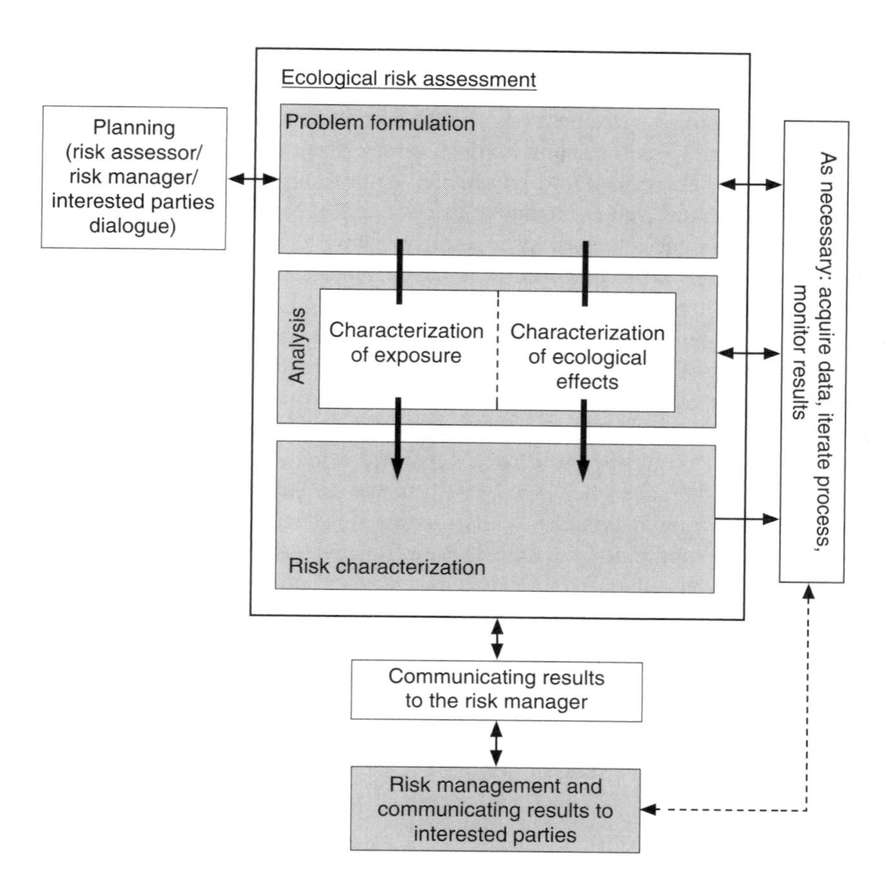

Figure 3.1 Framework for Ecological Risk Assessment (from USEPA[1])

both scientifically meaningful and socially relevant. Some ecologists have defined a parallel concept of *ecosystem health*,[10,11] but the appropriateness of this concept and the means to define and measure it are controversial among ecologists,[12–15] and there is no consensus among the general public about what constitutes ecological health or in which instances, or in what forms, it must be preserved.[16]

Planning

Lacking such a clearly defined reference point, ERA calls for an initial planning step that includes the explicit establishment of ecosystem management goals (Figure 3.1).[1] The planning process is a dialogue among risk assessors, risk managers, and, where appropriate, interested and affected parties ("stakeholders"), to determine the goals and scope of the assessment. However, according to USEPA,[1] planning should be separated from the scientific conduct of the risk assessment proper, to "ensure that political and social issues, though helping define the objectives for the risk assessment, do not bias the scientific evaluation of risk." This separation is consistent with a principle espoused by the National Research Council (NRC);[9] however, its appropriateness is explored further in the next section and in Chapter 9.

ERA planners seek agreement on (1) the decision context, (2) management goals and objectives, and (3) information needs. Characterizing the decision context entails understanding the decisions faced by officials, groups, or citizens regarding an environmental problem, as well as the public values; the legal, regulatory, and institutional factors; the geographic relationships; and the available risk management options that make up the context of those decisions. It also includes identifying risk assessors, risk managers, other specialists, and interested individuals and groups who should be involved in the planning process. Management goals are "general statements about the desired condition of ecological values of concern"[1] whereas management objectives are sufficiently specific to allow the development of measures.[17] Objectives must identify what matters given the decision context (in other words, what valued ecological characteristic should be protected), what protection requires, and what level of improvement, or direction of change, is to be achieved. Examination of informational needs entails determining whether an ERA is warranted and, if so, its scope, complexity, and focus.[17] Suppose, for example, there were concerns over the decline of a sport fishery in a reservoir influenced by municipal effluents and agriculture. Understanding the decision context may require listing the potential regulatory or restorative actions that could be taken by officials, farmers, reservoir users, and other citizens throughout the watershed; involving individuals representing each of those groups; and appreciating the values and the legal and economic interests held within each group. The management goal might be to maintain a viable sport fishery in the reservoir, and objectives might entail a listing of desirable species to be maintained.

Problem Formulation

The USEPA defines problem formulation as "a process of generating and evaluating preliminary hypotheses about why ecological effects have occurred, or may occur, from human activities."[1] It requires (1) the identification of assessment endpoints, (2) the development of one or more conceptual models, and (3) the development of an analysis plan. Assessment endpoints operationalize the valued ecological characteristics identified in the management objectives by identifying those that are both ecologically relevant and susceptible to human-caused stressors and by selecting specific ecological entities, and measurable attributes of those entities, to embody those valued characteristics in the analysis. For example, if a management objective was to maintain a viable fishery for a list of popular recreational species, then assessment endpoints might include population size, mean individual size, and recruitment for those species.

A conceptual model is "a written description and visual representation of predicted relationships between ecological entities and the stressors to which they may be exposed."[1] The visual representation usually takes the form of a box-and-arrow diagram illustrating hypothesized relationships between sources (human activities that produce stressors), stressors (chemical, biological or physical entities that can induce an adverse response), exposure pathways, and receptors (ecological entities that may be adversely affected). An example is presented in Chapter 11 (see Figure 11.2). Initial versions of the conceptual model for a complex problem may be overly detailed;

later versions can be simplified to emphasize only those pathways that figure importantly in the analysis plan.

The analysis plan identifies those hypotheses[a] that are believed to be important contributors to risk, or that can be feasibly reduced through management efforts. The plan specifies data needs, data collection methods, and methods for analysis of existing or newly collected data to confirm, or quantify, the underlying relationships and estimate risks.

Referring again to the reservoir fishery example, if fishery declines are hypothesized to result either from low dissolved oxygen (DO) concentrations caused by excessive nutrient inputs from municipal and agricultural sources or from agricultural pesticide use, diagrams (and accompanying text) would be produced illustrating these hypothesized sources and the pathways of pollutant transport to the lake. The ecological processes specific to each pollutant, nutrient effects on dissolved oxygen levels, pesticide effects on aquatic food webs, and the ultimate effects on the assessment endpoints would also be diagrammed. Following an evaluation of existing data, an analysis plan might call for the analysis of data on pesticide use in the watershed, municipal effluent characteristics, water quality in the lake and its tributaries, and fish populations.

Analysis

Analysis is "a process that examines the two primary components of risk, exposure and effects, and their relationships between each other and [with] ecosystem characteristics."[1] Exposure analysis describes sources of stressors, stressor transport and distribution, and the extent of contact or co-occurrence between stressors and receptors. Exposure analysis may be carried out using environmental measurements, computational models, or a combination of these. The product of exposure analysis is an exposure profile describing the intensity, spatial extent, and timing of exposure. In effects analysis, the effects that are thought to be elicited by a stressor are first identified. Effects of concern are then subjected to an ecological response analysis, which examines the quantitative relationship between the stressor and the response, the plausibility that the stressor may cause the response (causality), and links between particular measures of effect and the assessment endpoints. In the sport fishery example, exposure analysis would examine the magnitude, timing, and spatial dynamics of nutrient inputs; it would also characterize reductions in DO concentrations, since low DO constitutes a secondary stressor potentially affecting the assessment endpoints. Exposure analysis would also characterize the input, fate, and transport, and the resulting water concentrations of pesticides used in the watershed. Effects analysis would include a literature analysis to identify the kinds of effects potentially caused by these stressors and to determine whether exposure–response relationships had been estimated for the same or phylogenetically similar species. It would also evaluate the possibility that the primary effects of one of these stressors on the food base are causing secondary effects in the assessment endpoints. Effects analysis would also examine the relative timing of exposures and observed effects of concern to determine whether there is a causal relationship.

[a] Except as otherwise specified, *hypothesis* in this document refers to a *maintained hypothesis*, or statement thought to be true (i.e., an assumption).

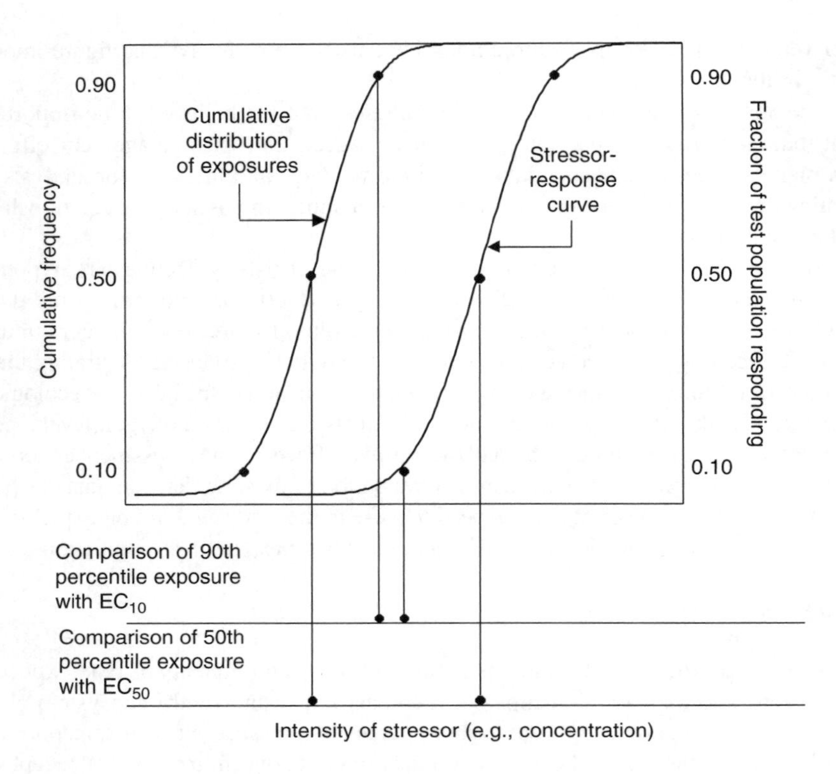

Figure 3.2 Estimation of risk by comparing a cumulative frequency distribution of exposure to a stressor and a stressor-response relationship; EC_X denotes stressor concentration affecting X% of test population (from USEPA[1])

Risk Characterization

Risk characterization is the process of uniting information about exposure and effects, to first estimate and then describe the likelihood of adverse effects of stressors. Risk estimates range in sophistication from simple, qualitative risk ratings (e.g., high, medium, or low), used when information is limited, to comparisons of point estimates of exposure and effective level, to comparisons of probability or frequency distributions of exposure and response.

Figure 3.2 illustrates the latter case. The intensity of exposure to a stressor varies across an assessed population of individuals, and this variability is expressed as a cumulative frequency (curve on left). The fraction of individuals in a tested population that responded to a given intensity of exposure also varied (curve on right). By aligning these curves on the same exposure axis, it is shown that median exposure is below the median level of sensitivity by a relatively large margin, and that 90% of individual exposures are below a level that caused a response in 10% of individuals, albeit by a smaller margin. These data would suggest a very low level of response is expected in the assessed population, assuming the test population adequately represents the assessed population.

Risk descriptions that accompany risk estimates should discuss the adequacy and quality of data on which the assessment is based, the degree and type of uncertainty associated with the evidence, and the relationship of the evidence to the hypotheses of the risk assessment. For example, the exposure and response distributions represented in Figure 3.2 may represent inherent uncertainty that cannot be further reduced, which is due to variability in the environmental distribution of the stressor and in the sensitivity of organisms tested. But there may be knowledge uncertainty associated with the data as well, if the number of exposure measurements or organisms tested was too low to adequately characterize the variability or if there were problems or biases associated with those measurements. There may be knowledge uncertainty concerning whether the response of the wild assessment population is similar to that of the test population, or whether the duration of the test and the endpoints examined were sufficient to characterize the possible effects. Risk descriptions should evaluate all lines of evidence, both supporting and refuting the risk estimates. They should also discuss the extent to which changes predicted in the risk assessment should be termed adverse, including the nature and intensity of expected effects, their spatial and temporal scale, and the potential of affected species or ecosystems to recover.

CRITIQUES OF ECOLOGICAL RISK ASSESSMENT

Using the steps of planning, problem formulation, analysis, and risk characterization, ERA seeks to construct a concise roadmap for science-based decision support — beginning with an inclusive, policy-informed discourse; proceeding through a rigorous process of hypothesis generation, data gathering, and evaluation; and leading to a set of carefully delimited statements about the probabilities of specific, adverse outcomes — to be provided to decision-makers. The process is intended to be flexible; it can employ tiers of increasing specificity (e.g., from screening-level to definitive), and sequences can be iterated as needed before proceeding to subsequent steps (see Figure 3.1).

Nonetheless, ERA has been subject to various criticisms. Some of these pertain to problems of application, others to methodology, and others to the premises underlying the role of science in decision-making. Many are centered on the treatment of scientific uncertainty, and several involve questions of whether science and policy can, or ought to, be separated. It is important to consider these issues openly when the use of ERA is contemplated for decision support, partly to be aware of the potential for misuse of the ERA process, and partly to acknowledge concerns that may be held by many stakeholders.

Some critics have charged that ecological risk assessors are prone to a rather sanguine view of the process, in which long-term laboratory tests of properly chosen sentinel species are assumed to yield results that are stable and adequately predictive of ecosystem responses (see Power and McCarty[18] and ensuing discussion).[19–21] They argue that the variability in stress-response among species and among field sites is sometimes ignored, and that biological regulatory mechanisms operating at the level of the field population or the ecosystem can confound the conventional interpretation of laboratory test results. These criticisms highlight the importance of using multiple lines of evidence (e.g., both field and laboratory observations) and making a full presentation of assumptions and uncertainties when characterizing risk, as called for in the ERA *Guidelines*.[1]

A common mistake in the analysis stage of ERA is ignoring statistical power, that is, the probability that a given experiment or monitoring study will detect an effect if it actually exists.[22–24] If hypothesis testing fails to reject the null hypothesis (no significant effect is detected), statistical power analysis determines the level of confidence that can be placed in the negative result; when power is low, a greater need for precaution is indicated.

Where the above criticisms pertain largely to ERA methods and applications, more fundamental issues have also been raised (see especially papers from a symposium held in 1994 entitled "Ecological Risk Assessment: Use, Abuse and Alternatives"[25] that call for the use of precautionary rather than risk-based approaches [e.g., the Wingspread Statement on the Precautionary Principle[26]]). Critics claim that (a) unintended ecological consequences of past actions demonstrate that ecosystems are too complex to be predictable under novel conditions, and (b) in view of these inherent uncertainties, it is immoral to rely upon the results of even a well-conducted risk assessment if alternative (albeit more costly) courses of action exist that appear to pose less hazard.[27,28] A related argument (see the Wingspread Statement) adds that the burden of removing uncertainty must lie with the proponent of any potentially risky action rather than with society at large. These arguments sometimes portray even the unbiased risk assessor as an enabling participant, who by virtue of his or her expertise lends a cloak of legitimacy to an intrinsically unjust process.[29] More often, the assessor is portrayed as biased (e.g., holding a narrowly reductionist worldview or having an organizational conflict of interest) or intentionally deceptive. In the end, according to this critique, ERA is at best unreliable for decision-making and at worst a tool to facilitate ecosystem exploitation.

Some of these criticisms pertain to governance structures themselves rather than to ERA per se. If indeed the validity of the governance structure underlying an environmental management effort is itself in dispute, then the trust that is necessary for an effective planning dialogue may be impossible to obtain, and ERA may be ineffective. In most cases, however, if an effective dialogue as described in the ERA *Guidelines* can be established, then many of the practical and fundamental issues that critics raise can be accommodated, even where deep-seated disagreements exist. As stated above, an effective planning dialogue clarifies the decision context, including participant values, burden of proof, institutional factors, and management alternatives, and it ensures that the assessment is not too narrowly conceived. Organizational interests and biases can be made clear at this stage as well. The *Guidelines* also state that the appropriateness of including stakeholders depends on the circumstances; in some cases, existing law and policy might narrowly prescribe the terms for conducting an assessment. However, it is unlikely that such a restriction ever is appropriate for assessment of problems in watersheds, where there are multiple sources and stressors, a variety of resources to protect, various regulatory authorities and incentive programs, and a need for broad community support.

In summary, through an inclusive planning dialogue and careful treatment of uncertainty, an ERA conducted according to the *Guidelines* can address many of the practical and philosophical criticisms that have been leveled against risk assessment. Further steps may need to be considered as well. Whereas the *Guidelines* argue for a strict delineation of policy and science — the planning process, where stakeholders may participate, remains "distinct from the scientific conduct of [the] risk assessment"[1] — other scientists have argued that the limits of science should be acknowledged not only at the planning stage but throughout the assessment. When risk assessors are forced to make judgments that go beyond the limits of the data, as they routinely do, they move from the realm of science into what Alvin Weinberg[30] has termed *trans-science*. These judgments reflect the knowledge, experience, and even cultural values of the assessor,[31] and they cannot, according to Weinberg, be viewed as free of bias. Funtowicz and Ravetz[32,33] likewise have suggested that as uncertainties, decision stakes, and urgency increase, problem-solving strategies correspondingly must progress from "applied science," to "professional consultancy," to "post-normal science." Post-normal science does not pretend to be value-free or ethically neutral, and it makes use of deliberation. The NRC[34] acknowledged that deliberation in the problem formulation stage of risk assessment that includes interested and affected parties can elicit insights that would not occur to assessors and managers alone. They thus called for deliberation involving decision-makers and interested and affected parties throughout the risk assessment process. The participation theme will be discussed further in Chapter 9.

WATERSHED APPLICATIONS OF ECOLOGICAL RISK ASSESSMENT

The use of the watershed as a geographic unit for the planning and management of water supply and flood control in the United States dates to the late nineteenth and early twentieth centuries, but its use for ecosystem protection is more recent (see Chapter 2). After the formation of the USEPA in 1970, the need for such an approach grew steadily as environmental regulatory programs proliferated and yet were spatially uncoordinated and lacked efficient mechanisms for sharing information. Also during this period, point-source pollution problems were beginning to be solved through the issuance of discharge permits, bringing to light the less tractable problems of nonpoint sources and habitat modification. Finally, in the 1990s environmental groups began to sue the USEPA over its failure to go beyond the source-by-source issuance of discharge permits in the thousands of cases where these had proved insufficient to rectify water quality impairment. Dozens of court actions, brought under the water quality standards (WQS) provision of the Clean Water Act of 1972 (CWA), required the states or the USEPA to determine, on a whole-water-body basis, the total maximum daily load (TMDL) allowable from all sources.

For these reasons, in the 1990s the USEPA began to encourage the use of a "watershed protection approach" (later termed simply the "watershed approach") for evaluating and managing threats to freshwater and estuarine ecosystems,[35–39] and they defined a framework for that process (a discussion of this and other frameworks is presented in Chapter 9). This approach provided an effective way of spatially delimiting ecological resources and the threats to those resources, engaging stakeholders in protection efforts, and promoting management actions that were concerted rather than piecemeal. Thus, the watershed protection approach focused on goal setting, partnerships, and management. Early USEPA guidance on the approach did not describe a role for ERA; there was an emphasis on procedures for calculating TMDLs,[40] but these were aimed at determining how to meet numeric WQS rather than at determining risks per se (see further discussion of WQS in Chapter 6). However, WQS do not address several aquatic ecological problems, including those due to hydrologic modification (e.g., water withdrawal, flow control, or development-related changes in runoff and recharge patterns), stream channel modification, removal of riparian vegetation, and introduction of nonnative species. Nor can they address chemicals for which no standards have been defined, indicate which of several pollutants may be causing an observed impairment, nor indicate whether a given protective or restorative measure, if implemented, will reduce the pollutant successfully. Nor can WQS adequately address problems whose severity is a function of spatial scale or the interactions of multiple stressors. Even motivated and involved teams of citizen and governmental partners can fail to achieve ecological improvements when risks in a watershed are not adequately understood. These are questions ERA is geared to address.

Therefore, ERA has a significant role to play as a tool for watershed management.[41] Five watershed ecological risk assessment (W-ERA) case studies were initiated by USEPA in 1993,[42,43] and results for several of these recently have been published.[41,44–48] The case studies were initiated to evaluate the feasibility of applying the ERA process to the complex context of watershed management. Watersheds were selected for study on the basis of data availability, identification of local participants, diversity of stressors, and significant and unique ecological resources. The watersheds selected were the Big Darby Creek in central Ohio; the Clinch River Valley in southwest Virginia and northeast Tennessee; the Platte River watershed in Colorado, Wyoming, and Nebraska, with special emphasis on the Big Bend Reach in south central Nebraska; the Middle Snake River in south central Idaho; and Waquoit Bay on the southern shore of Cape Cod in Massachusetts (Figure 1.1). These watersheds comprised different surface water types, stressors, scales, management problems, socioeconomic circumstances, and regions.

An initial review of the progress of these assessments through the problem formulation stage[43,49] found that ERA provided formal and scientifically defensible methods that were a useful contribution to a watershed management approach. They also found that the analyses in these five cases had not been as strongly linked to watershed management efforts as would be desired. However, subsequent experiences from these assessments have suggested that following W-ERA principles increases the likelihood that environmental monitoring and assessment data are considered in decision-making.[41,50,51] The three major principles that proved most beneficial were (1) holding regular meetings between scientists and managers to establish assessment goals and to share interim findings that could be of immediate value to managers, (2) using assessment endpoints and conceptual models to understand and communicate cascading effects and identify the most significant ecological concerns, and (3) combining data from many sources into an overall analytic framework, within which multiple stressor analysis is made feasible.[41] Chapters 10–12 will present the findings of economic studies that were funded in 1999 in three of those watersheds to further utilize the ERA results and extend their value for decision-making.

In addition to these USEPA applications of W-ERA, which sought to improve understanding of watershed condition and threats and, in a broad sense, to identify management needs, ERA has also been used by the U.S. Army Corps of Engineers (USACE) in the context of project planning for watershed restoration. While the principles and terminology are similar to those presented in this chapter, their adaptation to a planning process that explicitly includes the comparison of project alternatives is an important step toward the development of a framework that can encompass both ERA and economic analysis, since enabling that comparison is the crux of the integration problem. Therefore, the USACE approach is presented in detail in Chapter 4. While this volume does not include a case study applying the USACE approach, a hypothetical example illustrating its use is presented as Appendix 4-A.

REFERENCES

1. USEPA, Guidelines for ecological risk assessment, EPA/630/R-95/002F, Risk Assessment Forum, U.S. Environmental Protection Agency, Washington, D.C., 1998.
2. Cohrssen, J.J. and Covello, V.T., *Risk Analysis: A Guide to Principles and Methods for Analyzing Health and Environmental Risks*, U.S. Council on Environmental Quality, Washington, D.C., 1989.
3. PCCRARM, Framework for environmental health risk management, Presidential/Congressional Commission on Risk Assessment and Risk Management, Washington, D.C., 1997.
4. NRC, *Risk Analysis and Uncertainty in Flood Damage Reduction Studies*, National Research Council, Commission on Geosciences, Environment and Resources, Washington, D.C., 2000.
5. Morgan, M.G. and Henrion, M., *Uncertainty: A Guide to Dealing With Uncertainty in Quantitative Risk and Policy Analysis*, Cambridge University Press, Cambridge, UK, 1990.
6. Gentile, J.H., Harwell, M.A., Van Der Schalie, W.H., Norton, S.B., and Rodier, D.J., Ecological risk assessment: A scientific perspective, *J. Hazar. Mater.*, 35, 241, 1993.
7. Suter, G.W., *Ecological Risk Assessment*, Lewis Publishers, Boca Raton, FL, 1993.
8. USEPA, Framework for ecological risk assessment, EPA/630/R-92/001, Risk Assessment Forum, U.S. Environmental Protection Agency, Washington, D.C., 1992.
9. NRC, *Risk Assessment in the Federal Government: Managing the Process*, National Research Council, National Academy Press, Washington, D.C., 1983.
10. Rapport, D.J., What constitutes ecosystem health?, *Perspect. Biol. Med.*, 33, 120, 1989.
11. Schaeffer, D.J., Herricks, E.E., and Kerster, H.W., Ecosystem health I: Measuring ecosystem health, *Environ. Manage.*, 12, 445, 1988.
12. Simberloff, D., Flagships, umbrellas, and keystones: Is single-species management passe in the landscape era?, *Biol. Conserv.*, 83, 247, 1998.
13. Suter, G.W., A critique of ecosystem health concepts and indices, *Environ. Toxicol. Chem.*, 12, 1533, 1993.

14. Wicklum, D. and Davies, R.W., Ecosystem health and integrity?, *Can. J. Bot.*, 73, 997, 1995.
15. Lackey, R.T., Values, policy, and ecosystem health, *BioScience*, 51, 437, 2001.
16. Hood, R.L., Extreme cases: A strategy for ecological risk assessment in ecosystem health, *Ecosyst. Hlth.*, 4, 152, 1998.
17. USEPA, Planning for ecological risk assessment: Developing management objectives. External Review Draft, EPA/630/R-01/001A, Risk Assessment Forum, Office of Research and Development, U.S. Environmental Protection Agency, Washington, D.C., 2001.
18. Power, M. and McCarty, L.S., Fallacies in ecological risk assessment practices, *Environ. Sci. Technol.*, 31, 370A, 1997.
19. Mayer, F., Pittinger, C., Verstee, G.D., et al., Letter to the editor, *Environ. Sci. Technol. News Res. Notes*, 3, 116A, 1998.
20. Suter, G.W., Letter to the editor, *Environ. Sci. Technol. News Res. Notes*, 3, 116A, 1998.
21. Power, M. and McCarty, L.S., Authors' response, *Environ. Sci. Technol. News Res. Notes*, 3, 117A, 1998.
22. Peterman, R.M. and M'Gonigle, M., Statistical power analysis and the precautionary principle, *Mar. Pollut. Bull.*, 24, 231, 1992.
23. NRC, *Assessing the TMDL Approach to Water Quality Management*, National Research Council, National Academy Press, Washington, D.C., 2001.
24. Suter, G.W., Abuse of hypothesis testing statistics in ecological risk assessment, *Hum. Ecol. Risk Assess.*, 2, 331, 1996.
25. Mazaika, R., Lackey, R.T., and Friant, S.L., Special issue — ecological risk assessment: Use, abuse and alternatives, *Hum. Ecol. Risk Assess.*, 1, 1995.
26. Montague, P., Headlines: The precautionary principle, *Rachel's Environ. Hlth. Weekly*, 586, 1998.
27. O'Brien, M.H., Ecological alternatives assessment rather than ecological risk assessment: Considering options, benefits and hazards, *Hum. Ecol. Risk Assess.*, 1, 357, 1995.
28. O'Brien, M.H., *Making Better Environmental Decisions*, MIT Press, Cambridge, MA, 2000.
29. Pagel, J.E. and O'Brien, M.H., The use of ecological risk assessment to undermine implementation of good public policy, *Hum. Ecol. Risk Assess.*, 2, 238, 1996.
30. Weinberg, A.M., Science and its limits: The regulator's dilemma, *Issues Sci. Technol.*, 59, 1985.
31. Shabman, L.A., Environmental hazards of farming: Thinking about the management challenge, *South. J. Agric. Econ.*, 22, 11, 1990.
32. Funtowicz, S.O. and Ravetz, J.R., Risk management as a postnormal science, *Risk Anal.*, 12, 95, 1992.
33. Funtowicz, S.O. and Ravetz, J.R., A new scientific methodology for global environmental issues, in *Ecological Economics: The Science and Management of Sustainability*, Costanza, R., Ed., 1991, 10, 137.
34. NRC, *Understanding Risk: Informing Decisions in a Democratic Society*, National Research Council, National Academy Press, Washington, D.C., 1996.
35. USEPA, The watershed protection approach: An overview, EPA/503/9-92/002, Office of Water, U.S. Environmental Protection Agency, Washington, D.C., 1991.
36. USEPA, The watershed approach: Annual report 1992, EPA/849/S-93/001, U.S. Environmental Protection Agency, Washington, D.C., 1992.
37. USEPA, Watershed protection: A statewide approach, EPA/841/R-95/004, Office of Water, U.S. Environmental Protection Agency, Washington, D.C., 1995.
38. USEPA, Watershed protection: A project focus, EPA/841/R-95/003, Office of Water, U.S. Environmental Protection Agency, Washington, D.C., 1995.

39. USEPA, Watershed approach framework, EPA/840/S-96/001, Office of Water, U.S. Environmental Protection Agency, Washington, D.C., 1996.
40. USEPA, Draft guidance for water quality-based decisions: The TMDL process (Second Edition), EPA/841/D-99/001, U.S. Environmental Protection Agency, Office of Water, Washington, D.C., 1999.
41. Serveiss, V.B., Applying ecological risk principles to watershed assessment and management, *Environ. Manage.*, 29, 145, 2002.
42. CENR, Ecological risk assessment in the federal government, CENR/5-99/001, Committee on Environment and Natural Resources of the National Science and Technology Council, Washington, D.C., 1999.
43. Butcher, J.B., Creager, C.S., Clements, J.T., et al. Watershed level aquatic ecosystem protection: Value added of ecological risk assessment approach, Project No. 93-IRM-4(a), Water Environment Research Foundation, Alexandria, VA., 1997.
44. USEPA, Ecological risk assessment for the Middle Snake River, Idaho, EPA/600/R-01/017, U.S. Environmental Protection Agency, Office of Research and Development, National Center for Environmental Assessment, Washington, D.C., 2002.
45. USEPA, Clinch and Powell Valley Watershed ecological risk assessment, EPA/600/R-01/050, U.S. Environmental Protection Agency, Office of Research and Development, National Center for Environmental Assessment, Washington, D.C., 2002.
46. USEPA, Waquoit Bay Watershed ecological risk assessment: The effect of land derived nitrogen loads on estuarine eutrophication, EPA/600/R-02/079, U.S. Environmental Protection Agency, Office of Research and Development, National Center for Environmental Assessment, Washington, D.C., 2002.
47. Diamond, J.M. and Serveiss, V.B., Identifying sources of stress to native aquatic fauna using a watershed ecological risk assessment framework, *Environ. Sci. Technol.*, 35, 4711, 2001.
48. Valiela, I., Tomasky, G., Hauxwell, J., et al., Producing sustainability: Management and risk assessment of land-derived nitrogen loads to shallow estuaries, *Ecol. Appl.*, 10, 1006, 2000.
49. SAB, Advisory on the problem formulation phase of EPA's watershed ecological risk assessment case studies, SAB-EPEC-ADV-97-001, Science Advisory Board, U.S. Environmental Protection Agency, Washington, D.C., 1997.
50. Diamond, J.M., Bressler, D.W., and Serveiss, V.B., Diagnosing causes of native fish and mussel species decline in the Clinch and Powell River watershed, Virginia, USA, *Environ. Toxicol. Chem.*, 21, 1147, 2002.
51. USEPA, Report on the watershed ecological risk characterization workshop, EPA/600/R-99/111, Office of Research and Development, National Center for Environmental Assessment, Washington, D.C., 2000.
52. USEPA, Guiding principles for Monte Carlo analysis, Risk Assessment Forum, U.S. Environmental Protection Agency, 1997.

CHAPTER 4

A Framework for Risk Analysis for Ecological Restoration Projects in the U.S. Army Corps of Engineers

Heida L. Diefenderfer, Ronald M. Thom, and Keith D. Hofseth

CONTENTS

1-56670-639-4/05/$0.00+$1.50

INTRODUCTION

Ecosystem restoration is an increasingly important part of the civil works mission of the U.S. Army Corps of Engineers,[a] and restoration is often undertaken at the watershed scale[1,2] or at the scale of major watershed features such as delta marshes.[3] The effective planning of restoration projects requires the assessment of ecological risks.

This chapter is condensed from the U.S. Army Corps of Engineers' (USACE) Institute for Water Resources framework document that provides the general planner with a basic understanding of risk analysis in the USACE six-step ecosystem restoration planning process. The USACE objective in ecosystem restoration, one of the primary missions of the USACE Civil Works program, is to contribute to national ecosystem restoration by measurably increasing the net quantity and quality of desired ecosystem resources.[4] The focus of this chapter is on risk analysis: identifying the range of possible outcomes from alternative ecosystem restoration actions, assessing the potential for achieving the desired outcome, characterizing the likelihood of adverse consequences, and communicating these findings to stakeholders and decision makers.

Specifically, the USACE ecosystem restoration objective is "to restore degraded ecosystem structure, function, and dynamic processes to a less degraded, more natural condition."[4] This is further defined in USACE guidance, which states that "restored ecosystems should mimic, as closely as possible, conditions which would occur in the area in the absence of human changes to the landscape and hydrology. Indicators of

[a] Section 206 of the Water Resources Development Act of 1996.

success would include the presence of a large variety of native plants and animals, the ability of the area to sustain larger numbers of certain indicator species or more biologically desirable species, and the ability of the restored area to continue to function and produce the desired outputs with a minimum of continuing human intervention."[4]

In this chapter, a conceptual model of the site and landscape is advocated as a central organizing structure within the six-step process to achieve these objectives.[5] This is in response to USACE directives that restoration projects be conceived in a systems context[4] using an ecosystem or watershed approach.[6] The incorporation of ecological tools and concepts into the USACE planning process for ecosystem restoration is evolving.[7–9] The conceptual model delineates the empirical quantities to be addressed in risk analysis and modeling. Thus, this chapter describes an integration of concepts and tools from the science of ecological restoration with proven federal project planning processes. This integration, incorporating risk analysis into restoration planning, was called for by the USACE Evaluation of Environmental Investments Research Program.[10]

The risk analysis process requires planners to recognize and communicate the degree of uncertainty in each planning variable. The sharing of uncertainty information across a multidisciplinary planning team facilitates the identification of key variables affecting achievement of the planning objectives. The identification and inclusion of stakeholders further strengthens the knowledge base.[10] This process elevates risk management decisions from the sole province of the technical expert to the realm of the planning team and decision-makers.

Although the planning process is described in six distinct steps, these steps are iterative and often are carried out simultaneously in practice; the planning process is not linear. Planners and analysts work back and forth through the six steps until a comprehensive picture develops, which is communicated using the six steps as the reporting outline. Risk analysis within this context has the same character. The approach for incorporating risk analysis into the project-planning process provides direction intended to help the planner:

- Identify the acceptable levels of uncertainty at the start of the planning process.
- Use conceptual and numerical models to communicate the planning team's understanding of the ecosystem to others and to reduce the risk of misspecifying the system.
- Consider the uncertainty associated with the variables chosen to measure project effects.
- Use alternative designs to manage identified uncertainty.
- Use risk information to eliminate alternatives with unacceptable risk from consideration.
- Incorporate risk analysis into the USACE's four criteria of effectiveness, efficiency, completeness, and acceptability.
- Use an alternative's irreducible uncertainty as an attribute to be considered along with other attributes in the comparison of alternative plans.
- Use risk information in the final plan selection process.

The proposed approach is applicable to ecosystem restoration planning. The framework is sufficiently flexible to be scaled to projects of any size or budget; the degree of specification and data gathering can be tailored to the effort. The framework can be applied to studies of restoration, creation, reclamation, or protection alternatives.

This chapter makes simplifying assumptions to allow a focus on incorporating risk information into the planning and decision-making process. There are three other efforts associated with this framework document, which provide the technical detail needed to develop the necessary statistics. They offer information and guidance for incorporating risk assessment into cost-estimation and biological and hydrologic modeling. Reports on the latter two subjects have not yet been published. Three publications are available regarding cost-estimation: Noble et al.[11] is a postconstruction analysis comparing project expectations to outcomes, and Yoe's reports[12,13] provide guidance and demonstrate cost-estimation when there is uncertainty.

This chapter is divided into three sections. The first provides definitions and a brief background of risk analysis in USACE planning. The second develops the role of risk assessment in each of the six Planning Steps: (1) identifying problems and opportunities, (2) inventory and forecast, (3) plan formulation, (4) evaluation of plans, (5) comparison of alternatives, and (6) plan selection. The conceptual model is introduced in Planning Step 2, inventory and forecast. In Planning Step 3, plan formulation, habitat-modeling methods are detailed. The third section is a brief conclusion. An appendix (Appendix 4-A) provides a fully developed example of a tidal wetland restoration planning process, demonstrating the application of this approach.

BACKGROUND AND DEFINITIONS

Risk and Uncertainty

The term *risk* incorporates the notion of a negative or undesirable outcome.[14] This may take the form of actual harm to the environment, failure to achieve the planning objectives, or not using the most cost-effective measure to achieve a planning objective. The term *uncertainty* is used to describe "a lack of sureness about something or someone."[a] Uncertainty exists whenever there is doubt about an event, a piece of information, or the outcome of a process. Uncertainty can be attributed to two sources: (1) the variability of processes (inherent variability), or (2) incomplete knowledge (knowledge uncertainty). *Probability* is a way of quantifying uncertainty.[15]

Inherent variability is the ordinary variability in a system. In nature, it refers to the irreducible randomness of natural processes. In man-made systems it refers to the vagaries of the system; this randomness is irreducible from the perspective of the risk analyst. In the ecosystem restoration context, uncertainties related to inherent variability include things such as stream flow, assumed to be a random process in time; soil properties, assumed to be random in space; or the success rate of propagules deployed to revegetate a project area. Inherent variability is sometimes called *aleatory uncertainty.*

Knowledge uncertainty deals with a lack of understanding of events and processes, or with a lack of data from which to draw inferences; by assumption, such a lack of knowledge is reducible with further information. Knowledge uncertainty is sometimes called *epistemic uncertainty.*

[a] The definitions of uncertainty in this section are modified from the report of the National Research Council Commission on Geosciences, Environment and Resources.[48]

Many of the terms used to describe sources of error, uncertainty, and risk in the literature of risk analysis can be collapsed into the two previously named sources. For the purpose of risk analysis in USACE ecosystem restoration, the taxonomy used to describe the uncertainty is not as important as the source of the uncertainty.

Risk in USACE Ecosystem Restoration Projects

"First, do no harm," the principle of the medical doctor, is as applicable to the management of an ecological system as it is to the health of a human being. Identifying and documenting the structure and function of ecosystem components by creating a conceptual model elevates the understanding of causal relationships and thus reduces the risk of unintended consequences from management actions. The examination of restoration project scale, the size and pattern of a project relative to the landscape and active stressors, also helps to manage risk. By reducing the risk of unintended consequences, including ecological and economic consequences, risk analysis contributes to maximizing the cost-effectiveness of a restoration alternative.

The approach described in this report is not typical of health or ecological risk assessments, which typically focus on identifying the risks associated with sources of environmental stressors. Rather, the focus of risk analysis in the USACE planning context is to identify the range of possible and desirable outcomes of alternative environmental restoration actions, as well as the potential for any unwanted or adverse consequences from those actions.

Timing of Risk Analysis

Risk analysis is not something that is appended to the planning process or undertaken after a management measure has been selected. Effective risk analysis is woven into the iterative planning process, strengthening resulting decisions.

Iterative refinement of the problem statement and analysis[15] is key to restoration planning, from defining the objectives through the completion of post-construction monitoring. Morgan and Henrion state, "We cannot overemphasize the importance of the difference between this iterative conceptualization of the process of policy analysis and the simple linear approach."[15] New information informs redesign of the project or reconceptualization of the problem, and contributes to the body of knowledge that planners can apply to other projects.

Risk Analysis

Risk analysis includes three components: risk assessment, risk management, and risk communication.[16] Risk assessment is the process of identifying the variety of potential outcomes from a decision and the nature, likelihood, and uncertainty of adverse outcomes (see the "Planning" section in Chapter 3). Risk management describes the actions of decision-makers in response to the identified risk. Responses may range from doing nothing and accepting the chance of a negative outcome, to buying insurance to transfer the risk to others, to implementing a management measure to reduce or eliminate the risk. Risk communication is the vital step of informing the stakeholders and decision-makers of the risk assessment findings and risk management options.

It is important to understand that risk analysis — assessment, management, and communication — occurs throughout the planning process. Properly understood, risk analysis is an analytical approach to every decision.

PROCEDURES: THE PLANNING PROCESS

The risk analyses discussed in the following sections correspond to the key variables involved in each of the six steps of the USACE planning process for ecosystem restoration: identifying problems and opportunities, inventory and forecast, plan formulation, evaluation of plans, comparison of alternatives, and plan selection.

Planning Step 1: Identifying Problems and Opportunities

The problem — perceived degradation of ecosystem properties and reduction in related resources — must first be clearly stated. The first place to consider risk and uncertainty is in the selection of planning objectives. Planning objectives give a rational focus to the planning process. Optimal objectives for restoration projects will reflect a watershed, or other ecosystem, perspective.

Federal Objective

In ecosystem restoration, the federal objective is to "restore degraded significant ecosystem structure, function, and dynamic processes to a less degraded, more natural condition."[4] The federal objective becomes a goal for USACE planners in each ecosystem restoration planning study.[17] USACE planners then work with the stakeholders to develop planning objectives to meet this goal.

Planning Objectives (Outcomes) versus Design Objectives (Outputs)

USACE planners and stakeholders identify project goals and objectives as follows:

- **Goal** — A clear statement of the project targets, which are not necessarily measurable.
- **Planning Objective** — The outcome the project is designed to achieve; an index of accomplishment that may be nonmonetary or monetary. An example is increasing the population of an endangered species through a salt marsh habitat restoration. Planning objectives often are not measurable, and a lesser index of goal accomplishment is used at the risk of misrepresentation of goal accomplishment; the lesser index used generally is the design objective.
- **Design Objective** — The material output: the result of changes to physical, chemical, or biological components of the ecosystem caused by the project. An example is the set of more natural habitat features and processes required by an endangered species, including habitat connections to recolonization sources or migratory pathways.

Planning objectives for ecosystem restoration are analogous to the objectives defined in the planning phase of ecological risk assessment (see "Planning" in Chapter 3).

These objectives can take many forms, but regardless of the primary objective, an ecosystem or watershed perspective is needed to achieve them. Objectives may include, for instance, the restoration of populations of a single threatened or endangered species; a group of species, such as fish; or a community or ecosystem. One of the hallmarks of a good design objective is measurability. Setting the criteria to be measured will help to determine the requirements for risk analysis. If the measurement is the culmination of disparate estimates, it is analyzed and characterized best with risk analysis. Habitat units are often the measure used in USACE environmental studies. Estimates of habitat units are especially prone to variability and are well suited to risk analysis.

The objectives of a project will shape the conceptual models to be developed in Planning Step 2, inventory and forecast. The planning goal of the project, after the federal objective mentioned earlier, might be to restore a degraded significant salt marsh ecosystem's structure, function, and dynamic processes to a less degraded, more natural condition. Planning objectives might include reestablishing historical salt marsh areas and achieving sustainable populations of particular bird and fish species that use the habitat. Measurable design objectives might be the increased area of salt marsh; increased resting, feeding, and nesting areas for waterfowl; and increased feeding habitat and low tide refuge for fish. During Planning Step 1, when it is necessary to consider the criteria that will be established to assess the success or failure of the project, an ecosystem performance model can be developed to characterize potential progress toward the project goal.

Criteria and General Model for Ecosystem Performance

The general model for ecosystem performance (see Figure 4.1) provides the general direction with respect to structure and function that the ecosystem is expected to take on its trajectory toward meeting the project goal. Under a restoration scenario, the goal is to move the system from a degraded condition to one that is less degraded and more desirable. For management purposes, it is assumed that there is a positive relationship between the structure and function of an ecosystem. The natural structure of an ecosystem, habitat, or community has a corresponding functional condition, and to the extent that this is predictable, this information may be used to construct the ecosystem performance model.

Although the optimal condition is shown in the upper right corner of Figure 4.1, optimal functionality can occur in other system states. For example, the greatest net primary productivity of a marsh habitat may occur where only a subset of the species normally found in a climax community is present. Research has shown that a few species may account for the majority of productivity, and other species are redundant.[18] Hence, choosing the species to focus restoration on to restore optimal productivity is critical (see, for example, Hooper and Vitousek[19]). In addition, partial restorations, if done with the correct species, may produce optimal results. However, there is uncertainty as to which species have a significant impact on processes in which ecosystems.[20]

Moderate disturbances of a climax community may result in the greatest numbers of species as well as productivity.[21,22] This is because disturbance opens space for invasion of smaller, less competitive, but highly invasive species. Where sites may

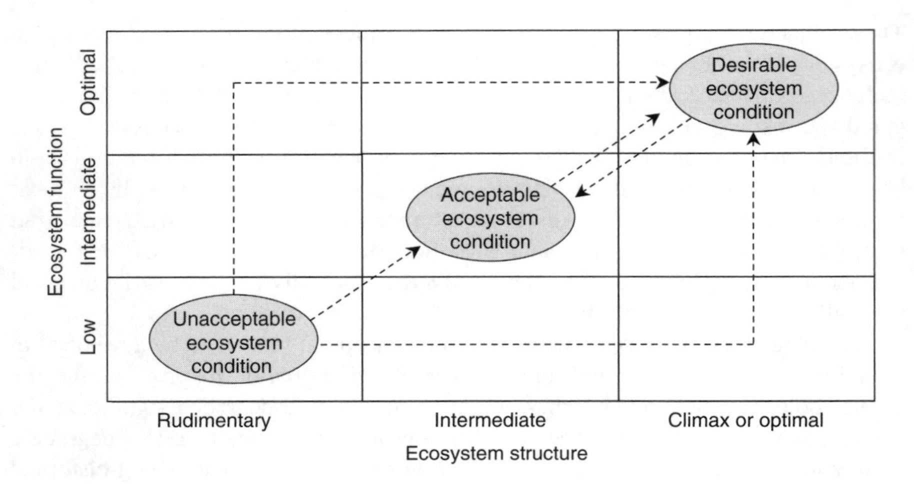

Figure 4.1 General model of ecosystem performance. An ecosystem or habitat that is in rudimentary condition with low functioning develops into a system with optimal structure and functioning. Development can take several pathways, and can oscillate between system states. (Modified from Thom,[8] with permission.)

be subjected to relatively regular and moderate disturbances, optimal conditions may be at a moderate level of structure but have high functionality. The difficulty and risk is that disturbances may not be predictable in time and intensity, and therefore the long-term fate of the restoration project may be more uncertain.

Figure 4.1 also indicates that a system may oscillate between states. This can be caused by stochastic processes such as human or natural disturbances, as well as stochastic climate-related forcing. This dynamic may be more pronounced in some system types than in others. It is important to recognize that the system can move between different structural and functional states and still maintain its long-term integrity.

Finally, and not explicit in Figure 4.1, is the fact that stability regimes are rarely ecosystem wide, but are limited to some fraction of the ecosystem.[23] This implies that if enough of an ecosystem is restored, sites within that system should support desirable resources.

If stressors are removed, the natural recovery (*passive restoration*) of ecosystems will tend to take place regardless of human intervention, but this may take a very long time — decades or centuries. *Active* restoration essentially means that humans act beyond stress removal to reduce the period of time required to improve ecosystem conditions, through a combination of physical intervention and natural recovery. At the desirable ecosystem condition, the system is fully functional, has an optimal structure, is resilient to disturbances, and is self-maintaining. However, the definition of *optimal* must be made with care and with relevance to the system under investigation. In the case presented here, it is assumed that optimal conditions are met with a natural climax community that, because of its persistence, is resistant and maintains itself through the ability to buffer changes. The term *optimal* implies a human value, and the optimal state represents what humans (i.e., restoration planners) view as the "best" condition for the system.

The levels of performance are divided into low, intermediate, and optimal in Figure 4.1. This essentially indicates that values (e.g., acreage) related to the structural conditions (e.g., the size of the pond–wetland interface) and the functional conditions (e.g., the annual duckling production per acre) can target a *range* (e.g., a ratio of wetland edge area to pond area between 1.0 and 2.0), rather than a single number. The ranges of values for structure and function are based on data from similar systems. An acceptable condition to project planners and stakeholders may be less than optimal. Hence, the goal for the project would be met if the system fell within a range of values for structure and function. Using a range of values acknowledges the following two sources of uncertainty:

- Uncertain understanding of relationships between ecosystem structure and function (knowledge uncertainty).
- Natural variability associated with the relationships between structural conditions and functional conditions (inherent variability).

Criteria for project success may also include constraints, or *zero impact thresholds*. For instance, the criteria for success in the Sagamore Marsh restoration study included no negative effect on the local water supply, wells, and area septic systems; no increase of residential flooding; and no negative impact on navigation in the Cape Cod Canal.[24] Another criterion of success was that the project would not harm existing eelgrass.

Planning Step 2: Inventory and Forecast

Introduction

To effectively assess the existing conditions of an ecosystem and define the problem in detail, planners must understand the physical, biological, and chemical conditions and processes at work. Developing a conceptual model can facilitate this understanding. The *conceptual model* is developed to ensure a shared vision of the planning context and the relationship among ecosystem components. The conceptual model is used as a formulation tool, a communication tool, and an assessment tool. Properly constructed, the conceptual model facilitates stakeholder participation, clarifies what ecological factors affect the achievement of project objectives, and minimizes the chance of misspecifying the system. Further, coupling the conceptual model with a decision process allows the planning team to deal with risk and uncertainty in a systematic way. The conceptual model identifies on paper the connections between the goal of the project and the physical and biological actions that need to be in place to achieve the project objectives.

Development of the Conceptual Model

What the Conceptual Model Provides

A conceptual model is a representation in words and illustrations of the major components and processes in a system and the linkages between them (see "Problem

Formulation" in Chapter 3). Although the conceptual model does not contain any quantitative information, it is frequently the first step toward developing a quantitative model, such as a numerical simulation model that mathematically links the components of a system. It illustrates how stressors affect or inhibit ecosystem functions and services (see Table 5.1), and helps identify sources of uncertainty and potential threats to reaching the goal.

The conceptual model is a rendering of the state of knowledge about the system under assessment. It is developed as a tool to describe the *existing condition* at a project site and to forecast the *without-project condition.* Under a no-action alternative, the ecosystem might remain the same, degrade further, or improve, depending on the degree of alteration and continued disturbance of the ecosystem. It is critical that the conceptual model be as complete and accurate as is feasible to reduce uncertainty about the ultimate ecological performance of the system. In later steps of the planning process, the conceptual model will be used to help predict the effects of various project alternatives. If a conceptual model cannot be developed for the habitat or species at a site, the project should not be attempted because the model provides a basis for management measures.

Conceptual Model Types

Several basic types of conceptual models are useful in ecosystem restoration planning. The first is the general model of *ecosystem performance* (see Figure 4.1). This model provides the general characterization of ecosystem output that is expected to take place on a trajectory toward meeting the project goal. The second model applies the principles of *landscape ecology* that define the required context for project success. The third model, the *ecosystem model*, provides a map of all of the important community and habitat components and processes of the ecosystem and the ways in which they interact. The fourth model is the *subsystem model*, which often takes the form of a species–habitat model in ecosystem restoration planning. Finally, a model coupling the biological outputs of the project with social and economic outcomes from the project is termed the *ecosystem services* model. All of these models can be developed at various scales.

Understanding the relationships among these types of conceptual models is useful when selecting and applying them for planning. For example, the ecosystem model for the project area is nested within the landscape model, geographically. Likewise, the species–habitat model is nested geographically within the ecosystem model. A temporal dimension is added to the ecosystem model to create the ecosystem performance model. The ecosystem performance model also incorporates species–habitat interactions and landscape dynamics. The economic outcomes represented in the ecosystem services model are affected by the outcomes of all other conceptual models described here, as well as by interactions such as those between the ecosystem and the landscape. For example, the restoration of one site may change the levels and impacts of recreation, hunting, or fishing in other areas, affecting the beneficial outcomes of the project.

General Form and Simple Example of a Conceptual Model

The general form of a conceptual model is shown in Figure 4.2a.

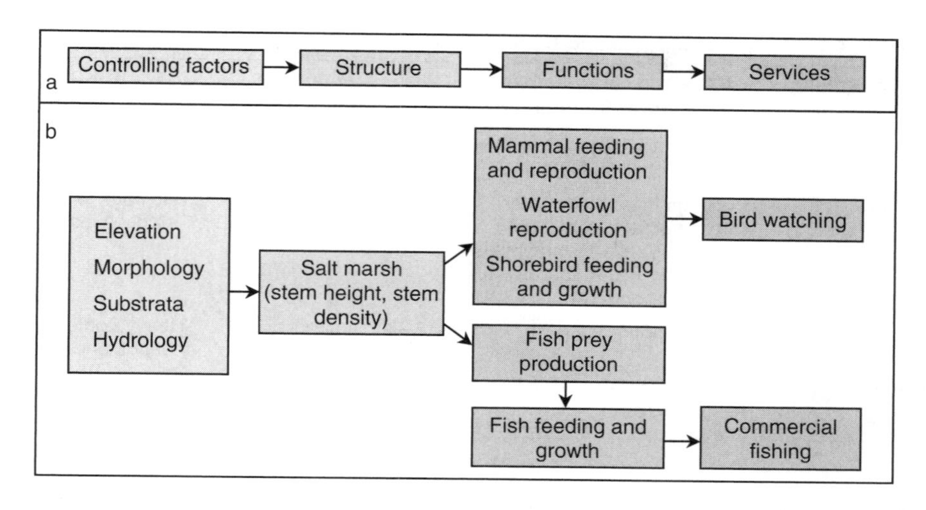

Figure 4.2 Generic structure of a conceptual model for ecosystem restoration (a) and simple conceptual model of a salt marsh system (b). General examples of (1) controlling factors are solar radiation, water depth, temperature, and nutrients; of (2) ecosystem structure are plant and animal biomass and diversity; of (3) functions are plant and animal production and diversification; and of (4) services are recreation, waste treatment, storm surge protection, and support for globally rare species.

The design objectives of a project may be structural (e.g., restore a marsh of a certain size and stem density) or functional (e.g., restore marsh primary production). The planning objectives, however, may include beneficial services (bird watching, flood control, recreation, aesthetics, fishing, or endangered species support). Of these services, only endangered species support is considered a *National Ecosystem Restoration* benefit; the others would be termed *National Economic Development* benefits, which are incidental in a restoration plan. To achieve any of these types of objectives, the factors that control the structure and function must be established, restored, or otherwise created at the site. If a certain objective is a priority for a particular site, the factors that need to be established to achieve this objective may be defined through the conceptual model.

Examples of controlling factors for a coastal marsh system are shown in a simplified conceptual model in Figure 4.2b, which for the purposes of this example does not attempt to completely list the controlling factors, structural features, functions, or ecosystem services of this system. Before the existing and future conditions of the ecosystem can be assessed, the ideal or preferred values of elevation, site morphology, substrata composition, and hydrology for this ecosystem type must be known. This information can be gathered through literature, discussions with experts, and new data collection. Current models using a hydrogeomorphic approach provide a good indication of some of the physical conditions and controlling factors for wetlands.[25]

Table 4.1 provides examples of the existing and preferred conditions in a hypothetical salt marsh. The information about present conditions at the salt marsh points to two major issues limiting functional performance: frequent disturbance by boat wakes, and smothering and fouling of habitat by trash. In addition, the site elevations

Table 4.1 Preferred and Existing Conditions in a Hypothetical Salt Marsh System

Controlling Factor	Preferred Range of More Natural Condition	Existing Site Conditions	Other Factors
Elevation	+10 to +13 feet relative to Mean Lower Low Water (MLLW)	+10 to +15 feet MLLW	
Site morphology	• Low gradient slope (<2%) from backshore to mouth • Semi-protected from waves • Free opening to larger estuarine system	• Eroded steep slope • Little protection from waves	Site open to large ship wakes
Substrata composition	Silt to medium sand	Medium sand and clay	Heavy floatable debris and trash
Hydrology	• Tidal inundation • Salinity 10-35 ppt	• Tidal inundation • Salinity 10-35 ppt	

are too great for full development of the salt marsh. One would predict that wakes and trash would continue to degrade the site and prevent recovery.

The level of certainty about the quantitative relationship between the controlling factors and the structure of the system enters into the assessment of the uncertainty of the project. If there is a high degree of uncertainty about the relationship between a controlling factor and the structure of the system, and this factor is believed to be highly important, there is good justification for directed research into the relationship to reduce the uncertainty. This principle applies to identifying potential causes for the degraded condition of the site. Site investigation might be needed to evaluate causes and to determine whether they are still contributing to the degraded condition. If they have been corrected, then there would be less uncertainty associated with restoration.

Landscape Model

The basic premise of a landscape model is that the flow of energy, animals, and materials within an ecosystem can be affected by the landscape.[26] Landscape features include the types, shapes, sizes, and distribution of habitat patches in the project-influential landscape of interest. Landscape ecology provides an integrated approach to understanding how large systems function to influence smaller subsystems. For project planning, the ecosystem targeted by the project area, such as a discrete wetland or river segment, is influenced by the functions of other ecosystems embedded in the landscape. For instance, the controlling factors in Table 4.1 can be controlled by site-specific management actions, but factors such as weather and disturbances upstream cannot be controlled and their influence should be considered in project planning. For many Corps projects, watershed boundaries define the landscape boundaries of interest. However, the influential landscapes of some projects may extend well beyond the influential watershed, such as the flyway landscapes of migratory birds. The principles of landscape ecology have direct applicability to the understanding, forecasting, planning, and design of ecosystem restoration projects.

Assessment on a *landscape scale* provides a wealth of data for understanding the relationship between changes such as landscape fragmentation and alterations in the populations of biological resources.[27] Removal or alteration of one or more elements may lead to the altered function of the remaining elements and ultimately result in degradation of significant resources in the project area. The U.S. Environmental Protection Agency specifically encourages resource managers responsible for wetlands cumulative impact assessment and protection to view wetlands in a landscape perspective.[28] In its review of aquatic ecosystem restoration, the National Research Council (NRC) made the following recommendations:[29]

> Wherever possible . . . restoration of aquatic resources . . . should not be made on a small-scale, short-term, site-by-site basis, but should instead be made to promote the long-term sustainability of all aquatic resources in the landscape. Whereas restoration on the large landscape scale is therefore definitely preferable to piecemeal restoration, small restoration efforts are not necessarily worthless or ineffective. Success in recreating a self-sustaining ecosystem is more likely, however, when the restoration is planned within the context of the target ecosystem's larger landscape.

Regardless of the magnitude of an ecosystem assessment, the understanding of landscape ecology is essential to the evaluation of the system's existing performance and future condition. The measurement of improved resource quality at a restoration site is qualified by landscape processes, which may reduce or negate site-specific effects. In addition to *scale*, considerations include landscape *structure, shape, linkages, configuration, number of patches,* and *disturbance*. For the salt marsh conceptual model example, the following questions at the landscape scale must be considered:

1. What is the quality of the landscape? If the salt marsh site is within a landscape that is highly degraded, the likelihood of the marsh's successful restoration diminishes. The landscape processes critical to the formation of habitats, such as sediment supply for marsh accretion, must be active at a level adequate to support development of the salt marsh without burying it during floods.
2. Are the target species present in sufficient numbers in the landscape? For example, if the goal of the project is to restore juvenile salmon habitat, but spawning habitat in the watershed has been eliminated, and no salmon exist within the landscape, the probability of the site meeting its goal is zero.
3. Is the site large enough to be used by the target species? As the size of the site increases, the probability that a species can find the site useful typically increases. Very small and isolated sites have a lower probability of use, and the residence time for migratory species is predictably greater in larger sites that provide a broad array of habitat types. For example, larger marshes contain a greater number of channel orders that vary in length, width, and depth, with deeper channels at the outer edges. At low tide, when much of the marsh is drained, fish use the deeper channels as a low-tide refuge, and when the tide floods the marsh, the fish occupy the smaller channels.
4. Are there adequate routes of access to the site? Linkages to other habitats must be large and arranged such that there is a reasonable likelihood that the target species will be able to reach the site.
5. How much disturbance will the salt marsh receive that might alter the controlling factor ranges? For example, will storms occur annually that flood the system and

deposit large volumes of sediment or erode the marsh? Frequent disturbances, especially during early establishment, can hamper system development.

Ecosystem Model

Ecosystem models come in a wide variety of forms, ranging from highly conceptual to fully integrated numerical simulation models. They generally include the atmospheric, geological, hydrological, and biological components of the ecosystem and attempt to describe all of the important connections among all of the major elements of each of these components. Compared with the relatively simple model shown in Figure 4.2b, ecosystem models are far more complex. Few, if any, fully integrated numerical ecosystem models have been developed for restoration purposes; however, models that simulate selected components (subsystems) of ecosystem models have been developed. For example, a simulation model of the hydrology of the Florida Everglades ecosystem was developed to provide specific information on the ways in which restoration of hydrology could improve ecological conditions in the ecosystem.[30] This large project required a level of modeling complexity greater than that needed for most projects.

Habitat or Subsystem Model

A habitat or subsystem model focuses on a particular component of the ecosystem. Figure 4.2b shows an example of a conceptual model for one habitat type, but there are many useful methods that generally can be called habitat models, some of which are semiquantitative. Well-known examples include the HydroGeoMorphic Approach (HGM) and the Habitat Evaluation Procedure (HEP).[25,31]

HEP assigns relative weights to the habitat quality and quantity provided to one or more selected species. For example, the salt marsh shown in Figure 4.2b may be marginally useful to shorebirds. Using HEP, the habitat quality would be given a score of 0.2 on a scale of 0 to 1, which represents its habitat suitability index (HSI). This index value is then multiplied by the geographical area of the habitat *potentially* used by shorebirds to calculate a weighted score in *habitat units* for the species–habitat subsystem. By incorporating more appropriate *natural* habitat(s) into the restoration plan for the marsh, one can elevate the shorebird score. According to the conceptual model, however, the natural condition that maximizes certain resources of significance, say shorebird use, may reduce the existing use of other significant resources, say rare waterfowl or fish. Although not detailed in Figure 4.2b, avifauna have different subhabitat requirements within the salt marsh so the distribution and quantity of habitat types will favor certain species or groups. These uses can be balanced through combining HEP analysis for selected species and good information on the controlling factors needed to result in the desired system state. According to the matrix in Figure 4.1, the desired system state must fit within the natural range of succession in the landscape to maximize the probability that it will develop to a state that is resilient to disturbances and is self-sustaining.

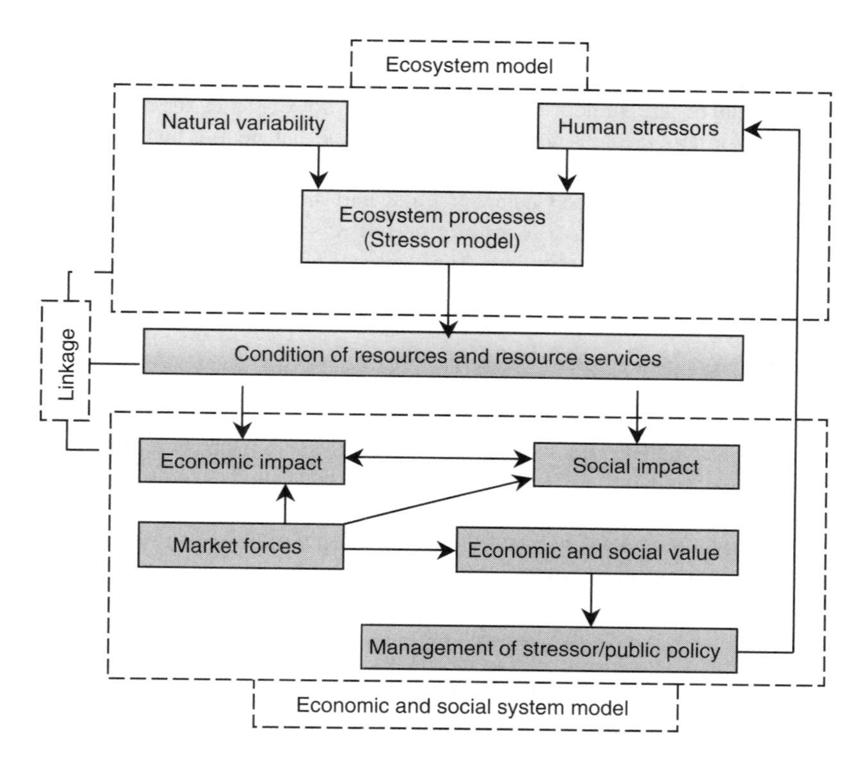

Figure 4.3 Conceptual model linking ecosystem model with social and economic model (From Thom and Wellman[32]).

Integrated Ecosystem Services Model

Integrated ecosystem services models link the ecosystem with related social and economic components. A simple integrated conceptual model for an aquatic ecosystem developed by Thom and Wellman[32] is shown in Figure 4.3. The primary assumptions in the model are that ecosystem processes provide resources that have economic and ultimately social benefits, and that human actions involved in using these resources have an impact on ecosystem processes. The model indicates that humans can make decisions that affect the ecosystem and thus affect the economic and social benefits accrued from the ecosystem. A more complex series of models, some of which attempt to integrate ecosystems with economic and social systems, were developed in an international effort to assess the effects of large-scale changes, such as climate, on the functions of coastal ecosystems: the Land–Ocean Interactions in the Coastal Zone program.[33]

The conceptual model from Figure 4.2b is linked to the economic and social system model by the functions and values produced by the salt marsh. To the extent that use values associated with these functions and values can be monetized (see the "Economic Value" section of Chapter 5, especially Table 5.3), they are considered National Economic Development benefits and are not the focus of the ecosystem restoration authority. Non-use values can be monetized as well, such as by using the contingent valuation method (see Appendix 5-A), but the USACE considers these values speculative and subject to biases and does not consider their use in project selection.[4] Estimating values for the many services that people benefit from more indirectly, such as water purification or preservation of genetic diversity, is also problematic; for example, see Costanza et al.[34] and the critique by Bockstael et al.[35] Therefore, the emphasis in Corps restoration projects is on establishing these social and economic linkages in qualitative terms; for example, more waterfowl production leads to more dollars spent on commerce and recreation.

Steps To Develop Conceptual Models

Models are often built from the ground up. For example, simple relationships are identified for a selected process, such as emergent marsh primary productivity. The model is then expanded to include system-scale effects, such as hydrodynamics, and the factors controlling these effects. As the model grows, it incorporates large-scale components including land-use and climate. Economic and social aspects may be included at the appropriate scale. It is advantageous to start simply at the smallest scale required to make decisions on the project. If the project is designed to provide nesting habitat for a selected bird species, for example, then the components of the habitat that meet nesting requirements, such as vegetation type and coverage, are the first elements specified in the model.

The steps to developing a conceptual model are described here for a model that focuses on restoration of a salt marsh habitat structure to support some of the marsh's native functions (Figure 4.2b).

1. Define the planning objectives of the restoration project. This is critical to developing the model. In this example, the planning objective is to restore the waterfowl and fish associated with the salt marsh.
2. Determine the aspects of the salt marsh that best meet the needs of the target species. For example, waterfowl may need ponds within the salt marsh for resting and feeding, and they may use vegetated islands in these ponds for nesting. Fish may need edges of the marsh within which to feed, and refuge areas where water remains during low tide. Verification or evaluation of these relationships can be carried out through directed field efforts or further literature reviews and discussions with experts in Planning Step 3. The initial model developed should identify critical data gaps and needs.
3. Determine the factors that control the development and maintenance of a salt marsh habitat necessary to meet the needs of the fish and birds, such as elevation, morphology, substrata, hydrology, and water quality.
4. Develop a diagram in the level of detail that accurately illustrates the connections among the controlling factors, the salt marsh structure, and the planning objectives of the project.

In practice, conceptual models are continuously updated throughout the planning process as information accumulates. In addition, postconstruction monitoring of restored sites often provides new insights into linkages and pathways that help refine the completeness and accuracy of the model.

Planning Step 3: Plan Formulation

Introduction

Conceptual and empirical models can be employed together to formulate management measures.

Systems Context

USACE ecosystem restoration planning seeks to reestablish a "self-regulating, functioning system."[4] A perspective focused on entire systems is an important means to this end. USACE planning guidance states:

> Restoration projects should be conceived in a systems context, considering aquatic (including marine, estuarine and riverine), wetland and terrestrial complexes, as appropriate, in order to improve the potential for long-term survival as self-regulating, functioning systems. This system view will be applied both in examination of the problems and the development of alternative means for their solution. Consideration should be given to the interconnectedness and dynamics of natural systems, along with human activities in the landscape, which may influence the results of restoration measures . . .
>
> The planning for ecosystem restoration objectives is essentially the same as for other water resources development purposes. However, there are some special considerations because of limitations in understanding the complex interrelationships of the components of ecological resources and services which are the focus of these studies, and because the environmental outputs considered in the evaluation process are typically not monetized. The consideration of significant resources and significant effects is integral to plan formulation and evaluation for any type of water resources development project. In ecosystem restoration planning, the concept of significance of outputs plays an especially important role because of the challenge of addressing non-monetized benefits.[4]

A *system view*, as described earlier, means consideration of the biological outputs of a restored system at a range of scales from the species to the community and from the patch to the landscape.

Conceptual and Empirical Models

Restoring an ecosystem is a complex undertaking. The better the understanding of the system components and controlling factors, the more likely project planning objectives will be achieved. Furthermore, having the best information from the start will result in better refinement of the physical actions that must take place, thus reducing costs associated with unnecessary activity. Ultimately, development of the conceptual model and its associated empirical models reduces the potential for error associated with the restoration actions and with the cost estimation for the project.

The conceptual model, with the identified pathways between chemical, physical, and biological components, can also provide insight into the changes that might be accomplished to restore the site. The general relationships represented by the conceptual model can be used as a guide for the application or development of empirical models in an effort to quantify these relationships.

The conceptual model is used together with applicable empirical models, such as hydrological models and habitat suitability indices (HSIs), to generate and evaluate a series of measures that would alter the system according to different design objectives that meet project-planning objectives. For example, factors such as inundation depth and duration, or components such as removal of an invasive species, could be considered for their effect on achievement of planning objectives.

In many cases, one or more species are emphasized. For example, if the project planning objectives included provision of habitat for the marsh wren (*Cistothorus palustris*), the HSI marsh wren model would be consulted.[36] The HSI model would rate the possible water-depth outcomes based on the empirical relationship between mean water depth and cover and reproduction suitability for the marsh wren. The conceptual model would provide information on variables influencing the mean water depth at the restoration site. Plans might then be formulated for different design objectives at the site, providing a range of suitability for the marsh wren based on mean water depth.

The conceptual model can also be used for planning at the community, ecosystem, and landscape levels. By delineating relationships among important components of the system, planners are able to envision the effects that changing variables have on groups of components, not only on single species. The conceptual model is a tool that helps planners apply the regulation cited in the previous section by using a system view to help them understand the problem and develop solutions.

The conceptual model provides an organization for the understanding of the habitat or subsystem upon which a quantitative model can be developed by adding mathematical links between subcomponents of the model. Although models such as the HSI are subsystem models, they can be linked with processes in the landscape that affect the subsystem. The reliability of the linkage will affect the ability to predict how changes in the landscape will affect changes in the habitat or subsystem. For simple projects like the one described in the example (Appendix 4-A), linkages between the subsystem and landscape processes can generally be articulated sufficiently for project planning using subsystem models. However, in very complex or large systems where a large investment is made, such as the Florida Everglades, subsystem models may not be adequate for evaluating sources of risk to the project. In such cases, spatially and temporally explicit process simulation models should be considered.

Landscape Variables

Several of the most important principles of landscape ecology relative to ecosystem restoration are considered in the plan formulation stage. Landscapes can be divided into matrices, patches, and corridors. A matrix is the area within which patches and corridors occur and represents the most extensive and most connected landscape element. The matrix, therefore, plays a dominant role in the functioning of the landscape. The works of Gosselink and Lee,[27] Forman and Godron[26] and Turner, Adger, and Lorensoni[33] among others, contribute to our understanding of the landscape ecology of the wetland, river, and coastal systems most associated with USACE restoration planning.

The scale at which humans perceive boundaries and patches in the landscape may have little relevance to the flows of energy and materials taking place there. Hence, investigations of the flows of energy and materials may be required to determine the appropriate scale at which to design restoration projects. In a practical sense, the scale of a restoration project is important for four reasons: (1) the project area needs to be large enough to limit any deleterious effects that boundary conditions may impose on interior aquatic functions; (2) project managers must be able to exert influence over zones in which major causes of ecological disturbance to the project are occurring; (3) the area needs to be sufficiently large to allow the monitoring and assessment of important effects of the project; and (4) the project should be of an affordable magnitude.[29]

The appropriate scale for restoration is likely to be defined by the requirements of target ecosystems and species groups. Allen and Hoekstra[37] suggest a hierarchical approach to restoration in which the prescribed size of the landscape to be restored determines which processes have sufficient space or time to be relevant in the restoration effort. Restoration, in this sense, is limited qualitatively, rather than quantitatively by area — that is, in consideration of the processes to be included or excluded, rather than by the acreage to be restored. Migratory waterfowl and shorebirds of Commencement Bay, for example, operate at the scale of the estuary and of the Pacific flyway. Thus, restoration projects for these birds must be sited within the flyway and designed to be of appropriate size and structure to both attract the birds and restore their numbers by providing available food, protection from predators, and resting areas in locations where those factors are limiting numbers. Attraction alone would simply move the birds from existing locations to new locations with no net gain in numbers or human benefit.

Management of natural wildlife reserves has supplied considerable information on scale issues affecting restoration. Based on island biogeography,[38] a well-established relationship between the number of species and habitat area has been used to predict the effect of habitat fragmentation on species in reserves. Hence, as a general rule, larger reserves are better than smaller to maintain high numbers of species. In an example presented by Gosselink and Lee,[27] one large reserve supported more native, interior species than two or more smaller reserves equal to the large reserve in total size. Although fragmentation and moderate levels of disturbance may increase the number of species in an area,[21] such an increase would be largely due to opportunistic species that generally do not require reserves for survival. The goal of reserves is generally to preserve species that would otherwise decline.

It is well established that the number of species within a patch is a function of a number of factors in addition to patch area, including within-patch heterogeneity, disturbance patterns, degree of isolation from sources of species, patch age, and matrix heterogeneity. Flow of species population members among patches is dependent on patch density and arrangement in the matrix and other factors that pose impediments to movement.

The NRC[29] recommended that historical records, including maps of resources, hydrology, and geomorphology, be consulted to determine an appropriate structure for target species groups. The primary reason for this recommendation is that the natural forces that resulted in the original structure of the landscape will still be active, or can be restored, and will tend to push the system toward its predisturbance structure once impediments are removed or circumvented. A secondary reason is that plant and animal communities have become adapted for optimally using the system in its natural state. A third and more pragmatic reason is that one cannot possibly model all of the chemical, physical, and biological interactions that shape the system and control its stability and functional performance; therefore, historical information on the site's natural state, particularly geography, provides the best information for a model upon which to base a restoration design.

The shape of a *patch*, or contiguous habitat, affects the type and number of species in the patch. Species may show preferences for either the edges or the interiors of patches. A quantitative metric of shape is the interior-to-edge ratio. For example, round patches have a large interior-to-edge ratio compared with that of very narrow, linear patches. Small patches may act only as edges, depending upon the size of the animal that potentially occupies them. The number of edge-dwelling or interior-dwelling species in a patch will be dependent to a certain extent on this ratio. Processes such as benthic productivity and nutrient flux will also be dependent on shape.

If they link two or more patches, certain habitat patches can act as corridors of movement for animals. For example, narrow strips of sedge marshes along riverbanks, ravines leading from uplands to intertidal areas, channels from deep areas to marsh flats, bands of eelgrass and kelp, and even drainage ditches represent landscape linkages for animals in estuarine environments in the Pacific Northwest. These linkages are generally rich in food resources or cover that increase the probability of an animal's successful use of two or more habitats in an estuary. Restored habitats must incorporate avenues of ingress and egress to be fully functional within the estuarine landscape. However, such corridors may also allow access to nonnative, invasive species.[39] If the restoration goal is to maintain native biodiversity and ecological integrity, optimum results are most likely when the natural connectivity among habitats in the landscape is restored and the connectivity of artificial habitats, such as clear cuts or agricultural fields, is minimized.[39]

Landscapes change naturally with time. Restoration must therefore account for aspects of landscapes that can evolve and influence long-term functional performance of constructed systems. The long-term viability of a restored system is dependent upon the landscape dynamics; a variety of factors inherent to the system, including stability, persistence, resistance, and resilience create a pattern of change over time. The stability of a system is the degree to which it displays long-term variability, or its tendency to convert from one type of system to another. Persistence refers to the period during which a certain characteristic of a landscape continues to be present. Resistance and resilience mean, respectively, the ability of a system to withstand and to recover from disturbances. In general, the goal of a restoration effort should be to create a system that is relatively stable, persistent, resistant, and resilient. The size of the habitat or ecosystem relative to the effect of a disturbance is a major concern. Protected ecosystem types must be sufficiently large to maintain viable populations of all native species.[39]

Examples — Marsh Wren and Mississippi Delta Marsh HSI Models

Examples of plan formulation for projects at regional and site-specific scales are the Mississippi River delta marshes restoration program, and the hypothetical restoration of marsh wren habitat, both of which incorporate an understanding of landscape-level processes.

Marsh Wren Example

This example demonstrates how a conceptual model and an empirical model can be employed to formulate management measures incorporating the landscape context. The approach employs information contained in the HSI marsh wren model,[36] guidance from Orth[40] on how to employ the HSI in USACE water resource planning, and a landscape ecology conceptual model. While it is unlikely that the marsh wren HSI would be used in restoration planning because this abundant bird is not considered an indicator of significance or of ecosystem condition, this relatively simple HSI is useful for demonstration.

The marsh wren breeds in freshwater and saltwater marshes. It feeds on a variety of insect species, and the food it selects depends on the size and age of the bird. It is important that the bird winters in warmer areas, such as Mexico and Florida, and accordingly, the bird depends not only on breeding habitat but also on the availability of productive and protective wintering habitat, which is often far from the breeding site. Nests are typically built in cattail (*Typha* spp.), emergent sedge (*Carex* spp.) or rush (*Scirpus* spp.) marshes, although other marsh habitats are also used. The wren nests in areas with standing water ranging from approximately 3 cm to 91 cm deep. Permanent standing water is generally required throughout the breeding season to supply a dependable food source and refuge from predators. In intertidal habitats, however, the marsh wren can nest where water is present only during high tides. Also important is the total area: marshes under 0.40 ha in size usually are not used by breeding marsh wrens. The HSI model applies primarily to the northern half and coastal areas of the U.S., the marsh wren's breeding range. It provides an estimate of the suitability of a habitat for breeding based on the controlling factors and structure. The HSI marsh wren model is as follows:

$$\text{HSI} = (\text{SIV1} \times \text{SIV2} \times \text{SIV3})^{1/3} \times \text{SIV4} \quad (4.1)$$

where the suitability indices are assigned based on data on controlling factors and structure, as follows:

SIV1 = the suitability index for the vegetation growth form
SIV2 = the suitability index for canopy cover of emergent herbaceous vegetation
SIV3 = the suitability index for mean water depth
SIV4 = the suitability index for canopy cover of woody vegetation

The value of an index ranges from 0.0 (completely unsuitable) to 1.0 (ideal condition). For example, the suitability index SIV1, based on data for the variable V1, the growth form of emergent hydrophytes, would be 1.0 if cattails, cordgrasses, or bulrushes dominated the site. In contrast, the value for mangroves would be 0.1. The indices can also be continuous. For example, the suitability index for a site would be 0.3 if the average water depth (SIV3) were 5 cm, or 1.0 if the depth were 15 cm or greater.

A conceptual model for marsh wren breeding can be created from the HSI information. It is important to include broader landscape aspects, as represented in

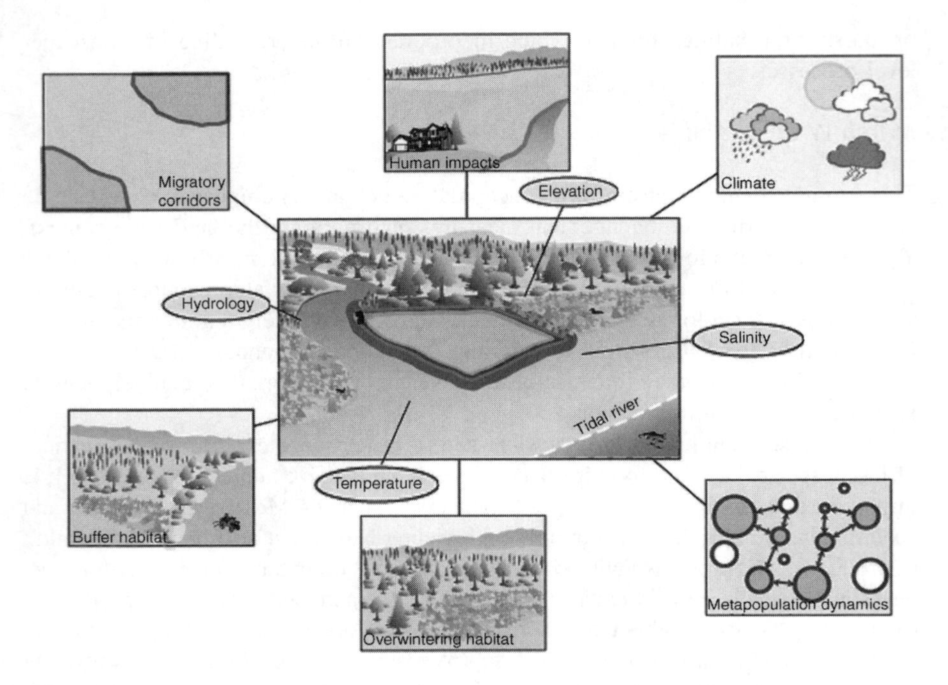

Figure 4.4 Site and landscape-scale conditions controlling habitat structure and function.

Figure 4.4, although they cannot be drawn from the HSI. The landscape aspects relevant to this example include, but are not limited to, the following:

- Presence and availability of adequate overwintering habitat.
- Presence of suitable migratory corridors between wintering and breeding habitats, providing resting and feeding sites.
- Adequate surrounding buffer habitats.
- Metapopulation dynamics (features of a series of populations at a landscape level that change over time, such as extinction and recolonization).
- Absence of human threats to survival, such as hunting and contaminants in the food web, over entire winter and summer range.
- Tolerable climate for marsh wren over the entire range.

These elements control the development of the structures essential for marsh wren habitat as detailed in the HSI model. Functions of such habitat include feeding, refuge from predators, and nesting success.

The HSI model does not explicitly treat all controlling factors at the site and landscape levels. Its starting point is habitat structure, itself an endpoint of other processes. If the HSI approach supports the conclusion that habitat is marginal, the cause may lie with landscape-level variables. The conceptual model is used to represent community, ecosystem, and landscape-level relationships and controlling factors. While the landscape-level assessment often is qualitative, empirical models of aspects, such as metapopulation dynamics, are sometimes developed as well. The landscape can be further examined by a newer approach, the HGM, which is described in the Mississippi Delta Marshes example in the following section.

Mississippi Delta Marshes Example

Restoration of the Mississippi River delta marshes of Louisiana provides an example of a large-scale, landscape-level restoration effort that is implemented under the Coastal Wetlands Planning, Protection and Restoration Act of 1990. At the heart of this program is a hydrologic basin-wide approach to restoring wetlands on the delta. Delta wetlands are being lost at a rate of about 65 km^2 (25 sq mi) annually. The loss not only threatens the wetlands and the fishery resources it supports, but also the infrastructure of towns and industry on the delta. A piecemeal, small-scale approach to restoring this system would not provide the scale of change required to slow or reverse the trend.[41]

The marshes are being lost through a combination of factors. A diagram of the major factors contributing to the loss is shown in Figure 4.5. Fundamentally, the marshes are eroding because they are not building elevation (accreting) to keep pace with relative sea-level rise. Subsidence, the combined effect of geological movement along faults and compaction of poorly consolidated sediments, is taking place on the delta.[42] Sediment input from the river to the marshes has largely been eliminated because of the levees. Although organic matter from marsh plants contributes to accretion, the buildup is not great enough to overcome the subsidence due to compaction of poorly consolidated clay and peat material. Salinity intrusion from the Gulf of Mexico affects the marsh vegetation and can contribute to changes in marsh production and organic matter deposition. Through channel dredging for oil and gas extraction, the hydrology of the delta has been drastically altered. The net result has been impedance of sheetflow across the marsh plain, accumulation of standing water, and drowning of the marsh interior. Storms and waves have eroded the marsh edges. The introduction of the nutria, a voracious herbivore of marsh plants, has created additional stress. At an even larger scale, the sediment supply has probably been altered throughout the watershed. The role that this alteration plays in delta marsh loss is unknown.

The understanding of each factor's role is key to designing the restoration plans. The primary strategy is to reintroduce sediment along with nutrients into the marshes to fuel marsh plant productivity and to build elevation. This strategy has led to the development of a project list that includes various methods to introduce sediment, stabilize and protect marsh edges, introduce nutrients, and recreate hydrology.

The delta marshes harbor an enormous abundance of bird species and a variety of economically and recreationally important aquatic animals. The production of the paneid brown shrimp (*Panaeus aztecus*) and to a lesser extent, the white shrimp (*Panaeus setiferus*) is dependent on a healthy marsh and adjacent submerged seagrass habitat. Therefore, shrimp habitat is a primary basis for delta marsh restoration. Increased shrimp production is believed to be linked to higher food resources and protection from predators afforded by marsh surfaces, channels, and rivulets, particularly the length of the land-water interface, or edge habitat.[41]

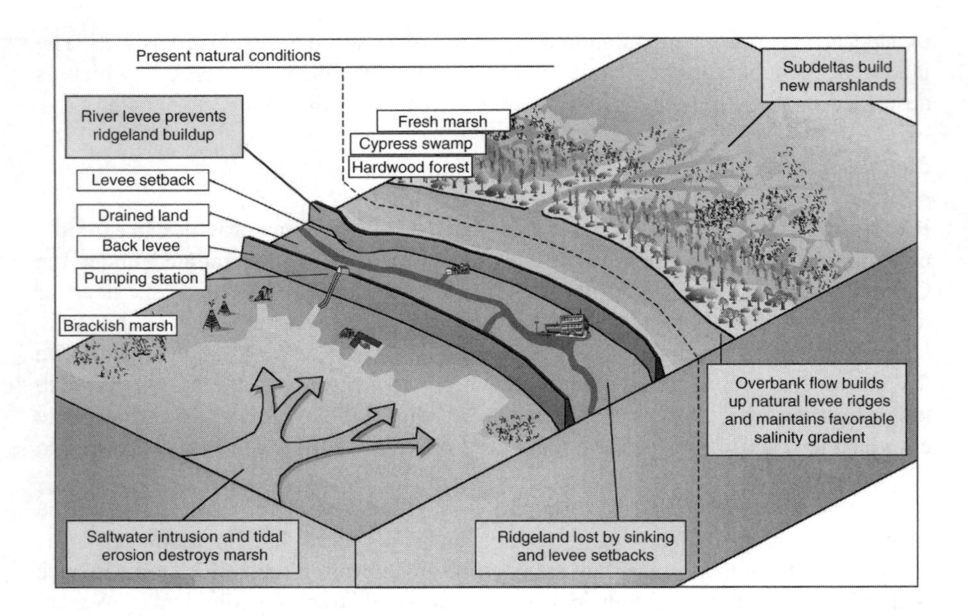

Figure 4.5 Natural and modified conditions along the Mississippi River corridor (Modified from Louisiana Coastal Wetlands Conservation and Restoration Task Force and Wetlands Conservation and Restoration Authority[42] and Gagliano[47]).

An HSI model for brown shrimp has been developed.[43] The model, which was developed before the edge-research findings cited previously, indicated that the suitability of the estuary increases with increasing coverage of the estuary by marsh, with a maximum suitability at 100%. The shrimp also prefer soft, peaty silt bottoms, an intermediate salinity range (10 ppt–20 ppt) during spring, and moderately warm spring temperatures (20°C–30°C). One index in the model incorporates variables for food and cover (*fc*), and one incorporates water quality variables (*wq*) as follows:

$$HSI_{fc} = (SI^2_{V1} \times SI_{V2})^{1/3} \tag{4.2}$$

$$HSI_{wq} = (SI_{V3} \times SI_{V4})^{1/2} \tag{4.3}$$

where the indices are:

SI_{V1} = the suitability index for marsh and seagrass cover
SI_{V2} = the suitability index for substrate composition
SI_{V3} = the suitability index for mean spring salinity
SI_{V4} = the suitability index for mean spring water temperature

SI^2_{V1} is squared to weight it more heavily in the composite index value. The HSI for a site is either HSI_{fc} (food, cover) or HSI_{wq} (water quality), whichever is lower.

Broader landscape conditions can have a strong influence on estuarine shrimp populations. Although no minimum habitat size is included in the HSI for brown shrimp, it is clear that increasing the area of highly suitable habitat would generally result in increased support for shrimp. The population levels would theoretically increase in concordance with increasing habitat area to a point at which other factors would limit shrimp populations. Because brown shrimp spawn in deep offshore areas, poor offshore spawning conditions may limit larval recruitment into the estuary. Heavy predation on adult shrimp offshore or heavy fishing pressure may also limit shrimp recruiting to estuarine habitats. Conversely, favorable offshore spawning conditions, removal of predators, and reduced fishing pressure may combine to produce massive recruitment to estuarine habitats.

The example of the Mississippi delta marsh restoration program illustrates the issues and uncertainties associated with restoration. First, the program rightly has determined that marsh restoration is dependent on major processes that are involved in delta formation (Figure 4.5). Restoration is focused on measures that can restore natural processes through interventions such as the introduction of sediment or the installation of wave protection. The program acknowledges that some processes, such as a weak zone in the Earth's crust, are not possible to address. Further, brown shrimp populations are dependent not only on habitat characteristics but on a variety of landscape-scale factors, such as spawning habitat conditions and fishing pressure. Hence, the success of the restoration in terms of brown shrimp is complex. Because of this complexity, and the limited understanding of their biology and natural variability, there is uncertainty regarding the response of shrimp to physical actions. Understanding the sources of these uncertainties represents a major step toward understanding potential outcomes from a restoration action.

Other Tools — HGM Models

The HGM, which is currently under development by the USACE, yields indices that can be used to evaluate wetland functions. It is not absolute but relative to regional reference sites and, like the HSI, is unitless. Guidebooks for its implementation in tidal fringe wetlands[44] and riverine wetlands[45] have been published. In comparison with the HSI models, which have a single-species focus, the HGM focuses on a suite of co-occurring species that inhabit a particular system. For example, among the functions that the HGM takes into account are not only *hydrogeomorphic* functions, but also *habitat-related* functions. Because the HGM relies heavily on data from regional reference sites to develop the rating for a restoration site, HGM guidebooks require extensive development and evaluation by region. There are presently only a few regions for which HGM models have been developed.

For the tidal fringe wetland HGM, four examples of functional capacity indices (FCI), are given in the following equations.[44] These equations combine subindices assigned to each variable after the variable is measured in the field. A published *functional profile*, to be developed from data from regional reference standard sites, is consulted to assign the subindex.

Hydrogeomorphic Functions

1. **Tidal surge attenuation** (TSA) — the capacity of a wetland to reduce the amplitude of tidal storm surges

$$TSA = (V_{DIST} + V_{ROUGH})/2 \tag{4.4}$$

where

V_{DIST} = variable subindex based on the distance that water must travel across an intervening tidal fringe wetland

V_{ROUGH} = variable subindex based on the potential effects of emergent vegetation, obstructions, and microtopographic features on hydrodynamics of tidal floodwaters

2. **Sediment deposition** (SD) — the deposition and retention of organic and inorganic particulates from the water column through physical processes

$$SD = (V_{PSC} + V_{FD} + V_{ROUGH})/3 \tag{4.5}$$

where

V_{PSC} = a variable subindex based on the proximity to the source channel

V_{FD} = a variable subindex based on flooding duration

Habitat Functions

3. **Plant community composition** (PCC) — the ability of the wetland to support a native plant community of characteristic species composition

$$PCC = (V_{COV} + V_{EXOTIC})/2 \tag{4.6}$$

where

V_{COV} = a variable subindex based on total percentage of vegetative cover

V_{EXOTIC} = a variable subindex based on percentage of vegetative cover by exotic or nuisance species

4. **Wildlife habitat utilization** (WHU) — the potential use of a marsh by resident and migratory avifauna, herpetofauna, and mammals

$$WHU = (V_{AE} + V_{UE} + V_{WHC})/3 \tag{4.7}$$

where

V_{AE} = a variable subindex based on the amount of edge between the intertidal vegetated, intertidal unvegetated, and subtidal areas

V_{UE} = a variable subindex based on amount and quality of upland edge

V_{WHC} = a variable subindex based on the wildlife habitat complexity

Both the FCI and the HSI are indices associated with approaches, the HGM and HEP approaches, respectively. The latter FCI, *wildlife habitat utilization*, is analogous to the HSI model presented earlier. However, the HGM addresses associations of species with common behaviors and habitat requirements. For a regional HGM, the marsh wren, for example, may be a major component of the avifauna, and therefore may influence the reference sites that are selected. If the reference sites selected are best for the marsh wren and species with similar habitat requirements, then the HGM functional capacity index for WHU will be a strong indicator of habitat quality for the individual species, as is the HSI.

In contrast to the HSI, the HGM is based on the recognition that wetland types are formed under hydrologic and geomorphic processes. For example, tidal fringe wetlands occur in a much different hydrological and geological regime than do organic soil flats. Therefore, maintenance of the larger regimes is important in assuring the continued development and functioning of the different wetlands that occur within them. This broader-scale setting requires that a restoration planner take a landscape or watershed view of the system within which he or she is working to reduce the uncertainty about the functional performance and long-term viability of a restored site.

Implications for Plan Formulation

Both HSI[40] and HGM models can be used with conceptual models for planning restoration projects. Because conceptual models illustrate scientific knowledge about the system or species, they provide both the basic understanding of the major aspects of the system and a pictorial description of how the system works. This is critical to any habitat-planning effort. Variables in the HGM models can be manipulated by several management measures to achieve an outcome; however, manipulations to improve conditions for one particular species would require special consideration. For example, habitat for the marsh wren could be enhanced through controlling water levels to improve intertidal and nontidal habitats, which would ultimately improve conditions for marsh wren nesting. The HSI models are conceptual, but they are partially parameterized by the assignment of quantitative values to at least some of the variables. The HSI models were designed for rapid, semiquantitative, onsite assessments of habitat conditions relative to selected species. There are two important assumptions to the HSI models:

- If the ideal habitat conditions are established, optimal functions will be realized.
- Broader landscape conditions are ideal or at least adequate.

There are, of course, significant uncertainties associated with each of these assumptions. These will be discussed later in this chapter. Nonetheless, management measures can be developed to arrive at site conditions that are at least conducive to a selected species' breeding.

Orth[40] provides a scenario for application of the HSI models in developing management measures. According to Orth, HEP and its HSI models were traditionally used to inventory resources (Planning Step 2) and to assess the effects of alternative planning formulations (Planning Step 4); the use of HSI models in Planning Step 3, plan formulation, has come into being only more recently in mitigation and restoration planning efforts. The goal for planning purposes is to achieve the highest total HSI value possible within physical and economic constraints. By adjusting the number, degree, and kind of management measures at a site (e.g., percentage of cover type flooded, average height of herbaceous canopy), one can see in relative terms the ecological outcome as a change in the HSI. In the restoration-planning context, HSI models are used to identify potentially useful management measures rather than to predict the effects of nonrestoration management measures.

The process described by Orth[40] involves examination of each of the habitat variables in relevant HSI models in light of four factors:

- Proximity to optimum conditions.
- Responsiveness to manipulation.
- Relative effect in calculating an HSI.
- Manipulation costs.

This process provides information that is particularly relevant to the individual species identified in the HSI models and is therefore most useful when the goals of a restoration project involve specific populations. Extrapolation from HSI results to restoration goals at the ecosystem level involves additional uncertainties. Orth cautioned that planning for target species may have undesirable effects on other species and suggested using community-based HSI models or performing the analysis on a suite of evaluation species to "balance their habitat needs."[40]

Because some of the HSI model variables can be manipulated in several ways to improve habitat quality, Orth noted that planners have several management options. For example, pipe irrigation systems, controlled flooding, and increased groundwater levels can manipulate the soil moisture regime. Each of these options has an associated cost and uncertainty, which can be evaluated as part of the planning process. However, not all variables are equally important in determining habitat quality, nor are they comparable at every unique site.

The conceptual understanding of the habitat from patch to landscape and from species to community can be used to conduct the risk assessment. The following are examples of questions that can be asked to link the project site to the landscape and to the conceptual model that represents it:

- Will there be adequate corridors?
- How will restoration for a species affect the community?
- Is the site large enough?
- How will the project's resistance and resilience change over time? Is this acceptable?
- What are the likely landscape changes over time? How will they affect the site?
- Does the design invite invasive species?
- How will the community affect the species of interest?
- What else could go wrong?

When risk is identified, the conceptual model can be used to develop risk management techniques, alternative plans, or construction methods. The project example at the end of this chapter illustrates these concepts.

Uncertainties Associated with the Models used for Management

Morgan and Henrion[15] point out that "Careful thought leads us to the following disturbing conclusion: *Every model is definitely false.*" Given that no model perfectly represents the real world, in this section we discuss the more salient uncertainties associated with the use of conceptual models, HSI models, and the HGM in restoration projects, and outline some considerations for reducing such uncertainties in practice.

Conceptual models are not quantitative; therefore the uncertainty directly associated with them derives from the possibility that relationships are depicted inaccurately. Relationships may be indicated where there are none, or cause-and-effect relationships may be implied inaccurately in the model. Conceptual models reflect the state of knowledge and the uncertainties of expert opinion. They are also based on quantitative data and are thus subject to the uncertainty associated with empirical measurements.

Both the HSI and HGM models have built-in uncertainties, as well as uncertainties associated with their application to plan formulation in ecosystem restoration projects. The built-in uncertainties of both models include those derived from their basis in expert opinion, which is to some degree subjective; incomplete data sources; and imperfect application of numerical equivalencies for habitat quality and functions. In the terminology of risk analysis, these are all forms of *knowledge uncertainty.* Sources of knowledge uncertainty in the HSI and HGM models include the standard sources of uncertainty associated with empirical quantities since the models are based on such data. These sources of uncertainty, some of which also derive from inherent variability, are further described in Morgan and Henrion:[15] statistical variation, systematic error, subjective judgment, linguistic imprecision, variability over time or space, inherent randomness, disagreement, and approximation by the model. These uncertainties can be reduced to some degree by studies directed at data gaps.

Another source of uncertainty associated with empirical quantities is called *randomness and unpredictability* by Morgan and Henrion[15] and is synonymous with the inherent variability of natural systems under study that was defined previously. When the HSI or HGM models are applied in plan formulation for specific natural or man-made systems, inherent variability may affect the outcome of the project in unpredictable ways. An example would be a major storm event occurring within months of a restoration planting effort, when plants are particularly susceptible to damage. As a general rule, it is possible to reduce knowledge uncertainty, but it is impossible to reduce such inherent variability.

A simple analysis of the HSI and HGM models shows that in restoration project planning applications, they are best applied at different scales. Because HSI is focused on the individual species, whereas HGM integrates landscape- and community-scale information and general habitat use, there are risks associated with the selection of one or the other of these model types to formulate alternative restoration plans. When HSI models are applied to plan formulation in ecosystem restoration projects, there is a risk that results will be extrapolated from the habitat of one species to another or from a species to objectives at a larger scale, such as the ecosystem. When HGM models are used in plan formulation, there is a risk that the unique requirements of certain species will not be adequately accommodated.

Design Risk

In the plan formulation step, design is used to reduce the risks associated with both knowledge uncertainty and inherent variability. Such approaches, however, can lead to overdesign. The costs of reducing knowledge uncertainty by using studies to fill data gaps can be directly compared with the costs of proposed design modifications suggested to manage the uncertainty. Designs proposed to manage engineering uncertainties, such as the level of tidal flooding that will result from opening a dike, need to be carefully weighed against their cost. Also, residual risk due to engineering limitations or engineering uncertainties should be documented for consideration when the alternative plans are evaluated in Planning Step 4 of the planning process.

Planning Step 4: Evaluation of Plans

It is important to identify all of the potential significant effects of each plan. Effects are usually evaluated relative to the planning objectives established in Planning Step 1. However, it is important to broaden the focus to the watershed or landscape level, using both the conceptual and empirical models to evaluate the system and examine the possibility of unintended or incidental consequences. When the variables being used to measure the effects are determined to encompass all significant effects, the evaluation process may proceed.

Five primary tasks are required to evaluate an alternative:

- Forecast the with- and without-project conditions.
- Compare the with- and without-project conditions.
- Assess the differences.
- Appraise plan effects.
- Qualify plans for further consideration.

In practice, forecasting the with-project conditions simply means projecting the state of the planning objectives into the future. The state or condition variables established in Planning Step 1 to measure the design objectives are forecast for each project alternative. The forecasts are made for each year of the period of analysis and are averaged to yield an average annual equivalent value. Comparisons and descriptions of with- and without-project conditions identify the net difference in the average annual equivalent values of the variables assessed relative to the planning objectives. Appraisal of a plan entails a subjective judgment of its effects: positive, negative, or neutral and important or not. Qualifying plans is the first screening of the alternatives; each alternative is assessed to determine its suitability for further consideration.

Alternative plans are qualified for further consideration using four criteria: *completeness, effectiveness, efficiency*, and *acceptability*.[4] One goal of this chapter is to explain how risk analysis relates to each of these criteria. Completeness is the measure "to which a given alternative plan provides and accounts for all necessary investments or other actions to ensure the realization of the planned effects."[4] Investments or other actions can include the design features and construction methods used to ensure the realization of the planned effects. Completeness is clearly a statement of risk assessment (i.e., the probability of project success). Effectiveness is "the extent to which an alternative plan alleviates the specified problems and achieves the specified opportunities."[4] Incorporating uncertainty into the analysis allows the analyst to measure and communicate the effectiveness of an alternative plan as a range of possible outcomes, along with their probability of occurrence. Efficiency is "the extent to which an alternative plan is the most cost-effective means of alleviating the specified problems . . . "[4] When uncertainty is incorporated into the cost estimate, cost-effectiveness is not only measured at the expected value, but also at the range of identified outcomes. Acceptability is "the workability and viability of the alternative plan with respect to acceptance by State [*sic*] and local entities . . . "[4] Explicitly incorporating risk analysis into the decision-making process allows stakeholders to consider their risk preferences when determining acceptability. These concepts are illustrated in Step 4 of the project example in Appendix 4-A.

Planning Step 5: Comparison of Alternatives

The surviving alternatives, which have been judged complete, effective, efficient, and acceptable, are compared to evaluate their relative contribution to the planning objectives, their cost, and other objectives that are important to the stakeholders. When metrics used to measure the effects of each alternative are incommensurate,[a] such as PCC habitat units and WHU habitat units, there is no objective way to choose between the alternatives. Value judgments must be used to make trade-offs between project cost and project outputs.[12]

Subjective values are often expressed as weights for each output measure. Many methods have been developed to elicit and quantify the relative weights of alternative outputs. They range from the simple assignation of weights to the complex pairwise comparison method used in the analytical hierarchy process (AHP).[46]

Once weights have been established, the alternatives can be ranked, the least-cost route to an outcome (the *efficient frontier*) identified, and incremental cost analysis conducted. However, decisions are often taken by imputing the relative value of each output without explicitly establishing weights. The sophistication of the methods for assigning weights to outcomes often outruns the subjective nature of the process.

The inclusion of uncertainty in the analysis calls for a comparison of the overlapping ranges of potential outcomes for each alternative, rather than a simple examination of the alternatives at their mean values. This allows alternatives that may be dominated by other alternatives at the mean to be included in the final set of choices.

[a] As explained under the heading "Development of the Conceptual Model," ecosystem restoration measures are formulated only for environmental benefits, and any National Economic Development benefits are incidental to the restoration plan.

Planning Step 6: Plan selection

Selection criteria should consider all important outputs and effects of the projects. This includes not only the design objective metrics assessed relative to the planning objectives, but also any other factors important to the stakeholders. At this point in a restoration study, all remaining plans, including the no-action plan, are equally rational choices given the metrics of the design objectives. In theory, the "best" choice is the one that maximizes general welfare. However, without commensurate outputs or market prices for such outputs, there is no objective way to identify this alternative. Therefore, one of the most important parts of plan selection is a convincing rationale.

CONCLUSION

The absence of complete information about ecosystems makes risk analysis an important part of any ecosystem restoration evaluation. The USACE six-step project planning process, augmented with concepts and tools from the science of ecological restoration, provides a framework for risk analysis in ecosystem restoration project planning. Conceptual models used in project planning include models of the ecosystem and ecosystem services, the landscape, and the habitat or subsystem. These models communicate to stakeholders our understanding of the system and reduce the chance of specification error. Empirical models, such as HGM and HSI, are known to have estimate error. The process presented here allows the analyst to explain risk information to the stakeholders for the purposes of decision-making. This procedure is illustrated in detail in Appendix 4-A, "Example Application of Risk Analysis to a USACE Ecological Restoration Project."

ACKNOWLEDGMENTS

The authors gratefully acknowledge the reviewers of this chapter for their valuable contributions: Dr. Richard Cole, Dr. David Moser, and Mr. Ken Orth of the USACE; Mr. Dick Ecker, Dr. Martin Miller, Dr. Walter Pearson, and Dr. Susan Thomas of Battelle; and Dr. Randy Bruins and Dr. Matt Heberling of the U.S. EPA. Graphics were prepared by Mr. Nathan Evans of Battelle. This research was supported by the U.S. Army Corps of Engineers Institute for Water Resources under a Related Services Agreement with the U.S. Department of Energy under Contract DE-AC06-76RLO1830.

REFERENCES

1. USACE (U.S. Army Corps of Engineers) and South Florida Water Management District, The Journey to Restore America's Everglades, Home Page, Dec. 17, 2003. Accessed Jan. 6, 2004 at http://www.evergladesplan.org/.

2. USACE, Upper Mississippi River — Illinois Waterway System navigation feasibility study; Draft integrated feasibility report and programmatic environmental impact statement, U.S. Army Corps of Engineers, Apr. 29, 2004. Accessed May 24, 2004 at http://www2.mvr.usace.army.mil/umr-iwwsns/.
3. USACE New Orleans District, Louisiana coastal area ecosystem restoration study Web site, U.S. Army Corps of Engineers, Apr. 23, 2004. Accessed May 24, 2004 at http://www.mvn.usace.army.mil/prj/lca/.
4. USACE, Planning guidance notebook, ER 1105-2-100, U.S. Army Corps of Engineers, Washington, D.C., 2000.
5. Thom, R.M. and Wellman, K.F., Planning aquatic ecosystem restoration monitoring programs, IWR Report 96-R-23, prepared for Institute for Water Resources, U.S. Army Corps of Engineers, Alexandria, VA and Waterways Experimental Station, U.S. Army Corps of Engineers, Vicksburg, MS, 1996.
6. USACE, Ecosystem restoration — Supporting policy information, EP 1165-2-502, U.S. Army Corps of Engineers, Washington, D.C., 1999.
7. Pastorok, R.A., MacDonald, A., Sampson, J.R., et al., An ecological decision framework for environmental restoration projects, *Ecol. Eng.*, 9, 89, 1997.
8. Thom, R.M., System-development matrix for adaptive management of coastal ecosystem restoration projects, *Ecol. Eng.*, 9, 219, 1997.
9. Yozzo, D., Titre, J., and Sexton, J., Eds., Planning and evaluating restoration of aquatic habitats from an ecological perspective, IWR Report 96-EL-4, submitted in association with the U.S. Army Corps of Engineers Waterways Experiment Station and PTI Environmental Services, Bellevue, WA and prepared for Institute for Water Resources, U.S. Army Corps of Engineers, Alexandria, VA, 1996.
10. Harrington, K.W. and Feather, T.D., Evaluation of environmental investments procedures: Interim overview manual, IWR Report 96-R-18, prepared for Institute for Water Resources, U.S. Army Corps of Engineers, Alexandria, VA, 1996.
11. Noble, B.D., Thom, R.M., Green, T.H., and Borde, A.B., Analyzing uncertainty in the costs of ecosystem restoration, IWR Report 00-R-3 (Draft), submitted in association with Battelle, Sequim, WA and prepared for Institute for Water Resources, U.S. Army Corps of Engineers, Alexandria, VA, 2000.
12. Yoe, C., Trade-off analysis planning and procedures guidebook, IWR Report 02-R-2, submitted in association with Planning and Management Consultants, Carbondale, IL and prepared for Institute for Water Resources, U.S. Army Corps of Engineers, Alexandria, VA, 2002.
13. Yoe, C., Risk analysis framework for cost estimation, IWR Report 00-R-9, submitted in association with Planning and Management Consultants, Carbondale, IL and prepared for Institute for Water Resources, U.S. Army Corps of Engineers, Alexandria, VA, 2000.
14. Males, R.M., Beyond expected value: Making decisions under risk and uncertainty, submitted in association with Planning and Management Consultants, Carbondale, IL and prepared for Institute for Water Resources, U.S. Army Corps of Engineers, Alexandria, VA, 2002.
15. Morgan, M.G. and Henrion, M., *Uncertainty: A Guide to Dealing With Uncertainty in Quantitative Risk and Policy Analysis*, Cambridge University Press, Cambridge, UK, 1990.
16. Yoe, C., An introduction to risk and uncertainty in the evaluation of environmental investments, IWR Report 96-R-8, submitted in association with Planning and Management Consultants, Carbondale, IL and prepared for Institute for Water Resources, U.S. Army Corps of Engineers, Alexandria, VA, 1996.

17. USACE and Institute for Water Resources, Planning manual, IWR Report 96-R-21, U.S. Army Corps of Engineers, Alexandria, VA, 1996.
18. Tilman, D., Knops, J., Wedin, D., et al., The influence of functional diversity and composition on ecosystem processes, *Science*, 277, 1300, 1997.
19. Hooper, D.U. and Vitousek, P.M., The effects of plant composition and diversity on ecosystem processes, *Science*, 277, 1302, 1997.
20. Loreau, M., Naeem, S., Inchausti, P., et al., Biodiversity and ecosystem functioning: Current knowledge and future challenges, *Science*, 294, 804, 2001.
21. Levin, S.A. and Paine, R.T., Disturbance, patch formation and community structure, *Proc. Natl. Acad. Sci.*, 71, 2744, 1974.
22. Grime, J.P., *Plant Strategies and Vegetation Processes*, John Wiley & Sons, Ltd., New York, 1979.
23. Connell, J.H. and Sousa, W.P., On evidence needed to judge ecological stability or persistence, *Am. Nat.*, 121, 789, 1983.
24. USACE, Sagamore marsh restoration study, Bourne and Sandwich, Massachusetts, EOEA File #10174, 1996.
25. Brinson, M.M., A hydrogeomorphic classification for wetlands, Wetlands Research Program Technical Report WRP-DE-4, U.S. Army Corps of Engineers, Waterways Experimental Station, Vicksburg, MS, 1993.
26. Forman, R.T.T. and Godron, M., *Landscape Ecology*, John Wiley & Sons, New York, 1986.
27. Gosselink, J.G. and Lee, L.C., Cumulative impact assessment in bottomland hardwood forests, *Wetlands*, 9, 89, 1989.
28. USEPA, A synoptic approach to cumulative impact assessment, EPA/600/R-92/167, U.S. Environmental Protection Agency, Corvallis, OR, 1992.
29. NRC (National Research Council), *Restoration of Aquatic Ecosystems: Science, Technology and Public Policy*, National Research Council, Commission on Geosciences, Environment and Resources, Washington, D.C., 1992.
30. Toth, L.A., Principles and guidelines for restoration of river/floodplain ecosystems — Kissimmee River, Florida, in *Rehabilitating Damaged Ecosystems*, 2nd ed., Cairns, J., Ed., Lewis Publishers, Boca Raton, FL, 1998, 49.
31. USFWS (U.S. Fish and Wildlife Service), Habitat evaluation procedures handbook, 870 FW-1, U.S. Fish and Wildlife Service, Washington, D.C., 1980.
32. Thom, R.M. and Wellman, K.F., Integrated ecosystem services model, unpublished.
33. Turner, R.K., Adger, W.N., and Lorensoni, I., Towards integrated modeling and analysis in coastal zones: Principles and practices, Land-Ocean Interactions in the Coastal Zone Reports and Studies No. 11, Land-Ocean Interactions in the Coastal Zone, Texel, The Netherlands, 1998.
34. Costanza, R., d'Arge, R., de Groot, R., et al., The value of the world's ecosystem services and natural capital, *Nature*, 387, 253, 1997.
35. Bockstael, N.E., Freeman III, A.M., Kopp, R.J., Portney, P.R., and Smith, V.K., On measuring economic values for nature, *Environ. Sci. Technol.*, 34, 1384, 2000.
36. Gutzwiller, K.J. and Anderson, S.H., Habitat suitability index models: Marsh wren, Biol. Rep. 82 (10.139), U.S. Fish and Wildlife Service, 1987.
37. Allen, T.F.H. and Hoekstra, T.W., Problems of scaling in restoration ecology: A practical application, in *Restoration Ecology: A Synthetic Approach to Ecological Research*, Jordan III, W.R., Gilpin, M.E., and Aber, J.D., Eds., Cambridge University Press, New York, 1987.
38. MacArthur, R.H. and Wilson, E.O., *The Theory of Island Biogeography*, Princeton University Press, Princeton, NJ, 1967.

39. Noss, R.F., Landscape connectivity: Different functions at different scales, in *Landscape Linkages and Biodiversity*, Hudson, W.E., Ed., Island Press, Washington, D.C., 1991.
40. Orth, K., Using habitat suitability index models in plan formulation, CEWRC-IWR-P, U.S. Army Corps of Engineers, Alexandria, VA, Apr. 14, 1993.
41. Boesch, D.F., Josselyn, M.N., Mehta, A.J., et al., Scientific assessment of coastal wetland loss, restoration and management in Louisiana, *J. Coastal Res.*, 20, pp. 15–53, 1994.
42. Louisiana Coastal Wetlands Conservation and Restoration Task Force and Wetlands Conservation and Restoration Authority, Coast 2050: Toward a sustainable coastal Louisiana, Louisiana Department of Natural Resources, Baton Rouge, LA, 1998.
43. Turner, R.E. and Brody, M.S., Habitat suitability index models: Northern Gulf of Mexico brown shrimp and white shrimp, FWS/OBS-82/10.54, U.S. Fish and Wildlife Service, 1983.
44. Shafer, D.J. and Yozzo, D.J., National guidebook for application of hydrogeomorphic assessment to tidal fringe wetlands, Wetlands Research Program, Tech. Report WRP-DE-16, U.S. Army Corps of Engineers, Waterways Experiment Station, Vicksburg, MS, 1998.
45. Brinson, M.M., Rheinhardt, R.D., Hauer, F.R., et al., A guidebook for application of hydrogeomorphic assessments to riverine wetlands, Wetlands Research Program, Tech. Report WRP-DE-11, U.S. Army Corps of Engineers, Waterways Experiment Station, Vicksburg, MS, 1995.
46. Saaty, R.W., The analytical hierarchy process, what it is and how it is used, *Math. Model.*, 9, 161, 1987.
47. Gagliano, S.M., Controlled diversions in the Mississippi River deltaic plain, Proceedings of International Symposium, Wetlands and River Corridor Management, Charleston, South Carolina, Association of Wetland Managers, July 5, 1989, 257.
48. NRC (National Research Council), *Risk Analysis and Uncertainty in Flood Damage Reduction Studies*, National Research Council, Commission on Geosciences, Environment and Resources, Washington, D.C., 2000.

APPENDIX 4-A

Example Application of Risk Analysis to a USACE Ecological Restoration Project

CONTENTS

1-56670-639-4/05/$0.00+$1.50

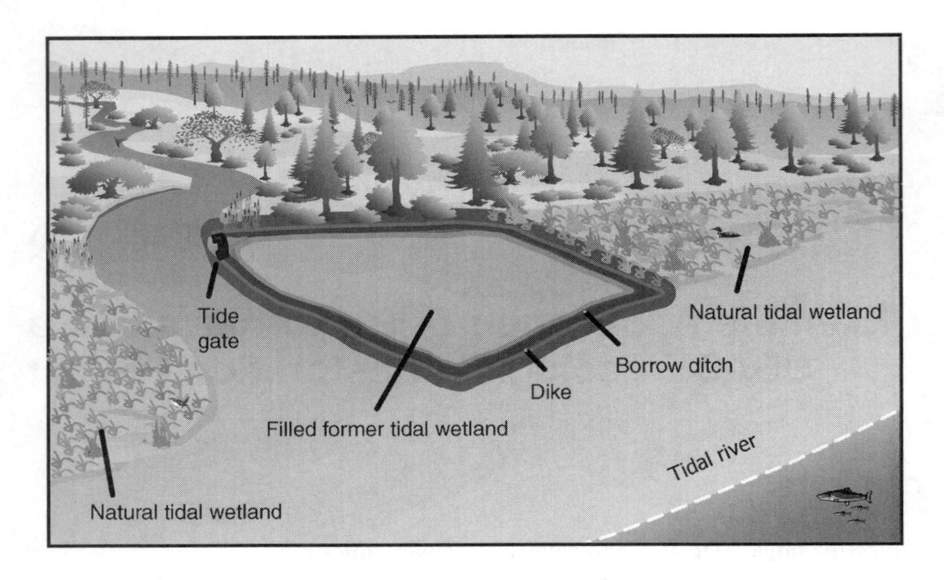

Figure 4-A.1 Diked and filled former marsh and adjacent habitats.

The following example employs the six planning steps described in Chapter 4 to create and select an approach for restoring the functions of a degraded tidal wetland. The planning *goal* of the project is to restore degraded significant salt marsh ecosystem structure, function, and dynamic processes to a less degraded, more natural condition.

PROJECT EXAMPLE STEP 1 — PROBLEMS AND OPPORTUNITIES: SETTING PLANNING OBJECTIVES

The tidal wetland is located adjacent to a river with a navigation channel (Figure 4-A.1), which requires periodic dredging to maintain authorized depths. Until the late 1970s, clean, dredged material was discharged into a portion of the tidal wetland that is contained within a dike. A small tide gate allows water to drain from the filled area back into the river. The diked area of approximately 25 acres is bordered on the downstream side by a relatively undisturbed wetland, on the upland side by naturally vegetated upland habitat, and on the upstream side by a small stream with an adjacent tidal wetland.

Diking and filling have cut off the hydrodynamic processes critical to maintenance of a viable wetland habitat. Prior to diking, water flushed over the site at a minimum of 12 times per month. Since tidal flushing was stopped, precipitation is the only source of water to the site. Hence, in addition to the physical loss of wetland habitat by fill, any remaining habitat is now freshwater/terrestrial, and completely different from the natural conditions. The changes have resulted in wetland habitat losses, which have had an effect on wildlife and habitat quality and functions. Removal of fill and breaching of the dike to restore natural tidal processes are the key actions required to restore the system.

The following *planning objectives* were established by USACE planners and stakeholders in response to the need to restore the site:

- Restore tidal marsh wildlife habitat.
- Restore natural tidal marsh wetland vegetation communities.

The planning objectives will be measured by proxy through design objectives. Tidal wildlife habitat and tidal wetland vegetation communities will be quantified using the HSI model for the marsh wren along with the wildlife utilization and plant community composition components of the HGM model. Together, these models account for species as well as for the general conditions for wildlife at the site. While it is unlikely that the abundant marsh wren would ever be the target of restoration planning, this relatively simple model is useful as an example. The HGM model provides a simple and direct measure of the amount of native vegetation cover relative to nonnative vegetation cover.

Given the inherent variability of natural systems, the planning team established the *design objective* that a viable alternative would be required to have at least a 50% chance of achieving a marsh wren habitat unit score of 9 or above.

There is uncertainty associated with the project. The risk of failure to meet the planning objectives is assessed through evaluation of the distribution of index values for the with- and without-project conditions. Yoe[1] presents a method to estimate the range of index values, which results in a minimum, median, and maximum value for an index. Yoe recommends that the values for variables be treated as ranges rather than point estimates. The purpose is to recognize and communicate to the extent possible the degree of uncertainty the study team associates with the values.

For example, the variable for cover of suitable vegetation type in the HSI model for the marsh wren can have a point estimate of 50%. However, uncertainties associated with cover estimates result in a range of 30% to 70%. In such a case, Yoe[1] suggests that the range be used for the analysis; if the distribution of the values were known, then the range could be based on the 95% confidence limits. The HSI index would be calculated using a range of values for each variable. Minimum, maximum, and mean HSI index values would result from the series of calculations using all combinations of values. In a similar fashion, the minimum, maximum, and mean index values can be determined for with-project conditions. By comparing the with- and without-project results, one can determine whether — at a minimum, maximum, and on average — the with-project conditions of the HSI index will be improved over the without-project conditions.

PROJECT EXAMPLE STEP 2 — INVENTORY AND FORECAST

Conceptual Model

The conceptual model of the landscape and controlling factors for marsh wren habitat (Figure 4.4) provides a basis for understanding the existing conditions and their relationship to the status of wildlife and wetland habitat functions. This model is based on information summarized from the HSI report for the marsh wren and an understanding

Table 4-A.1HGM and HSI Indices for Existing Conditions Analysis

Index	Minimum	Median	Maximum
Marsh wren HSI	0.00	0.10	0.20
WHU HGM	0.00	0.10	0.20
PCC HGM	0.00	0.05	0.10

of landscape-level processes that can affect the formation and maintenance of marsh wren habitat.[2] Using the model as a guide, the relevant processes and structures that have been affected are tidal hydrology, standing water, and marsh cover featuring cattails, sedge, and rush. Some of the broader landscape-level features that have been changed are the presence of surrounding buffer habitats and potential human threats.

Existing Conditions Analysis

The HSI marsh wren model (Equation 4.1) indicates the quality of existing ecosystem conditions. Because the HSI model is not necessarily inclusive of tidal marsh wildlife, the HGM tidal marsh models provide additional guidance to the design.[3,4] For example, the HGM model for WHU (Equation 4.7) includes factors related to edge between intertidal vegetation and unvegetated edge, as well as upland edge and habitat complexity. The HGM model for native vegetation (Equation 4.6) indicates only that the highest scores are assigned to systems entirely containing native vegetation species. Hence, potential threats for invasion of the system by nonnative vegetation must be considered in the assessment of alternatives.

Without modification of hydrology and removal of fill, the existing conditions will remain indefinitely. There is no expected improvement or reduction in overall habitat quality or functioning. The existing conditions are poor for wildlife and vegetation, as indicated by the calculations of HSI and HGM indices (Table 4-A.1).

PROJECT EXAMPLE STEP 3 — PLAN FORMULATION

The conceptual model and the HSI and HGM models reveal that restoration of hydrology and vegetation features is required to restore the desired functions expressed as planning objectives. Evaluation of the HSI model (Equation 4.1) indicates that the percentage of canopy cover of woody vegetation is the most responsive variable; this is also a variable in the HGM PCC model (Equation 4.6). Mean water depth is a limiting factor; there cannot be emergent herbaceous vegetation without standing water. An obvious measure would be to remove the entire facility (dike and fill) and allow the area to be flooded again. Plant recruitment is expected to occur naturally by airborne, waterborne, and bird dispersal mechanisms.

What Could Go Wrong?

1. As plant recruitment mechanisms potentially limit both the number and rate of species recruitment to the site, invasive exotic species may become established and dominate the site. To reduce this risk, desirable species could be planted in the project area.

2. Storm waves could destroy immature vegetative stands or prevent natural vegetation from becoming established. This suggests an alternative that leaves the dike in place until the vegetation is mature. The existing tide gate could be used to regulate the water depth behind the dike.
3. Grazing activity by species such as the Canada goose may delay colonization of the site. Various techniques can be used to reduce this risk.
4. Protected vegetation may never develop the root system necessary to withstand storm waves, resulting in the loss of tidal vegetation when the dike is removed. Allowing limited exposure of the vegetative stand by partial removal of the dike may reduce this risk.

The following alternatives will be carried forward in this example.

1. Removal of dike and fill.
2. Removal of dike and fill with planting of desirable species and grazing prevention.
3. Removal of fill only with planting of desirable species and grazing prevention. The tide gate will be modified to regulate water depths.
4. Removal of fill, planting of desirable species, grazing prevention, and partial removal of the dike to allow a robust environment for the vegetative stand to develop.
5. Removal of fill, planting of desirable species, prevention of grazing, and removal of the dike in segments over time to allow a robust environment for the vegetative stand to develop and achieve a natural tidal wetland environment.

Potential Sources of Errors in the Analysis

There are several sources of errors and uncertainties in the previous example. For instance, it is assumed that if the habitat conditions were built, the marsh wren would occupy the site. However, the wren may not have sufficient numbers in the region to colonize the site, an uncertainty that can be investigated by consulting experts. Elevations of the site may be difficult to determine and to excavate precisely during construction, resulting in another source of uncertainty. In a microtidal marsh system, a few inches of change in elevation can significantly affect the type of plant community that develops. Further, the fill and dike material are assumed to be clean and disposable without hazardous material costs, an uncertainty that needs to be addressed through assessment of the materials and careful control during construction.

PROJECT EXAMPLE STEP 4 — EVALUATION OF ALTERNATIVE PLANS

Initial Appraisal of the Alternatives

Five alternatives were identified during the plan-formulation phase. The first step in forecasting the with-project conditions is to consider these alternatives in the context of the landscape. Given the limited scale of the project, it is assumed that most effects of the alternatives would be limited to the project area, and that all significant effects would be adequately measured by the variables selected in Planning Step 1: the HSI for the marsh wren, and the HGM scores for tidal marsh. It is necessary to convert these values to habitat units for comparison because not all alternatives have the same spatial footprint.

The habitat value scores for each alternative are estimated using the method described in Yoe.[1] Briefly, this method takes a distribution, rather than a point estimate, as the input for each variable in the model. A Monte Carlo process is used, in which the input distributions for each variable are convolved into the distribution of habitat units score. For this example, triangular distributions are used to describe each variable.

The cost estimates used in this evaluation also include uncertainty. Using the method outlined in Yoe[5] and Yoe,[1] the cost estimates account for both unit cost and unit quantity uncertainty. This method also uses a Monte Carlo process, and it produces a distribution of total cost for each alternative.

Alternative 1: Removal of Dike and Fill

With removal of the dike and fill, the system should tend toward development of marsh vegetation, and, ultimately, of ponding. However, without planting or grazing prevention and with uncertainty in the proximity of a seed source, the most likely values for HSI model vegetation variables SIV1, SIV2, and SIV4 are limited (Table 4-A.2). The minimum values for these vegetation variables are very low, due to the possibility for system-wide sediment deposition and erosion with complete dike removal. This possibility for disturbance also tends to keep the most likely and minimum values for SIV3, mean water depth, relatively low. Uncertainty in construction techniques also affects this variable — specifically, bulldozing to an HSI target for the marsh wren that is measured in centimeters. Distributions such as these typically are developed by a panel of experts and are subject to error based on the experience and perspective of the participants.[6]

The total percentage of vegetation cover, which is a variable in the PCC score in the HGM model, is affected by uncertainties in several factors: grazing, natural colonization, the proximity of seed sources, and disturbance, as discussed earlier (Table 4-A.3). The other component in the PCC score, the percentage of cover of exotic or nuisance species, is affected by the proximity of seed sources and by grazer preferences.

The qualities of edge and habitat complexity that drive the WHU indices are positively affected by removal of the dike, which restores intertidal–subtidal and upland edges. Without planting and grazing prevention, however, the vegetation and ponding will develop slowly and limit the development of associated edge habitats. Uncertainties in the quality of postconstruction edges will also exist without the use of planting as a restoration tool. Wildlife habitat complexity is positively affected by the presence of stream and subtidal habitats, but vegetation and ponds will develop relatively slowly under this alternative, keeping the most likely and maximum values low (Table 4-A.4).

Using the process outlined by Yoe,[5] the input distributions for Alternative 1 resulted in a distribution of habitat unit scores for each model, and a distribution of estimated costs (Figure 4-A.2). Including uncertainty in the analysis allows the

Table 4-A.2 HSI Input Variable Scores (Average Annual Values)

	Alternative 1 Index			Alternative 2 Index			Alternative 3 Index			Alternative 4 Index			Alternative 5 Index		
Variable[a]	most likely	min	max	most likely	min	max	most likely	min	max	most likely	min	max	most likely	min	max
SIV1	0.80	0.40	0.99	0.70	0.20	0.99	0.90	0.40	0.99	0.80	0.40	0.99	0.90	0.50	0.99
SIV2	0.80	0.40	0.99	0.70	0.20	0.99	0.90	0.40	0.99	0.80	0.40	0.99	0.90	0.40	0.99
SIV3	0.85	0.30	0.99	0.80	0.20	0.99	0.90	0.40	0.99	0.85	0.30	0.99	0.90	0.40	0.99
SIV4	0.80	0.40	0.99	0.70	0.20	0.99	0.90	0.40	0.99	0.80	0.40	0.99	0.90	0.40	0.99

[a] SIV1 = vegetation growth form
SIV2 = % canopy emergent herbaceous vegetation
SIV3 = mean water depth
SIV4 = % canopy woody vegetation

Table 4-A.3 PCC Input Variable Scores (Average Annual Values)

	Alternative 1 Index			Alternative 2 Index			Alternative 3 Index			Alternative 4 Index			Alternative 5 Index		
Variable[a]	most likely	min	max	most likely	min	max	most likely	min	max	most likely	min	max	most likely	min	max
Vcov	0.90	0.20	0.99	0.85	0.10	0.99	0.95	0.40	0.99	0.90	0.20	0.99	0.95	0.40	0.99
Vexotic	0.20	0.01	0.99	0.20	0.01	0.99	0.20	0.01	0.99	0.20	0.01	0.99	0.20	0.01	0.99

[a] Vcov = total % vegetative cover
Vexotic = % vegetative cover by exotic/nuisance species

Table 4-A.4 WHU Input Variable Scores (Average Annual Values)

	Alternative 1 Index			Alternative 2 Index			Alternative 3 Index			Alternative 4 Index			Alternative 5 Index		
Variable[a]	most likely	min	max	most likely	min	max	most likely	min	max	most likely	min	max	most likely	min	max
Vae	0.70	0.50	0.80	0.80	0.50	0.90	0.50	0.30	0.60	0.70	0.50	0.80	0.85	0.60	0.95
Vue	0.00	0.00	0.00	0.90	0.80	1.00	0.00	0.00	0.00	0.00	0.00	0.00	0.90	0.80	1.00
Vwhc	0.60	0.30	0.80	0.70	0.10	0.99	0.70	0.50	0.80	0.60	0.30	0.80	0.90	0.60	0.99

[a] Vae = amount edge between the intertidal vegetated, intertidal unvegetated, and subtidal areas
Vue = amount and quality of upland edge
Vwhc = wildlife habitat complexity

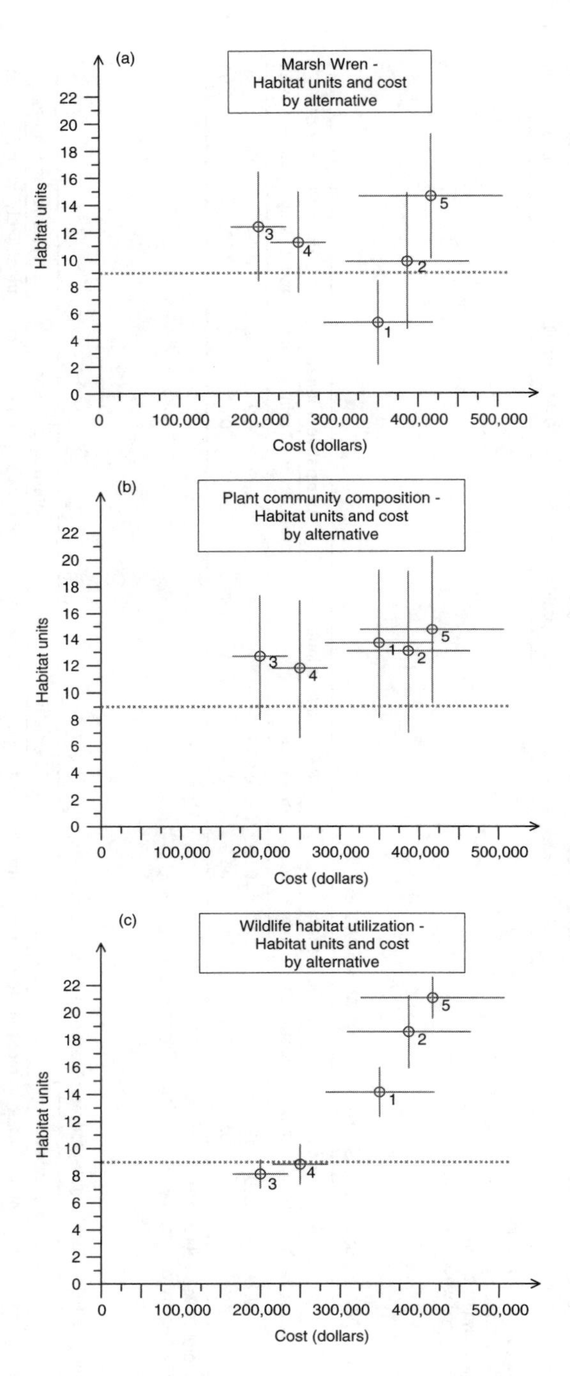

Figure 4-A.2 Habitat units and cost by alternative for three models: marsh wren habitat suitability index (a), plant community composition (b), and wildlife habitat utilization index (c). The mean score and the fifth and ninety-fifth percentiles are shown for habitat units and cost.

analyst to communicate to other decision-makers a degree of belief about the expected outcome of the plan.

Alternative 2: Removal of Dike and Fill with Planting of Desirable Species and Grazing Prevention

This alternative, like Alternative 1, produces a system that is exposed to sediment deposition and erosion because of dike removal, but in addition, desirable species are planted and stressors are removed by grazing prevention. This tends to increase the most likely and maximum values for the vegetation variables in both the HSI model (Table 4-A.2) and the PCC score (Table 4-A.3), while keeping the minimum values low, due to the possibility of disturbance. The quality of vegetative cover is likely to be improved by giving desirable species a head start over invasive exotics, and therefore the most likely percentage of cover by exotic and nuisance species is lowered. Maximum possible values for this variable remain high because an invasion is always possible, depending on the proximity of seed sources and dispersal vectors. Mean water depth, SIV3, is unchanged.

The accelerated succession of plant communities due to grazing prevention and planting tends to hasten the development of ponds and thus to increase the WHU scores for edge in the HGM model (Table 4-A.4). As with Alternative 1, restoration of both intertidal-subtidal and upland edges is begun with complete dike removal. Uncertainties remain in the quality of post-construction edge, but these can be minimized by planting. Wildlife habitat complexity is increased by the presence of more developed plant communities and ponds, although the minimum possible value remains low due to the exposure to sedimentation and erosion in the early stages of marsh development.

Alternative 3: Removal of Fill Only with Planting of Desirable Species, Grazing Prevention, and Modification of the Tide Gate to Regulate Water Depths

Alternative 3 affords a very protected start for plants, and therefore increases the most likely and maximum values for the HSI and PCC vegetation variables, with the exception of exotic and nuisance species cover; the most likely and maximum values for this variable decrease because desirable species are planted in relatively protected conditions (Tables 4-A.2 and 4-A.3). Reducing the exposure of the system to major sediment deposition or erosion also increases the most likely and maximum values for mean water depth. In general, minimum values increase under this alternative as well, due to the low likelihood of catastrophic disturbance.

Without dike removal, however, the number of edge habitats in the system remains very low. No upland edge or intertidal–subtidal edge habitat is restored, and therefore the most likely and maximum values for these variables in the HGM model are low or zero (Table 4-A.4). Some edge habitats are created by the accelerated development of plant communities in this protected environment, leading to the establishment of ponds. Although the protected development of vegetation in this system will create some wildlife habitat complexity, the most likely and maximum values for this variable are not high because the dike impedes habitat connectivity; the protection from disturbance keeps the minimum value relatively high.

Alternative 4: Removal of Fill, Planting of Desirable Species, Grazing Prevention, and Partial Removal of the Dike

A portion of the dike on the stream side of the project site is removed in this alternative, increasing the possibility of local disturbance. This uncertainty reduces the most likely values of the vegetation variables in the HSI model and the PCC score to slightly below the levels for Alternative 3 (Tables 4-A.2 and 4-A.3). The most likely and minimum values for mean water depth are also reduced, due to the possibility of localized deposition or scour in the area of the dike breach. Overall, this alternative provides a relatively protected environment for the development of plant communities to target levels. The PCC values for exotic and nuisance vegetation are unchanged, with invasions relatively well controlled.

Alternative 4 is an improvement over Alternative 3 in providing some intertidal–subtidal edge habitat in the area of the dike breach, which increases the edge scores in the WHU module. However, because the purpose of the dike breach is to provide tidal inflow to the project site, there is no construction activity on the upland side, and no upland edge is restored; the score for the amount and quality of upland edge remains zero, as in Alternative 3 (Table 4-A.4). Wildlife habitat may reach the level of complexity estimated for Alternative 3, but there is also the possibility of disturbance slowing community development or affecting the mean water level.

Alternative 5: Removal of Fill, Planting of Desirable Species, Grazing Prevention, and Removal of the Dike in Segments over Time

Overall, this alternative provides a protected start for plants and ultimately should achieve a restored ecosystem without remnants of the dike limiting habitat continuity. The possibility of disturbance during the most vulnerable postplanting period is kept low by removing the dike in segments, which increases the minimum values for the vegetation variables in the HSI model and for total percentage of cover in the PCC score (Tables 4-A.2 and 4-A.3). The most likely and maximum values for these variables are high, equal to those of Alternative 3. The PCC values for exotic and nuisance vegetation remain unchanged.

The design of the project phases is critical to the values assigned to the variables in the WHU model. The upland portion of the dike should be removed in the first phase of construction, when the dike is first breached to permit tidal inundation, to allow habitat connectivity to develop across the upland–marsh boundary. With this specification, the most likely and maximum values for the edge and complexity variables in the HGM model are the highest of any alternative (Table 4-A.4). Intertidal–subtidal edge is also created as the dike is removed in segments, and ponding begins as the plant communities develop rapidly from the initial planting in a relatively protected environment. Minimum values for these variables are also the highest of any alternative, due to the low probability of disturbance.

Qualifying Plans for Further Consideration

Having appraised the effects of each plan, the analyst can now qualify each plan for further consideration based on the four criteria: completeness, effectiveness, efficiency, and acceptability.[7] Completeness or actions to ensure the realization of the planned effects can now be judged by the uncertainty associated with each plan. Completeness was defined in Planning Step 1 in terms of uncertainty; to avoid unacceptable risk, viable alternatives were required to have at least a 50% chance of achieving a marsh wren habitat unit score of 9 or above.

The evaluation results for all alternatives are summarized in Figure 4-A.2. Examining these graphs of habitat unit scores, the following observations can be made: the identified outcomes of Alternative 1 are clearly below the planning requirement of 50% over a marsh wren habitat score of 9 (Figure 4-A.2a). Therefore, Alternative 1 fails the criterion of completeness. Regarding effectiveness, a perfect habitat would have a score of 25 — the higher the habitat unit score, the more effective the measure. The inclusion of uncertainty in the analysis allows the analyst to determine whether an alternative is more effective over the entire range of outcomes or simply at the average value. In our example, Alternative 5 is the most effective plan by all three measures; however, there is considerable overlap between Alternative 5 and Alternative 2. Efficiency, or cost-effectiveness, is also revealed in the analysis. For example, the Alternative 3 score for marsh wren habitat units exceeds scores for Alternatives 1, 2 and 4, which cost more and yield less in terms of marsh wren habitat.[a] Alternatives 3 and 4 do share some solution space, as do Alternatives 2 and 5. Using the PCC, Alternative 3 scores exceed those for Alternative 4, and Alternative 1 scores exceed those for Alternative 2 (Figure 4-A.2b). With respect to the WHU, all of the plans are efficient (Figure 4-A.2c), in that a more costly alternative is never less effective.

From an efficiency standpoint, the analyst could qualify the following for further consideration: all plans, based on the WHU scores; Alternatives 1, 3, and 5, based on the PCC scores; or only Alternatives 3 and 5, based on the Marsh Wren HSI scores. The final decision will depend on the relative weight assigned to each model score. Acceptability can also be assessed by the risk information revealed in the analysis. The local sponsor may be unwilling to participate in an alternative that costs too much. In this example, if the local sponsor were limited to $450,000, Alternative 5 could be unacceptable because it represents some risk of exceeding that cost.

The plans that qualify for further consideration in this example are Alternatives 2, 3, and 5. The marsh wren habitat score for Alternative 1 presents too great a risk of failure. This alternative is incomplete. Alternative 4 was judged inefficient in that Alternative 3 produces more for less cost in terms of both the PCC score and the marsh wren score. The final three alternatives will be compared in Step 5.

[a] Orders of stochastic dominance, correlation of outcomes, and dependencies are ignored to simplify the example.

PROJECT EXAMPLE STEP 5 — PLAN COMPARISON

In this example, the measures used to evaluate the alternatives are incommensurate; that is, marsh wren habitat units cannot be added to PCC habitat units or wildlife habitat units. Comparison must proceed either by examining each output separately or by establishing weights for each. If weights are established, the outputs can be summed and cost-effectiveness and incremental cost analysis used to compare plans. This example proceeds by examining each output separately.

The remaining alternatives in our example are the no-action plan and Alternatives 2, 3, and 5. Based on Figure 4-A.2, the following observations can be made: the common metric to each output is cost. Comparing the cost of each alternative shows that Alternative 3 will always cost less than Alternatives 2 or 5. Alternative 5 costs more than Alternative 2 at the mean; however, the cost estimates have considerable overlap. This overlap indicates that Alternative 5 could end up costing less than Alternative 2.

Because Alternative 3 is the least-cost with-action alternative, it is considered the "first added," and the others are compared incrementally with it. Alternative 2 contributes a clear incremental increase of 10.5 WHU habitat units over Alternative 3. However, PCC units are only marginally improved at the mean, 0.4 habitat units, and have a greater degree of dispersion, with some potential outcomes lower in PCC by as much as 2 habitat units. The output of Marsh Wren habitat units for Alternative 2 is lower at the mean and the fifth and ninety-fifth percentiles than for Alternative 3. Given these outcomes, it is only Alternative 2's contribution to WHU that gives it potential merit. It will be the subjective evaluation of WHU relative to marsh wren habitat and PCC that determines the strength of this advantage.

Alternative 5 makes incremental contributions to all three measures for its additional cost, whether it is considered added to Alternative 3 or to Alternative 2. For example, at the mean, Alternative 5 contributes an additional 2.2 marsh wren habitat units over Alternative 3. The incremental cost of Alternative 5 from Alternative 3 at the mean is $217,000. Subjective judgment will be needed to answer the incremental cost analysis question: "Is it worth it?"

Depending on the weights given to each output, it is possible to construct a scenario in which all three alternatives remain on the efficient frontier. However, in this example, we assume the contribution of an additional 10.5 WHU habitat units by Alternative 2 is not sufficient to offset the mean loss of 2.6 marsh wren habitat units incurred by selecting Alternative 2 over 3. Therefore, Alternative 2 is removed from further consideration.

PROJECT EXAMPLE STEP 6 — PLAN SELECTION

The final choice set in our example consists of Alternatives 3 and 5 and the no-action plan. All three options can be defended. The no-action plan has low habitat values, but the decision-makers may think the restoration project is too expensive for the outcomes achieved. Alternative 3 has similar accomplishments on the marsh wren and PCC metrics for less than half the cost of Alternative 5. Alternative 5 makes a significant contribution to the WHU index. The subjective values brought to the analysis by the decision-makers will determine the final selection.

What are the other criteria to consider in the final plan selection? One of the risks identified earlier in this report was the risk that the metrics used to measure the contribution of a plan to the planning objectives (the marsh wren, PCC, and WHU habitat units in our example) do not fully reflect the stated goals of the planning effort. One goal of the USACE restoration program is to reestablish natural, self-regulating ecosystems. In contrast to Alternative 5, Alternative 3 leaves in place the man-made levees and tide gate. This is reflected in the indices, in that the number of acres restored is less, and thus the habitat unit scores will be less for an equal HSI score. However, this numerical difference may not adequately express the importance of this outcome to the stakeholders.

What is the convincing argument for selecting a single alternative from the three remaining? Planners soon learn that what convinces some does not convince others. One example assumes that Alternative 5 is selected because of the significant contribution it makes to the WHU index and because it completely restores the site to a more natural condition. The cost for this is more than double the cost of Alternative 3. However, some decision-makers may think this additional cost is not worth the incremental gain and that the difference may be better spent restoring a different site. This is when the public debate begins in earnest. At this stage, after the planning team has made its recommendation, it should step back and let the decision-makers, such as USACE leadership, local sponsors, and stakeholders, make the final judgment.

REFERENCES

1. Yoe, C., An introduction to risk and uncertainty in the evaluation of environmental investments, IWR Report 96-R-8, submitted in association with Planning and Management Consultants, Carbondale, IL and prepared for Institute for Water Resources, U.S. Army Corps of Engineers, Alexandria, VA, 1996.
2. Gutzwiller, K.J. and Anderson, S.H., Habitat suitability index models: Marsh wren, Biol. Rep. 82 (10.139), U.S. Fish and Wildlife Service, 1987.
3. Shafer, D.J. and Yozzo, D.J., National guidebook for application of hydrogeomorphic assessment to tidal fringe wetlands, Wetlands Research Program, Tech. Report WRP-DE-16, U.S. Army Corps of Engineers, Waterways Experiment Station, Vicksburg, MS, 1998.
4. Brinson, M.M., Rheinhardt, R.D., Hauer, F.R., et al., A guidebook for application of hydrogeomorphic assessments to riverine wetlands, Wetlands Research Program, Tech. Report WRP-DE-11, U.S. Army Corps of Engineers, Waterways Experiment Station, Vicksburg, MS, 1995.
5. Yoe, C., Risk analysis framework for cost estimation, IWR Report 00-R-9, submitted in association with Planning and Management Consultants, Carbondale, IL and prepared for Institute for Water Resources, U.S. Army Corps of Engineers, Alexandria, VA, 2000.
6. Ayyub, B.M., Methods for expert-opinion elicitation of probabilities and consequences for Corps facilities, IWR Report 00-R-10, prepared for Institute for Water Resources, U.S. Army Corps of Engineers, Alexandria, VA, 2000.
7. USACE (U.S. Army Corps of Engineers), Planning guidance notebook, ER 1105-2-100, U.S. Army Corps of Engineers, Washington, D.C., 2000.

CHAPTER 5

Introduction to Economic Analysis in Watersheds

Matthew T. Heberling and Randall J.F. Bruins

CONTENTS

This chapter discusses economic analysis in relationship to watershed management. As with the introduction to ecological risk assessment that was presented in Chapter 3, the goal here is to provide sufficient background to make the succeeding chapters understandable to the noneconomist, rather than to provide a comprehensive introduction to the topic. First, it describes welfare economics as the foundation of environmental and natural resource economics, and the related concept of economic value. Next, this chapter introduces some tools that are used for the valuation of environmental goods and services, and some watershed-related applications of those tools. Then it presents game theory, a set of approaches for modeling decisions that are based on economic theory. Finally, it discusses ecological economics, an emerging field that has criticized the mainstream economic paradigm, and its potential contribution to the practice of watershed analysis.

1-56670-639-4/05/$0.00+$1.50

Table 5.1 Definition of Ecosystem Services and Related Terminology[a]

Ecological functions or processes — the characteristic physical, chemical, or biological activities that influence the flows, storage, or transformations of materials or energy within and through ecosystems.
Example: the uptake of nitrogen from soil by vegetation.

Ecosystem services —

- Ecological functions or processes that directly or indirectly enhance human well-being.
  Example 1: the production by an estuary of seafood harvested for human consumption.
  Example 2: the uptake of nitrogen by riparian and wetland vegetation in a watershed, which supports downstream seafood production, recreational opportunities, and aesthetic qualities by preventing excess nutrient loading to estuarine waters.
- Ecosystem attributes from which people derive amenities, such as serenity, beauty and cultural inspiration, and recreational opportunities.
  Example: the autumn foliage in a mixed, temperate forest on a complex landscape, which people currently enjoy or would enjoy viewing.
- The preservation by ecosystems of the potential to sustain or fulfill human life in ways that are as yet unrecognized.
  Example: the maintenance by diverse ecosystems of a reservoir of biologically active chemical compounds whose uses as pharmaceuticals might be determined in the future.

Ecosystem goods — physical outputs produced by ecosystems that enhance human well-being. (Note that ecosystem functions or processes that support the production of ecosystem goods are ecosystem services.)

Ecological benefits — in general, any contributions to human well-being derived from ecosystems. In this book, the term applies specifically to *net improvements* in human well-being that result from *changes* in the flow of ecosystem goods and services resulting from environmental management actions. Since changes in human well-being are often difficult to measure or estimate, changes in ecosystem goods and services *per se* also are considered to be ecological benefits as long as the relationship to human well-being is evident.
Examples:

- The contributions to human well-being from the protection or restoration of riparian and wetland vegetation, such as increased seafood production and improved recreational and aesthetic quality.
- The contributions to human well-being from the reduction of concentrations of airborne particulate matter, such as the enhanced aesthetic qualities of the surrounding landscape.
- The potential for future improvements in human well-being that results from actions to preserve species whose pharmaceutical uses might be recognized and developed in the future.

[a]Definitions developed in discussions of the USEPA Ecological Benefits Assessment Strategic Plan Workgroup, with reference to terminology of Freeman,[49] Daily,[50,51] King[52] and Whigham.[53]

WELFARE ECONOMICS

Economists study the allocation of scarce resources across competing uses. Aquatic ecosystems provide many goods and services to humans (Table 5.1 and Table 5.2). Like time and money, the allocation of environmental goods and services entails important choices because all wants cannot be satisfied.

Welfare economics is the study of agents who are making choices, under the given assumption that they are trying to maximize their well-being (i.e., their *welfare* or satisfaction, also termed *utility*). Economists focus on choices made by agents such as individuals or firms. They assume individuals are rational — that

Table 5.2 Daily's Classification of Ecosystem Services with Illustrative Examples

Production of Goods	• food (terrestrial animal and plant products, forage, seafood, spice) • pharmaceuticals (medicinal products, precursors to synthetics) • durable materials (natural fiber, timber) • energy (biomass fuels, low-sediment water for hydropower) • industrial products (waxes, oils, fragrances, dyes, latex, rubber, etc., precursors to many synthetic products) • genetic resources (intermediate goods that enhance production of other goods)
Regeneration Processes	• cycling and filtration processes (waste detoxification and decomposition; soil fertility generation and renewal; air and water purification) • translocation processes (dispersal of seeds necessary for revegetation; pollination of crops and natural vegetation)
Stabilizing Processes	• coastal and river channel stability • compensation of one species for another under varying conditions • control of the majority of potential pest species • moderation of weather extremes (such as of temperature and wind) • partial climate stabilization • hydrological cycle regulation (mitigation of floods and droughts)
Life-Fulfilling Functions	• cultural, intellectual, and spiritual inspiration • aesthetic beauty • existence value • scientific discovery • serenity
Preservation of Options	• maintenance of the ecological components and systems needed for future supply of these goods and services and others awaiting discovery

(Adapted from Daily, GC. *Environ. Sci. and Policy,* 3, 333, 2000 and as cited in USEPA, Planning for Ecological Risk Assessment: Developing Management Objectives, External Review Draft, EPA/630/R-01/001A, June 2001.)

is, they make choices that maximize their well-being subject to constraints on time and money — and that firms maximize profits subject to technology or resources. These decisions are examined through *marginal analysis* — that is, by determining how beneficial or costly one additional unit of a good or service would be to the agent.

In an ideal market, agents' decisions will lead to an efficient outcome, or one in which all mutually beneficial trades have been made. In other words, under conditions of economic efficiency, also termed *Pareto efficiency*, the distribution of resources among agents is such that no one can be made better off without making someone else worse off.[a] Rarely, however, do markets achieve efficient outcomes for environmental goods and services.[1] More often, characteristics of the market or of the goods and services make trade in the marketplace difficult. Economists describe these as situations of market failure, and they may attempt to identify social arrangements, including policies and institutions, for adjusting the distribution of resources to improve efficiency.

Some ecosystem goods and services, like hydropower or bottled water, are traded in markets, yet imperfections in these markets may lead to inefficiency and degradation.

[a] It should be noted that efficient outcomes are not always fair. The concept of equity is discussed later in the section titled "Complementary Analyses."

Others, including public goods such as recreational fishing sites and ecological services such as aesthetics or groundwater recharge, may lack markets entirely; economists refer to these as nonmarket goods and services.

Further inefficiencies in the market exist because aquatic ecosystems have been used as waste receptacles; third parties are "external" to these market transactions, although they are affected by them. Consider, for example, pollutant discharges by a firm into a river that is used by downstream households for recreation; regular markets provide no mechanism to compensate these third parties for the effects of these externalities and are therefore inefficient.

A final type of market failure occurs when the economic agents have incomplete information, or differing information, about a good or service.[2] Information may be incomplete because not all the relationships within an aquatic ecosystem are fully known; for example, decisions to pollute or to develop may be made without a full understanding of the consequences.[2] Asymmetric information may lead to strategic interaction among those involved, rather than straightforward responses based on supply and demand.[3,4]

Recognition of these kinds of market failure has led to the development of natural resource and environmental economics as specialized subfields of welfare economics. Natural resource economics examines the optimal allocation of scarce resources over time, including both nonrenewable resources (e.g., minerals) and renewable resources (e.g., fisheries and water resources).[1] Environmental economics tends to focus on two main issues: regulating pollution or damages as an externality, and valuing nonmarket goods.[5]

ECONOMIC VALUE

At this point it is necessary to provide a clear definition of economic value. Freeman[6] defines economic value within the welfare economic framework. Because each individual is considered to know how well off he or she is in a given situation, and because each individual's well-being depends on both private and public goods, the economic value of any particular good should be based on the associated changes to individuals' well-being. In some cases, markets help define economic value, but in the absence of markets, or in cases of market failure, other techniques are needed.

In either situation, economic value is defined as the maximum of something someone is willing to give up to get something else.[6] It does not need to be measured in dollars (e.g., an individual may be willing to give up the usefulness of a dam to obtain an increase in water quality and better fishing), but the dollar metric allows economists to compare trade-offs to all other goods. *Willingness to pay* (WTP) is a monetary measure of a welfare change or economic value; it is the maximum amount a consumer would pay to obtain or avoid a particular change. An alternative measure to WTP is *willingness to accept* (WTA), the minimum amount of money an individual is willing to take to give up some change. Both WTP and WTA measure value, but they are likely to differ for a number of reasons.[6,7] For example, they use different starting points for the initial levels of well-being (for an improvement, WTP is measured by starting at the individual's level of well-being before the improvement and WTA is calculated by starting at the individual's level of well-being after the improvement). Also, WTP is constrained by income, while WTA has no upper constraint. Economists typically use WTP to value benefits because it is easier to estimate.[7]

Economic value for environmental goods and services has been separated into use and nonuse value. *Use value* applies when people get some satisfaction from personal utilization of environmental goods and services; use can be direct or indirect. An example of direct use is enjoying the woods while hiking. To one who enjoys fishing for smallmouth bass, indirect use may mean valuing crayfish because smallmouth bass eat them. The idea of *nonuse value*, first introduced by Krutilla,[8] comes from the notion that individuals can value environmental goods and services regardless of whether they use the resource.[a] For example, individuals in the United States are willing to devote resources to protecting Brazilian habitat for the endangered golden tamarind monkey, even though they do not ever expect to visit the protected area or to see the species. The total economic value for a nonmarket good or service is the aggregate of these categories of values.

Economists have developed a variety of methods for estimating nonmarket values.[6] The methods can be categorized according to how the data are generated (based on observed or hypothetical behavior).[9] Observed-behavior approaches, referred to as *revealed preference methods*, infer values from data on actual market choices related to the public good.[6,7] Table 5.3 briefly describes four revealed preference approaches. Revealed preference approaches require market data, which limits the kinds of environmental goods that can be valued. The assumptions on which these approaches rely also affect the results. The hedonic price method, which examines the effect of differences in environmental quality on, for example, housing or job markets, assumes that all buyers in the market perceive these environmental characteristics.[10]

Hypothetical approaches, called *stated preference methods*, use data generated by placing individuals in hypothetical choice settings. These methods are needed when no behavior can be observed (or no other market data exist to infer value), such as to estimate nonuse values or to value changes that have not yet occurred. These approaches typically use surveys that determine WTP or WTA; Table 5.3 describes two such approaches. Stated preference methods typically require more time and cost to develop and implement than revealed preference approaches, and they can be subject to bias. These biases can create uncertainty about whether respondents would actually pay the amounts they indicate.

Benefit transfer is an alternative to either stated or revealed preference methods.[7] This method estimates the value of environmental goods and services by transferring the results of previous studies at different locations.[11] For example, the value of clean water in Ohio could be approximated using a number of different studies that estimate the value of reducing excess nutrients in Pennsylvania waterways. Like stated preference methods, it can be used in the absence of market data, but it is less expensive to implement. However, many factors need consideration to determine whether benefit transfer will provide adequate information.[7]

[a] Nonuse value is sometimes termed *passive use value*.

Table 5.3 Methods for Estimating Values of Environmental Goods and Services

Method	Description	Examples
Revealed Preference Methods (Can Estimate Use Values Only)		
Market	When environmental goods are traded in markets, their value can be estimated from transactions.	The benefits of an oil spill cleanup that would result in restoration of a commercial fishery can be projected from changes in markets for fish, before and after the spill, and their effects on fishermen and consumers.
Production function	The value of an environmental good or service can be estimated when it is needed to produce a market good.	If an improvement in air quality would lead to healthier crops, the value of the improvement includes, e.g., the reduction in fertilizer costs to produce the same amount of agricultural crops.
Hedonic price method	The value of environmental characteristics can be indirectly estimated from the market, when market goods are affected by the characteristics.	If an improvement in air quality affects housing prices in a region, the value of the improvement can be measured by statistically estimating the relationship between house prices, house characteristics, and air quality.
Travel cost method	The value of recreational sites can be estimated by examining travel costs.	The value of a recreational fishing site to those who use it can be estimated by surveying visitors to determine the relationship between the number of visits and the costs of travel.
Stated Preference Methods (Can Estimate Both Use and Nonuse Values)		
Contingent valuation method	Individuals are surveyed regarding their willingness to pay for a specifically described nonmarket good.	In a telephone survey, respondents are directly asked their willingness to pay, via a hypothetical tax increase, for a project that would reduce runoff, improving the health of a particular stream.
Conjoint analysis	Survey respondents evaluate alternative descriptions of goods as a function of their characteristics, so the characteristics can be valued.	In a mail survey, hypothetical alternative recreational fishing sites are described by type of fish, expected catch rate, expected crowding, and round-trip distance; respondents' preferences are used to calculate value for changes in each of the characteristics.

To summarize, the choice of valuation technique depends on the values individuals have for the good or service (i.e., use and nonuse), the availability of appropriate data, the researcher's constraints (e.g., time and money), and the ability to minimize biases. For more detail on revealed preference methods, stated preference methods, and benefit transfer approaches (such as the theory, analysis, and steps), the reader is referred to Freeman,[6] Hanley and Spash,[10] and Desvousges et al.[12] For additional information on estimating ecological benefits, the reader should see USEPA.[7,13]

Three of the case studies presented in later chapters of this book used stated preference techniques (Table 5.3). Chapter 10 explores the use of a contingent valuation method (CVM) model to value alternative development approaches in the Big Darby

Table 5.4 Structure of a Cost–Benefit Analysis

1. Definition of project and policy alternatives.
2. Identification of project and policy impacts.
3. Which impacts are economically relevant?
4. Physical quantification of relevant impacts.
5. Monetary valuation of relevant impacts.
6. Discounting of costs and benefit flows.
7. Applying the net present value test.
8. Sensitivity analysis.

Source: Hanley and Spash.[10]

Creek watershed of central Ohio, and Chapter 11 presents a study of the use of conjoint analysis (CA) to study social trade-offs among development policies in the Clinch Valley of southwestern Virginia and northeastern Tennessee. Chapter 14 discusses a method that uses a type of CA to scale potential restoration projects as part of natural resource damage assessment. To prepare the reader unfamiliar with those methods, Appendix 5-A discusses their differences more at length.

COST-BENEFIT ANALYSIS

Cost–benefit analysis (CBA) is the process of summing the value of the individual welfare changes, present and future, associated with a project or policy. The purpose is to assess all changes that can be feasibly measured to determine whether society gains more than it loses. If the benefits exceed the costs so that the gainers could potentially compensate the losers — this is termed the *potential Pareto criterion* — the project or policy is said to improve efficiency.[7,10] Under this criterion, it is considered irrelevant whether compensation actually occurs. The procedure may be used prospectively, in planning, or retrospectively, to determine whether planned goals were met. CBA was originally developed to assess the net economic value of public works projects, the outputs of which usually were market goods, and the goal of which was to produce net social benefit.[6,10] Some of the earliest examples of its use were for water resource projects in the United States,[10,14] so the relationship between CBA and watershed management is longstanding.

Hanley and Spash[10] describe eight stages of CBA (Table 5.4). The first stage defines what is to be analyzed to reveal how the project or policy will cause change. The next stages identify the relevant impacts and their physical characteristics, including applicable time horizons, as necessary for economic comparison. For example, if stream restoration is undertaken to improve stream ecological communities, then the time necessary to plant the riparian zone; the duration of required maintenance; the lag period for fish population response; and the type and magnitude of the response need to be determined. The process of economic valuation is next. Its purpose is to express all changes in the common metric of dollars. Where market prices of goods and service do not exist, or do not capture the full value, corrected or "shadow" prices are calculated, as discussed later. Negative effects of the project are estimated as opportunity costs, or the lost value of a resource that cannot be used because of the project.[7,15] For example, if a firm chooses to pollute a river, an opportunity cost might be the lost value of recreational fishing. The sixth step, discounting of cost and benefit flows, is necessary to translate all values into present value when benefits and costs occur at different times. Discount rates represent the time value of money and scale down future values (i.e., the higher the discount rate, the lower the present value of future streams of benefits and costs).[a] Present values can be compared; if the net present value is greater than zero, the project or policy is said to improve efficiency. If more than one project or policy is being compared, the one with the largest net present value is said to be the most efficient or to provide the largest improvement in social welfare. The final stage is sensitivity analysis, which examines the uncertainty of the relevant impacts and discount rate.

COMPLEMENTARY ANALYSES

Traditionally, a complete economic analysis is comprised of three techniques: CBA, economic impact analysis (EIA), and equity assessment.[7,16] Where CBA provides information about economic efficiency, the other two techniques examine resource distribution. These two latter types of analysis are briefly discussed in this section, as well as cost-effectiveness analysis (CEA) and natural resource damage assessment (NRDA) as they relate to CBA.

Tietenberg[15] defines impact analysis, whether environmental or economic, as a process to quantify the consequences of various actions. By this definition, it is similar to CBA and CEA; however, rather than transforming all changes into a single (dollar) metric, it simply organizes a large amount of information for decision support. The USEPA[7] defines EIA as a process to examine the distribution of impacts (both positive and negative), usually by examining economic changes across a variety of economic sectors.

Fair distribution is an important goal in both welfare and ecological economics, and equity and efficiency are sometimes traded off. Because it relies on the potential Pareto criterion, CBA is not concerned with whether the potential compensation actually takes place; therefore a project by which society as a whole benefits may cause transfers of wealth, creating winners and losers. Equity assessment allows economists to understand changes in distribution of wealth due to a policy or project. According to the USEPA's economic guidance,[7] the first step is to identify potentially-affected subpopulations; next steps may involve determining each subpopulation's net benefits or the distribution of the net benefits among the subpopulations. Most often, however, equity has not been a decision criterion in water resource projects, since as long as net benefits over society as a whole exceed zero, those subpopulations experiencing positive net benefits theoretically could compensate the

[a] While the practice of discounting the future is well supported based on observation of individuals' choices, the selection of a discount rate for any particular analysis of public goods, such as infrastructure, health or the environment, frequently is controversial. Applying discount rates to decisions that affect the well-being of future generations is especially problematic, since the potential Pareto criterion cannot be reasonably applied across generations.[6]

others. Research to investigate how winners could compensate losers may be needed to better ensure equitable outcomes.[17]

CEA resembles CBA but considers only costs. It may be used in situations where the estimation of benefits is infeasible (e.g., because of time or budget constraints) or too uncertain.[7,15] It is also used when a specific target — such as a pollution level to be achieved, or an area of new habitat to be created — has been established by policy and is not subject to an efficiency test. Given such conditions, CEA can help sort out the management alternatives. It compares alternatives by calculating a benefit–cost ratio for each, where benefits may be measured in nonmonetary units, though costs are measured in dollars. The benefits may also be the achievement of a prespecified environmental target. Alternatives with high benefit–cost ratios would be preferred to those with low ratios. Note that because benefits and costs are not measured in the same units, only a ratio can be calculated, not a difference. This also suggests that only CBA provides information about economic efficiency.

NRDA is a process for economic analysis established under the Clean Water Act, the Comprehensive Environmental Response, Compensation, and Liability Act of 1980 (CERCLA) and the Oil Pollution Act of 1990.[18] These Acts hold liable for damages those who release hazardous substances and oil into the environment,[19,20] and they establish trustees (i.e., officials who act on behalf of the public) responsible for recovering the damages. Damages comprise the cost to restore the injured natural resources to their baseline condition, compensation for interim losses pending recovery, and the cost of damage assessment.[18,21] Assessment methods are used to scale the magnitude of the environmental restoration effort to offset the magnitude of the harm (more detail is provided on the broad classes of these assessment methods in Chapter 8; case studies of these particular assessment methods are in Chapter 14 and Chapter 15). Some of the techniques involved are similar to those employed in CBA for valuing ecological benefits; in this sense, NRDA requires a retrospective application of the assessment methods, whereas project evaluation, the more routine use, normally is prospective.[6]

GAME THEORY

Game theory, like other subfields of economics, is concerned with human behavior and can examine individuals interacting within a market, or in situations of market failure. Gibbons[22] defines game theory as the study of decision problems when multiple entities are involved. Varian[4] simply calls it the study of interacting decision-makers. This discipline provides a theory of strategic behavior where an outcome depends on many individuals' strategies and the current conditions of the situation. Most commonly, games include three elements: players, strategies, and payoffs.[3,22,23]

Game situations can be modeled as cooperative or noncooperative. In cooperative games, players can make binding agreements affecting the overall objective of those involved.[3,23] Most environmental applications deal with noncooperative games, in which such agreements cannot occur.[3] Strategic interaction in games can be modeled as dynamic (changing over time) or static. Game theory has played an important role in analyzing externalities, bargaining, free-riding behavior (the reaping of benefits of a public good without paying), and principal-agent problems (situations where parties have incentives to hide information or actions).[3]

For example, if a government wants to regulate a firm's ability to pollute, where only the firm knows its emissions and abatement costs, the government wants to determine how the firm will react to the environmental policy. Game theory can help design a system with incentive compatibility, i.e., one in which individuals will provide truthful estimates.[3] International externalities, where one country's action affects another's welfare (e.g., greenhouse gas emissions or water diversion), are modeled as noncooperative games if no jurisdiction exists to enforce agreements.[3] On the other hand, for interstate water disputes, enforceable agreements are possible. Chapter 12 discusses the use of cooperative game models to inform an interstate water negotiation in the Platte River watershed of Colorado, Wyoming, and Nebraska.

ECOLOGICAL ECONOMICS

A relatively new paradigm has developed out of the controversies of welfare economics. Many point out that the assumptions used to develop the utilitarian perspective (e.g., preferences are fixed, agents are rational and self-interested) are not always accurate.[24–26] As Gowdy[27] elaborates,

> . . . the focus of most economists on markets and market solutions, to the exclusion of the behavior of actual ecosystems and actual human behavior, has been at least partially responsible for a variety of wildlife policy failures, from forestry to fisheries to the protection of endangered species.

Ecological economists contend that a transdisciplinary approach, spanning economics and natural science, is needed to address environmental problems.[28] Costanza describes opposing assumptions about economic growth: technological optimism (which he represents as the economists' perspective) assumes unlimited economic growth because of human ingenuity; technological pessimism (the ecologists' view) holds that technology cannot abrogate the constraints of resources and energy and that economic stagnation is inevitable.[29] He concludes that technologically pessimistic policies should be pursued, and that ecological economic research should compare pessimistic to optimistic policies and work to reduce uncertainty regarding the effects of technology. Sahu and Nayak suggest that ecological economics is not constrained by the mechanistic assumptions of welfare economics, but uses a systems approach.[30] Whereas environmental and natural resource economists define the environment as a part of the economy (the environment is an asset), ecological economists define the economy as a part of the environment.[31] Therefore, ecological economists may use the tools of conventional economics, but they also believe that new approaches are needed to answer some environmental questions.[28] In valuation, for example, ecological economists place more emphasis on the physical characteristics and ecological health of the system, which may not be captured by values elicited from individuals.[32]

Daly[31] defines the goals of ecological economics as efficient allocation, equitable distribution, and sustainable scale. Two of these goals have been mentioned previously as important to welfare economics. The third relates the "physical volume of throughput," to "the carrying capacity of the environment over time;"[31] it acknowledges that excessive economic growth can cause environmental destruction.[32]

According to Tacconi,[33] the driving force of ecological economics is analysis related to describing and achieving sustainability. Toman et al.[34] define the central issue of sustainability as the well-being of future generations, subject to constraints imposed by the functioning of the natural environment.

APPLICATIONS OF ECOLOGICAL ECONOMICS

Ecological economics does not offer a single, theoretically integrated, and widely accepted analytic paradigm similar to CBA. Much of its contribution has been in the form of a wide-ranging critique of welfare economics, some of it aimed at establishing an alternative moral-philosophical framework having sustainability (rather than social welfare, as earlier defined) as the objective. At the same time, the techniques of environmental and resource economics are not necessarily rejected. At some risk of over-generalization, the ecological-economic critique may be summarized as calling for the increased use of participative processes; a greater focus on equity; the integration of multiple scientific paradigms and methods; the evaluation of multiple objectives; and explicit recognition of biophysical processes and limits.

Several analytic techniques that employ a biophysical constraint have been, or could be, applied in a watershed setting. A broad family of methods has treated energy as limiting for all meaningful work and therefore useful as a biophysical and economic least-common-denominator. This premise has been applied with various forms of energy: (a) the energy used to produce goods and services in national economies (*embodied energy*);[35] (b) the solar energy that is captured by living and nonliving earth systems, is transformed (intensified) by physical or biological processes, and represents the "real wealth" of both ecosystems and human economies (*emergy*);[36] and (c) the energy that is available (*exergy*) or unavailable (*entropy*) to do work in a process or system.[37,38] Each of these approaches assumes the goal is to maximize useful energy rather than welfare or utility. Their proponents suggest that policies diminishing available energy should be avoided, even if they appear to increase social welfare. Their appeal is that energy, unlike utility, can be comprehensively estimated, even for nonmarket goods, and is subject to accounting under the laws of thermodynamics. However, these approaches have long been criticized by welfare economists as not being able to address the scarcity of some resources (e.g., primary minerals)[39] and as simply ignoring the supply and demand principles by which economic systems really operate.[40]

Ecological footprint methods consider the ecological services provided by earth's terrestrial or marine ecosystems as necessary to support life and economic activity but also as limited by the total area of those ecosystems.[41,42] These techniques examine the areas required to support existing patterns of resource use (including, e.g., the appropriation of productive areas in poorer countries to support the consumption patterns of richer countries) to determine whether uses are equitable and sustainable, and to compare alternative resource-use scenarios. Critics have described the method as powerfully illustrative, since the ecological demands of economic activity often go unrecognized, but also as overly simplified, technologically pessimistic, and biased against trade.[43] These criticisms suggest that biophysically based methods, while providing useful insights, should not be used uncritically as a substitute for welfare-based methods.

Other approaches have not sought to establish an alternative economic basis but rather to link abiotic, biotic, and economic models to simulate feedbacks within ecological–economic systems. Biophysical constraints are achieved only by ensuring that the future welfare effects of anticipated ecological changes can be taken into account when policies are designed. For example, Costanza and coworkers[44–47] developed the Patuxent Landscape Model to determine the effects of societal activities, especially land use and agriculture, on aquatic biological endpoints and land values in the Patuxent River watershed. This effort has attempted to integrate models of land use conversion, agricultural practice, stream hydrologic and ecological processes, and ecological succession and habitat type, in a spatially explicit simulation framework. Approaches of this type inform welfare-based analysis with the best-available ecological modeling methods. An ecological–economic case study that integrates economic, land use, and ecological models to analyze development is presented in Chapter 13.

CONCLUSION

This chapter has summarized important economic concepts related to watershed management. The case studies presented in Part II use a variety of economic methods to analyze management alternatives in support of decision-making. Because there are many such methods, any generalized conceptual approach for integrating ecological risk assessment with economic analysis requires substantial flexibility. This chapter illuminates some of the possibilities.

Implicit in this discussion is the requirement that, to study the trade-offs, economists must accurately comprehend the relevant effects of any potential decision. Ecological risk assessment is one plausible tool that can deliver information on the ecological outcomes (e.g., Suter[48]). Chapters 6 and 7 describe the somewhat disjointed manner in which ecological risk assessment and economics are currently used in the development and modification of water quality standards, and both chapters discuss possibilities for improved integration.

ACKNOWLEDGMENTS

Definitions in Table 5.1 were developed through discussions with Steve Newbold, Rich Iovanna, Sabrina Ise-Lovell, Wayne Munns, Margaret McVey, Nicole Owens, T.J. Wyatt, and Lynne Blake-Hedges.

REFERENCES

1. Fullerton, D. and Stavins, R., How economists see the environment, in *Economics of the Environment*, Robert Stavins, Ed., W. W. Norton and Company, New York, 2000.
2. Hurley, T. and Shogren, J., Environmental conflicts with asymmetric information: Theory and behavior, in *Game Theory and the Environment*, Hanley, N. and Folmer, H., Eds., Edward Elgar Publishing, Northampton, MA, 1998.
3. Folmer, H. and de Zeeuw, A., Game theory in environmental policy analysis, in *Handbook of Environmental and Resource Economics*, van den Bergh, J.C.J.M., Ed., Edward Elgar Publishing, Northampton, MA, 1999.
4. Varian, H., *Microeconomic Analysis*, W.W. Norton and Company, New York, 1992.
5. Cropper, M. and Oates, W., Environmental economics: A survey, *J. Econ. Lit.*, 30, 675, 1992.
6. Freeman III, A.M., *The Measurement of Environmental and Resource Values: Theories and Methods*, Resources for the Future, Washington, D.C., 1993.
7. USEPA, Guidelines for preparing economic analyses, EPA/240/R-00/003, U.S. Environmental Protection Agency, Washington, D.C., 2000.
8. Krutilla, J., Conservation reconsidered, *Am. Econ. Rev.*, 57, 777, 1967.
9. Mitchell, R.C. and Carson, R.T., *Using Surveys to Value Public Goods: The Contingent Valuation Method*, Resources for the Future, Washington, D.C., 1989.
10. Hanley, N. and Spash, C.L., *Cost–Benefit Analysis and the Environment*, Edward Elgar Publishing Company, Northhampton, MA, 1993.
11. Boyle, K. and Bergstrom, J., Benefit transfer studies: Myths, pragmatism, and idealism, *Water Resour. Res.*, 28, 657, 1992.
12. Desvousges, W., Johnson, F.R., and Banzhaf, H.S., *Environmental Policy Analysis With Limited Information*, Edward Elgar Publishing, Northampton, MA, 1998.
13. USEPA, A framework for the economic assessment of ecological benefits, Science Policy Council, U.S. Environmental Protection Agency, Washington, D.C., Feb. 1, 2002.
14. Dasgupta, A.K. and Pearce, D.W., *Cost–Benefit Analysis: The Theory and Practice*, MacMillan Press Ltd., London, 1978.
15. Tietenberg, T., *Environmental and Natural Resource Economics*, Addison-Wesley Longman, Inc., New York, 2000.
16. Hanley, N., Cost–benefit analysis of environmental policy and management, in *Handbook of Environmental and Resource Economics*, van den Bergh, J.C.J.M., Ed., Edward Elgar Publishing, Northhampton, MA, 1999.
17. Zilberman, D. and Lipper, L., The economics of water use, in *Handbook of Environmental and Resource Economics*, van den Bergh, J.C.J.M., Ed., Edward Elgar Publishing, Northhampton, MA, 1999.
18. Mazotta, M., Opaluch, J., and Grigalunas, T., Natural resource damage assessment: The role of resource restoration, *Nat. Resour. J.*, 34 (Winter), 153, 1994.
19. Kopp, R. and Smith, V., *Valuing Natural Assets: The Economics of Natural Resource Damage Assessment*, Resources for the Future, Washington, D.C., 1993.
20. Unsworth, R. and Bishop, R., Assessing natural resource damages using environmental annuities, *Ecol. Econ.*, 11, 35, 1994.
21. Penn, T. and Tomasi, T., Environmental assessment: Calculating resource restoration for an oil discharge in Lake Barre, Louisiana, USA, *Environ. Manage.*, 29(5), 691, 2002.
22. Gibbons, R., *Game Theory for Applied Economists*, Princeton University Press, Princeton, NJ, 1992.
23. Nicholson, W., *Microeconomic Theory*, The Dryden Press, New York, 1992.

24. Norton, B., Costanza, R., and Bishop, R., The evolution of preferences: Why "sovereign" preferences may not lead to sustainable policies and what to do about it, *Ecol. Econ.*, 24, 193, 1998.
25. Frank, R.H., *Passions Within Reason: The Strategic Role of the Emotions*, Norton, New York, 1988.
26. Kolstad, C., *Environmental Economics*, Oxford University Press, New York, 2000.
27. Gowdy, J., Terms and concepts in ecological economics, *Wildl. Soc. Bull.*, 28(1), 26, 2000.
28. Costanza, R., Daly, H.E., Norgaard, R., Cumberland, J., and Goodland, R., *An Introduction to Ecological Economics*, St. Lucie Press, Boca Raton, FL, 1997.
29. Costanza, R., What is ecological economics?, *Ecol. Econ.*, 1, 1, 1989.
30. Sahu, N. and Nayak, B., Niche diversification in environmental/ecological economics, *Ecol. Econ.*, 11, 9, 1994.
31. Daly, H.E., Allocation, distribution, and scale: Towards an economics that is efficient, just, and sustainable, *Ecol. Econ.*, 6, 185, 1992.
32. Turner, R.K., Environmental and ecological economics perspectives, in *Handbook of Environmental and Resource Economics*, van den Bergh, J.C.J.M., Ed., Edward Elgar Publishing, Northhampton, MA, 1999.
33. Tacconi, L., *Biodiversity and Ecological Economics: Participation, Values and Resource Management*, Earthscan Publications Ltd., London, 2000.
34. Toman, M., Pezzey, J., and Krautkraemer, J., Neoclassical economic growth theory and "sustainability," in *The Handbook of Environmental Economics*, Bromley, D.W., Ed., Blackwell Publishers, Malden, MA, 1995.
35. Costanza, R., Embodied energy and economic valuation, *Science*, 210, 1219, 1980.
36. Odum, H.T., *Environmental Accounting: Emergy and Environmental Decision Making*, John Wiley & Sons, New York, 1996.
37. Jørgensen, S.E., *Integration of Ecosystem Theories: A Pattern*, Kluwer, Dodrecht, The Netherlands, 1997.
38. Faber, M., Manstetten, R., and Proops, J., *Ecological Economics: Concepts and Methods*, Edward Elgar Publishing, Northampton, MA, 1996.
39. Heuttner, D.A., Net energy analysis: An economic assessment, *Science*, 192, 101, 1976.
40. Shabman, L.A. and Batie, S.S., Economic value of natural coastal wetlands: A critique, *Coast. Zone Manage. J.*, 4, 231, 1978.
41. Rees, W.E. and Wackernagel, M., Ecological footprints and appropriated carrying capacity: Measuring the capital requirements of the human economy, in *Investing in Natural Capital: The Ecological Economics Approach to Sustainability*, Jannson, A.M., Hammer, M., Folke, C., and Costanza, R., Eds., Island Press, Washington, D.C., 1994, 362.
42. Folke, C., Kautsky, N., Berg, H., Jansson, Å., and Troell, M., The ecological footprint concept for sustainable seafood production: A review, *Ecol. Appl.*, 8, S63, 1998.
43. Costanza, R., The dynamics of the ecological footprint concept, *Ecol. Econ.*, 32, 341, 2000.
44. Costanza, R., Wainger, L., and Bockstael, N., Integrated ecological economic systems modeling: Theoretical issues and practical applications, in *Integrating Economic and Ecologic Indicators: Practical Methods for Environmental Policy Analysis*, Milon, J.W. and Shogren, J.F., Eds., Praeger Publishing, Westport, CT, 1995, 45.
45. Reyes, E., Costanza, R., Wainger, L., Debellevue, E., and Bockstael, N., Integrated ecological economic regional modelling for sustainable development., in *Models of Sustainable Development.*, Faucheux, S. et al., Eds., Edward Elgar Publishing, Aldershot, UK, 1996, 253.
46. Costanza, R. and Ruth, M., Using dynamic modeling to scope environmental problems and build consensus, *Environ. Manage.*, 22, 183, 1998.

47. Geoghegan, J., Wainger, L.A., and Bockstael, N.E., Spatial landscape indices in a hedonic framework: An ecological economics analysis using GIS, *Ecol. Econ.*, 23, 251, 1997.
48. Suter, G.W., Adapting ecological risk assessment for ecosystem valuation, *Ecol. Econ.*, 14, 137, 1995.
49. Freeman III, A.M., *The Measurement of Environmental and Resource Values: Theory and Methods*, Resources for the Future, Washington, D.C., 2003.
50. Daily, G.C., Introduction: What are ecosystem services?, in *Nature's Services: Societal Dependence on Natural Ecosystems*, Daily, G.C., Ed., Island Press, Washington, D.C., 1997, 1, 1.
51. Daily, G.C., Management objectives for the protection of ecosystem services, *Environ. Sci. Policy*, 3, 333, 2000.
52. King, D.M., Comparing ecosystem services and values, National Oceanic and Atmospheric Administration, Damage Assessment and Restoration Program, Jan. 12, 1997. Accessed Mar. 4, 2004 at http://www.darp.noaa.gov/pdf/kingpape.pdf.
53. Whigham, D.F., Ecosystem functions and ecosystem values, in *Ecosystem Function and Human Activities: Reconciling Economics and Ecology*, Simpson, R.D. and Christensen, N.L., Eds., Chapman & Hall, New York, 1997, 10, 225.

APPENDIX 5-A

Discussion of Stated Preference Methods Used in Three Case Studies

This appendix discusses the differences between two stated preference methods used for valuing environmental goods, the contingent valuation method (CVM) and conjoint analysis (CA). CVM was used to value alternative development scenarios in the Big Darby Creek watershed of central Ohio (Chapter 10), while CA measured the trade-offs among development policies in the Clinch Valley of southwestern Virginia and northeastern Tennessee (Chapter 11) and scaled potential restoration scenarios for Lower Fox River and Green Bay, Wisconsin and Michigan (Chapter 14).

CVM measures value directly by asking respondents' their willingness to pay, using a specified payment vehicle (e.g., a change in the electric bill or in taxes), to avoid or obtain a particular change. The question format could be open-ended (i.e., how much are you willing to pay . . . ?) or dichotomous-choice (i.e., would you be willing to pay $X amount: yes or no?). Mitchell and Carson[1] describe CVM as a "versatile tool for directly measuring a range of benefits for a range of goods consistent with economic theory." Unlike revealed preference techniques, which are limited to valuing existing goods at existing quantity and quality levels, CVM can be used to measure both use and nonuse values of goods that may not presently exist. As a result of compensation claims associated with the Exxon Valdez oil spill in Prince William Sound, Alaska, the National Oceanic and Atmospheric Administration (NOAA) convened a panel to conduct hearings on the validity of CVM.[2] The panel established rigorous guidelines for legally admissible studies. Nonetheless, the method remains controversial among some economists because of its hypothetical nature. Several potential biases have been identified,[3,4] and CVM models have had a mixed performance when subjected to internal and external validity tests.[5,6]

Whereas CVM typically measures the value of a good as a whole, CA induces respondents to evaluate alternatives as a function of their attributes, so that the attributes can be individually valued.[7–9] For example, a respondent may be asked to state a preference (and perhaps to rate the strength of the preference) between alternative streams for fishing. The streams are said to vary as to the type of fishing, expected catch rate, expected crowding, expected weather, and round-trip distance.[10] One attribute, in this case driving distance, usually is either a cost or a proxy for cost to allow estimation of WTP. By choosing one alternative, the respondent reveals a (strength of) preference for that particular bundle of attribute values vis-à-vis the others presented. By presenting a series of choice sets in which these attribute values are varied, respondent preferences can be disaggregated and the contribution of each attribute to the combined preference determined.

Environmental management alternatives and their fiscal, social, and ecological results also occur as bundles in the real world and can be analyzed using CA. The technique has been used in environmental applications where attributes are cardinal (e.g., travel distance) or class (e.g., terrain type) variables associated with the economic and environmental elements of a choice, such as a choice of recreational opportunity[11–13] or electricity generation scenario.[14] If the key features, both ecological and nonecological, that define each alternative can be expressed by the selected attributes, then CA can be used to quantify the key sources of stakeholder preference and to inform the design of an optimal alternative.

The multiattribute choice process employed in CA could avoid or reduce certain biases associated with the bid process in CVM, especially if the choices presented were meaningful and plausible to survey respondents.[15] Potential difficulties with such an application include: (1) the difficulty of constructing choice sets that encompass the needed range of potential management options and outcomes; (2) the potential for confusing or fatiguing respondents if too many attributes or choice sets are presented; and (3) a lack of experience in applying the method to evaluate indirect or nonuse values.

CA is similar to CVM, and therefore some of the same benefits apply, including the ability to value goods that have not been observed yet (e.g., impacts of global climate change), as long as they can be described adequately to the respondent. But whereas CVM results apply only to the scenarios or goods described, CA results can be extrapolated to any good within the range of attribute values used, even a good that was not specifically tested. It also avoids some of the problems of dichotomous-choice CVM, such as yea-saying (i.e., bias toward agreement). However, CA has not been subjected to the same scrutiny as CVM, questionnaire design is difficult, and the optimal design is unsettled.[15]

REFERENCES

1. Mitchell, R.C. and Carson, R.T., *Using Surveys to Value Public Goods: The Contingent Valuation Method*, Resources for the Future, Washington, D.C., 1989.
2. Arrow, K.J., Solow, R., Portney, P.R., et al., Report of the NOAA panel on contingent valuation, *Fed. Re.g.,* 58, 4601, 1993.
3. Diamond, P.A. and Hausman, J.A., Contingent value: Is some number better than no number?, *J. Econ. Perspect*, 8, 45, 1994.
4. Desvousges, W.H., Hudson, S.P., and Ruby, M.C., Evaluating CV performance: Separating the light from the heat, in *The Contingent Valuation of Environmental Resources*, Bjornstad, D.J. and Kahn, J.R. Eds., Edward Elgar Publishing, Cheltenham, UK, 1996, 117.
5. Hanneman, W.M., Valuing the environment through contingent valuation, *J. Eco Persp.*, 8, 19, 1994.

6. Bjornstad, D.J. and Kahn, J.R., Characteristics of environmental resources and their relevance for measuring value, in *The Contingent Valuation of Environmental Resources: Methodological Issues and Research Needs*, Bjornstad, D.J. and Kahn, J.R. Eds., Edward Elgar Publishing, Cheltenham, UK, 1996, 3.
7. Louviere, J.J., Conjoint analysis modeling of stated preferences: A review of theory, methods, recent developments and external validity, *J. Trans. Eco. Pol.*, 22, 93, 1988.
8. Louviere, J.J., Relating stated preference methods and models to choices in real markets: Calibration of CV responses, in *The Contingent Valuation of Environmental Resources*, Bjornstad, D.J. and Kahn, J.R. Eds., Edward Elgar Publishing, Cheltenham, UK, 1996, 167.
9. Kahn, J.R., *The Economic Approach to Environment and Natural Resources*, Harcourt Brace/Dryden Press, Fort Worth, TX, 1998.
10. Heberling, M., Valuing Public Goods Using the Stated Choice Method, PhD Dissertation thesis, The Pennsylvania State University, University Park, 2000.
11. Adamowicz, W., Louviere, J., and Williams, M., Combining revealed and stated preference methods for valuing environmental amenities, *J. Env. Econ. Manage.*, 26, 271, 1994.
12. Adamowicz, W., Perceptions versus objective measures of environmental quality in combined revealed and stated preference models of environmental valuation, *J. Env. Econ. Manage.*, 32, 65, 1997.
13. Roe, B., Boyle, K.J., and Teisl, M.F., Using conjoint analysis to derive estimates of compensating variation, *J. Env. Econ. Manage.*, 31, 145, 1996.
14. Johnson, F.R. and Desvousges, W.H., Estimating state preferences with rated-pair data: Environmental, health and employment effects of energy programs, *J. Env. Econ. Manage.*, 34, 79, 1997.
15. Hanley, N. and Wright, R.E., Using choice experiments to value the environment, *Env. Res. Econ.*, 11, 413, 1998.

CHAPTER 6

Ecological and Economic Analysis for Water Quality Standards

Randall J.F. Bruins and Matthew T. Heberling[a]

CONTENTS

Mechanisms for safeguarding aquatic ecological resources under the Clean Water Act (CWA) are aimed at providing "protection" wherever it is "attainable."[b,1] Ecological risk assessment (ERA) procedures (see Chapter 3) have been used to determine what measures are protective and whether they are physically attainable, whereas economic analyses (see Chapter 5) have been used primarily to determine what is cost effective and financially attainable. CWA language seemingly has left little room for weighing the benefits of protection against its costs, and therefore the integration of ecological and economic analyses in CWA programs has been limited.[c] However, there have been recent calls for increased flexibility in the use of benefits analysis (see NRC[2] and Chapter 7 of this volume).

Water quality standards (WQS) underpin several important regulatory protections for U.S. waters. In addition to the WQS program itself, the CWA authorizes regulatory programs for the establishment of national effluent guidelines for specific industries,

[a] The views expressed in this chapter are those of the authors and do not necessarily reflect the views or policies of the U.S. Environmental Protection Agency.
[b] Section 101(a) (2) of the CWA
[c] Exceptions include regulatory development in support of effluent guideline limitations (e.g., the development documents for the Final Rule on Concentrated Animal Feed Operations).[27]

1-56670-639-4/05/$0.00+$1.50

facility-specific effluent permits, and water-body specific total maximum daily loads (TMDLs). Effluent guidelines are designed based on available pollution control technology and its cost-effectiveness, but permit programs rely on water quality–based limits, in addition to technology-based limits. Effluent guidelines set the minimum performance in permits for a large number of industrial facilities.

For facilities not covered by effluent guidelines, permits are based on technology performance, usually evaluated by the permit writer's best professional judgment. However, if WQS are not met, these permits can then be tightened, or TMDLs can be required. Therefore the ecological and economic bases of WQS are worth examining.

WATER QUALITY STANDARDS AND ECOLOGICAL RISK ASSESSMENT

Under the CWA, states, tribes, and territories with approved WQS programs must establish *designated uses* for their water bodies. Uses are designated — such as use by aquatic life, use for fishing and fish consumption, use as a drinking water supply, use for full or limited body contact recreation — for which water and habitat quality are expected to be suitable. Designation for use by aquatic life, which requires ensuring conditions suitable for "the protection and propagation of fish, shellfish, and wildlife,"[1] normally is required for all waters, and subcategories may be established where different aquatic communities have differing water quality or habitat requirements. For example, cold water communities typical of some higher altitude or headwater streams have more stringent requirements for dissolved oxygen (DO), temperature, and suspended solids than nearby warm water communities, and thus subcategories for warm water and cold water aquatic life use are often established. Uses that are designated must include any *existing uses*, defined as uses actually attained at any time since 1975, and they should be considered presently attainable (as is more fully discussed in the following section).[a]

Ambient water quality criteria (AWQC) for specific pollutants, which are scientifically derived criteria, are then adopted to protect the designated uses. The designated uses and corresponding criteria, taken together with provisions that prevent the degradation of high quality waters, constitute the WQS for a given water body. WQS are proposed by states and approved by the USEPA. They are used as a basis for setting allowable pollutant levels for point-source discharges, such as from publicly owned treatment works (POTWs), industries, combined sewer overflow (CSO) outfalls, and concentrated animal feeding operations (CAFOs) exceeding a certain size. WQS and criteria are considered to be, respectively, the regulatory and scientific foundations of programs established under the CWA to protect U.S. waters.

Even if individual dischargers are substantially in compliance with discharge limits, WQS may be violated if there are unregulated point sources or nonpoint sources, such as urban or agricultural runoff (which typically are not regulated), or if the accumulation of permitted unregulated and background pollutant loads exceeds a water body's assimilative capacity. If regular violations for one or more pollutants

[a] This is according to criteria and definitions in 40 CFR 131.10 (g) and 131.10 (d).

cause the water body to be listed as impaired, the state or USEPA must conduct a study of the drainage area of the affected water body, determine the TMDL from all sources that can be assimilated without a violation of standards, and then revise all discharge permits accordingly (or possibly require control of nonpoint sources). WQS also play an important role in nonregulatory CWA programs; impairment as evidenced by WQS violations is taken into account in the targeting of federal funds used by states or tribes for water quality improvement projects.[3,4]

AWQC, the scientific component of WQS, are established or recommended by the USEPA but may be modified by the states to reflect site-specific conditions; separate AWQC are designed for protection of aquatic ecosystems and human health. Those intended for ecosystem protection include aquatic life criteria for many toxic contaminants, clarity, DO, and nutrients.[5] They also include biological criteria (or *biocriteria*), which evaluate the condition of aquatic ecological communities. The principles of ERA play an important role in the determination of each; some are based on a characterization of stressor-response relationships and others on statistical comparison to a reference condition.

Using the terminology of ERA, AWQC derivation procedures for toxic contaminants and DO have an implicit management goal of protecting aquatic communities from the adverse effects of specific chemical stressors and a management objective of limiting those stressors so as to prevent the occurrence of acute or chronic effects in 95% of aquatic taxa.[6,7] The assessment endpoint therefore is the viability of 95% of species; measurement endpoints include survival, growth, biomass, and fecundity of individuals of the species tested, typically in laboratory exposures. Stressor-response assessment procedures for AWQC involve constructing *species-sensitivity distributions* (SSDs) of the results of acute or chronic tests for a variety of fish and invertebrate taxa to estimate the concentration corresponding to the fifth percentile of tested species' responses.

Biological criteria evaluate aquatic communities themselves rather than aquatic stressors. Typically they consist of multimetric indices, such as the Index of Biotic Integrity (IBI)[8] or Invertebrate Community Index (ICI),[9] which are adjusted to fit regional conditions. Biocriteria are calculated from biological survey data to evaluate the integrity of a water body's fish or macroinvertebrate assemblages through comparison to a *reference* or minimally impacted condition.[10] The indices are aggregates of individual metrics that quantify the presence, abundance, or condition of particular species or groups of species that have been found to be either sensitive to or tolerant of various classes of stressors. Reference conditions are usually determined by identifying reference sites (sites judged to be minimally impacted) for a given region. Biocriteria calculation methods are regionally adjusted so as to maximize their ability to discriminate reference sites from sites in the same region that are known to be impacted by human influence; sites scoring lower than a reference score are assumed likely to be suffering one or more of those impacts.

The management goal implicit in the use of biocriteria in WQS is the protection of aquatic communities from any human-induced stressors, and the management objective is the prevention of human-induced impacts on fish or invertebrate community integrity. The assessment endpoint therefore is community integrity, which may be defined as degree of similarity to "the most robust aquatic community to be expected in a natural condition."[10] Measurement endpoints are the component metrics, determined using standard biosurvey methods.[10] As a result of the use of indices such as IBI and ICI in WQS programs, their inclusion in water quality monitoring programs is becoming more widespread, and the data frequently are available for use in W-ERA. Because IBI and ICI are important in the case studies presented in Chapters 10, 11, and 13, additional information on their derivation is presented in Appendix 6-A.

Nutrient criteria may be based on stressor-response relationships, similarity to reference conditions, or a combination of the two. The USEPA derived a set of recommended nutrient criteria for regions of the United States[11] by statistically defining a reference condition, but the Agency also suggested using stressor-response relationships as an alternative basis.[12] For example, the Ohio EPA employed an IBI score as the response endpoint in stressor-response analyses for total nitrogen (TN) and total phosphorus (TP) that formed the basis for recommended nutrient criteria[13] and for the TP target of a draft TMDL.[14]

When WQS are incorporated into discharge permits, permit limits are designed using exposure assessment methods to ensure that ambient exposures, beyond an immediate mixing zone, will be low enough to avoid acute or chronic toxicity, except under extreme low-flow conditions.[15] TMDL targets frequently are based on AWQC that have been derived from stressor-response analyses. Waste-loading allocations for particular sources are derived using principles of exposure assessment so as to achieve the target (i.e., attain acceptable risk) under design conditions (such as high-flow or low-flow, depending on the nature of the source).

WATER QUALITY STANDARDS AND ECONOMIC ANALYSIS

In general, WQS are based on a level of water quality that provides for the protection of aquatic life (i.e., propagation of fish, shellfish, and wildlife) wherever it can be reasonably attained,[1] not wherever it can be shown to provide net economic benefit. The CWA provides limited basis for economic analysis in conjunction with WQS programs.[1,16] If all technology-based effluent limits for point sources as well as cost-effective and reasonable best management practices for nonpoint sources have been fully implemented and a water body still has not attained the designated use, a state or tribe may conduct a use attainability analysis (UAA) to determine whether the use should be removed or a variance (i.e., a temporary suspension of a water quality standard without removing the use) should be granted.[1] The UAA must show that the designated use is not attainable based on any of several grounds, one of which is that the installation of further controls (i.e., going beyond the technology standard) would result in "substantial and widespread economic and social impact."[a] USEPA guidance recommends that the determination of "substantial" impact be based on the financial burden to affected households (for a facility that is publicly owned) or to private-sector entities of installing additional pollution controls; "widespread" impacts are those involving relatively large changes in socioeconomic conditions throughout a community or surrounding area.[16] Conversely, where water quality is

[a] See 40 CFR 131.10 (g).

higher than required to meet designated uses, CWA antidegradation provisions prevent the issuing of any permit that would result in a significant lowering of water quality unless necessary to allow an "important" economic or social development; an "important" development is one that would have "significant" and "widespread" impacts if foregone.[16]

The USEPA also performs economic analyses of WQS under other (i.e., non-CWA) authorities. Cost analyses of federally implemented regulations are required under Executive Order 12866[17] and the Unfunded Mandates Reform Act,[18] and depending on the magnitude of the federal action, a cost–benefit analysis (CBA) may also be presented. For example, the USEPA performed an economic analysis of the California Toxics Rule that established numeric water quality criteria for toxic pollutants necessary to meet the requirements of the CWA.[19] Even in so doing, however, the USEPA does not make the promulgation of its WQS-related rules subject to an economic efficiency test (i.e., a determination of whether benefits exceed costs), nor have states, tribes, or territories relied on such a test for WQS. A 1983 proposed revision of WQS regulations that would have allowed CBA to serve as a basis for changes in designated uses was discarded following public comment. In spite of previous regulatory language that required states to "'. . . take into consideration environmental, technological, social, economic, and institutional factors' in determining the attainability of standards for any particular water segment," the agency recognized "inherent difficulties" in balancing costs or benefits with achievement of CWA goals.[20] The USEPA *Interim Economic Guidance for WQS* allows CBA to be presented as part of an economic impact analysis for UAA but suggests that the determination for assessing benefits be coordinated with USEPA regional offices.[16]

Comprehensive efforts to integrate ecological and economic analyses have been rare due, in part, to existing policy. In most cases, ecological analyses determine what measures are protective and physically attainable, and separate economic analyses determine only what is financially attainable. For example, where designated uses are not being attained, stakeholders may be engaged in seeking least-cost mechanisms for meeting a TMDL target, but stakeholder preferences with respect to the ecological or other benefits of attainment normally do not play a role in identifying the target, or in downgrading the use. However, the NRC[2] has criticized this approach to WQS as "narrowly conceived" and has suggested that a "broadened socioeconomic benefit–cost framework" be employed for use designation. Novotny et al.[21] recommended the use of CBA in UAA in cases where "the nonmarket impacts (especially on water quality benefits) are likely to be large or the costs of incremental benefit very large," in spite of a lack of guarantees that USEPA reviewers would accept such an analysis as persuasive.

Furthermore, stakeholder preferences come into greater play wherever the protection of water quality is dependent on the integrity of riparian systems and adjacent uplands — especially in headwater systems. The CWA affords little federal authority for controlling the physical modification (other than dredging or filling) of streams or riparian systems or for the control of nonpoint source (NPS) pollution resulting from upland land uses. Headwater systems, including intermittent or ephemeral streams, while of critical ecological importance,[22] are also very numerous and highly subject to disturbance and may need to be protected through approaches involving public cooperation and evaluation of benefit. For example, the Kansas Legislature[23] has mandated that certain types of low-flow or intermittent streams be entirely exempted from CWA requirements except on those stream segments where the economic efficiency of regulation can first be demonstrated. In Ohio, although the applicability of the CWA to headwater streams has not been questioned, a need for stakeholder input as to the appropriate level of protection is acknowledged.[24]

THE NEED FOR INTEGRATION

Risk and economics are unavoidably linked. In the post-*Silent Spring* era, U.S. society entered into a number of social contracts that arguably combined elements of bold foresight and naïveté — foresight with regard to the importance of reducing ecological risks, but naïveté with regard to scientific nuance and cost. The 1973 Endangered Species Act required federal agencies to "insure that any action . . . is not likely to jeopardize the continued existence of any endangered species or threatened species . . ." before the sheer numbers of endangered and threatened species and their potentially overwhelming habitat protection or restoration costs were well understood. Consider, for example, the substantial costs and far-reaching social disruption that would be required to restore some endangered salmon runs in the Pacific Northwest.[25] Similarly, the 1972 CWA established a goal "to restore and maintain the chemical, physical, and biological integrity of [the] Nation's waters" and called for achieving a level of water quality that provides for the protection and propagation of fish, shellfish, and wildlife, and recreation in and on the water, "wherever attainable,"[a] well before TMDL lawsuits would require that longstanding water quality impairments be addressed and the lion's share of blame would shift from big industry and sewage treatment plants to agriculture and urban sprawl. There is now a wider recognition that reducing ecological risks is quite costly, and that its costs are paid not only by big, discrete polluters but also by society at large.

Moreover, risks, as humans define them, have an economic dimension. This does not imply that ERA should be limited by economics or serve only as input to economic analyses, but rather that any risk humans can recognize has economic implications. By definition, a risk entails a probability of an "adverse" effect, or an effect that is contrary to what is desired. Therefore, risk is defined with respect to human preference. Even in those cases where norms or standards have been established by statute or regulation, subjective interpretation is often needed. As stated earlier, it is not possible to precisely define terms such as *integrity* with reference to ecosystems, and the *attainability* of a level of water quality is usually a function of cost. In many instances, USEPA regulatory programs have been required to codify a particular interpretation of these normative terms. Whenever it is allowable and practicable, however, determining the preferences of interested and affected individuals can be a means to identify the best alternatives and to ensure broadly based support for management efforts.

Since people's information about risks is usually incomplete, technical information about risk plays an important role in informing those preferences. Furthermore, the form

[a] 33 USC 1251 (a)

of the technical information is critical. Compendia of monitoring data, problem reports, or expert opinions can all prove misleading because they do not provide for the rigorous and systematic determination of, for example, objectives, causative agency, and the probabilities and uncertainties associated with projected outcomes. Risk characterization, the last step in ERA, links each of these elements in careful, informative statements. ERA is needed if economic analysis of complex ecological problems is to be done.

Just as risks have an economic or preferential dimension, so decisions about actions to reduce risks always entail trade-offs. This interrelationship of information, preferences, and effective management argues for the thorough integration of ERA and economic analysis. It should be obvious, furthermore, that an approach in which the disciplines are compartmentalized rather than integrated will invariably lead to an analysis of poorer quality. Such an approach would assume that the natural and social sciences do not bring differing lenses to the understanding of goals and problems, and that the analytical requirements of each are mutually grasped without difficulty. In fact, the fundamental relationship between the social and natural sciences has long been a matter of philosophical dispute,[26] and while dialogue between economists and ecologists has dramatically increased in recent years, it still must be assumed that, in any new circumstance, conscious effort will be required to establish mutual understanding between the disciplines and a concerted approach to environmental problem-solving.

When is integrated analysis needed? ERA often is needed to determine the likely ecological responses to proposed management actions, and economic analysis often is needed to interpret those ecological changes, and other changes, in terms of human well-being so that decisions are effective and beneficial. Whenever both ERA and economic analysis are needed to address a watershed management problem, the analytic processes should be undertaken in an integrated fashion. Here, the term *integrated* does not necessarily imply that any distinction between the respective sciences is erased, or that either loses its essential character. It does imply that these analytic processes will be mutually informed and fully coordinated. The alternative, a piecemeal or haphazard process, is unlikely to serve decision-makers or stakeholders as well.

To accomplish integration in practice, extensive interaction is needed between the disciplines as well as with others who have relevant knowledge or a stake in solving the problem. The form these interactions should take will vary according to the circumstances, but experiences from the field of environmental management can be drawn upon to identify certain principles, and sequences of events, that help determine success. Chapter 7 further examines the problems inherent in analyses conducted to support the regulatory management of water quality in the United States and proposes an integrated approach to the development of water quality standards.

REFERENCES

1. USEPA, Water quality standards handbook, second edition, EPA/823/B-94-005a, U.S. Environmental Protection Agency, Office of Water, Washington, D.C., 1994.
2. NRC, *Assessing the TMDL Approach to Water Quality Management*, National Research Council, National Academy Press, Washington, D.C., 2001.

3. USEPA, Integrated planning and priority setting in the clean water state revolving fund program, EPA-832-R-01-002, U.S. Environmental Protection Agency, Office of Water, Washington, D.C., 2001.
4. USEPA, Nonpoint source program and grants guidance for fiscal year 1997 and future years, U.S. Environmental Protection Agency, Office of Water, Washington, D.C., 1996.
5. USEPA, National recommended water quality criteria-correction, EPA/822/Z-99/002, U.S. Environmental Protection Agency, Office of Water, Washington, D.C., 1999.
6. Stephan, C.E., Mount, D.I., Hansen, D.J. et al., Guidelines for deriving numerical national water quality criteria for the protection of aquatic organisms and their uses, NTIS PB85-227049, U.S. Environmental Protection Agency, Office of Research and Development, Duluth, MN, 1985.
7. USEPA, Final water quality guidance for the Great Lakes System; final rule, *Fed. Reg.,* 60, 15365, 1995.
8. Karr, J.R., Assessment of biotic integrity using fish communities, *Fisheries*, 6, 21, 1981.
9. DeShon, J.E., Development and application of the Invertebrate Community Index (ICI), in *Biological Assessment and Criteria: Tools for Water Resource Planning and Decision Making*, Davis, W.S. and Simon, T., Eds., Lewis Publishers, Boca Raton, FL, 2004, 217.
10. USEPA, Biological criteria: Technical guidance for streams and small rivers. Revised edition, EPA/822/B-096/001, U.S. Environmental Protection Agency, Office of Water, Washington, D.C., 1996.
11. USEPA, Ambient water quality criteria recommendations: Information supporting the development of state and tribal nutrient criteria for rivers and streams in nutrient ecoregion VI: Corn belt and Northern Great Plains, EPA/822/B-00/017, U.S. Environmental Protection Agency, Office of Water, Washington, D.C., 2000.
12. USEPA, Nutrient criteria technical guidance manual: Rivers and streams, EPA/822/B-00/002, U.S. Environmental Protection Agency, Office of Water, Washington, D.C., 2000.
13. OEPA, Association between nutrients, habitat, and the aquatic biota in Ohio rivers and streams, Technical Bulletin MAS/1999-1-1, Ohio Environmental Protection Agency, Division of Surface Waters, Columbus, OH, 1999.
14. OEPA, Total maximum daily loads for the Upper Little Miami River, draft report, Ohio Environmental Protection Agency, Division of Surface Water, Columbus, OH, 2001.
15. USEPA, U.S. EPA NPDES permit writers' manual, EPA/833/B-96/003, U.S. Environmental Protection Agency, Office of Water, Washington, D.C., 1996.
16. USEPA, Interim economic guidance for water quality standards, EPA/823/B-95/002, U.S. Environmental Protection Agency, Office of Water, Washington, D.C., 1995.
17. The President, Executive Order 12866 of September 23, 1993; regulatory planning and review, *Fed. Reg.,* 58, 1993.
18. USEPA, OAQPS economic analysis resource document, U.S. Environmental Protection Agency, Office of Air Quality Planning and Standards, Innovative Strategies and Economics Group, Research Triangle Park, NC, 1999.
19. USEPA, Economic analysis of the California Toxics Rule, EPA Contract No. 68-C4-0046, U.S. Environmental Protection Agency, Washington, D.C., 1999.
20. USEPA, Water quality standards regulation, *Fed. Reg.,* 48, 51400, 1983.
21. Novotny, V., Braden, J., White, D. et al., A comprehensive UAA technical reference, Project 91-NPS-1, Water Environment Research Foundation, Alexandria, VA, 1997.
22. Zale, A.V., Leslie, D.M., Jr., Fisher, W.L., and Merrifield, S.G., The physicochemistry, flora, and fauna of intermittent prairie streams: A review of the literature, Biological Report 89(5), Fish and Wildlife Service, U.S. Department of the Interior, Washington, D.C., 1989.
23. Kansas Legislature, Kansas Legislature, Substitute Senate Bill 204, 2001.

24. OEPA, Fact sheet: Clean rivers spring from their source: The importance & management of headwater streams, Division of Surface Water, Ohio Environmental Protection Agency, Columbus, OH, 2000.
25. Lackey, R.T., Salmon policy: Science, society, restoration, and reality, *Environ. Sci. Policy*, 2, 369, 1999.
26. Little, D. E., *Beyond positivism: Toward a methodological pluralism in the social sciences*, Delittle@Umd.Umich.Edu, 2002. Accessed Dec. 9, 2002 at http://www-personal.umd.umich.edu/~delittle/BEYPOSIT.PDF.
27. USEPA, Environmental and economic benefit analysis of final revisions to the National Pollutant Discharge Elimination System regulation and the effiuent guidelines for concentrated animal feeding operations. EPA 821-R-03-003, U.S. Environmental Protection Agency, Office of Water, Washington, DC., 2002.

APPENDIX 6-A

Using Multimetric Indices to Define the Integrity of Stream Biological Assemblages and Instream Habitat

CONTENTS

To determine whether a stream provides a suitable environment for a robust biological community, measurement of a set of chemical and physical water quality parameters (e.g., toxics, dissolved oxygen, temperature) is not sufficient. A chemical that is present, but not on the monitoring list, may be affecting the stream community, and episodic exposures, which are difficult to detect without continuous sampling, can also cause long term effects. Even high quality water can fail to support robust communities if other factors affect the stream environment. The physical habitat of the stream may have been altered (such as by channelization) in a way that removes instream cover or substrate needed by organisms, and barriers such as low-head dams may prevent migration or recolonization. Changes in stream hydrology that result from watershed development or flow diversion can create flow conditions that degrade the instream environment as well.

A goal of the Clean Water Act is "to restore and maintain the chemical, physical, and biological integrity of the Nation's waters."[a] The term *biological integrity* implies a concept of wholeness that encompasses more than water quality alone. Indeed, to determine whether a stream biological community is flourishing as expected, it makes sense to measure the community itself. Because biological communities are

[a] 33 USC 1251 (a).

1-56670-639-4/05/$0.00+$1.50

both complex and variable over space and time, however, the list of aspects that could be measured is long, and the measurements are not meaningful without interpretation. To establish an operational definition, various aggregate indices have been designed that measure selected ecological parameters and express some aspect of *integrity* (e.g., of the fish or invertebrates assemblages present, or of the instream habitat) on a simple numerical scale. While the concept itself remains controversial[1] and some argue that, in the aggregation of measures, information useful for assessment is lost rather than gained,[2] the approach has gained sufficient acceptance to become widely used in environmental monitoring and regulation.[3] Since watershed ecological risk assessments often must rely on the data that are available, whether or not they are ideal, their application in stream assessments is common. Four such indices that are referred to in later chapters of this book are briefly described here. Methods used for computing an index vary regionally; they are modified to fit regional ecological conditions. This description relies on methods used by the Ohio Environmental Protection Agency (OEPA);[4] methods applied in other locations, while not identical, are similar.

It should be noted that indices of biotic integrity are not necessarily useful for the study or management of rare species. Although Karr and Chu state that the explicit inclusion of threatened or endangered species in an index can improve their management,[1] bioassessments that are conducted for routine monitoring of stream condition may not have the spatial or temporal intensity needed to detect them. Therefore, the indices, like those used by OEPA, may not be designed to respond to the presence or absence of rare species. Furthermore, a low score on one metric that is due to the absence of a rare species could be masked by high scores on other metrics.

Another potential weakness of integrity indices is that the choice of sampling techniques may be taxonomically limiting. For example, the Invertebrate Community Index, described later, relies heavily on artificial substrates, and its metrics mainly reflect organisms that colonize those substrates. As a result, the presence and diversity of noninsect taxa such as crustaceans and mollusks, many of which are sensitive to human disturbance (see Chapters 10 and 11), are poorly reflected in the index.

INDEX OF BIOTIC INTEGRITY (IBI)

The IBI, originally developed by Karr,[5] expresses the status of the stream fish assemblage in a given location at the time of sampling. A stream reach of a given length is sampled by electrofishing techniques, and captured fish are identified to the species level.[6] To compute a set of 12 metrics, species are categorized into various groupings, including taxonomic family, tolerance to pollution, feeding type, breeding type, and whether indigenous or exotic (Table 6-A.1). Visible skin or subcutaneous disorders are also recorded; these include deformities, eroded fins, lesions and ulcers, and tumors. For each metric, a score of 5, 3, or 1 is assigned according to whether the sample approximates (5), deviates somewhat from (3), or strongly deviates from (1) the reference value, or that value expected under minimally impacted conditions. For most metrics, the reference value is scaled according to drainage area (i.e., the area of the watershed above the point sampled) since fish assemblages in larger streams tend

Table 6-A.1 Individual metrics constituting two indices of biological integrity used by the Ohio Environmental Protection Agency

Metric #	Index of Biotic Integrity (IBI)[a]	Invertebrate Community Index (ICI)
1	Total number of indigenous fish species	Total number of taxa
2	Number of darter species (Percidae)	Number of mayfly taxa (Ephemeroptera)
3	Number of sunfish species (Centrarchidae)	Number of caddisfly taxa (Trichoptera)
4	Number of sucker species (Catostomidae)	Number of true fly taxa (Diptera)
5	Number of pollution-intolerant species	Percent mayflies (Ephemeroptera)
6	Percent abundance of tolerant species	Percent caddisflies (Trichoptera)
7	Percent abundance of omnivores	Percent Tanytarsini midges
8	Percent abundance of insectivores	Percent other true flies and non-insects
9	Percent abundance of top carnivores	Percent pollution tolerant organisms
10	Total number of individuals	Number of EPT taxa[b]
11	Percent lithophils (species requiring clean gravel or cobble for spawning)	
12	Percent with deformities, eroded fins, lesions, and tumors	

[a]Metrics listed are for wadable, nonheadwaters sites. For other sites, some metrics differ.
[b]EPT = Ephemeroptera (mayflies), Plecoptera (caddisflies) and Tricoptera (stoneflies). Index is determined only from sampling of natural, not artificial, substrates.

naturally to be more diverse. The index is a sum of scores of the individual metrics, with a maximum score of 60. The interquartile range (25th percentile–75th percentile) of IBI for wadable, warm-water reference sites in Ohio is 38–50.[a,4]

MODIFIED INDEX OF WELL-BEING (MIWB)

The Index of Well-Being, developed by Gammon[7] and modified by the OEPA,[4] also expresses the status of fish assemblages. It uses the same sampling data as required for the IBI but also requires determination of the total weight of each species in the sample. The index is computed as follows:

$$\text{MIwb} = 0.5 \ln N + 0.5 \ln B + \overline{H}(\text{no.}) + \overline{H}(\text{wt.})$$

where:

N = relative numbers of all species, excluding species designated "highly tolerant"

B = relative weights of all species, excluding species designated "highly tolerant"

[a] Wadable streams are those that can be sampled by personnel walking in the streams, but do not include headwaters streams (drainage area < 20 mi^2). Warm water streams, which include most streams in Ohio, are those not capable of supporting cold water fauna such as trout.

$\overline{H}$ (no.) = Shannon diversity index based on numbers

$\overline{H}$ (wt.) = Shannon diversity index based on weight

The Shannon diversity index is computed as:

$$\overline{H} = -\Sigma \frac{(n_i)}{N} \ln \frac{(n_i)}{N}$$

where:

n_i = number or weight of the *i*th species

N = total number or weight of the sample

The interquartile range of MIwb for wadeable, warm-water reference sites in Ohio is 8.3–9.4.[4]

INVERTEBRATE COMMUNITY INDEX (ICI)

The ICI was developed by DeShon and others to determine the condition of the benthic, or bottom-dwelling, invertebrate assemblage.[4,6,8] Where there is sufficient stream flow, a device consisting of a series of hardboard plates, spaced along an eyebolt, is submerged in the stream and allowed to be colonized for a period of six weeks during the summer months. It is then collected for laboratory enumeration and identification of the attached organisms. To augment observations from the artificial substrates, a net is used to sample organisms occurring on natural substrates. Where the artificial substrates cannot be used, the natural substrates are sampled more extensively. When possible, collected individuals are identified to species, but sometimes identification is only to the genus or a higher level. As with IBI, species are categorized into groups for calculation of the index. The ICI is composed of 10 metrics (Table 6-A.1) that are scored as either 6, 4, 2, or 0 according to a relationship that varies with drainage area. These relationships are more complex than those for fish. For example, diversity of certain groups first increases and then decreases as drainage area increases. Like the IBI, the highest possible score is 60. The interquartile range of ICI for reference sites in Ohio where artificial substrates could be used is 36–48.[4]

QUALITATIVE HABITAT EVALUATION INDEX (QHEI)

The QHEI evaluates physical characteristics of stream habitats that are important to fish and invertebrate communities.[6,9,10] Six principal metrics compose the index, each having two to five constituent measures (Table 6-A.2). The metrics describe the material covering the stream bottom (*substrate*), areas where fauna can hide (*cover*), complexity and stability of the stream channel (*channel quality*), naturalness and stability of the streamside environment (*riparian/erosion*), variety of instream habitat

Table 6-A.2 Primary and secondary metrics constituting the Qualitative Habitat Evaluation Index (QHEI) used by the Ohio Environmental Protection Agency

Metric	Score	Subscore
Substrate	≤ 20	
Type		0–20
Quality		–5–3
Instream Cover	≤ 20	
Type		0–9
Amount		1–11
Channel Quality	≤ 20	
Sinuosity		1–4
Development		1–7
Channelization		1–6
Stability		1–3
Riparian/Erosion	≤ 10	
Width		0–4
Floodplain quality		0–3
Bank erosion		1–3
Pool/Riffle	≤ 20	
Max depth		0–6
Current available		–2–4
Pool morphology		0–2
Riffle/run depth		0–4
Riffle substrate stability		0–2
Riffle embeddedness		–1–2
Gradient	≤ 10	
Total Score	≤ 100	

Source: Rankin.[9]

types such as riffles, runs, and pools (*pool/riffle*), and steepness of the stream in the direction of flow (*gradient*). The maximum score is 100. The interquartile range of QHEI for wadable, warm-water reference sites in Ohio is 68–78.

REFERENCES

1. Karr, J.R. and Chu, E.W., *Restoring Life in Running Waters: Better Biological Monitoring*, Island Press, Washington, D.C., 1999.
2. Suter, G.W., *Ecological Risk Assessment*, Lewis Publishers, Boca Raton, FL, 1993.
3. USEPA, Biological criteria: Technical guidance for streams and small rivers. Revised edition, EPA/822/B-096/001, U.S. Environmental Protection Agency, Office of Water, Washington, D.C., 1996.

4. OEPA, Biological criteria for the protection of aquatic life. Volume II: Users manual for biological field assessment of Ohio surface waters, WQMA-SWS-6, Ohio Environmental Protection Agency, Columbus, OH, 1987.
5. Karr, J.R., Assessment of biotic integrity using fish communities, *Fisheries*, 6, 21, 1981.
6. OEPA, Biological criteria for the protection of aquatic life. Volume III: Standardized biological field sampling and laboratory methods for assessing fish and macroinvertebrate communities, WQMA-SWS-3, Ohio Environmental Protection Agency, Columbus, OH, 1987.
7. Gammon, J.R., The fish populations of the middle 340 km of the Wabash River, Tech. Rep. 86, Purdue University Water Resources Center, Lafayette, IN, 19976.
8. DeShon, J.E., Development and application of the Invertebrate Community Index (ICI), in *Biological Assessment and Criteria: Tools for Water Resource Planning and Decision Making*, Davis, W.S. and Simon, T., Eds., Lewis Publishers, Boca Raton, FL, 2004, 217.
9. Rankin, E.T., The Qualitative Habitat Evaluation Index (QHEI): Rationale, methods, and application, Ohio Environmental Protection Agency, Columbus, OH, 1989.
10. Rankin, E.T., Habitat indices in water resource quality assessments, in *Biological Assessment and Criteria: Tools for Risk-based Planning and Decision Making*, Davis, W.S. and Simon, T., Eds., Lewis Publishers, Boca Raton, FL, 1995, 181.

CHAPTER 7

Decision-Making and Uncertainty in Ambient Water Quality Management

Leonard A. Shabman

CONTENTS

INTRODUCTION

The Federal Water Pollution Control Act Amendments of 1972, as modified by the Clean Water Act of 1977 and the Water Quality Act of 1987, are the foundation for the nation's water quality management program. Section 101(a) of the Act (hereafter referred to as the Clean Water Act — CWA) set a goal for the nation of ". . . restoring the chemical, physical and biological integrity of the nation's waters," with an interim goal of securing "fishable and swimmable" waters. The CWA expects the states, with U.S. Environmental Protection Agency (USEPA) oversight, to translate these narrative goal statements into measurable ambient water quality

1-56670-639-4/05/$0.00+$1.50

standards (WQS).[a] The states then assign one or more of the standards to each water body in their jurisdiction and monitor those waters to determine whether the assigned standards are being met.[1]

After 1972, the primary strategy for meeting ambient standards focused on reducing the concentrations of a limited number of "criteria" pollutants in the wastewater effluent of point source dischargers.[b] Effluent standards appropriate to different industry groups were set at the national level in consideration of technical feasibility and reasonableness of costs.[c] States (in most cases) issue National Pollutant Discharge Elimination System (NPDES) permits that limit the allowable concentration of the pollutant in the effluent. Over the past 30 years, pollutants discharged by industry and municipal wastewater treatment plants have declined relative to the growth in population and economic growth, and the ambient quality of the nation's lakes, rivers, reservoirs, groundwater, and coastal waters has generally improved.

Nonetheless, states' monitoring programs report that ambient WQS in many places are not being met.[d,2] In some places, unregulated (under the CWA) nonpoint sources are the cause of the violation. As a general matter, mandating further reductions of criteria pollutants only from point sources will not secure the interim or restoration goals of the CWA in many watersheds.[e] The result has drawn attention to a long-dormant provision in Section 303 of the CWA. The so-called Total Maximum Daily Load (TMDL) program calls for watershed-scale analysis to identify all the pollutants from all sources responsible for WQS violations. The analysis is expected to estimate the pollutant load consistent with securing those standards, and allocate that allowable load to both point and nonpoint sources.[3]

In many watersheds, alteration of land cover, riparian habitat, flow, or channel geomorphology may be responsible for standards violations. In fact, the CWA draws a distinction between pollutants and pollution, with the term *pollution* referring to factors other than chemical discharges that affect the condition of a water body.[4] If the control of pollutants alone cannot secure the assigned ambient WQS, the standard can be reconsidered to reflect this reality. On the other hand, a comprehensive approach to water quality management to include pollutants and pollution may be adopted. A "watershed approach" to water quality management has been advanced as a way to integrate controls on pollutants into a broader program of aquatic restoration.[f,5] A more comprehensive analysis in support of the watershed approach

[a] As will be discussed later in this chapter, a standard includes both a designated use for the water body (lake, river, stream, or estuary) and an associated measurable criterion that is a surrogate for the use. In addition, a water quality standard must include an antidegradation provision to ensure that no pollutants will be discharged that compromise the existing designated uses.[6]

[b] See Houck[25] for a concise history.

[c] CWA § 304(b)(1)(B) calls for a consideration of costs against benefits in setting effluent standards to assure that costs are reasonable and not wholly disproportionate to the benefits received. However, both legislative history and case law limit the reliance on such economic calculations as a primary consideration in setting limits.

[d] The technical validity of state water quality monitoring reports has been questioned and improvements have been proposed.[5,36–38]

[e] The CWA does include a provision for requiring water-quality based, as opposed to technology based, NPDES effluent concentration limits in places where ambient WQS are not being met.[6]

[f] One of the nation's most ambitious aquatic ecosystem restoration efforts is in South Florida (the Everglades), where controls on phosphorus pollutant loads are being pursued in combination with restoration of wetlands and historic hydrologic flow patterns.[39]

would inform decisions on assigning standards, on the factors that contribute to standards violations, and on water quality management strategies.[a]

Integrated watershed assessment (IWA), as this more complete analytical system might be called, would draw from hydrology, geomorphology, ecology, and water chemistry, as well as the social sciences, to build models for informing water quality management decisions (see Chapter 9). Movement toward IWA will pose analytical challenges, but our understanding of watershed processes continues to grow and the data and computing power to use those data continue to advance.[b,2] Nonetheless, modeling limitations mean that IWA must recognize and report analytical uncertainties. The margin of safety (MOS) calculation required in the TMDL calculation is one acknowledgement of this reality.[4]

The implications of analytical uncertainty for applying IWA to water quality decision-making are the focus of this chapter. The chapter is organized into five sections. The next section ("Analytical Uncertainty and Water Quality Management") includes a conceptual discussion of the sources of analytical uncertainty and the implications for IWA. The section concludes with an endorsement of an approach to analysis and decision-making called *adaptive implementation* (AI).[4] The "Adaptive Implementation and the Clean Water Act" section argues that the continuing planning process called for by the CWA supports IWA and AI. Because IWA and AI include economic analysis, the "Water Quality Standards, Economics, and Adaptive Implementation" section describes an approach to economic analysis that is both consistent with the CWA and supports AI. The chapter concludes with a discussion of some of the challenges for carrying out AI.

ANALYTICAL UNCERTAINTY AND WATER QUALITY MANAGEMENT

Water quality management begins with the assignment by states of *designated uses* for each waterbody, subject to USEPA approval. The Agency's policy is that states should establish uses that encompass the protection or restoration of biological integrity,[6] even though the precise meaning of the term *integrity* remains a matter

[a] As a matter of practice, the watershed approach remains closely tied to the TMDL program and its focus remains limited to pollutants as stressors. Existing USEPA guidance places management strategies to address pollution outside the scope of the TMDL program (40 CFR Part 130, Section 130.7). In fact, the pending revision to the TMDL rule has been termed the *watershed rule*.[40]

[b] An ambient focus was the foundation for the nation's water quality management strategy prior to the 1972 amendments.[25] States were expected to identify sources of pollutants that were causing ambient WQS violations and then require controls at the sources most responsible for the water quality violation. Ideally, the most cost-effective water quality management strategy would be defined and then implemented.[41,42] However, in even modestly complex watersheds there were multiple sources of pollutants. Neither the available monitoring data nor the existing water quality models available at the time allowed the states to unambiguously determine which sources were responsible for the standards violation and then defensibly mandate differential load reduction requirements. Also, potentially regulated sources resisted requirements to limit their discharges, often making cost or fairness-of-burden arguments. These technical and administrative difficulties in securing reductions were the motivation for the nationally uniform, technology-based effluent standards requirements called for by the 1972 amendments. Based on this history there are those who favor extending the technology-based effluent standards approach to other sources of pollutants over time. They find the logic of, and attention to, the TMDL program and the watershed approach a step backward in securing the CWA goals.[43,44]

of legal ambiguity[7] and scientific dispute.[8,9] The interim goal of the CWA is swimmable and fishable waters, again terms whose definition have been subject to different interpretations.[10]

In practice, measurable criteria that serve as the surrogates for *integrity, swimmable*, and *fishable* are also a part of the standard and are the operational focus for water quality planning and decision-making. Criteria may be some acceptable frequency, duration, and magnitude of the presence of a pollutant (sediment concentration, bacterial colonies) or a waterbody condition (chlorophyll *a* or dissolved oxygen concentrations). More recently, indices of algal, fish, or (most often) benthic species or communities have been employed as water quality criteria.[a] Criteria can be defined as highly discrete measures. As one example, depending on the designated use, the acceptable frequency of violation of a dissolved oxygen criterion may be stated along a continuum from zero percent of the time to some positive number. In addition, the acceptable frequency of criterion violation might differ by waterbody, even for the same broad designated use.

In its most analytically complete form, an IWA would employ a set of linked mathematical models from multiple disciplines to inform decisions on the assignment of uses to a water body, on the surrogate criteria to represent the uses, and on the investment and regulatory actions to meet the criteria.[b] More generally, for decision support IWA serves the two purposes of forecasting and backcasting. *Forecasting* (prediction) uses models to conduct policy experiments on paper before committing significant private and public resources to an action. Models are asked "what-if" questions to help in deciding whether a predicted watershed outcome from an action (for example, how limits on a pollutant will affect compliance with a water quality criterion) is worth the prospective costs of the action. Another use of IWA models is backcasting. *Backcasting* is the opposite of prediction and might be thought of as using a model for explanation. In this use, the condition of a waterbody would be monitored (the monitoring would focus on whether the water quality criterion is being met), and a model might be used to explain the factors that led to the measured condition. For example, a model would be used to explain how pollutant sources and other factors cause a violation of a water quality criterion.[c]

Figure 7.1 is adapted from the conceptual layout of a model that was used to predict (and explain) multiple cause-and-effect relationships between concentration of nutrients and biological response for a Neuse estuary in North Carolina.[11] The layout recognizes that the Neuse River receives nutrient loads from a number of

[a] The capabilities of the monitoring program should be considered when choosing criteria. For example, a criterion defined in terms of frequency, duration, and magnitude is meaningless if the available monitoring data will assess only the frequency with which the criterion is exceeded. In addition, because water quality assessment is a sampling and statistical inference procedure, the acceptable level of the criterion should be stated with confidence limits.[38]

[b] IWA takes a significant step beyond ecological risk assessment by adding explicit consideration of socioeconomic values and costs to the analysis. This added dimension can increase the contribution of analysis to environmental decision-making.[9]

[c] Models can include intuitively held mental constructs, as well as highly complex mathematical representations that are solved by computer algorithms and rely on access to extensive data sets. The existing scientific knowledge of the cause and effect relationships; the data, staff, and budget resources available for conducting analysis; and other considerations should influence the complexity of the IWA model. (See Chapra[45] for a discussion of model complexity.)

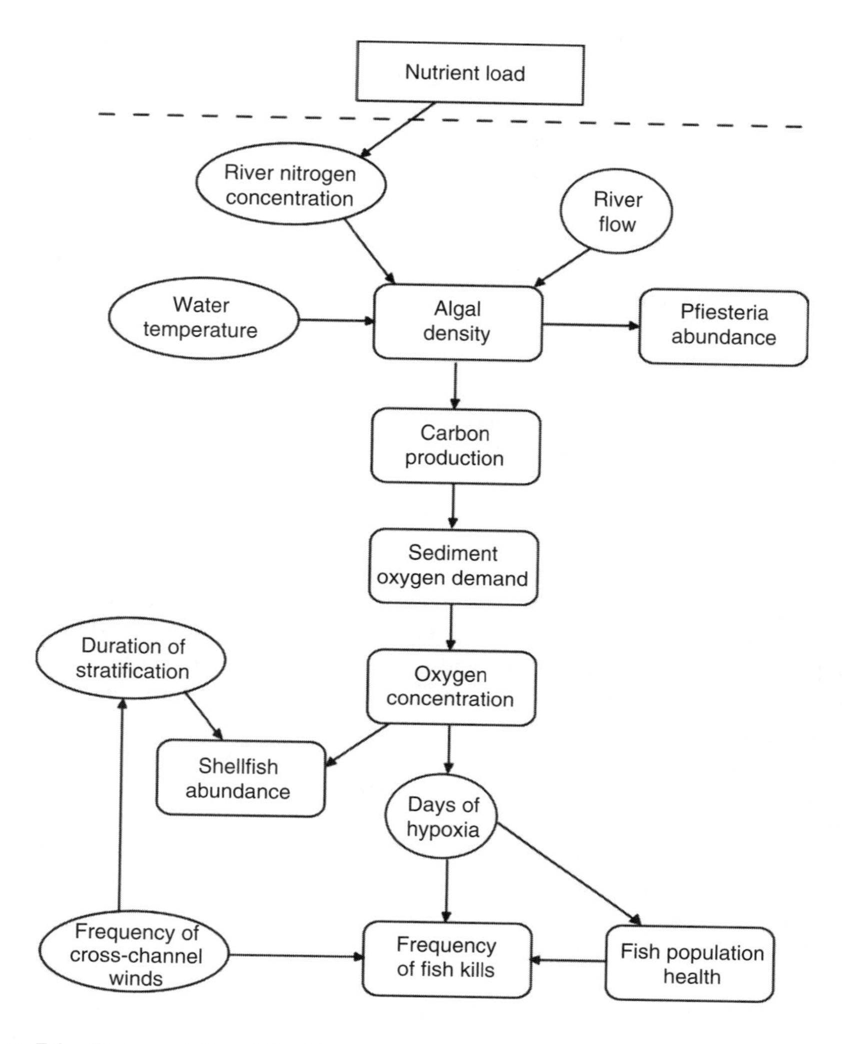

Figure 7.1 Representation of the Neuse network model.

different sources that result in a nutrient concentration. This concentration, in concert with natural phenomena such winds and tides, affects the algal community (chlorophyll *a*) and then end conditions (uses), such as shellfish abundance. The conceptual layout in Figure 7.1 guided the construction of a computational model of the Neuse estuary. Each of the general relationships in the causal chain depicted in Figure 7.1 was represented as an equation, and data were collected to quantify the specified relationship. In addition, the modeling effort explicitly accounted for uncertainty in the specified relationships and from the limited data that were available to quantify those relationships.[a,11] This was one of three models that were employed in developing

[a] A Bayesian perspective means that there is no need to make a distinction between risk and uncertainty as sources of prediction error.

the Neuse River TMDL plan.[12] Use of multiple models is a recognition that no single model can be relied upon to give a certain answer.

The Neuse River team recognized that while models and data have improved dramatically over time, model prediction error is unavoidable. There are a number of reasons: (1) the inherent randomness of natural systems, making any prediction of a single outcome suspect; (2) limits on conceptual understanding of the watershed stressors and responses to stressor reduction in the waterbody of interest (for example, variable lag times between an action and a system response); (3) the necessity of simplifying the model of the system to make the model tractable; and (4) limits of data to parameterize the inevitably simplified model for the water body.[4,13,14]

In addition, the prediction error is not uniform throughout the model. Refer again to Figure 7.1. If the criterion selected for measuring water quality use attainment is closer to the control actions on nutrient loads (not depicted in Figure 7.1) — for example, river nitrogen concentration — then model predictions of the effect of a control action (wastewater treatment plant controls) on the surrogate will be more certain. However, even in this case, cause-and-effect predictions are likely to have error bands around them. If the criterion is closer to the designated use (for example shellfish abundance, which is a surrogate for the designated use of shellfish harvest), the uncertainty in the prediction of the effect of wastewater treatment controls is likely to be greater than for the nutrient concentration measure. In short, model prediction error cannot be avoided and is likely to become greater as we measure water quality program success by criteria that are the closest surrogates for designated uses.

A probability network model was used as the solution process for the Neuse estuary analysis. The model relationships were stated in probabilistic terms (using a Bayesian framework), and the uncertainties were propagated through the solution to report the resulting prediction uncertainties. Because of the Bayesian orientation, this modeling approach incorporated judgmental as well as statistically derived estimates of uncertainties.[11]

Ecological risk assessment, the foundation for IWA, recognizes the need to explicitly account for uncertainty (see Chapter 3). However, the manner in which uncertainty is accommodated in the decisions made depends on whether decision-making is static or adaptive.[15] Generally, static decision-making occurs when there is a single decision point and there is little expectation of future monitoring and study to revise or modify the model or the decision made. Introduction of uncertainty is accomplished by assigning probabilities to certain critical relationships or data and then making calculations that propagate and report the resulting prediction uncertainty to decision authorities. The "best" actions are chosen in consideration of the predicted outcomes and the decision-makers' attitude toward risk.[15] Modeling typically done for the TMDL program implicitly assumes a static decision-making setting, but explicit reporting of uncertainty is not the norm. Instead the modelers are instructed to assign a MOS to the TMDL calculation.[4]

Static decision-making is not appropriate for larger-scale watershed management efforts, such as that in the Neuse. At the same time, the use of the MOS calculation in the TMDL program has been a focus of criticism.[4] The alternative to static decision-making and best professional judgment for MOS assignment can be called *adaptive decision-making*.[15] Adaptive decision-making (learning while doing through feedback) means that watershed management includes learning as a goal and that management choices are made over time in response to new knowledge about the system being managed. Adaptive decision-making adds a dynamic component to IWA.

Anderson et al.[15] describe different approaches to adaptive decision-making: passive adaptive management, active adaptive management, and evolutionary problem-solving. The distinctions between passive and active adaptive management are generally well understood. Passive adaptive management is less experimental than active and less costly. Passive adaptive management might involve monitoring outcomes of actions taken to see whether goals were achieved, but it would add little to understanding of why a goal was or was not met. In the end, the models used to inform future decision-making may not be significantly improved through passive adaptive management.

In active adaptive management, the learning occurs through (1) continued monitoring of the water body to determine how it responds to actions taken and (2) carefully designed experiments in the watershed. More generally, learning can occur through any process associated with scientific investigations and can include literature reviews as well as laboratory experimentation. The expectation in active adaptive management is that the learning will be used to improve the IWA models. In this sense, the motivation behind active adaptive management is to build better models over time as a way to better inform future decision-making.

The distinction between either form of adaptive management and evolutionary problem-solving turns on the responsiveness of goals to learning. Adaptive management is goal seeking, and evolutionary problem-solving is goal discovering. In the context of the CWA, the NRC[4] used the term *adaptive implementation* (AI) to describe evolutionary problem-solving, drawing a distinction between adaptive management and adaptive implementation. For water quality management under the CWA, model prediction uncertainty means that future water quality criteria cannot be assigned without knowing how the waterbody will respond to actions taken to change the water from its current condition. Different uses and associated criteria might be assigned once we know more about the watershed — and once we learn more about the costs of attaining different WQS.[a]

Models in support of AI must characterize and report on inevitable modeling uncertainties. The probability network computational algorithm used in the Neuse River case was well suited to supporting AI because it highlighted the areas of uncertainty that need the greatest attention to reduce future model prediction error. The approach helps to target the adaptive management studies for model improvement for the next round of decision support. In summary, while the purpose of water quality management is to resolve a water quality problem, water quality management also is a learning process with the goal of improving our understanding of a watershed to improve the models that will inform subsequent rounds of decision-making.[b]

[a] The role of cost in assigning uses and associated criteria is discussed later in this chapter under "Water Quality Standards, Economics, and Adaptive Implementation" and "Decision-Making Challenges to Adaptive Implementation."

[b] It would be cost prohibitive to apply AI in its most developed form for many smaller watersheds where water quality criteria are not being met. A pilot watershed approach[4] could be developed. A number of pilot watersheds would be selected for extensive modeling and model refinements over time. The resulting models would then be used to support decision-making in watersheds with conditions similar to the pilot areas.

ADAPTIVE IMPLEMENTATION AND THE CLEAN WATER ACT

Section 303e of the CWA authorizes a continuing planning process (CPP) for watershed water quality management.[a] The steps in the CPP are iterated over time. The CPP begins with the assigned ambient standards and associated criteria for the waterbody. Whether oriented towards the long term or interim goals of the CWA, the statement of watershed goals as represented in the WQS (uses and criteria) is the foundation for the CPP. Water quality monitoring is conducted to determine whether the criteria are being met.[b] If the criteria are met, then (ideally) there is an ongoing monitoring program that will continue to assess the waterbody condition on a defined schedule.

If the criteria are violated, a plan to reduce the stressors must be developed, or a review of the standard (designated use or criterion) can be undertaken.[c] In the situation where a standards review is not initiated, a suite of models are used to calculate the maximum pollutant load consistent with meeting the water quality criterion and to predict the effect of load reduction alternatives at point and nonpoint sources on the water quality criterion. The expected result is a plan for an initial sequence of regulatory and budgetary actions to secure required load reductions and (presumably under the CPP) implementation of the actions.[d] Ideally, the watershed and water quality modeling will recognize other factors that affect water body condition, such as flow and habitat alteration. If multiple factors were considered, other watershed management actions related to flow or habitat would become a part of the overall plan for securing the designated use. Future monitoring, or other means of water quality assessment,[e] establish whether the criteria have been met after the initial actions in the plan are taken. If the criteria are not met, then the CPP anticipates either that, based on what has been learned, additional stressor reduction actions will be taken or that the uses and criteria will be adjusted.

The CPP can accommodate the need for water quality analysis and management decision-making to recognize and accommodate multiple sources of uncertainty. The CPP can be structured as an adaptive decision-making process so that stressor reduction actions and water quality uses and criteria will be revised over time with experience and further analysis.

[a] See 40 CFR Part 130.5.

[b] In fact, determining whether a criterion is met is a matter of much controversy and has been the focus of recent USEPA guidance.[5]

[c] The processes for a review of standards are discussed in detail in Sections 7.4 and 7.5.

[d] Adaptive implementation should not be confused with phased implementation. Phased implementation[1] of TMDL plans assumes (implicitly) that both the assignment of a future designated use (and criteria) as well as the strategies to meet the criteria can be determined at the time the original plan is developed. The adjective *phased* suggests that implementation is not immediate, but only waits until implementation funds accumulate and necessary regulatory authorities are secured.[1]

[e] If there are lag times between the actions taken and the response of the water body, then other ways to track progress may be used. For example, inventories of implemented nonpoint source control actions may be combined with the results from studies of the "edge of field" load reduction effectiveness and pollutant load transport and fate to build models that predict water quality conditions. The model-predicted water quality criteria, as opposed to actual monitoring data, would be used to measure attainment.

WATER QUALITY STANDARDS, ECONOMICS, AND ADAPTIVE IMPLEMENTATION

The assignment of designated uses and criteria to water bodies recently has become a matter of intense public debate[16,17] and USEPA attention,[18] focusing on how, and by whom, uses and criteria are defined and then assigned to particular water bodies. This section begins with a discussion of the CWA process for assigning water quality uses and criteria to water bodies and then proposes a role for economic analysis and adaptive implementation in the use and criteria assignment process.

Water Quality Standards: Establishment and Assignment to Water Bodies

States, subject to USEPA review, define a range of designated uses for their waters and then establish water quality criteria as measurable surrogates for each use. In turn, the states are expected to assign one or more such uses to each water body and then monitor them to determine whether the criteria for that use are being met. In executing this responsibility, a state may use the system of tiered standards. For example, the "swimming" use might distinguish between seasonal use, primary water contact recreation (i.e., full immersion), and secondary water contact recreation (e.g., canoeing or fishing). Then, microbial bacteria criteria for each use tier would vary according to the likelihood of water body contact expected for the population.

As another example, *biological integrity* indices (see Appendix 6-A for examples) might be used to develop tiered standards. The definition of biological integrity has been equated with a reference condition where an unimpaired stream is defined as a condition where human disturbance is at a minimum. The overall goal of establishing a reference condition is to describe the natural potential of the water body and habitat types characteristic of the region, independent of the extent of human degradation. However, it does not follow that this undisturbed reference condition needs to become assigned as the WQS. Indeed, in areas where human activities have significantly altered the landscape, a biological criterion that results in a WQS that only can be achieved independent of past and prospective future human disturbance may be unattainable. A tiered approach to biological criteria (biocriteria) development for watersheds has been employed in several states to accommodate this reality. Specifically, indices of biological condition that are deemed attainable in any watershed would be assigned in recognition of the development history and future economic activity in that watershed. For example, stream use in Ohio is classified according to different levels of aquatic life support potential, and then index values (biocriteria) are associated with the different use designations.[19] Then one of these uses is assigned to (deemed attainable in) each water body in the state.

Recently, the USEPA has promoted *use attainability analysis* (UAA) as the analytical foundation for assigning attainable WQS for a watershed.[20] A UAA employs both technical and socioeconomic considerations for evaluating whether a use and its associated criteria are attainable.[a] The principal lesson to be drawn from the current interest in the UAA process is that the CWA goals define a direction for water quality improvement, but the goals are not required end states for all watersheds in all places. Nonetheless, even if a tiered standard is adopted, still further refinement of the uses and associated criteria for any specific location can be considered over time through the AI process.[b] In fact, given modeling uncertainty, there is a scientific imperative to systematically revisit WQS within the AI process.

Economics and the Assignment of Water Quality Standards

An economic framework can organize information about choices and trade-offs that must be made when determining attainability. At the conceptual level, an economic analysis framework considers incremental gains in relation to costs from achieving different levels of use as reflected by different levels of use attainment. In the stylized representation of Figure 7.2, levels of a criterion — μ — for measuring use are arrayed along the horizontal axis. The criterion might be expressed as the frequency with which dissolved oxygen is above a given concentration, the frequency that a chlorophyll *a* concentration is not exceeded, or the level of an index of biological integrity.[c] Movement from left to right represents a change toward the restoration goal stated in the CWA. At the far right of the figure, restoration approaching a predisturbance condition is attained for the watershed, shown as R; also shown is CC, the current condition of the water body. Marginal benefits and marginal costs of securing different levels of a water quality criterion are shown as solid lines in Figure 7.2.[d] The water quality management challenge is to assign an attainable water quality use and associated criterion for a water body; the attainable level may be less

[a] The uses may be modified if the presence of natural conditions or irrevocable human conditions makes attainment impossible. In addition, a use may be modified if attainment would result in widespread adverse social and economic impact. A further discussion of UAA is included under "Decision-Making Challenges to Adaptive Implementation."

[b] The standard setting decision has two related parts: assigning an attainable use and defining a surrogate criterion for that use. Sec. 304(a)(1) of the CWA limits the selection of criteria to "scientific knowledge," and so would not expect economic arguments to be a part of criteria selection. As will be noted, however, there is room for economic arguments in assigning uses, and there can be much variation in the levels of use within a broad use category, such as water contact recreation. As these uses vary, so too will the criteria that are derivative of the use. In effect the decisions on both use and criteria selection are inextricably linked. In this paper a standard is described as "a use and its associated criterion."

[c] A composite metric representing multiple uses might be developed. For example, see Griffiths.[46]

[d] Economic logic asks whether the added costs for each additional movement toward R are warranted by the additional benefits realized. However, some have noted that economic logic can be in tension with legal interpretations of laws that call for the meeting of clear standards. U.S. Supreme Court Justice Stephen Breyer has described the tension in this way: "Economic reasoning is often difficult to reconcile with bright-line rules. Economics often concerns gradations, with consequences that flow from a little more or a little less. But the law, at least in a final appeals court, often seeks to create clear distinctions of kind."[47]

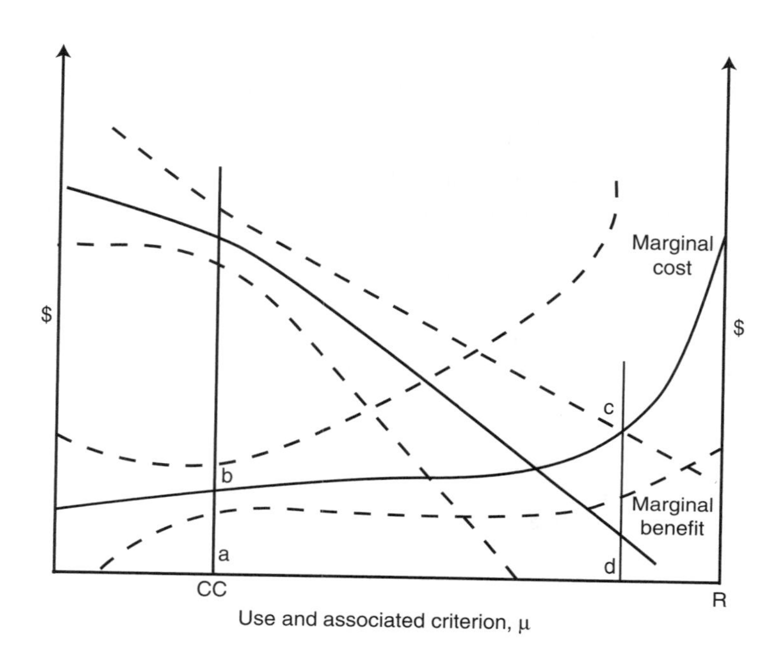

Figure 7.2 Conceptual economic framework.

than R.[a] In the conceptual IWA framework, the attainable level is where incremental benefits are equal to incremental costs.[b]

The marginal cost relationship shows that the costs incurred for increasing the controls on point and nonpoint pollutant loads, as well as for reducing other barriers to complete restoration (restoration of hydrology, for example), grow at an increasing rate as controls become more stringent. In Figure 7.2, costs grow at an increasing rate with movement from CC to R. The marginal benefit curve suggests that the value to people of moving the assigned criterion from CC toward R increases, but at a decreasing rate. The theory of value in Figure 7.2 is economic; that is, value is defined in terms of satisfaction of people's preferences — here for different levels of the water quality criterion. The empirical approach to economic valuation — as some professional economists practice the art — is to calculate money-equivalent measures of individual people's preferences and then aggregate those measures into a benefit estimate. Valuation, as an empirical exercise, rests on the argument that choices individuals make in market exchange provide data that analysts can use to translate people's preferences into money terms. The logic of the argument is

[a] In concept, the selected level for the water body may be less than CC, but antidegradation requirements in WQS would prohibit this choice. In fact, under the CWA, a use that was actually attained at any time since 1975 is considered to be an "existing use," even if that use is later precluded by water quality conditions; an existing use is not subject to change and must be retained in the standard.[6]

[b] The states are allowed to consider environmental, technological, social, economic, and institutional factors in assigning attainable uses with associated criteria to a water body. In effect, there is room for the application of economic logic. However, the CWA does not allow states to exclusively rely on calculations of net benefits to determine attainability.[27]

straightforward. In market exchange, money income is sacrificed (a price is paid) to secure any good or service. By arguing that preferences guide market choices, analysts conclude that the money value of a good or service is at least equal to the amount of income a person spends to obtain the service. Thus, market prices are the raw data for preference measurement. The often-explicit premises of this revealed choice framework are that individuals know their preferences for goods and services (states of the world) before being confronted with a choice, that people are willing to pay to satisfy those preferences, and that whatever an individual chooses is in the interests of that individual.[21] It is the benefit–cost analyst's responsibility to measure those preferences in money terms.[22] Valuing people's preferences for environmental services (such as improved water quality) has been an interesting research puzzle for many economists, who seek to obtain data on revealed choices from other markets or from choices made in hypothetical markets (see Chapter 5 for a general discussion and Chapters 10, 11, and 14 for examples).

Returning to Figure 7.2, the reality of model prediction error is represented by error bands surrounding the benefit and cost relationships. These error bands are shown as the dashed lines surrounding the marginal benefit and cost curves. There are several reasons for these error bands. First, it analytically follows that uncertainties in benefit and cost estimates represented in Figure 7.2 are simply propagated from prediction errors in the underlying hydrologic, chemical, engineering, and biological models on which they rely.

However, the uncertainties in the benefit and cost relationships have other sources, stemming from uncertainties in human behavior and choice. In the case of cost estimation, realized costs will depend on the institutional form of regulation (technology based, performance based, market-like) and general market conditions.[23] In fact, one of the principal economic arguments for regulatory reform is that the costs of compliance and government program administration will be different under different regulatory structures.[24]

Error bands that surround benefit measures might simply be ascribed to limited data or technical barriers to preference measurement. However, a more fundamental methodological reason for the error bands is that the preferences (the source of value) are not stable and are not revealed in the choices people make. Economic theorizing in the tradition of the Austrian and American Institutionalist schools of thought argues that preferences are not given, something to be recalled when making choices. Instead, preferences are in flux and are formed and then revised in the process of making choices.[a] If preferences are not recalled but instead are formed as future choices are made, then no matter what tools are used or what data are available, benefit estimates for future water quality conditions cannot be known in advance even to individuals themselves, but instead are discovered over time as water quality conditions change. This perspective is consistent with the basic premise of AI that stresses learning over time because learning affects future choices.

Also, note in Figure 7.2 that the error bands widen when moving from CC toward R. The error in predicting the costs of securing a criterion level grows as R is approached because we have less certainty about possible behavioral and technological

[a] See Shabman and Stephenson[48] for a review of this literature.

responses to meeting this more stringent criterion. Also, because preferences are discovered in consideration of new circumstances and in consideration of realized costs, the benefits for different watershed conditions are less certain the further one gets from the current condition. As a result, the further from CC, the wider the range of possible "correct levels" of a future water quality criterion. Calculations made in an initial time period offer little guidance for choosing an appropriate water quality criterion at a future date.

It follows from these arguments that making economics part of AI does not mean that calculations of net benefits are necessary or even desirable.[a] Instead it is useful to think of AI as a public decision-making process for discovering marginal costs and benefits (preferences) over time.[b] The analysis to inform this public learning process might be called *proximate knee of the cost curve* analysis. This approach to analysis is what it means to bring economics to water quality management decision-making.

Beginning at CC in Figure 7.2, a decision would need to be made on whether to move in the direction of improved water quality — toward point R. One of the economist's contributions to IWA is to help measure marginal implementation costs as well as can be done,[c] and to report uncertainties in the estimates. Actions to move toward R would be chosen for implementation until a marked increase in the slope of the cost curve is encountered. The logic is that water quality management strategies beyond that point might not be pursued in the initial time period, unless there is a strong social consensus that the incremental gains are likely to be of significant value. The term *proximate* recognizes the uncertainties in the cost and effectiveness estimates underlying the marginal cost curve. Even if there is no distinct point of change in the predicted marginal cost curve, there can be significant uncertainty in the estimate. In that case, a decision would be made to move forward with those activities that show the most certain cost and effectiveness while awaiting the learning that comes from the AI process before determining what the next increment, if any, should be.

Some could argue that the proximate knee of the cost curve argument is not consistent with the CWA as codified in USEPA regulations. They would cite the integrity goal (R in Figure 7.2) as the requirement of the CWA, only to be modified when there is compelling documentation that the use is not attainable.[25] By focusing water quality decision-making on whether R is attainable, the burden of proof falls on those responsible for the pollutants or pollution to show why actions to secure R impose unreasonable costs. If CC is the starting point for defining what is attainable, as argued in Figure 7.2, the burden of proof is on the public to show that the costs of the improvements beyond CC are not unreasonable (i.e., are attainable). This logic is consistent with the way effluent standards are set under the CWA, where limits are imposed until the costs are deemed unreasonable. In addition, there is an analytical problem with starting at R. The predicted costs and benefits of

[a] In fact, excessive reliance on such calculations is not permissible under the CWA.[6]

[b] The challenges to ensuring that such a process is representative and secures results that serve public values, as opposed to only the interest and values of the participants, are recognized. However, for purposes of this discussion these issues need not be addressed.

[c] The calculation of costs will be of limited scope and detail. The focus is likely to be on predicting private and public expenditures, and perhaps on the changes in net income of regulated party. While approaches to more comprehensive cost analysis can be described, such cost estimation can, as noted, be fraught with uncertainties. See USEPA[23] for a discussion of cost analysis.

changes away from R will have significant and irreducible error bands. Water quality improvement actions whose outcomes are too uncertain ought not to be considered "attainable." In fact, as a practical matter, analysis and decisions on regulations and on investments to secure water quality changes can be made only by considering the expected costs of moving away from CC.[a]

In fact, knee of the curve analysis is not an unfamiliar concept within the USEPA. For example, the Combined Sewer Overflow (CSO) Control Policy of 1994 requires that marginal costs of control strategies be compared to the incremental benefits before implementing CSO controls.[b]

> The permittee should develop appropriate cost/performance curves to demonstrate the relationship among a comprehensive set of reasonable control alternatives that correspond to the different ranges specified in Section II.C.4. This should include an analysis to determine where the increment of pollution reduction achieved in the receiving water diminishes compared to the increased costs. This analysis, often known as knee of the curve, should be among the considerations used to help guide selection of controls.[26]

Designing CSO controls to avoid discharges during unusual wet-weather events to achieve the goal of fishable–swimmable water may not be justified if non-CSO sources dominate the loading of the pollutant or if water contact uses are unlikely during wet-weather events or during certain seasons. In such cases, the state may change the use and associated criteria for a particular water body, following a UAA. In fact, the CSO guidance on coordinating CSO long-term control plans with WQS reviews was an effort to integrate CSO planning with the UAA. It states that:

> If chemical, physical or economic factors appear to preclude attainment of use, the data collected during the planning process may be used to support revisions in water quality standards. These revisions could include adoption of uses that better reflect the water quality standards that can be achieved with a level of CSO control that does not cause substantial and widespread economic and social impact[10]

Unlike the CSO Control Policy, the UAA application is not organized around consideration of marginal benefits and costs, even though the UAA process is used to evaluate how far to move along the cost curve (to the right) in Figure 7.2. Nonetheless, the interim economic guidance for WQS explicitly states, "Benefit–cost analysis is not required to demonstrate substantial and widespread effects under the federal water quality standards regulation."[27] Instead, the UAA applies an affordability test to determine the attainable level of the standard. Hence, to change a designated use and associated criterion, states must demonstrate that implementing the controls to achieve a certain WQS would result in substantial and widespread economic and social impacts.

Substantial impact refers to the financial costs of controls necessary to achieve a WQS, given the socioeconomic situation of a community. In determining whether

[a] The AI approach expects that WQS will be systematically revisited over time. Therefore, the possibility of moving further toward R is always under consideration.

[b] This policy was codified by reference in the Wet Weather Quality Act of 2000 that amended the act to include Section 402(q)(1).

the costs are substantial, the annualized per-household cost of sewer services, including existing and new costs, is divided by the median household income to calculate the municipality preliminary screener (MPC). If the result is less than 1 percent, the state could decide that the annualized costs are minimal, and the financial and economic impacts do not warrant revising the standard. If the MPC is between 1 and 2 percent or greater than 2 percent, a secondary score is calculated based on the evaluation of debt, socioeconomic, and financial management indicators.[a] The results of the secondary score and the MPC are summarized in a matrix, from which states must determine whether the impacts are substantial.[b] In addition, just demonstrating substantial impacts is not a sufficient condition to change the designated use of a water body. The entity must also demonstrate widespread economic impacts. Widespread impacts are a change in socioeconomic conditions, determined by comparing the financial burden with a series of indicators, for a defined, affected area.[c]

In the UAA analysis, a total expenditure that the community can afford is calculated, based on guidance that sets limits on financial burden. Therefore, the expectation is that this amount of expenditure will be made on water quality improvements.[d] In Figure 7.2, the area abcd represents the expenditure that results from the affordability calculation and that must be made, without regard to considerations of incremental costs and benefits.[e]

To add the necessary economic content to IWA, the current UAA affordability calculations need to be conditioned with the logic of proximate knee of the curve analysis. In fact, the implicit UAA premise that there should be a national requirement to spend a certain amount of a community's budget on water quality (or any other item), without regard to relative costs and benefits, might not have been the intent of the CWA. Such a change would bring the UAA process more in line with the NPDES permitting process. Specifically, the stringency of an NPDES permit limit is governed by the application of a "reasonableness test" of the costs and a determination that the costs are not wholly disproportionate to the environmental benefit received. Posing the question, "are the costs wholly disproportionate to benefits received" (where benefits are in biological terms) is to apply net-benefit logic and to organize that logic around knee-of-the cost-curve analysis.[f] The application of this NPDES logic recognizes that at some point incremental costs may

[a] The indicators for public entities include bond rating and overall net debt as a percent of the full market value of taxable property (debt indicators), unemployment rate and median household income (socioeconomic indicators), and property tax revenue as a percent of full market value of taxable property and property tax collection rate (financial management indicators). In the case of private entities, these include indicators for liquidity, solvency, and leverage.

[b] The justifications for the recommended percent cutoffs are not documented (J. Keating, personal communication, 2004).

[c] Indicators include median household income, unemployment, overall net debt as a percentage of the full market value of taxable property, percent of households below the poverty line, impact of community development potential, and impact on property values. In the case of a private entity, widespread impacts include the decrease in taxes as a result of a decrease in profits, and jobs lost due to relocation of industry.

[d] In spending this sum, there is an opportunity to allocate the spending so that the marginal costs across all improvement strategies are equated.[49]

[e] If the uncertainty bands in Figure 7.2 were considered, the budget would "buy" a wide range of possible criterion levels.

[f] This approach to economic analysis in other venues is common. See for example Shabman and Stephenson.[48]

exceed incremental benefits, but the NPDES review relies on the regulatory and deliberative processes among stakeholders and not on calculations from economists to determine when incremental costs are "too high" in relation to incremental benefits.[a] Likewise, proximate knee-of-the-curve analysis does not answer the "how much is enough" question, even in an initial time period. Instead, through public decision-making, agencies and stakeholders would, informed by the cost analysis, select a near-term standard and the strategies to meet that standard. This process, repeated over time in recognition of new information, is the process by which the marginal benefits (values) are discovered.[b] Such a process must be what engaging stakeholders in watershed management decisions means. The next section discusses some of the challenges of meshing AI with stakeholder involvement, with illustrations drawn for the Neuse River, North Carolina TMDL planning process.

DECISION-MAKING CHALLENGES TO ADAPTIVE IMPLEMENTATION

Several challenges must be met if AI is to be successfully applied. The first challenge will be to select the initial regulatory and investment actions expected to improve water quality. In making that selection, IWA models could be used for backcasting (as described earlier) to provide the analytical support for considering alternative use and associated criteria levels, alternative reductions in loads and other factors to meet the criterion, and the costs and effectiveness of the regulatory and investment strategies that will achieve each level, recognizing uncertainty. Based on this evaluation, the first increment of actions would be implemented.[c]

For the Neuse River situation, the decision by the State of North Carolina, with USEPA concurrence, was to seek a 30% reduction in delivered nutrient loads to the Neuse estuary from all sources. The decision process was informed by both the results from technical modeling and a formally convened panel of stakeholders.[28] The several models used in developing the initial Neuse River water quality improvement strategy had suggested a wide range of load reductions, and in some cases the individual modeling efforts calculated and reported errors in the estimates. The WQS that governed the Neuse planning process required that a chlorophyll *a* criterion of 40 mg/L be met 90% of the time.[29,30] One model result suggested that a 30% load reduction might result in a violation of the criterion more than 10% of the time, but the model prediction error was also reported. Meanwhile, and equally important, there was significant disagreement over whether the criterion had a sound scientific

[a] Generally speaking, this permitting process is best characterized as static and not adaptive decision-making, although permit limits may be adjusted over time in response to new information.

[b] If part of the analytical foundation for decision-making is to poll the public for its preferences, there is no substitute for direct survey information designed to measure and represent those preferences. One particular method of eliciting preferences across various attributes is the use of conjoint analysis (see Appendix 5-A and Chapters 11 and 14). However, the purpose of such analysis is not to develop a money measure of the benefit function in Figure 7.2.

[c] In terms of the TMDL process, the water body would be impaired until the specified use and criteria were met. Once met, further improvements would be considered through the CPP. As long as R is not secured, the possibility of moving toward R would be open for consideration.

basis, given the origin of the measure,[31] and over whether the criterion was a compelling surrogate for the desired uses.[28]

In recognition of sharply increasing predicted costs of seeking more than a 30% reduction, in recognition of the uncertainties of the predicted criterion response,[a] and in recognition that the criterion itself was in need of review, a 30% reduction was selected as the first increment. Practically, it was possible to secure agreement only on the 30% reduction, but not more. In fact, the state recognized that failure to reach agreement on an acceptable first increment of load reduction could have resulted in court challenges and delay of any reductions taking place. Instead of delaying water quality improvement actions by seeking an exact and ultimately elusive answer to the correct criterion and long-term strategy, the Neuse River plan is now being implemented with an understanding that the plan will be revisited over time.[b]

A second challenge is to establish the times when the necessity and merits of additional actions will be evaluated. Without such a schedule, there will be no assurance of "reasonable progress" toward implementing an effective water quality strategy. The length of time between program reevaluations should be governed by at least two considerations. If there is to be learning before taking the next step, then there must be adequate time for that learning to occur and be incorporated in improved models. For example, if there is reason to expect lag times between implementing actions and water quality response, then there must be time to allow the responses to materialize. Also, there needs to be some predictability of the regulatory requirements to allow for reasoned investment decision-making by the regulated parties who will make investment and management decisions that affect the operation of their facilities. It would seem that no less than five years (an NPDES permit cycle) should pass before new increments of reduction would be considered. In the case of the Neuse River, there is no schedule defining when such reevaluation would take place.[c]

A third challenge is ensuring that the IWA models will be improved over time. Such improvement requires monitoring of the whole watershed and of practices put in place, conscious experimentation, active tracking of the scientific literature in multiple disciplines, and application of the new data and knowledge to refine the IWA models. This AI effort requires a commitment of staff and budget, as well as maintenance of the collaborative stakeholder process. The details of setting knowledge enhancement priorities and securing adequate study resources are not

[a] The general conclusion was that, based on the weight of multiple model results, there was a 50–50 chance of meeting the WQS with a more stringent 45% load reduction (Roessler, personal communication, 2004).

[b] Roessler, personal communication, 2004. In fact, one TMDL plan states the following[12]: "It should be acknowledged from the outset that though the predictions and decisions contained in this document are based on the best currently available information, there is substantial uncertainty in them. For this reason, DWQ intends to follow an adaptive approach to managing the estuary. In other words, DWQ will use the models to guide decision making, but continuing observation of the watershed and estuary, as nitrogen controls are implemented (i.e., Neuse Rules, and other measures such as wetlands restoration and establishment of conservation easements), is expected to be our best approach for determining the appropriate level of management."

[c] Triennial reviews of a state's system of standards are required by Section 303(c)(1) of the CWA. It is at this review that changes in standards are proposed and that the USEPA must approve such changes. However three years may not provide adequate time for the adaptive implementation cycle to be completed.[6]

the focus here.[a] It is worth noting that the costs for any AI effort will be significant, but that the budgets available for watershed planning in general, and the TMDL program in particular, remain limited.[b]

The Neuse River plan does not include a commitment by the state or the USEPA to future model improvements and so cannot be described as an example of an AI program. In fact, the lead staff person for the Neuse TMDL study from the State of North Carolina has been designated to work on other issues and in other watersheds.[c] The Neuse adaptive process may move forward, however, because university researchers in North Carolina secured a USEPA grant to continue experimentation, monitoring, and model development in support of future decision-making for the watershed.[32]

While an AI process can serve the nation's water quality management programs, a fourth and central challenge will be clarifying the decision-making roles and responsibilities for defining appropriate criteria and making regulatory and investment obligations over time. The USEPA, the states' water quality management agencies, and local governments all will have responsibilities. At the same time, the roles of regulated parties, citizens, and interest groups such as environmental organizations will need to be recognized. There is a large and growing literature of the promise and the challenges of collaborative decision-making for environmental management.[33–35]

For watershed-based water quality management, the general commitment to stakeholder involvement and collaborative decision-making will need to be given a specific form that is sensitive to CWA processes but also pays attention to the economic, as well the legal, chemical, biological, and engineering dimensions of decisions on WQS assignment and the actions to secure the standards. Maguire,[28] in summarizing the stakeholder decision process for the Neuse River TMDL, offers a general lesson for the watershed approach to water quality management:

> Perhaps the biggest constraint imposed on stakeholder-science interactions by the regulatory process was the narrow casting of the TMDL issue for the Neuse. Impaired water quality in the Neuse was defined by failure to meet a standard for chlorophyll concentration. Not only was this standard defined without regard for the cost effectiveness tradeoffs at the forefront of most stakeholders' minds, it also failed to connect in a meaningful way with the features of water quality that most stakeholders are worried about: fish kills, noxious algal blooms, contaminated shellfish and turbid water. More broadly than that the whole TMDL process fails to address [the] issues that are important to stakeholders: . . . Are costs worth the benefits, in both economic and environmental terms? Have the difficult tradeoffs among competing values been adjudicated in a manner that is consistent with democratic ideals? . . . Failure of the Neuse TMDL process to address most stakeholder concerns . . . stems from an overly narrow construction of regulatory decision making, where the scientific basis for regulation is limited to biophysical concepts . . . rather than the full range of biological, economic, social and cultural elements that should inform regulatory decisions.

[a] See Reckhow[30] on priority setting.

[b] For cost reasons AI is not practical for small watersheds. See previous note, Section 7.2.

[c] Roessler, personal communication, 2004.

CONCLUSION

Proximate knee-of-the-cost-curve analysis is an analytical framework that will support adaptive implementation in water quality decision-making and is consistent with the CWA and its decision-making requirements. The proposed approach in this paper calculates the incremental costs, and associated uncertainties, of increasingly stringent pollutant and pollution controls necessary to secure different levels of one or more water quality uses and associated criteria. The proposed approach anticipates that a collaborative decision process will judge whether those added costs are justified by the predicted changes in the uses and associated criteria (benefits). The incremental cost analysis focuses a public discussion over "How clean is clean enough?," both inside the regulatory agencies and in the open public process around estimates of incremental costs. In addition, the proposed process is a continuing one that systematically revisits both standards and water quality management strategies over time.

ACKNOWLEDGMENTS

This work has been partially supported by the USEPA, Grant #83088301, "Adaptive Implementation Modeling and Monitoring for TMDL Refinement." The assistance of Puja Jawahar of Resources for the Future is gratefully acknowledged.

REFERENCES

1. USEPA, Guidance for water quality based decisions: The TMDL process, EPA440/4-91/001, NTIS PB92-231620, U.S. Environmental Protection Agency, Washington, D.C., 1991.
2. USEPA, Water quality conditions in the United States: A profile from the 2000 National Water Quality Inventory, EPA/841/R-02/003, U.S. Environmental Protection Agency, Washington, D.C., 2002.
3. USEPA, Draft guidance for water quality-based decisions: The TMDL process (Second Edition), EPA/841/D-99/001, U.S. Environmental Protection Agency, Office of Water, Washington, D.C., 1999.
4. NRC, *Assessing the TMDL Approach to Water Quality Management*, National Research Council, National Academy Press, Washington, D.C., 2001.
5. USEPA, The twenty needs report: How research can improve the total maximum daily load (TMDL) program, EPA/841/B-02/002, U.S. Environmental Protection Agency, Washington, D.C., 2002.
6. USEPA, Water quality standards handbook, second edition, EPA/823/B-94-005a, U.S. Environmental Protection Agency, Office of Water, Washington, D.C., 1994.
7. Adler, R.W., The two lost books in the water quality trilogy: The elusive objectives of physical and biological integrity, *Environ. Law*, 33, 29, 2003.
8. Karr, J.R., Bioassessment and non-point source pollution: An overview, Second National Symposium on Water Quality Assessment, Washington, D.C., U.S. Environmental Protection Agency, Office of Water, 4-1.
9. Lackey, R.T., Values, policy, and ecosystem health, *BioScience*, 51, 437, 2001.

10. USEPA, Guidance: Coordinating CSO long-term planning with water quality standards reviews, EPA/240/R-00/003, U.S. Environmental Protection Agency, Office of Water, Washington, D.C., 2001.
11. Borsuk, M.E., Stow, C.A., and Reckhow, K.H., Integrated approach to total maximum daily load development for Neuse river estuary using Bayesian probability network model (Neu-BERN), *J. Wat. Resour. Plan. Manage.*, 129, 271, 2003.
12. North Carolina Department of Environment and Natural Resources, Phase II of the total maximum daily load for total nitrogen to the Neuse River Estuary, North Carolina, NC Department of Environment and Natural Resources Division of Water Quality, Dec., 2001. Accessed May 13, 2004 at http://h2o.enr.state.nc.us/tmdl/Docs TMDL/ Neuse%20TN%20TMDL%20II.pdf.
13. Sarewitz, D.R., Pielke, R.A., and Byerly, R., *Prediction: Science, Decision Making, and the Future of Nature*, Island Press, Washington, D.C., 2000.
14. Sarewitz, D. and Byerly, R., Prediction in policy: A process, not a product, *Nat. Haz. Observer*, 23, 1, 1999.
15. Anderson, J.L., Hilborn, R.W., Lackey, R.T., and Ludwig, D., Watershed restoration: Adaptive decision making in the face of uncertainty, in *Strategies for Restoring River Ecosystems: Sources of Variability*, Wissmar, R.C. and Bisson, P.A., Eds., American Fisheries Societies, Bethesda, MD, 2003.
16. GAO, Water quality: Improved EPA guidance and support to help states develop water quality standards that better target cleanup efforts, GAO-03-308, General Accounting Office, Washington, D.C., 2003.
17. U.S. Congress, The need to update water quality standards to improve Clean Water Act programs, Committee on Transportation and Infrastructure, The Subcommittee on Water Resources and Environment, 108th Congress, 2nd Session, June 19, 2003.
18. USEPA, Guidance for 2004 assessment, listing and reporting requirements pursuant to sections 303(d) and 305(b) of the Clean Water Act, U.S. Environmental Protection Agency, 2003. Accessed Jan. 5, 2004 at http://www.epa.gov/owow/tmdl/tmd10103/index.html.
19. OEPA, 3745-1-07 Water use designations and statewide criteria, Ohio Environmental Protection Agency, Division of Surface Water, Dec. 30, 2002. Accessed Feb. 5, 2004 at http://www.epa.state.oh.us/dsw/rules/01-07.pdf.
20. USEPA, Strategy for water quality standards and criteria: Setting priorities to strengthen the foundation for protecting and restoring the nation's waters, EPA/823/R-03/010, U.S. Environmental Protection Agency, Washington, D.C., 2003.
21. Randall, A. and Peterson, G.R., Valuation of wild and resource benefits: An overview, in *The Valuation of Wildland Resource Benefits*, Randall, A. and Peterson, G.R., Eds., Westview Press, Boulder, CO, 1984.
22. Randall, A., Why benefits and costs matter, *Choices*, 14, 38, 1999.
23. USEPA, Guidelines for preparing economic analyses, EPA/240/R-00/003, U.S. Environmental Protection Agency, Washington, D.C., 2000.
24. Shabman, L., Stephenson, K., and Shobe, W., Trading programs for environmental management: Reflections on the air and water experiences, *Environ. Pract.*, 4, 153, 2001.
25. Houck, O.A., *The Clean Water Act TMDL Program: Law, Policy and Implementation*, Environmental Law Institute, Washington, D.C., 1999.
26. USEPA, Combined sewer overflow (CSO) control policy; notice, EPA/830/Z-94/001, NTIS PB95-156840, U.S. Environmental Protection Agency, Washington, D.C., 1994.
27. USEPA, Interim economic guidance for water quality standards: Workbook, EPA/823/B-95/002, U.S. Environmental Protection Agency, Washington, D.C., 1995.

28. Maguire, L.A., Interplay of science and stakeholder values in Neuse river TMDL process, *J. Wat. Resour. Plan. Manage.*, 129, 261, 2003. Excerpt reprinted with permission of the publisher, ASCE.
29. Borsuk, M.E., Stow, C.A., and Reckhow, K.H., Predicting the frequency of water quality standard violations: A probabilistic approach to TMDL development, *Environ. Sci. Technol.*, 36, 2109, 2002.
30. Reckhow, K.H., Assessment of the value of new information for adaptive TMDL, unpublished.
31. Water Resources Research Institute of the University of North Carolina, The Chlorophyll *a* standard: A primer, WRRI News, 238, Mar. 1, 2001. Accessed Jan. 30, 2004 at http://www2.ncsu.edu/ncsu/CIL/WRRI/news/328.html.
32. Reckhow, K. H., Shabman, L.A., Roessler, C., et al., Adaptive implementation modeling and monitoring for TMDL refinement. EPA Grant Number R830883, June 1, 2003 through May 31, 2006, National Center for Environmental Research, USEPA, Jan. 29, 2004. Accessed Jan. 29, 2004 at http://cfpub.epa.gov/ncer_abstracts/index.cfm/fuseaction/display.abstractDetail/abstract/6139/report/0.
33. Daniels, S.E. and Walker, G.B., *Working Through Environmental Conflict: The Collaborative Learning Approach*, Praeger, Westport, CT, 2001.
34. O'Leary, R., Bingham, L.B. Eds., *The Promise and Performance of Environmental Conflict Resolution*, Resources for the Future, Washington, D.C., 2003.
35. Webler, T. and Tuler, S., Integrating technical analysis with deliberation in regional watershed management planning: Applying the national research council approach, *Pol. Studies J.*, 27, 530, 1999.
36. GAO, Water quality: Key EPA and state decisions limited by inconsistent and incomplete data, RCED-00-54, General Accounting Office, Washington, D.C., 2000.
37. Smith, E., Ye, K., Hughes, C., and Shabman, L., Assessing violations of water quality standards under section 303(I) of the Clean Water Act, *Environ. Sci. Technol.*, 25, 606, 2001.
38. Shabman, L. and Smith, E., Implications of applying statistically based procedures for water quality assessment, *J. Wat. Resour. Plan. Manage.*, 129, 330, 2003.
39. USACE (U.S. Army Corps of Engineers) and South Florida Water Management District, The Journey to Restore America's Everglades, Home Page, Dec. 17, 2003. Accessed Jan. 6, 2004 at http://www.evergladesplan.org/.
40. USEPA, Status proposed watershed rule, U.S. Environmental Protection Agency, Mar. 19, 2003. Accessed May 13, 2004 at http://www.epa.gov/owow/TMDL/watershedrule/.
41. Dorfman, R., Jacoby, H.D., and Thomas, H.A., *Models for Managing Regional Water Quality*, Harvard University Press, Cambridge, MA, 1972.
42. Farrow, D.R. and Bower, B.T., Towards more integrated management of watersheds: Some past efforts, present attempts, and future possibilities, *Wat. Resour. Update*, 93, 13, 1993.
43. Houck, O.A., The clean water act TMDL program V: Aftershock and prelude, *Env. Law Rep. News Anal.*, 32, 10385, 2002.
44. Malone, L.A., The myths and truths that threaten the TMDL program, *Environ. Law Rep. News Anal.*, 32, 11133, 2002.
45. Chapra, S.C., Engineering water quality models and TMDLs, *J. Wat. Resour. Plan. Manage.*, 129, 247, 2003.
46. Griffiths, C., Multi-criteria cost effectiveness using a water quality index, USEPA National Center for Environmental Economics, Symposium on Cost-Effectiveness Analysis for Multiple Benefits, Sept. 22, 2003. Accessed Jan. 8, 2004 at http://yosemite.epa.gov/ee/epa/eed.nsf/webpages/CEAforMB.html.

47. Breyer, S., Economic reasoning and judicial review, AEI-Brookings Joint Center 2003 Distinguished Lecture, 2003. Accessed Jan. 5, 2004 at http://www.aei.org/news/newsID.19559.filter./news_detail.asp.
48. Shabman, L. and Stephenson, K., Environmental valuation and its economic critics, *J. Wat. Resour. Plan. Manage.*, 382, 2000.
49. Mann, J.Z., Economic infeasibility and EPA's 1994 CSO policy: A successful solution in Massachusetts still leaves a turbid understanding between state and federal officials, *Boston Coll. Env. Affairs Law Rev.*, 26, 857, 1999.

CHAPTER 8

Scaling Environmental Restoration to Offset Injury Using Habitat Equivalency Analysis

P. David Allen II, David J. Chapman, and Diana Lane

CONTENTS

1-56670-639-4/05/$0.00+$1.50

INTRODUCTION

Chapters 6 and 7 have discussed the ecological and economic analyses needed to guide attainment of the water quality goals of the Clean Water Act. In this chapter, we introduce the concept of scaling ecological restoration, not to achieve a prospective goal but to offset a past environmental harm, which may have continuing consequences. *Ecological restoration* is defined herein as the process of assisting the recovery of an ecosystem that has been degraded, damaged, or destroyed.[1] We provide an overview of the laws that make environmental restoration an important currency for offsetting environmental harm. We introduce the metrics that are used to measure environmental harm and restoration and the techniques that are used to compare the amount of restoration with the amount of environmental harm. Finally, we discuss, in detail, the most common scaling technique, *habitat equivalency analysis* (HEA).

Environmental Degradation Is Often Addressed via Restoration

In the United States, environmental protection is accomplished via a complex web of law, funding, and agency and private effort. The legal framework for environmental protection includes the common law principles of nuisance, trespass, toxic tort, negligence, public trust, and *parens patriae* (parent of the country), as well as numerous local, state, and federal statutes, regulations, and ordinances. Although the mechanisms for environmental protection are myriad and complex, most can be categorized into three broad themes: (1) development and dissemination of information, including monitoring, planning, reporting, research, and public participation; (2) prevention of harm, including permitting, standards, natural resource preservation, and land acquisition and management; and (3) restoration, including incident response, remediation, habitat improvement, land management, and recovery of compensatory damages for restoration. Even looking only at federal environmental statutes as a potentially representative subset of the available mechanisms to protect the environment, a dizzying array of environmental provisions has been enacted into law. Table 8.1 lists the common name and citation of most environmental federal statutes, along with each statute's principal purposes corresponding to the three themes listed earlier.

As can be seen in Table 8.1, restoration is an important goal throughout federal law. Furthermore, restoration is becoming an increasing priority for federal, state, and local governmental agencies, and for nongovernmental organizations and the general public. In part, this is due to public participation becoming almost routine for most environmental actions. For instance, Environmental Assessments and Environmental Impact Statements have become the norm for all major environmental actions by the federal government, pursuant to the National Environmental Policy Act; public agencies issue public comment periods on most environmental administrative actions; and third parties routinely litigate over both procedural and substantive issues under a wide variety of statutes. In addition, the most obvious and easiest opportunities to prevent harm have already been exploited. For instance, National Pollutant Discharge Elimination System permits and Construction Grants,

Table 8.1 U.S. Environmental Statutes at the Federal Level

Name	Citation	Primary Purpose[a]
Atomic Energy Act	42 U.S.C. §§2011-2286i, 2296a-2296h-13	Information, prevention
Bald and Golden Eagle Protection Act	16 U.S.C. §§668-668d	Protection
Clean Air Act	42 U.S.C. §§7401-7671q	Information, prevention
Coastal Barrier Resources Act	16 U.S.C. §§3501-3510	Prevention, **restoration**
Coastal Zone Management Act	16 U.S.C. §§1451-1465	Prevention, **restoration**
Comprehensive Environmental Response, Compensation, and Liability Act	42 U.S.C. §§9601-9675	**Restoration**
Emergency Planning and Community Right-To-Know Act	42 U.S.C. §§11001-11050	Information
Endangered Species Act	16 U.S.C. §§1531-1544	Information, prevention, **restoration**
Energy Reorganization Act	42 U.S.C. §§5801-5879	Information
Environmental Quality Improvement Act	42 U.S.C. §§4371-4375	Information
Federal Insecticide, Fungicide, and Rodenticide Act	7 U.S.C. §§136-136y	Information, prevention
Federal Land Policy and Management Act	43 U.S.C. §§1701-1785	Information, prevention, **restoration**
Federal Water Pollution Control Act (Clean Water Act)	33 U.S.C. §§1251-1387	Information, prevention, **restoration**
Food Security Act (Swampbuster)	16 U.S.C. §§3821-3822	Prevention, **restoration**
Forest Service Organic Legislation	16 U.S.C. §§473-482, 551	Information, prevention, **restoration**
Geothermal Steam Act	30 U.S.C. §§1001-1027	Information, prevention
Hazardous Materials Transportation Act	49 U.S.C. §§5101-5127	Prevention
Low-Level Radioactive Waste Policy Act	42 U.S.C. §§2021b-2021j	Information, prevention
Marine Mammal Protection Act	16 U.S.C. §§1361-1421h	Information, prevention, **restoration**
Marine Protection, Research, and Sanctuaries Act	16 U.S.C. §§1431-1447f	Information, prevention, **restoration**
Migratory Bird Treaty Act	16 U.S.C. §§703-712	Protection
Mineral Lands Leasing Act	30 U.S.C. §§181-287	Information, prevention, **restoration**
Mining and Minerals Policy Act	30 U.S.C. §21a	Prevention, **restoration**
Multiple-Use Sustained Yield Act	16 U.S.C. §§528-531	Information, prevention, **restoration**
National Environmental Policy Act	42 U.S.C. §§4321-4370d	Information
National Forest Management Act	16 U.S.C. §§1600-1687	Information, prevention, **restoration**
National Wildlife Refuge System Administration Act	16 U.S.C. §§668dd-668ee, 715s	Information, prevention, **Restoration**

(*Continued*)

Table 8.1 U.S. Environmental Statutes at the Federal Level (*Continued*)

Name	Citation	Primary purpose[a]
Nuclear Waste Policy Act	42 U.S.C. §§10101-10270	Information, prevention
Oil Pollution Act	33 U.S.C. §§2701-2761	Prevention, **restoration**
Outer Continental Shelf Lands Act	43 U.S.C. §§1331-1356	Information, prevention
Pollution Prevention Act	42 U.S.C §§13101-13109	Information, prevention
Public Rangelands Improvement Act	43 U.S.C. §§1901-1908	Information, prevention, **restoration**
Refuse Act	33 U.S.C. §407	Prevention
Resource Conservation and Recovery Act	42 U.S.C. §§6901-6992k	Prevention, **restoration**
Safe Drinking Water Act	42 U.S.C. §§300f-300j-26	Information, prevention
Submerged Lands Act	43 U.S.C. §§1301b, 1311, 1312	Prevention
Surface Mining Control and Reclamation Act	30 U.S.C. §§1201-1328	Information, prevention, **restoration**
Taylor Grazing Act	43 U.S.C. §§315-315o-1	Information, prevention, **restoration**
Toxic Substances Control Act	15 U.S.C. §§2601-2692	Information, prevention, **restoration**
Transportation Equity Act	Public Law No. 105-178	Information, prevention, **restoration**
Uranium Mill Tailings Radiation Control Act	42 U.S.C. §§7901-7942	**Restoration**
Water Resources Development Act	42 U.S.C. §1962	**Restoration**
Wild and Scenic Rivers Act	16 U.S.C. §§1271-1287	Information, prevention
Wild Free-Roaming Horses and Burros Act	16 U.S.C. §§1331-1340	Prevention, **restoration**
Wilderness Act	16 U.S.C. §§1131-1136	Prevention

[a] *Information* includes monitoring, planning, reporting, and research. *Prevention* includes permitting, standards, preservation, and land ownership. *Restoration* (in bold) includes response, remediation, and habitat improvement. Nonenvironmental purposes such as development are not listed.

pursuant to the Federal Water Pollution Control Act or Clean Water Act (CWA), have already controlled most point sources to surface water, and many of the most popular, significant, and available lands have already been incorporated into the National Park Service, the National Wildlife Refuge System, and state and local equivalents. However, many impacts persist, and restoration provides a potential currency for compensating and offsetting continuing environmental harm.

Techniques for Scaling Restoration to Balance Environmental Impacts

For environmental restoration to work as a currency for offsetting harm, rather than simply as a qualitative goal, techniques are required to (1) measure the amount of harm, (2) measure the amount of restoration, and (3) establish proportionality between measurements of harm and measurements of restoration. The various scaling techniques are essentially accounting methods to balance past and future restoration gains with past and future impact losses, using discount rates[a] if desired to convert the losses and gains to the same point in time. The techniques are deceptively simple in concept: add up all of the losses, determine the amount of gain per unit of restoration, and divide the total losses by the per-unit gains to determine the total amount of restoration required.

However, realistic measurement of restoration gain in the field can be quite difficult because of the many species and ecological services that can be positively or negatively affected by restoration at varying spatial and temporal scales. Therefore, various scaling techniques based on different kinds of metrics have been devised to simplify the comparison between environmental restoration and harm. Examples of metrics that can be used to compare restoration gains with impact losses include numbers, density, or biomass of organisms; acreage of habitat; amount of ecological services (see "Welfare Economics" in Chapter 5) provided by natural resources or habitats; and the value or cost of natural resources and services. Therefore, the various restoration-to-impact comparison techniques are often classified as resource-to-resource, service-to-service, or value-to-value scaling.

For instance, when the number of organisms lost and gained can be counted, a resource equivalency analysis (REA)[b] can be conducted using resource-to-resource scaling. For this technique to work, one must discern which organisms are lost to a particular impact and which are gained by a particular restoration. Here, the primary challenge is to differentiate the environmental impact losses and restoration gains of interest from population fluctuations caused by other factors such as immigration, emigration, competition, and other ecological constraints.

Similarly, if the acreage of habitat lost and gained can be measured, then an HEA[2,3] can be undertaken. However, most impacts do not completely eliminate habitat, and most restorations do not create completely new and functioning habitat. Therefore, common practice includes estimating the percent loss and the percent gain of (a) particular ecosystem service(s)[c] to allow service-to-service scaling. Here, ecological services are defined as the physical, chemical, and biological processes through which natural ecosystems support and sustain all life, including human life.[4,5] Measuring these services can also be quite difficult in the field because

[a] *Discount rate* refers to the rate at which dollars or other valued items or services being provided in different time periods are converted into current time period equivalents. Typically in the environmental scaling context, a social discount rate is used. The *social discount rate* is the rate at which society as a whole would be willing to trade off the consumption or use of goods and services in the current period until some future period. However, some economists and ecologists have noted the difficult issue of future generations being unable to negotiate with the current generation about how to allocate today's scarce resources.

[b] *Resource equivalency analysis* or *REA* is usually thought of as a special type of HEA, and many HEAs conducted by trustees would fit into the REA classification of this chapter. The NOAA regulations at 15 C.F.R. § 990.53(d)(2) describe resource-to-resource scaling, which would include REAs.

[c] Choosing which service or services to use in an HEA can be based on: (1) the ability of (a) service(s) to represent the habitat as a whole; (2) the particular importance of (a) service(s) because of legal protections, high public values, or particular agency interest; or (3) the service(s) lost and gained can be readily measured or estimated for the habitats being analyzed. Choosing the "right" service(s) for a habitat is not always straightforward, and different choices can lead to very different results in terms of the amount of needed restoration that is indicated by the HEA calculation.

ecosystem functions are typically quite complex and poorly understood. Therefore, calculating the total services provided by a habitat may be so imprecise that percent loss and gain estimates may be too variable to allow practical estimates of the amount of habitat actually needed.

Finally, value equivalency analysis (VEA), using value-to-value scaling, can be applied in a variety of situations that are not well-suited for REA and HEA; for example, (1) proposed restoration projects provide different natural resources, habitats, or services than those lost to impacts; (2) organism numbers, habitat acreage, or important services (as defined by ecosystem experts or the general public) cannot be measured accurately in the restorations or impacts;[a] or (3) important differences exist between impact losses and restoration gains that are not reflected in the metrics that can be conveniently used to determine scaling.[b] Chapter 14 presents an example of a VEA (there called a *Total Value Equivalency* study).

The value-to-value approach compares the utility or monetary value of the losses with the increase in utility or monetary value of restoration actions. This comparison uses *utils*[c] or dollars as the unit of measure. The service-to-service and resource-to-resource approaches are analogous to the value-to-value approach, except that the unit of measure is a specific service or resource rather than utils or money. In the next section, we focus on the service-to-service comparisons typically done in HEA. However, much of this discussion applies to habitat-to-habitat and resource-to-resource scaling as well. Under specific conditions, discussed under "Quantifying Ecosystem Services," the value-to-value approach to scaling reduces to the service-to-service approach and is thus equivalent.

No absolute conditions dictate when either the value-to-value or service-to-service approach is appropriate for scaling compensatory restoration projects. However, some guidelines[6] have been developed to help determine which approach might be preferred for any given situation. When evaluating the applicability of the service-to-service approach, the analysis should consider the following question: Are restoration projects available that will provide resources of a similar type, quality, and comparable value to those resources injured?

[a] For instance, in the Green Bay NRDA (see Chapter 14), known injuries included microscopic precancerous lesions in walleye livers, low but statistically significant increases in deformity rates and decreases in reproductive rates in several bird species, and advisories against consuming birds and fish. However, the trustees doubted that these injuries could be converted credibly into numbers of lost organisms or habitat-based service metrics. Instead, they designed surveys to elicit people's preferences between eliminating the known injuries (without measurement of or conversion to numbers of organisms lost or the percentage of habitat-based services lost) under differing timeframes versus implementing various scales of restoration programs.

[b] This can be particularly important when the only impacts that are easy to measure are of low value, and practical-but-expensive restoration is available to address the more valuable but harder-to-measure impacts. For instance, the aesthetic value to a coastline community of preventing construction of a confined disposal facility next to their beach would probably be much higher but harder to measure than, say, the cost of creating an equivalent (compared to the confined disposal facility) acreage of lake bottom, volume of lake water, or even amount of secondary productivity by excavating an inland pond nearby (i.e., the pond is not equivalent, even if its easily measured area, volume, and secondary productivity are).

[c] "Utils" are the unit of measure of Utility, in economic terms, a measure of satisfaction. Underlying most economic theory is the assumption that people do things because doing so gives them utility. Utility is typically translated into dollar terms when trying to compare the level of utility across individuals.

The term *comparable value* can have various interpretations depending on the prevailing paradigm of the analysis. To an economist, this term would most likely be interpreted to mean utils or monetary value — the per-unit dollar value of a habitat. To a biologist or ecologist, this term might be thought of as an ecological function, such as primary productivity. The exact interpretation of the term matters only in relation to the end goal of the restoration project. If the restoration project were undertaken to compensate people for losses, as in the case of natural resource damage claims, then monetary value or utils would likely be an appropriate interpretation. However, under other circumstances, the analysis may not be concerned with compensating people, but rather trying to ensure that the net quantity of a specific habitat is not reduced. Furthermore, the analyst may be concerned that conversion of losses and gains into monetary measures might "wash away" information about the injuries and restorations that are critical to public or agency understanding or acceptance. Under those conditions, ecological function might be the appropriate definition. The ultimate use of the analysis should guide the interpretation chosen.

Examples of Environmental Impacts and Restoration Opportunities Scaled under Federal Programs

Environmental impacts caused by human activities are so widespread and complex that many have not yet been properly described, let alone covered by environmental laws and programs. Nevertheless, many federal statutes address impacts by scaling restoration. One of the most common impacts is the release of hazardous substances and oil. The Comprehensive Environmental Response, Compensation, and Liability Act (CERCLA), the Oil Pollution Act (OPA), and the CWA provide for response to ongoing spills and releases, remediation of contamination in the environment, and compensation through restoration for public losses caused by releases. The last of these three provisions, compensation, is the basis for natural resource damage assessment (NRDA) programs and legal claims, and the heart of an NRDA is determining the amount of restoration needed to compensate for public losses caused by the release of hazardous substances and oil. Therefore, natural resource trustees[a] often use REA, HEA, and VEA techniques to scale restoration programs such as habitat preservation and restoration, species introductions and controls, elimination of non-point source runoff and other stressors, and enhancement of human recreational opportunities related to natural resources (see Chapter 14).

Another widely known federal program that scales restoration is wetland mitigation.[7] Section 404 of the CWA regulates dredging and filling in navigational waters. These provisions have led to programs to prevent the net loss of wetlands. Where wetlands are lost or impacted, HEA is often used to determine how many wetlands, and of what type and quality, must be restored as mitigation.

[a] CERCLA and OPA designate the federal government, states, and Indian tribes as natural resource trustees. The National Contingency Plan at 40 C.F.R. § 300.600 delegates federal trusteeship to federal departments, such as Interior and Commerce, authorizes Governors to delegate state agencies, and specifies that federally recognized tribes are trustees. Delegated agencies are the only plaintiffs that can bring natural resource damage claims on behalf of the public, pursuant to CERCLA, OPA, and the Federal Water Pollution Control Act.

In addition, many federal statutes have sufficient flexibility to create permitting and settlement opportunities that include scaled restoration. Permitting discretion, enforcement discretion, and the authority to compromise claims allow agencies (often departments of justice) to include environmental restoration actions as alternatives to traditional permit requirements, fines, or penalties. Such actions have been successfully pursued by the U.S. Environmental Protection Agency (USEPA) under the CWA and the U.S. Fish and Wildlife Service (USFWS) under the Migratory Bird Treaty Act, and creative settlements that restore the environment directly are becoming more popular throughout many permitting and enforcement programs. In these cases, the amount of restoration has been determined more often through negotiations and estimates of litigation risk than by balancing of losses and gains of natural resources, habitat, services, or values. Nevertheless, credible scaling techniques can help adversaries find a common ground that is more likely to be accepted by the courts, stakeholders, the scientific community, and the general public.

Finally, land management agencies can use restoration-scaling techniques to help determine the type and amount of action required to address various ongoing impacts to natural resources. This can be particularly important where restorative actions on agency lands (or on lands with conservation easements or other similar protections) are being used to compensate for impacts that occur beyond agency lands. Similarly, private landowners can sometimes use protection or management of their own lands to offset impacts for various actions permitted by environmental agencies. Conservation banking, for example, allows permitted "incidental takes" of endangered species or permitted habitat loss to be offset by purchasing credits in land banks that are protected and managed to benefit particular species or habitat types.[8] In these cases, scaling techniques can be used to determine what actions are sufficient and affordable, both for purposes of landowner planning and for convincing permitting authorities.

Methods for Restoration Scaling

Once the resource-to-resource, value-to-value, or service-to-service approach has been determined, the analyst is faced with the choice of which method to implement. If the value-to-value approach is taken, a number of methods are available to calculate the losses and gains in value for the scaling exercise. A thorough description of the methods available to implement the value-to-value approach is beyond the scope of this chapter. However, the short outline later in this chapter provides the context. Where the analysis is being conducted to determine the amount of compensation due to the public for a loss of natural resources, the methods available to determine such compensation are from *nonmarket economic valuation*. Nonmarket valuation focuses on measuring the value of goods and service that are not commonly traded in economic markets. Economists have developed, or adapted, a number of methods to measure the value of nonmarketed goods and services, including travel cost, hedonic analysis, contingent valuation method (CVM), and conjoint analysis (choice modeling; see Chapter 5, especially Table 5.3).

Travel cost, an approach developed in the 1940s, uses the cost the individual is willing to pay (endure) to travel to a recreational site as a proxy for a market price.

This method has been used to estimate the value of a multitude of recreational activities. Under conditions where the injury primarily causes a reduction in recreational activity, and the proposed restoration action entails the creation or improvement of recreational facilities or access, this method can be used to estimate both the value lost because of injury and the value gained because of the proposed restoration projects.

Hedonic analysis builds on the understanding that often the value for a good can be divided into component parts. A common example is the value of a house. Everything else held equal, a house with more square footage, or more bedrooms or bathrooms, would cost more than a smaller house, or one with fewer bedrooms or bathrooms. The hedonic method uses the fact that environmental quality is another factor influencing the value of homes. All else held equal, a home near a contaminated site will most likely cost less than one some distance away. When done appropriately, the difference in those values is the estimate of the loss in value due to the contamination. Under these conditions, the value of various environmental disamenities could be estimated, and the value of a restoration project that reduces one or a number of these conditions could be scaled to the reduction in value on home prices caused by the contaminated site.

Both the travel cost and hedonic methods use information about people's behavior (revealed preference) to estimate value. The contingent valuation method (CVM) and conjoint analysis are stated preference methods that value nonmarket goods and services (see Appendix 5-A). In the context of restoration scaling, CVM directly questions individuals about how they would value the prevention of a specific environmental injury such as that under consideration by the analysis, and alternatively how they would value the proposed restoration projects. Conjoint analysis or choice modeling is often thought of as an extension of CVM. Where a CVM study would often provide one alternative scenario to individuals to determine their value for the scenario, or at most a few scenarios, conjoint analysis provides individuals with a number of choices between alternative injury or restoration scenarios. Each of these four methods has been used to scale restoration projects in a value-to-value framework for NRDAs.

HEA is currently the only method available for the service-to-service approach. HEA is based on the idea that natural resources can be thought of as an asset that provides a flow of services. These services can be to humans or to other ecological resources.[a] Examples of services to humans include fishing, hunting, boating, hiking, bird watching, flood control, shoreline storm protection, and enjoyment of a healthy and functioning natural environment. Services to other ecological resources include habitat for food, shelter, and reproduction; organic carbon and nutrient transfer through the food web; energy transfer through the food web; biodiversity and maintenance of the gene pool; food web and community structure; prevention of the spread of exotic or disruptive species; and natural succession processes. REAs, HEAs, and VEAs can be combined in many ways, depending on the nuances of environmental impacts and practical restoration opportunities at any given site and

[a] Most authors define *ecosystem services* as services provided to humans. NOAA's[6] distinction of services provided to humans and those provided to other ecological "resources" is not inconsistent since the term *resources* implies a value to humans, even if that value is passive or indirect.

the context of applicable law. Chapter 15 presents a habitat-based replacement cost (HRC) analysis that combines HEA and REA to evaluate the cost of environmental restoration sufficient to offset losses of aquatic organisms to impingement and entrainment of aquatic organisms at power plant facilities, pursuant to Section 316(b) of the CWA. Chapter 14 presents a total value equivalency (TVE) study, which is a VEA, that was used to compare the value to the general public of different cleanup scenarios and other restoration programs, and their willingness to pay for these scenarios and programs, pursuant to the NRDA provisions of CERCLA. Of all of the scaling techniques, though, HEA is probably the most popular, and its popularity among natural resource and environmental protection agencies seems to be increasing. Therefore, the rest of this chapter focuses on restoration scaling using HEA.

INTRODUCTION TO HABITAT EQUIVALENCY ANALYSIS

Background

HEA is a method for determining the amount of restoration needed to compensate for losses of natural resources, typically resulting from oil spills, hazardous substance releases, or physical injuries such as vessel groundings. Restoration is scaled so that the human or ecological service gains provided at compensation sites equal the cumulative service losses at the injured site, where ecological services are defined as the physical, chemical, or biological functions that one natural resource provides for another.[2] Thus, HEA is used to determine the amount of restoration that is required to compensate for past, current, and future injuries.

The HEA model is used to calculate the service losses from past, ongoing, and future injuries (the debit side of the model) and the service gains in the future from proposed restoration (the credit side of the model). Restoration is usually a separate activity from remedial actions to be performed at the site. Even if remedial activities restore services at the site to 100% of baseline levels (i.e., conditions absent the impact), restoration is still required to compensate for past and interim losses. The HEA model incorporates a discount rate (typically 3%) to account for the fact that ecological services gained from restoration conducted in the future are less valuable to the public than ecological services available today. Figure 8.1 presents a graphical representation of the debit calculations from an HEA model, and Figure 8.2 presents a representation of the credit calculations from an HEA model.

Advantages of HEA

One of the key benefits of HEA is that it allows trustees (or regulators) and potentially responsible parties (or the regulated community) to bypass the evaluation of economic damages resulting from natural resource injuries and to proceed directly to scaling and planning restoration. Using HEA can reduce the cost and time required for an assessment and can focus efforts directly on restoration. In practice, HEA calculations are often used as a tool in settlement discussions. HEA calculations

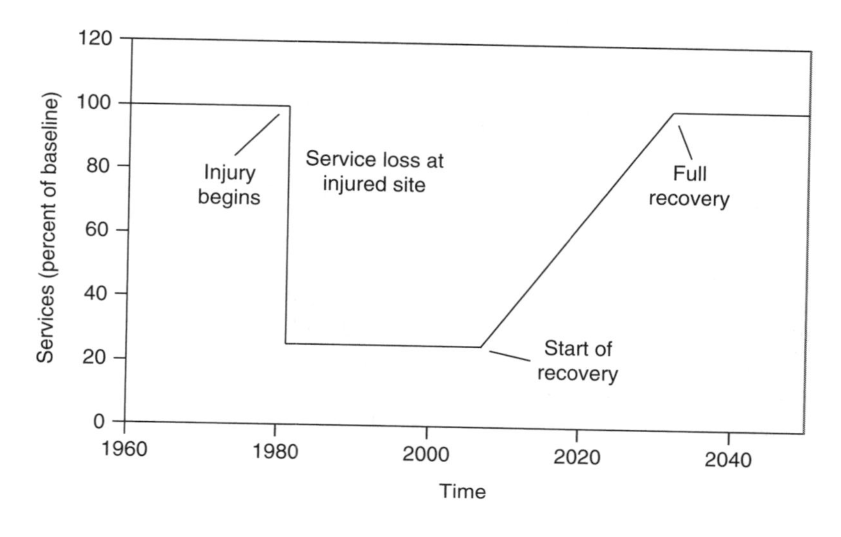

Figure 8.1 Total service loss at a hypothetical injured site, from an HEA model. The effect of a nonzero discount rate on service flows over time is not shown.

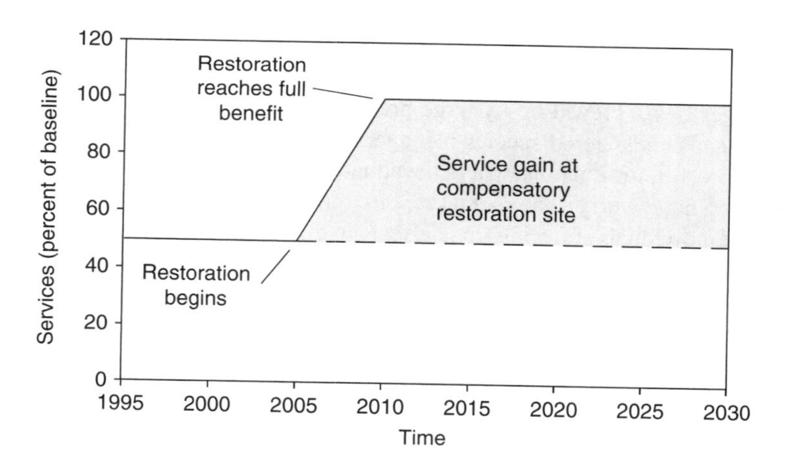

Figure 8.2 Total service gain at a hypothetical restored site, from an HEA model. The effect of a nonzero discount rate on service flows over time is not shown.

also have been used as the basis for compensation claims in litigated cases involving physical destruction of sea grass by grounded vessels.[a,b]

Another benefit of HEA is that it explicitly creates a connection between units of services lost because of injury and units of services gained through restoration. The connection provides a clear demonstration to the public that the trustees have fulfilled their mandate of compensating the public for losses of natural resources

[a] United States v. Great Lakes Dredge and Dock Co., 259 F.3d 1300.
[b] United States v. Melvin A. Fisher et al., 174 F.3d 201.

and their services. The implicit assumption of HEA is that the public can be compensated with direct service-to-service scaling when the services provided by proposed restoration actions are of similar type, quality, and value as the services lost due to injury.[2]

Limitations of HEA

HEA is considered to be an appropriate tool for scaling compensatory restoration only when (1) a common metric can be defined for natural resource services that captures the level of services provided by injury and replacement habitats, (2) the landscape context of the injury and replacement habitats provides similar opportunity to supply the relevant ecological services, and (3) sufficient data on HEA input parameters exist or are cost-effective to collect.[2] When these conditions are not fulfilled, the HEA process is unlikely to result in the "right" amount of restoration. As with all models, a lack of appropriate inputs necessarily limits the validity of the outputs.

HEA would be difficult to apply in a situation where a change in services might have a very positive effect on the ecological function of a habitat or ecosystem, but a neutral or negative effect in how humans value those services. For example, constructing artificial island habitat may increase the population of colonial nesting birds. However, such increases could be viewed simultaneously as a success and a failure by different groups. Perhaps fish and wildlife managers hope that increased numbers of herons and cormorants will help stabilize populations of small fish that are no longer being preyed on by large predatory fish wiped out by local contamination. Perhaps endangered species managers hope to reestablish extirpated species of gulls. However, in situations where the aquatic food chain is contaminated, NRDA practitioners may worry that increased bird populations could lead to increased numbers of individuals suffering toxic effects (e.g., cancer and endocrine disruption) and to consumption advisories for waterfowl hunters. Fishery scientists may disagree that birds are an adequate substitute for predatory fish in the aquatic food web. Furthermore, local fishermen may view cormorants and herons as competitors, and local residents may view gulls as a nuisance. Reaching a consensus between these groups on what ecosystem services should be used to balance restoration gains with impact losses could be quite difficult, yet it would be critical for determining both the type and scale of proposed restoration actions.

Selecting a Metric

Selecting appropriate assumptions for use in the HEA model is a key element of the process. HEA has no objective standard for determining which metric (or metrics) should be used to estimate service losses at injured sites and replaced services at restoration sites. Metrics used for capturing service losses are often based on ecological attributes of a site that are assumed to correlate with services: standing above-ground biomass in a grassland could be used as a surrogate for net primary productivity; abundance of keystone species could be used as a surrogate for ecosystem health. For example, at the Blackbird Mine site in east-central Idaho, numbers of naturally spawning spring and summer chinook salmon were used as the metric for service loss from

Panther Creek. This species was chosen because of its economic importance and its role in providing a nutrient base for the stream.[9] The service gains calculated for proposed restoration projects also were based on the degree of recovery of salmon.

For the Pecos Mine, the New Mexico Office of Natural Resources Trustee used a series of decision rules based on data collected as part of a risk assessment to estimate service losses to habitats injured by metals released from an abandoned mine.[10] Service losses to aquatic habitat were estimated based on the ability of the Pecos River to provide suitable habitat for aquatic biota. The frequency and extent of exceedences of aquatic life criteria and evidence of adverse effects on fish and aquatic invertebrate communities, such as fish kills and population shifts or reductions, were used to categorize service loss within reaches of the river.

For the North Cape oil spill off of Rhode Island, NOAA developed HEAs using metrics for American lobsters,[11] bivalves (focusing on surf clams),[12] endangered piping plovers,[13] a variety of seabirds,[14] and a variety of organisms in a representative food chain.[15] For these HEAs, scaling was based on lobster egg or juvenile equivalents in a demographic population model (scaled to purchase and re-release of lobsters), bivalve productivity (scaled to hatchery stocking), oiling of piping plover nesting areas and resultant reductions in productivity (scaled to protection of additional breeding sites), direct mortality and reduced fledging productivity of seabirds (scaled to protection and management of nesting sites for loons and eiders), and trophically equivalent production based on a food chain transfer model (scaled to eelgrass bed protection).

HEA in NRDAs and in the Courts

The early development and first applications of HEA were in NRDAs. One of the first applications of HEA was an assessment of lost wetland services resulting from the 1990 *M/T BT Nautilus* oil spill in the Kill Van Kull of New York/New Jersey Harbor.[a] The earliest application in a Superfund setting was the contamination at the Great Swamp National Wildlife Refuge.[3] The initial Superfund application was developed not as a substitute for a full damage assessment but rather as a method to develop a damages claim in a timely and cost-effective manner, due to pending bankruptcy proceedings. The early applications were innovative developments in the NRDA arena in two main ways: they extended and formalized the conceptualization by Freeman[16] of the environment as an asset that provides a flow of services, and they focused the measure of damages as the scale of restoration projects necessary to compensate for interim lost uses. CERCLA requires that recovered damages be used to "… restore, replace, or acquire the equivalent of such natural resources …"[b] However, until the development of HEA in these early cases, the trustees had not based their measure of damages on the cost of conducting restoration projects scaled to the injury. Previously, interim lost use damages had been developed using non-market valuation techniques to estimate the interim lost use value. That monetary value was then collected and used to undertake restoration.

[a] Consent Decree in the Matter of the Complaint of Nautilus Motor Tanker Co. Ltd., Owner of the *M/T B.T. Nautilus* for Exoneration from or Limitation of Liability. 1994. Civ. No. 90 CV2419 (D.N.J. April 8).
[b] 42 U.S.C. §9607(f)(1).

In the early 1990s, HEA was applied to a number of sea grass injury claims in the Florida Keys National Marine Sanctuary. Two of these assessments[a,b] were challenged in court on the applicability of HEA as a reliable method to be used in NRDA. In these cases, the defendants challenged the overall validity of the HEA method, the detailed data inputs underlying the specific analysis, and the underlying data on admissibility grounds under the U.S. Supreme Court's Daubert standards[c] of testability, known error rate, general acceptance in the profession, and peer review of the method.

In each of these areas, the court's rulings refuted the defendants' claims and supported the admissibility of HEA as a reliable method to assess natural resource damages. Regarding testing, the court found that HEA is not a scientific principle subject to testing in the conventional sense, but rather a mathematical equation subject to the limitations of the input data. On the contention of a known error rate, the court determined that this argument is not applicable to HEA, since it is just a mathematical equation. The issue of error rate may apply to the underlying data, but not to HEA itself. Although the specific analysis challenged in the court was not published in peer-reviewed journals at the time, the court recognized that the specific analysis had undergone peer review by the trustees. On the final issue of general acceptability, the court ruled that although HEA had not had time to truly gain general acceptance, the relative "youth" of a scientific technique does not make it any less valid. The conclusions from both court rulings strongly supported the admissibility of HEA as an appropriate method to determine compensatory restoration project scale when the primary category of lost on-site services pertains to the ecological and biological function of an area; when feasible restoration projects are available that provide services of the same type, quality, and comparable value to those that were lost; and when sufficient data to perform the HEA are available or cost effective to collect.

Subsequently, HEA has been used to scale compensatory restoration for dozens of NRDAs across the nation in such diverse habitats as Florida coral reefs, salmon habitat in the Northwest, and estuarine wetlands in south Texas. REA is a recent evolution of HEA where the basic unit of measure is individuals of a given species, such as chinook salmon or marbled murrelets, rather than habitats or their services. The underlying concepts or types of data necessary for REA are completely analogous to HEA.

HOW THE HEA TECHNIQUE WORKS

Quantifying Ecosystem Services

Conducting a specific HEA involves three steps: document and quantify lost services, identify and evaluate replacement project options, and scale the replacement project to compensate for the lost services over time. As discussed previously, scaling is the process by which the size of a specific restoration project is compared to the

[a] United States v. Great Lakes Dredge and Dock Co., 259 F.3d 1300.
[b] United States v. Melvin A. Fisher et al., 174 F.3d 201.
[c] Daubert v. Merrell Dow Pharmaceuticals, Inc., 509 U.S. 579.

amount of environmental harm, and a determination is made as to whether the restoration project provides sufficient compensation to offset the losses. Equation 8.1 formalizes the concept of value-to-value scaling of restoration actions to compensate for losses due to injury or destruction of resources.[a]

$$\left[\sum_{t=0}^{B} V^j * \rho_t * \left\{\left(b^j - x_t^j\right) / b^j\right\}\right] * J = \left[\sum_{t=1}^{L} V^p * \rho_t * \left\{\left(x_t^p - b^p\right) / b^j\right\}\right] * P \quad (8.1)$$

where t refers to time (in years):

- $t = 0$; the injury occurs
- $t = B$; the injured habitat recovers to baseline
- $t = C$; time the claim is presented
- $t = I$; habitat replacement project begins to provide services
- $t = L$; habitat replacement project stops yielding services

and where:

- V^j is the annualized per unit value of the services provided by the injured habitat (without injury)
- V^p is the annualized per unit value of the services provided by the replacement habitat
- x_t^j is the level of services per acre provided by the injured habitat at the end of year t
- b^j is the baseline (without injury) level of services per acre of the injured habitat
- x_t^p is the level of services per acre provided by the replacement habitat at the end of year t
- b^p is the initial level of services per acre of the replacement habitat
- ρ_t is the discount factor, where $\rho_t = 1/(1 + r)^{t-C}$, and r is the discount rate for the time period
- J is the number of injured acres
- P is the size in acres of the replacement project that equates the losses with the gains from restoration

This formulation recognizes two important points about the comparison of the types of services lost through the injury and those gained from the restoration actions: service losses and gains may occur over different time periods, and the value of those services may vary. Balancing the losses and gains in terms of their present discounted value requires that both potential differences be addressed. It is often the case that the injury causes a loss in services closer to the period of analysis than restoration, or past losses may need to be compared to future restoration. It is common to adjust for these differences in time periods using discounting. Accounting for the relative value of the services affected by the injury and restoration is a more complex matter.

[a] For consistency with the literature, this formulation follows that presented in NOAA's Overview of Habitat Equivalency Analysis with slight modifications.

Under certain restrictive conditions, where $V^j = V^p$, the value-to-value approach can be reduced to the service-to-service equivalence. The assumption that injured and restored resources are equal in terms of per-unit monetary and ecological values is not often obviously correct. If the per-unit monetary values are equal, then the value-to-value equation reduces to the service-to-service equation. Under those conditions, one can then compare the ecological values of the services lost and gained. To better understand where the equal ecological services assumption may hold, King[17] developed a series of evaluations — capacity, opportunity, payoff, and equity (C.O.P.E.) — as a way to structure the comparison between the ecological services lost due to injury and those gained from restoration.

Capacity relates to the biophysical characteristics of the specific habitat, and it evaluates whether the created habitat has the biophysical capacity to provide the ecological function provided by the lost habitat. Opportunity evaluates the ecological landscape of the habitat, and whether the habitat is provided in a similar landscape context such that the function capacity can be realized. Payoff relates to the ability of the given habitat with capacity and opportunity to actually realize the ecological function required. Finally, equity compares the habitat in terms of societal trade-off. Even though a given habitat may have the capacity, opportunity, and payoff to provide the necessary ecological function, if it is provided in a way such that there is a societal inequity, then the habitat in question may not be of comparable value.

Under conditions where the per-unit monetary value meets the necessary equality requirements and an evaluation of C.O.P.E. is sufficiently satisfied, then one could determine that the restoration habitat is of equal value, and HEA is an appropriate method for implementing the service-to-service approach. Unfortunately, circumstances that satisfy all of these various requirements are relatively uncommon,[18] and analysts must resist the temptation to apply HEA, perhaps because of its ease of use, where the requirements are not met.

Quantifying Injury

Quantifying lost services focuses on measurement of the physical area of injured habitat and an evaluation of the severity of the injury (percent service loss per area), as determined using the chosen metric. While there is fundamentally no reason why multiple measures of the injured habitat cannot be evaluated, for use in HEA those multiple measures are typically aggregated into a single measure that relates to the overall percent service loss due to the injury. Once a metric or aggregate metric is developed, it becomes the basis for estimating b^j and x_t^j on the injury side and b^p and x_t^p on the restoration side.

The period of time for which services are below baseline can vary dramatically. Injury of a lightly oiled, high-energy sandy beach may persist for months or seasons. Tropical sea grasses may take years to recover from physical destruction, whereas coral may take decades or longer to fully recover from destruction. In some cases, the habitat may never fully recover the level of functionality that would have existed absent the injury. When determining the duration of injury, the analyst must incorporate the effects of any primary (on-site) restoration actions undertaken to limit injury. Primary restoration actions may return injured habitats to baseline sooner or

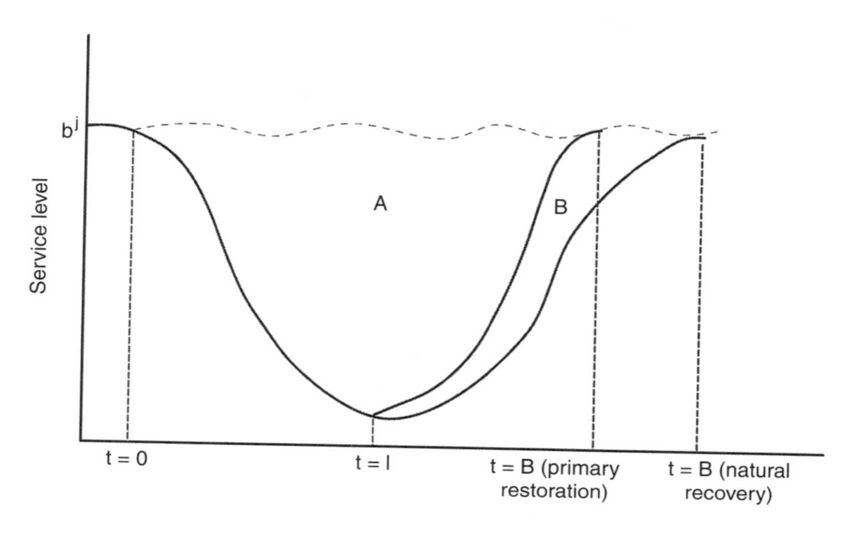

Figure 8.3 Graphical representation of injury quantification. Area A+B represents the total services lost if no action is taken and recovery occurs naturally. Area A represents the total services lost if on-site restoration partially fixes the injury. Variables are as defined in Equation 8.1.

allow for recovery to higher ecological function than would be achieved if no primary restoration were undertaken. Finally, the analyst must determine the expected path of recovery. Figure 8.3 is a graphical representation of the injury quantification.

The two main sources of data available to quantify injury are existing literature and original studies. For many habitats, there is an extensive literature that can be tapped to develop the necessary inputs. For those cases where no literature currently exists or the range of uncertainty is larger than the analyst considers appropriate, original studies would need to be conducted. Original studies should be designed to ensure that quantitative measures of the chosen metric can be developed.

Quantifying Restoration

Quantifying the benefits of specific restoration projects for scaling requires developing a similar set of information as for injury quantification. The analyst must identify and evaluate habitat restoration options to ensure that they provide services of comparable value to those that were lost. The relative productivity of the restoration actions must be evaluated in terms of the chosen metric and relative to the baseline services of the injured resources. The period of time required to plan and implement the restoration projects must be determined. Following implementation, the analyst must determine how long the restoration project will take to reach its maximum function. Finally, the analyst must determine the ultimate functionality of the restored habitat in terms of the baseline conditions of the injured area.

The concluding step in HEA is determining the size of the restoration project so that the total discounted flow of services provided by the newly restored habitat is equal to the discounted flow of services lost due to the injury. To accomplish this, the analyst must choose an appropriate discount rate. Discounting enables comparison of two flows of services that occur in different time periods. When the HEA is developed to address lost public resources over time, it is correct to apply the *social discount rate*. This social discount rate is the rate at which society as a whole would be willing to trade off services across different time periods. The literature regarding the applicability of different discount rates is extensive and beyond the scope of this chapter.

Scaling Restoration to Match Injury

The final step in implementing HEA is determining the scale, *P*, of the restoration project(s) so that, over time, the discounted flow of services is equal for restoration and injury. In the HEA framework, this would involve identifying the area necessary to generate the appropriate services. In the REA framework, this would involve determining the total quantity of resources (birds, fish, etc.) necessary to generate the service flow. In practice, the scale of the necessary restoration projects is determined by first identifying the per-unit net present value of the restoration project, and then dividing the total discounted lost services by this amount.

Equation 8.2 shows the basic HEA formula for the case $V^j = V^p$, rearranged to solve for the quantity of services, *P*, it is necessary to create to just offset the lost services, *J*:

$$P = J * \frac{\left[\sum_{t=0}^{B} \rho_t * \left(b^j - x_t^j\right) / b^j\right]}{\left[\sum_{t=I}^{L} \rho_t * \left(x_t^p - b^p\right) / b^j\right]} \tag{8.2}$$

The numerator on the right-hand side is the sum of the annual discounted percent service loss due to the injury. The denominator is the sum of the discounted percent service gain, in terms of the baseline level of services of the injured habitat. *J* is the total area of the injured site. Solving the equation, we find *P*, the area of habitat necessary to be created or improved such that the discounted quantity of services gained just equals the discounted quantity of services lost. Normalizing the percent service loss and gains in terms of b^j, the baseline (uninjured) level of services at the injury site, ensures that the relative percent service changes will be equal. An important note on HEA is that, as recognized in the court rulings, it is fundamentally a simple mathematical equation and not a "model" in the sense of a groundwater transport or air circulation model. No unique set of parameters is used across all applications.

DISCUSSION

In terms of integrating ecological risk assessment and economics benefits analysis, HEA has an important role. American law recognizes restoration as an important goal, and environmental practice increasingly turns to restoration to offset or compensate the public for harm that cannot be easily prevented. Transforming the qualitative restoration goal into a quantitative currency to offset environmental impacts requires scaling methods. Therefore, methods have been developed to compare restoration with impacts by measuring resource numbers, habitats, services, and values. HEA deals with adverse environmental impacts by scaling the services provided by habitats that have been injured and restored.

HEAs are increasing in popularity and use because they are easy to understand, and under certain conditions, they do not require site-specific studies or extensive involvement by economists, so scaling can be accomplished through straightforward trades of habitats and services by whatever staff are available. The inputs for an HEA, however, particularly estimates of the percentages of total services lost or gained, require a thoughtful consideration of whether total services can be estimated or whether certain key services can be singled out as most important. For any particular site and circumstances, this consideration requires an understanding of the legal context for restoration, the ecological functions and services provided by injured and restored habitats, and the human interests in those services, whether by ecologists, other stakeholders, or the general public.

Therefore, determining whether to use HEA rather than REA or VEA requires an integrated understanding of ecological impacts and risks as well as economic analysis. Furthermore, choosing a subset of ecological services for use in a particular HEA entails an inherently value-laden judgment that may require insights into the ecological functions and services of greatest concern for ecological management as well as an understanding of the desires of key stakeholders, legislative intent, or the general public. Finally, where restoration gains and injury losses are not conducive to direct comparison as measured by services or resource quantity, economic analyses of the values being gained and lost may be necessary.

In conclusion, HEAs provide an important opportunity for integration of information from ecology and risk assessment on the one hand and economic analysis on the other. In some cases, such integration can lead to straightforward scaling of obvious habitats and services. In other cases, detailed analysis of ecological functions, services, and values may be required to ensure that HEA (or REA or VEA) outputs are relevant and valid.

REFERENCES

1. Society for Ecological Restoration and Science and Policy Working Group, The SER primer on ecological restoration, Society for Ecological Restoration, 2002. Accessed Mar. 4, 2004 at http://www.ser.org.
2. NOAA (National Oceanic and Atmospheric Agency), Habitat equivalency analysis: An overview, Prepared by the Damage Assessment and Restoration Program, Silver Spring, MD, Oct. 4, 2000.
3. Unsworth, R.E. and Bishop, R.C., Assessing natural resource damages using environmental annuities, *Ecol. Econ.*, 11, 35, 1994.
4. Daily, G.C., Ed., *Nature's Services: Societal Dependence on Natural Ecosystems*, Island Press, Washington, D.C., 1997.
5. Daily, G.C., Alexander, P.R., Ehrlich, P.R., et al., Ecosystem services: Benefits supplied to human societies by natural ecosystems, *Issues Ecol.*, 2, 1997.
6. NOAA (National Oceanic and Atmospheric Agency), Scaling compensatory restoration action: Guidance document for natural resource damage assessment under the Oil Pollution Act of 1990, National Oceanographic and Atmospheric Agency, 1997. Accessed June 1, 2004 at http://www.darp.noaa.gov/pdf/scaling/pdf.

7. USACE (U.S. Army Corps of Engineers), USEPA (U.S. Environmental Protection Agency), NRCS (Natural Resources Conservation Service), USFWS (U.S. Fish and Wildlife Service), and NOAA (National Oceanic and Atmospheric Agency), Federal guidance for the establishment, use and operation of mitigation banks, *Fed. Re.g.,* 60, 58605, 1995.
8. USFWS (U.S. Fish and Wildlife Service), Guidance for the establishment, use, and operation of conservation banks. Memorandum to regional Directors, Regions 1–7, and to the Manager, California Nevada operations, U.S. Fish and Wildlife Service, 2003. Accessed June 1, 2004 at http://endangered.fws.gov/policies/conservation-banking.pdf.
9. Chapman, D., Iadanza, N., and Penn, T., Calculating resource compensation: An application of the service-to-service approach to the Blackbird Mine, hazardous waste site, Technical Paper 97-1, National Oceanic and Atmospheric Administration, Damage Asssessment and Restoration Program, Oct. 16, 1998.
10. Hagler Bailly Services, Pecos Mine operable unit: Natural resource damage assessment, Prepared for New Mexico Office of Natural Resources Trustee and the New Mexico Environment Department by Hagler Bailly Services, Inc., Boulder, CO, Mar. 5, 1998.
11. McCay, D.P.F., Gibson, M., and Cobb, J.S., Scaling restoration of American lobsters: Combined demographic and discounting model for an exploited species, *Mar. Ecol. Prog. Series*, 264, 177, 2003.
12. McCay, D.P.F., Peterson, C.H., DeAlteris, J.T., and Catena, J., Restoration that targets function as opposed to structure: Replacing lost bivalve production and filtration, *Mar. Ecol. Prog. Ser.*, 264, 197, 2003.
13. Donlan, M., Sperduto, M., and Hebert, C., Compensatory mitigation for injury to a threatened or endangered species: Scaling piping plover restoration, *Mar. Ecol. Prog. Ser.*, 264, 213, 2003.
14. Sperduto, M.B., Powers, S.P., and Donlan, M., Scaling restoration to achieve quantitative enhancement of loon, seaduck, and other seabird populations, *Mar. Ecol. Prog. Ser.*, 264, 221, 2003.
15. McCay, D.P.F. and Rowe, J.J., Habitat restoration as mitigation for lost production at multiple trophic levels, *Mar. Ecol. Prog. Ser.*, 264, 233, 2003.
16. Freeman, A.M., *The Measurements of Environmental and Resource Values: Theory and Methods*, Johns Hopkins University Press, Baltimore, MD, 2003.
17. King, D.M., Using ecosystem assessment methods in natural resource damage assessment, National Oceanic and Atmospheric Administration, Damage Assessment and Restoration Program, Jan. 21, 1997.
18. Dunford, T.C.G. and Desvousges, W.H., The use of habitat equivalency analysis in natural resource damage assessments, *Ecol. Econ.*, 48, 49, 2004.

CHAPTER 9

Conceptual Approach for Integrated Watershed Management

Randall J.F. Bruins, Matthew T. Heberling, and Tara A. Maddock[a]

CONTENTS

[a] The views expressed in this chapter are those of the authors and do not necessarily reflect the views or policies of the U.S. Environmental Protection Agency.

1-56670-639-4/05/$0.00+$1.50

In Chapter 6 a rationale was presented for integrating ecological risk assessment (ERA) and economic analysis in watershed management: (a) that both risks and actions to reduce risk have an economic dimension because they invoke preferences and tradeoffs; (b) that technical information about risks, as is provided by ERA, is necessary for the formation of informed preferences; and (c) that the compartmentalization of disciplinary efforts leads to a poorer quality of analysis. It was recommended that whenever both ERA and economic analysis are needed to address a watershed management problem, they should be undertaken in an integrated fashion, which means that they should be mutually informed and fully coordinated. The next chapters discussed integration approaches specific to the development of water quality standards (Chapter 7) and the determination of compensation for ecological injury (Chapter 8). The goal of this chapter is to develop a more generalized, conceptual approach for achieving ERA–economic integration in a watershed management context. This chapter first examines existing frameworks that have been used for watershed management, then considers some guiding principles, and finally presents a new conceptual approach that incorporates ERA into a well-integrated management process.

EXISTING FRAMEWORKS FOR WATERSHED MANAGEMENT

Various frameworks, emanating from the fields of risk assessment, environmental monitoring, project planning, environmental regulation, and natural resource management, have been applied to watershed management processes, but none has addressed specifically the ERA–economic integration problem. Review of these frameworks reveals several characteristics by which they differ, which will be seen later to have bearing on the integration problem. The first of these is comprehensiveness with respect to the management process. Some frameworks address only *monitoring* or *assessment*, stopping short of decisions, whereas others are for *planning and management* as a whole, including decisions (and often, implementation, evaluation, and adaptation). The second has to do with the intended use. Some can be termed *situational*, or responding to the advent of a problem or opportunity; others are for ongoing management and may be termed *regular*. The third characteristic is disciplinary breadth. Some frameworks are focused within the *natural sciences*, whereas others emphasize both the *natural and social sciences*. The final characteristic is the degree to which the process is open to stakeholders, ranging from no explicit role to a role that entails negotiation rights. These four characteristics have been used to create an illustrative typology of some existing frameworks (Table 9.1). A discussion of each of these frameworks, in relation to the typology, is presented in Appendix 9-A.

GUIDING CONSIDERATIONS FOR AN INTEGRATED MANAGEMENT PROCESS

Given the existing frameworks, what considerations should guide the design (via borrowing and adaptation) of an approach for ERA–economic integration? According to the U.S. Environmental Protection Agency's (USEPA's) Science Advisory Board,[1] the processes used should have the following characteristics: they should be transparent

Table 9.1 Typology of Frameworks that have been Applied to the Processes of Watershed Assessment and Management[a]

	Monitoring and Assessment	Planning and Management
Situational: For project design or problem response	EMAP (Environmental Monitoring and Assessment Program) indicator design[34]**[0]**	Society for Environmental Toxicology and Chemistry's ecological risk management framework[37]**[1]**
	DPSIR (Driving forces, Pressures, State, Impacts, Response) indicator framework design[35]**[0]**	Framework for environmental health risk management[17]**[2]**
	Guidelines for ecological risk assessment[12]**[1]**	*U.S. Army Corps of Engineers project planning*[22]**[2]**
	Framework for the economic assessment of ecological benefits[36]**[1]**	*World Commission on Dams planning and project development framework*[38]**[3]**
		USEPA's watershed project guidance[39]**[3]**
Regular: For ongoing management of watershed resources	Monitoring program with cyclical redesign[40]**[0]**	Clean Water Act watershed management cycle[23]**[2]**
		2000 U.S. Forest Service land and resource management planning framework[41,42]**[2]**

Bold, bracketed numbers indicate degree of stakeholder integration in the process.[b]

Italics indicate an emphasis on the integration of natural and social sciences.[c]

[a]See Appendix 9-A for description of cited frameworks.

[b]Bold, bracketed numbers are further explained as follows:

[0] — No explicit stakeholder role; process may be amenable to stakeholder involvement, but such involvement is not described.

[1] — Stakeholder-informed process; stakeholder involvement occurs primarily at the outset, as part of goal-setting.

[2] — Stakeholder-engaging process; stakeholder involvement is sought throughout the process.

[3] — Stakeholder-empowering process; process occurs at the initiative of stakeholders themselves, or framework deals explicitly with issues of "power" and assigns specific rights to stakeholders.

[c]Integration of social sciences denotes the use of scientific methodologies, not stakeholder inclusion alone. It includes economics and the decision sciences.

(clearly understandable) to all parties; flexibly applied; dynamic (interconnected and iterative); open and cooperative; informed by many different sources and disciplines; and they should reflect holistic, systems thinking. Bellamy et al.[2] comment on the tendency for natural resource management efforts to fail to develop clear goals, achieve an integrated perspective, match actions to objectives, and evaluate outcomes. They develop a broad set of criteria for evaluating efforts that have been implemented. These criteria are useful prospectively as well and are presented here as relevant to the development of an integrated process. They state that an effective process

> (a) addresses evaluation from a systems perspective, (b) links objective to consequence, (c) considers the fundamental assumptions and hypotheses that underpin core

Table 9.2 Important Considerations in Framework Design, and Resulting Design Elements

Consideration	Specific Points	Framework Design Element
Unique value of ERA	• Ecologically informed, biophysical in nature • Structured, deductive process proceeding from goal —> objectives —> hypotheses —> analyses —> risk characterization	• Scientific character of analysis is not compromised • Retain core of ERA process
Sensitivity to critiques of ERA	• Some stakeholders may perceive overall process as unfair • Assess broad range of alternatives (not constrained set) • Acknowledge limits of science throughout assessment	• Acknowledge potential "winners" and "losers"; extend negotiating rights • Comparative assessment of alternatives • "Deliberation" by "extended peer community" throughout process
Key aspects of economic thought	• Individual preferences and trade-offs are essence of value • Citizen sovereignty (as constrained by mandate of representative government)	• Comparative assessment of alternatives • Stakeholders in process; analysis of preferences • Risk communication is inside process
Methodological pluralism	• Neither ERA nor economic analysis ascendant • Both deliberative (constructive) and logico-deductive processes can inform decisions	• Extended peer community • Decision is based on input from multiple disciplines
Importance of adaptive management	• Costs and uncertainties often high; politics may not bear full implementation • Assume incremental, negotiated decisions; include analysts here and in subsequent steps	• Negotiation part of decision process • Adaptive management integral, not accessory
Linkage of situational and regular management processes	• Both types of processes are needed • Should be mutually supporting	• Linked cycles

policy or program objectives, (d) is grounded in the natural resource, policy/institutional, economic, socio-cultural and technological contexts of implementation in practice, (e) establishes practical and valid evaluation criteria by which change can be monitored and assessed, (f) involves methodological pluralism including both quantitative and qualitative methods to ensure rigor and comprehensiveness in assessment, and (g) integrates different disciplinary perspectives (i.e. social, economic, environmental, policy and technological).

Based on these ideas and the examination of other frameworks, a set of considerations that address watershed management generally and are also specific to the ERA–economic integration problem are listed in Table 9.2. These considerations, and the design elements resulting from each, are summarized later in this chapter.

As was emphasized in Chapter 3, ERA has unique value as an ecologically informed process that conceptually defines the ecological system at hand and the anthropogenic forces acting upon it and that progresses, in structured and logical fashion, from ecosystem management goals to the characterization of risks affecting those goals. An integrated framework should retain the processes composing the analytic core of ERA, and the essentially scientific character of the analysis should not be compromised. At the same time, to secure broad participation leading to robust solutions, there must be sensitivity to the critiques of ERA discussed in Section 3.2, particularly that ERA can be too narrowly focused — bearing the mantle of "science" yet serving particular interests[3] or lacking a clear link to management efforts.[4] These criticisms may be answered by an approach that emphasizes the comparative assessment of a range of management alternatives; that identifies stakeholder groups that are likely to bear the respective risks and benefits of the alternatives; and that sees negotiation among these groups as legitimate. Where uncertainties and decision stakes are high, the approach should acknowledge the limits of science by accommodating "deliberation"[5] by "extended peer communities"[6] throughout the process. Scheraga and Furlow[7] coined the term *policy-focused assessment* to describe a scientific process that is constantly engaged with stakeholders and decision-makers so that the results will be relevant to policy.

The incorporation of economics into the process implies there will be an increased emphasis on the measurement of individual preferences, expressed as the willingness to make trade-offs. This dynamic reaffirms the importance of a comparative assessment of alternatives. It also implies that risk communication, necessary for informing preferences, is an essential component of an integrated process (whereas it is accessory to ERA) and that stakeholder preferences will be analyzed in some form.[a]

Methodological pluralism[8] is a relevant goal because the salient attributes of environmental management problems are not adequately modeled by any single disciplinary paradigm. The extended peer community should include multiple disciplines;[8] both qualitative and quantitative data collection methods may be needed, and deliberative as well as deductive processes may be relevant. In the decision-making phase, it may not be possible to reduce all relevant factors to a single dimension; multiple objectives may need to be treated.

Adaptive management has been described as a "learn-by-doing" approach to decision-making, in which both goals and approaches are subject to revision over time.[9] When the process is applied to the implementation of a plan or policy, rather than the ongoing management of a resource, the term *adaptive implementation* (AI) may be used.[10] Analytical frameworks often treat AI as an accessory process — a postanalytic feedback loop that acknowledges that uncertainty and complexity may prevent us from precisely hitting the target on the first try. Experience, however, suggests something more. Where costs of remediation or restoration are high, the political will to take fully responsive actions may be lacking, even where scientific knowledge is relatively adequate. Interested parties might first negotiate a less costly interim decision. AI could then constitute an indispensable learning process through which a community gradually acquires the willingness to take more vigorous steps.

[a] For guides on sharing environmental information with the public, refer to USEPA;[44,45] for useful information on terminology for communicating ecological concepts, see Schiller et al.[15] and Norton.[46]

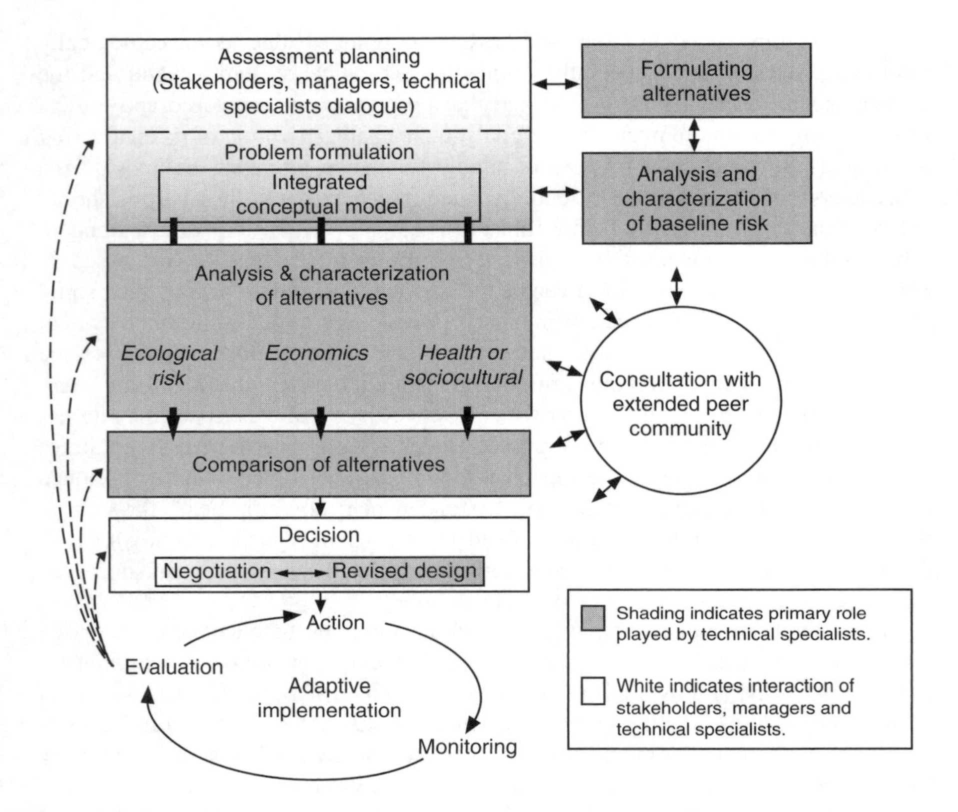

Figure 9.1 A conceptual approach for the integration of ecological risk assessment and economic analysis in watershed management.

As Holling et al.[11] put it, "managers as well as scientists learn from change," and the same can be said for other stakeholders. If so, it would be a mistake to view negotiation as a purely nonscientific process taking place after the specialists have "had their say." Rather, technical specialists should participate in the design of an incremental process that yields information and employs evaluation criteria at each step. They should be expected to play a supporting role during negotiations and to be actively engaged through the adaptive implementation process.

Finally, because environmental management entails both regular and situational processes, it may be important to examine how the problem-oriented process of ecological–economic analysis, decision-making, and AI that is being developed herein can best interact with ongoing resource management requirements.

DIAGRAMMING AN INTEGRATED MANAGEMENT PROCESS

Figure 9.1 diagrams a conceptual approach that addresses each of the guiding considerations listed in Table 9.2 and, in so doing, responds to each of the SAB and Bellamy criteria. The major components are discussed in the succeeding sections.

In many respects the approach is similar to the *ERA Framework*. However, ERA estimates only the likelihood of adverse ecological effects, and it assumes that economic analysis, if needed, will be able to use the assessment results. This approach modifies the *ERA Framework* at every stage of risk assessment, beginning with the planning process, to ensure compatibility. In so doing, however, the core scientific character of ERA is not compromised. The scope of planning and problem formulation are broadened, but the key steps of articulating ecological values, goals, objectives, and endpoints are still carried out. Analysis and characterization of ecological risks is carried out in a scientific manner as part of the analysis of management alternatives and sometimes also as part of an assessment of baseline risks.

This conceptual approach would be placed in the upper right cell of the typology presented earlier in Table 9.1; that is, it is a situational process rather than an ongoing one, triggered by need. However, it includes an adaptive implementation phase, which may continue, and it can be linked to or used within an ongoing watershed management cycle. It is a planning and management approach that includes decision-making and implementation; it is not limited to providing information for decision-support. It generally assumes that stakeholders and decision-makers will be involved in the initial stages and will remain engaged at some level throughout the process, such as through consultations with an extended peer community, but that analysis and characterization will be conducted by technical specialists. Depending on the decision context, stakeholders may be empowered to participate in or to make decisions (i.e., it would be scored as **[2]** or **[3]** in Table 9.1). Each of these aspects is further discussed in the following sections.

The sequence of discussion is not necessarily that in which the process will occur. The process may begin with assessment planning, initiated because a problem or opportunity has been recognized. On the other hand, a proposal for one or more actions may have been formulated that now requires full evaluation, or a study of baseline risks (i.e., present and future risks, if no new action is taken) may have been conducted that demonstrates a need for actions to be formulated and comparatively assessed. A separate step for the study of baseline risks is not needed at all if the analysis of alternatives includes the no-action alternative. However, assessment planning, problem formulation, and formulation of alternatives all should be completed prior to the assessment of alternatives and subsequent steps (although the reiteration of these steps may be necessitated by later findings, or by intervening events).

Assessment Planning

Assessment planning is analogous to *planning* in ERA and to *identifying problems and opportunities* in the U.S. Army Corps of Engineers (USACE) project planning process (see Chapter 4); it is here termed *assessment planning* to distinguish it from the more encompassing terms *project planning*, used by the USACE, and *resource management planning*, used by the U.S. Forest Service (see Appendix 9-A). It is a stage that emphasizes discussions among analysts of multiple disciplines (i.e., ecological, economic, and others as needed), risk managers and, where appropriate, stakeholders about values and goals. It is conducted as described in the "Planning" section of Chapter 3, except on three major points. First, the identification of the decision context is somewhat expanded. Besides identifying the decisions to be made and determining their context, assessment planners must also determine who has the authority to make the decisions and what criteria they expect to use. These are critical factors for the characterization and comparison of alternatives; analysts need to know how the decision-makers view the decision situation so their comparisons comprise all the needed elements. For example, decision-makers may be specifically constrained to consider, or not to consider, particular factors such as cost, equity, or threatened and endangered species, or to prioritize some factors vis-à-vis others.

Second, the scientific disciplines needed to address all important dimensions of the problem should be represented in assessment planning. Besides ecology and economics, which are the focus of this book, the watershed management problem may have implications for human health, requiring the involvement of health risk assessors. In addition, sociocultural issues such as environmental justice concerns or threats to cultural artifacts could require the parallel involvement of additional disciplines (geography, cultural anthropology, archeology, etc.), here and throughout the assessment process (see Appendix 9-B). These various analysts should help decision-makers elucidate their time horizon of concern. Decisions have both short- and long-term consequences, and ecological and economic timeframes of analysis will need to acknowledge the time horizons of the relevant processes involved, the decision-makers, and the other disciplines.

Third, not only must interested and affected parties be identified, but the ways in which they may be benefited or harmed by the alternatives under consideration should be indicated because, depending on the legal context, it may be necessary or advisable to accord them negotiating rights, or to address compensation issues, in the decision process. This information will also be useful if the negotiation process is to be modeled (e.g., using game theoretic techniques; see Chapter 5).

Problem Formulation

In the *ERA Guidelines*,[12] problem formulation is a scientific process that is kept separate from planning (see Figure 3.1, and refer to the discussion of problem formulation in Chapter 3). As shown in Figure 9.1, however, it is separated from assessment planning by a dashed line to indicate the tendency for these two steps to be closely associated in practice. For example, conceptual models produced in problem formulation diagrammatically illustrate for stakeholders and decision-makers the complex causes, nature, and ramifications of ecological problems in watersheds,[13] as is necessary for assessment planning.

The distinction between these two steps is further reduced here because of the need to broaden conceptual models and assessment endpoints to include socioeconomic as well as ecological impacts — an exercise that is likely to rely on repeated discussions with interested and affected parties. In ERA, risk hypotheses, which are proposed explanations of relationships between sources, stressors, exposure pathways, receptors, and ecological effects, are the basis of conceptual models (see "Problem Formulation" in Chapter 3). To include socioeconomic impacts, risk hypotheses must be extended to include the changes in ecosystem services (see Table 5.2) that will be associated with the changes in those endpoints. Finally, since the evaluation of alternatives is also required for an integrated assessment, risk management hypotheses are needed as well; these are proposed explanations of how management alternatives will affect sources, exposures, effects, and services.

Chapter 3 used the example of the decline of a hypothetical reservoir fishery to illustrate the components of ERA. The "Problem Formulation" section listed population size, mean individual size, and recruitment of popular angling species as appropriate ecological assessment endpoints, and it stated that conceptual models should diagram the ecological processes whereby the stressors suspected of causing the decline, in this case agricultural pesticides and municipal and agricultural nutrients, were thought to exert effects. Continuing that example, the integration of economics at the problem formulation stage would require adding management alternatives to the conceptual model. In this example, suppose that a baseline risk assessment (see the next section of this chapter) had identified nutrient loadings to the reservoir as the actual cause of the decline, and that risk management alternatives to be studied (see "Formulation of Alternatives" in this Chapter) included restricting further sewerage connections to the municipal treatment plant, upgrading the treatment plant, instituting an incentives program for riparian zone restoration, and conducting an outreach program to encourage conservation tillage. Extending the conceptual model would require adding each of these alternatives to the diagram and illustrating their expected effects on the ecological processes relevant to the endpoints. Additional effects that might have ecological relevance would be diagrammed as well, such as important beneficial or detrimental effects on species that were not the original subject of the assessment. These might require defining additional ecological assessment endpoints.

Economic effects of the alternatives must also be added to the model. Since the ecological assessment endpoints in this example (fish species, population, size, etc.) are not directly valued, the link to ecosystem services such as fishing success must be included in the diagram, and assessment endpoints corresponding to the service changes (for example, value to recreational users) must be added. Other economic effect pathways, such as the effects of plant upgrade costs or land use changes on the local economy, also need to be included. Finally, other kinds of changes expected to result from the alternatives, such as changes in human health or quality of life, should also be indicated. Complete risk management hypotheses will consist of a causal chain that extends from a given management alternative to each of the applicable ecological and economic assessment endpoints.

The analysis plan, which is the final product of problem formulation, must include procedures for evaluating the risk management hypotheses, including the efficacy of proposed management actions and the relationship between ecological responses and ecosystem services. The plan must include quantification of the spatial and temporal extent of endpoint changes.[14] (In the reservoir example, ecosystem service improvements resulting from a management action would depend on the size of the area over which the fishery was improved and the time required to effect the improvement.) The plan must also include proposed methods for the comparison of alternatives that closely reflect the needs of decision-makers, as determined during assessment planning (see "Comparison of Alternatives" in this chapter for further discussion). Finally, the analysis plan and other products of problem formulation (assessment endpoints and conceptual models) must be verified with managers and stakeholders as being not only technically accurate but also well targeted to the most important concerns. If members of these groups have been engaged throughout assessment planning and problem formulation, they may have acquired in the process sufficient technical knowledge to understand these products. If not, or if the economic methods to be used later will require surveys of a broader audience or the general public, then careful work will have to be done at this stage to build a risk communication capability. Steps may include developing common-language terminology to express key ecological concepts,[15] and using focus groups to refine this lexicon and verify assumptions about the values held by the public or stakeholder groups.

Analysis and Characterization of Baseline Risk

If preexisting information is not sufficient, a separate study of baseline risks may be conducted prior to the formulation of alternatives. Although definitions can vary slightly, baseline risks are defined as the present and future risks to ecosystems or human health that would occur if no new action were taken.[16] Baseline risk assessment is a formal part of environmental impact assessments conducted under the National Environmental Policy Act (NEPA) and site characterizations conducted under the Comprehensive Environmental Response, Compensation and Liability Act (Superfund). Since NEPA requirements are invoked only when an action is proposed, the action alternative and no-action alternative are assessed in the same stage of environmental impact assessment, and baseline assessment as a separate step is not needed. Under Superfund, on the other hand, baseline assessment is needed to characterize the risks prior to remedial action design. In watershed management, a separate baseline assessment as shown in Figure 9.1 may be required if the kind of management action needed, or the need for any action at all, is unclear.

Characterizing baseline risks may also require characterization of harms that have already occurred. Risks to economic well-being or sociocultural values may also form part of this analysis, but these risks are more easily addressed in comparative than absolute terms and are therefore likely to receive limited attention at this stage. Methods for analysis and characterization of ecological risk were discussed in Chapter 3; methods for the assessment of health risks are presented elsewhere.[17,18] Determining the magnitude and severity of ecological or health effects helps determine the need for management actions. Determining causality and pathways of exposure provides information useful in the design of management alternatives. Developing models of exposure and response, and risk characterization approaches, establishes the methods that will be used in the comparative analysis of management alternatives.

The generation of exposure scenarios may be an important part of baseline risk assessment. Scenarios are often used to describe alternative circumstances for which risk will be estimated. In some instances they help describe the range of the expected exposure conditions; for example, an assessment of pesticide impacts on watershed resources may require setting up a range of use scenarios to cover the different types of practices actually occurring in the watershed. Exposures resulting from all

Table 9.3 Categories (and some Examples) of Watershed Management Measures

Control of point sources (source reduction, waste recycling, waste pretreatment, or improvement of waste treatment infrastructure).
Control of urban or agricultural nonpoint sources (land use changes, runoff detention structures, improved waste management, educational outreach programs).
Contaminant remediation (chemical spill cleanup, acid mine drainage treatment).
Stream channel and riparian restoration (tree planting, instream structures).
Wetland restoration (removal of berms or drains, planting of wetland vegetation).
Species management (habitat creation, control of nonnatives, reintroductions).
Water resource development (irrigation, hydropower, recreation).
Improvement of other use values (access).
Strategies for adaptation to global change (land use changes to accommodate sea-level rise).

scenarios would then be used in the full characterization of baseline risk. In other cases, scenarios result from alternative assumptions about an unknown future; for example, alternative CO_2 emission assumptions and global climate models are being used to establish alternative future climate scenarios for watershed risk assessment.[19] These scenarios are part of baseline assessment if they do not correspond to designed policies or alternative management actions but rather form a positive basis for design of management actions. On the other hand, some future scenarios are explicitly policy based. For example, Coiner et al.[20] developed future scenarios for the Walnut Creek watershed of Iowa based on alternative policies that respectively prioritized agricultural production, water quality, and biodiversity; and Hulse et al.[21] developed scenarios for the Muddy Creek watershed of Oregon reflecting different policies with respect to development density and conservation. Policy-based future scenarios, which enable a normative comparison of policy outcomes, would be developed as part of the next stage, "Formulation of Alternatives."

Formulation of Alternatives

This phase entails the development of alternative action plans for achieving the watershed management objectives. Depending on the nature of the watershed problems and the management goals, there is a wide array of management actions that may be considered at this step (see Table 9.3). The planning process may include engineering design or policy development; the discussion of specific techniques is beyond the scope of this chapter. Details of processes that can be used for developing alternative plans are presented elsewhere.[22–26] While actions to reduce ecosystem risks are emphasized in this chapter, actions designed to reduce human health risks, improve socioeconomic well-being, or preserve sociocultural values may cause ecological changes and therefore may also need to be evaluated according to the procedures in this chapter.

To avoid a bias toward preselected solutions, planning objectives and constraints should be clearly established in advance,[22] and a broad range of alternatives should be examined (see "Critiques of Ecological Risk Assessment" in Chapter 3). A given alternative should not comprise just the design of management actions such as those listed in Table 9.3; long-term success depends on establishing a planned system that also includes implementation tools (such as permits, incentives, and information) and institutional and organizational arrangements (such as extension services).[26]

Consultation with Extended Peer Community

Funtowicz and Ravetz[6] describe *extended peer communities* as including scientists outside the specific discipline or practice at hand and others lacking formal knowledge but possessing practical, including local, knowledge. The term used here includes interested and affected parties and decision-makers, in addition to scientific peers. *Consultation* does not apply to the assessment planning phase, where interested and affected parties are already an integral part of the process. It applies rather to components such as analysis and characterization that are explicitly scientific. Consultation recognizes on the one hand that these steps must be carried out by analysts with specialized knowledge, and on the other that risk assessment often requires judgments that go beyond strict inference and are therefore susceptible to bias. Consultation is a process in which technical information from the assessment is discussed with the extended peer community for purposes of (a) identifying issues or deficiencies in the assessment and (b) keeping interested and affected parties engaged during what can be a lengthy process. It is equivalent to the term *deliberation* as used by the National Research Council (NRC).[5]

Analysis and Characterization of Alternatives

In this stage the alternatives are assessed from the perspective of various disciplines including ERA, economics, and possibly others such as human health risk or sociocultural assessment, depending on the situation. A brief summary of sociocultural assessment methods is presented as Appendix 9-B.

In Figure 9.1 the disciplines are shown as jointly conducted, indicating at least an exchange of information and at best an integrated analytic approach. However, it is by no means a requirement that the disciplines depart from their characteristic approaches, as long as they are mutually informed. Since ecological and economic time frames of analysis may differ, the time frame for each should be made explicit.

Analysis of alternatives is guided by the risk management hypotheses, indicating which exposures and responses are likely to be affected by risk management. Those not expected to be affected remain part of the baseline risk but are not included in the alternatives analysis. The ecological risk component estimates the changes in exposure profiles likely to result from each management alternative. Where management alternatives create new exposures, i.e., to stressors that were not originally present (such as to sediments from project construction or to pesticides used to control invasive species), additional exposure profiles and exposure-response relationships beyond those of the baseline assessment must be developed.

Ecological risk characterization describes probabilities, magnitudes, and severities of effects on ecological assessment endpoints. These should be described both in absolute terms and as changes with respect to baseline. Uncertainties in the effect estimates must be characterized as well, and the uncertainties, as well as the other parameters, must be carried forward into the economic analysis.[14]

The economic component analyzes costs (including financial and opportunity costs) and benefits associated with the management alternatives. This includes, to the extent practicable, the changes (with respect to the no-action baseline) in ecosystem services (see Table 5.2) that are associated with changes in the ecological assessment endpoints. This especially includes services that can be quantified objectively, such as biophysical services (e.g., the production of food, fiber, or other goods, and regeneration and stabilization processes) and services that are quantifiable by revealed preference methods (e.g., many forms of recreation). It may also include life-fulfilling functions (including functions corresponding to nonuse values), if these can be quantified by benefit transfer methods. The use of stated-preference or other subjective methods to quantify these services is not ruled out at this stage, but for pragmatic reasons such efforts may best be carried out as part of the subsequent comparison phase. For example, if a stated preference questionnaire were to be designed and administered, it may be possible, and therefore cost-effective, to do so in such a way as to effect a multifactoral comparison, as described in the next section.

Comparison of Alternatives

This step is included in the conceptual approach based on the assumption that not all factors important for decision-making can be objectively reduced to a single vector and that the comparison step itself therefore is both subjective and nontrivial. Even if net economic benefit to society, as determined by cost–benefit analysis, is an important criterion, there will usually be other ecological, moral, political, or legal factors that it cannot adequately encompass. Comparison is the step in which these various factors are arrayed in terms as amenable as possible to those of the legitimate decision-maker, be it an agency official, the collective of residents of a jurisdiction, or individual landowners. Any process used to assign subjective weights to the factors, or to enable individuals or groups to systematically compare the alternatives (based on information about these factors and their subjective judgment) is considered to be part of the comparison phase. Methods may include stated preference analyses (appropriate for large groups of individuals) or decision-analytic approaches in which factors are weighted by technical experts, or by representatives of interested and affected parties, acting either as individuals or within consensus-seeking groups (see Morgan[27] for a useful summary of nonmonetary, multicriteria evaluation methods). On the other hand, if the ultimate decision will be reached by negotiation among parties with divergent interests, the comparison methods used might seek to identify the alternative that the parties believe is the best they can hope to obtain, rather than the one with optimal overall utility. The comparison process is carried out according to agreements made during the assessment planning phase (in which the decision context, including decision-makers and decision factors, was described) and the problem formulation phase (in which the comparison methods to be used were verified).

Decision

Because environmental management problems in watersheds usually are multidimensional, it is unlikely that a problem can be solved based on the actions or authority of any single entity. Therefore, the decision process is likely to involve multiple parties. In spite of the findings of the analysis and characterization of alternatives, or because of the associated uncertainties, the parties may hold divergent beliefs about expected outcomes of a given alternative, or even if they agree on technical issues, they may have divergent incentives or expectations regarding compensation. They may also have divergent interpretations of legal constraints on the decision process. Therefore, a decision may entail less a consensus selection among the alternatives than a negotiated redesign. Where implementation cost is a predominant factor, negotiation may entail scaling back on a design or agreeing to a provisional schedule of incremental implementation, conditioned on verification that performance criteria are being met. Technical specialists therefore may be called on to assist the negotiation process.

Adaptive Implementation

Because achieving agreement can be difficult, a provision for AI may therefore be indispensable to reaching a decision: it can provide a middle-ground approach that satisfies no one but provides a respite until confirmatory data are available. However, a flexible or incremental approach does not constitute adaptive management unless several criteria are met. Holling et al.[11] recommend that experimental perturbations be designed to evaluate specific questions. Walters[28] emphasizes that perturbations need to be great enough to probe system responses across domains of interest; cautious incrementation may not produce any usable information. The NRC[9] stated that adaptive management must not only generate useful information but must specify the mechanisms by which the information will be translated into policy and program redesign. Depending on the findings and the nature of the agreement, evaluation of the data could lead to further action or could trigger renewed negotiation; it could also invalidate certain assumptions of planning, problem formulation, or analysis, indicating that earlier stages of the process need to be reiterated. The possibility of revisiting earlier steps in the assessment as more information is learned is indicated by broken lines in Figure 9.1.

Linkage to Regular Management Cycles

The process described here for integrated assessment is situational; that is, it should not be thought of as a cyclical process that can never be completed. By contrast, resource management is ongoing, and the two processes can be mutually supportive. For example, the rotating basin approach to Clean Water Act (CWA) management (see Appendix 9-A) identifies priorities and needed actions, which may call for a detailed, integrated assessment in situations where needed actions are unclear or where regulatory approaches are insufficient. Stakeholder processes that may have been established as part of that cycle can be drawn upon for the integrated assessment.

The rotating basin approach also establishes a long-term water-quality and biological monitoring database that can establish temporal trends and correlations in stressors and biological response that can be useful in establishing causation, exposure profiles, and stress-response relationships. The management alternatives to be considered in the integrated analysis can include (among other measures) regulatory and incentive mechanisms provided for under the CWA, to be implemented and monitored as part of the regular management cycle. Similarly, some watershed resources (e.g., forest resources) are adaptively managed in an ongoing fashion (see Appendix 9-A, Figure 9-A.5). Integrated assessments can link effectively to an adaptive management cycle.

EXAMPLES OF ANALYSIS AND CHARACTERIZATION FOLLOWED BY COMPARISON OF ALTERNATIVES

Planning and problem formulation (together with baseline ERA and formulation of alternatives) lay the groundwork for a successful integrated analysis, but the technical aspects of integration are encountered in the analysis and characterization of alternatives and in the comparison step that follows (see Figure 9.1). Because there are a variety of ecological and economic analytic tools that could be applied in these stages, the specific elements of these steps will also vary. This section provides examples to illustrate how ecological and economic techniques might interact.

Example 1: Cost–Benefit Analysis of All Changes that Can Be Monetized, with Qualitative Consideration of Other Changes

Cost–benefit analysis (CBA; see Chapter 5) is commonly used where decision-makers are concerned about the net economic benefit to society of a given action (that is, to determine whether economic efficiency is increased).[29] As discussed in Chapter 6, CBA is required for certain federal actions. In an integrated assessment where changes in economic efficiency will be a key factor in the decision, the process may occur as diagramed in Figure 9.2. For each management alternative, ecologists would quantify the changes expected in each ecological assessment endpoint. Changes that could not be quantified would be characterized qualitatively. Other analysts might examine quantitative or qualitative effects on health or quality of life, as needed. Economists would look first at the financial costs of the alternatives and any effects that could be determined from markets (for example, opportunity costs of land taken out of agricultural production). Economists would then seek to monetize the effects estimated from the ecological or other analyses, using revealed preference or benefit transfer methods wherever possible (see Table 5.3). Due to their required time and cost, stated-preference techniques would be used only if other methods were unsatisfactory (that is, if nonuse values are important or reliable studies from similar settings are lacking). Based on this information, economists would analyze economic efficiency, equity, and impacts. This information and information about effects that either could not be quantified or could not be monetized would be carried forward into the comparison step.

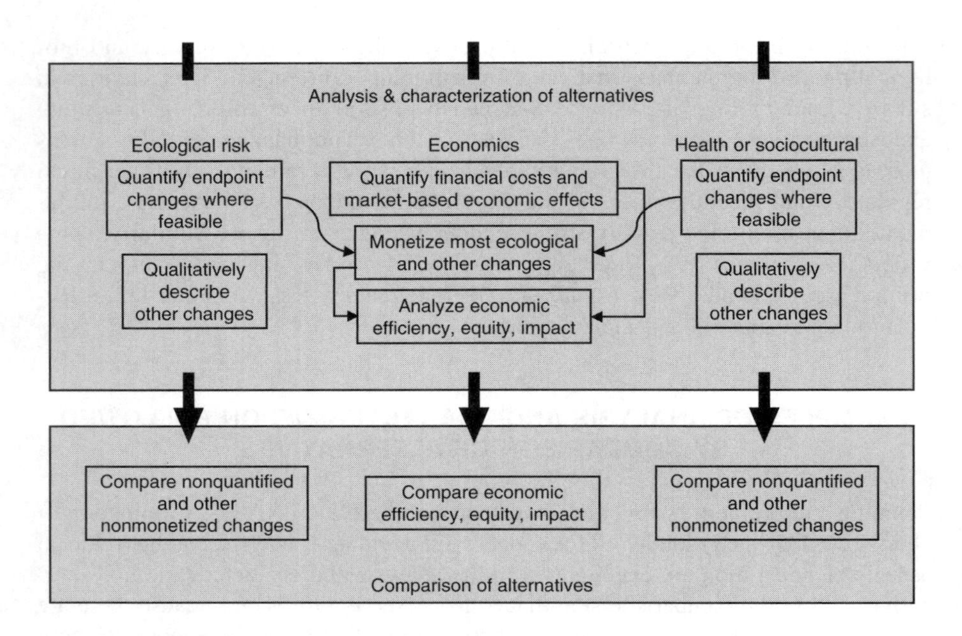

Figure 9.2 Analysis and characterization of alternatives, followed by their comparison, example 1: CBA of all changes that can be monetized, with qualitative consideration of other changes.

Example 2: Use of Stated Preference Techniques to Effect Integration of Ecological, Economic, and Other Factors

In the previous example, stated preference methods, if used at all, would monetize the ecological changes associated with one or more management alternatives. Figure 9.3 diagrams the use of stated preference methods to achieve a more broadly based comparison, such as one that includes the ecological, health, quality-of-life, equity, and impact dimensions of a choice. For example, this could be accomplished using a contingent valuation method (CVM, see Appendix 5-A) survey that explains the effects of the management alternatives (i.e., that "frames" the alternatives) in each of those dimensions before asking individuals about their willingness to pay (WTP) or to accept (WTA) (see Chapter 5). To design such a survey, each of those dimensions would first need to be analyzed and characterized, with all effects quantified to the extent possible. The technical findings would then need to be refined (such as through the use of focus groups) into a format that highlighted only the most important factors and used commonly understood language.[15] A broadly framed CVM approach that was similar to this in certain respects was employed in the Big Darby Creek watershed case study presented in Chapter 10.

A broad comparison could also be accomplished using a choice modeling method such as conjoint analysis (CA, see Appendix 5-A). In this approach, focus groups would again be used to identify the most important factors across those dimensions and to establish common terminology. Survey design would entail transforming those dimensions into choice attributes so that respondents' choices would reveal

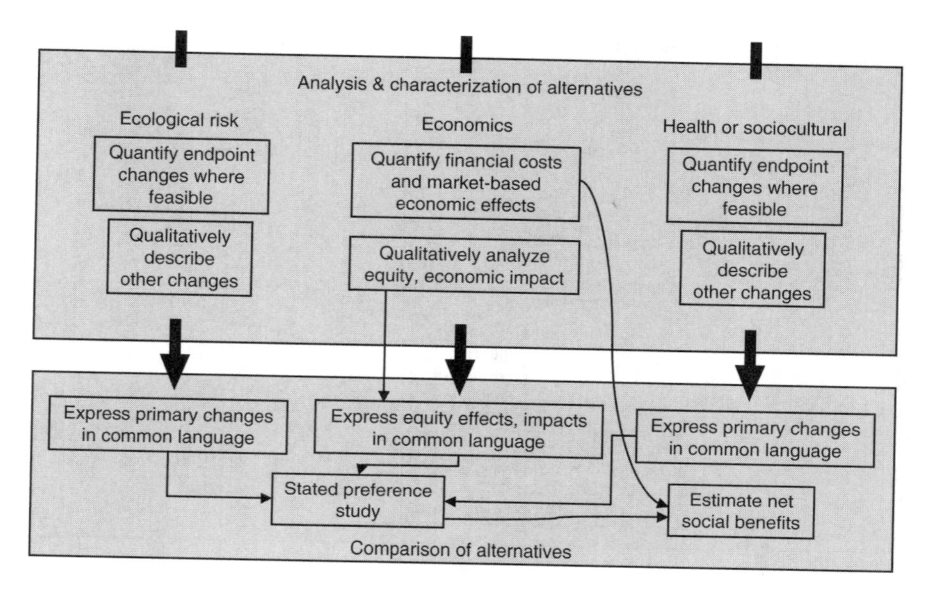

Figure 9.3 Analysis and characterization of alternatives, followed by their comparison, example 2: Use of stated preference techniques to effect integration of ecological, economic, and other factors.

how the various dimensions contributed to WTP or WTA. A method of this type was used in both the Clinch Valley case study presented in Chapter 11 and the Lower Fox River and Green Bay study presented in Chapter 14.

Example 3: Use of Linked Ecological and Economic Models to Dynamically Simulate System Feedbacks and Iteratively Revise Management Alternatives

A disadvantage of sequentially integrated assessments, in which ecological changes are estimated and then economically evaluated, is that there is no opportunity to simulate dynamic interaction between economic and ecological processes.[30,31] In cases where the economic effects of changes in ecosystem quality (such as effects on housing, recreational, or agricultural values[32,33]) will have an important influence on land-use decisions and ecosystem quality, an integrated system that models these feedbacks may enable a better understanding of the behavior of the real systems. In Figure 9.4, models of the ecological processes affecting the assessment endpoints are linked to a regional economic model in a manner that allows parameter feedbacks over time. Once such a modeling system is established, management alternatives can be simulated and iteratively revised to optimize their design according to a variety of criteria, such as cost-effectiveness, equity, and ecological risk. The example pictured in Figure 9.4 arbitrarily assumes a case where ecological and economic models are linked and that other effects (e.g., on health or quality of life) are estimated using other methods. The example further assumes that it may be difficult to estimate net social benefit from such a modeling approach, since WTP or WTA

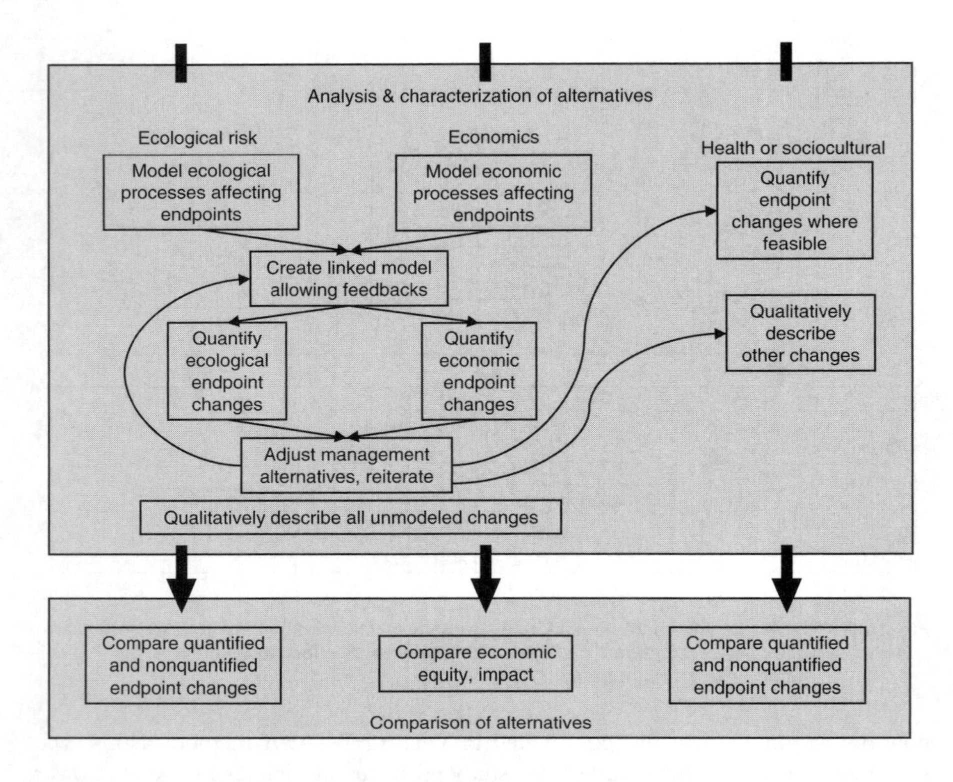

Figure 9.4 Analysis and characterization of alternatives, followed by their comparison, example 3: Use of linked ecological and economic models to dynamically simulate system feedbacks and iteratively revise management alternatives.

for nonuse values is not estimated, although in theory an appropriate benefit transfer module could be added to the model. In the comparison step, the modeling results for the various management alternatives or for different optimization criteria could be described, along with qualitative discussion of any effects that could not be quantified by the modeling effort.

CONCLUSION

This conceptual approach does not represent a fundamental departure from existing practice. Its steps correspond in large part to those of other frameworks (Table 9.4); they differ as needed to emphasize the ERA–economic integration problem. However, the incorporation of multiple disciplines into an integrated assessment process may create significant challenges of communication, coordination, and funding. Therefore the use of this approach is not appropriate in all instances where ERA alone is called for. However, if decisions need to be informed on the basis of both ecological risks and economics, an integrated approach, while more demanding, is more likely to provide coherent information.

Table 9.4 Rough correspondence between the components of the conceptual approach for ERA-economic integration and other selected watershed management frameworks[a]

Component of Conceptual Approach for ERA-Economic Integration (Figure 3.1)	Corresponding Component			
	Framework for Ecological Risk Assessment[12]	Framework for Integrated Environmental Decision Making[1]	Watershed Management Model[43]	U.S. ACE Six-Step Planning Process
Assessment Planning	Planning	Phase I: Problem Formulation	Phase I: Assessment/Problem Identification	Identifying Problems and Opportunities
Problem Formulation	Problem Formulation			
Analysis and Characterization of Baseline Risk	Analysis Risk Characterization	Phase II: Analysis and Decision Making		Inventorying and Forecasting Conditions
Formulation of Alternatives	NA[b]		Phase II: Planning	Formulating Alternative Plans
Analysis and Characterization of Alternatives	Analysis (reiteration) Risk Characterization (reiteration)			Evaluating Alternative Plans
Comparison of Alternatives				Comparing Alternative Plans
Decision	NA			Selecting a Plan
Adaptive Implementation	NA	Phase III: Implementation and Performance Evaluation	Phase III: Implementation Phase IV: Evaluation	NA

[a]See Appendix 9-A for discussion of other watershed management frameworks.

[b]Not applicable.

This conceptual approach is used in Part II of this book as a vantage point from which to analyze six case studies. Each case study has limitations. As was mentioned in Chapter 1, in the first three case studies (Chapters 10–12) the involvement of economists came well after ERA had been initiated, and in one case ERA was never completed. Furthermore, none of the case studies has encompassed the full span of management activities, from assessment planning to adaptive implementation. Nonetheless, the conceptual approach helps to illustrate how the methodological advances and insights from each case study could be used to fullest advantage, both in the watersheds that were studied and in other settings where similar methods could be applied.

REFERENCES

1. SAB, Toward integrated environmental decision-making, EPA-SAB-EC-00-011, U.S. Environmental Protection Agency, Science Advisory Board, Integrated Risk Project, Washington, D.C., 2000.
2. Bellamy, J.A., Walker, D.H., McDonald, G.T., and Syme, G.J., A systems approach to the evaluation of natural resource management initiatives, *J. Environ. Manage.*, 63, 407, 2001. Excerpt reprinted with permission from Elsevier.
3. Pagel, J.E. and O'Brien, M.H., The use of ecological risk assessment to undermine implementation of good public policy, *Hum. Ecol. Risk Assess.*, 2, 238, 1996.
4. Butcher, J.B., Creager, C.S., Clements, J.T., et al., Watershed level aquatic ecosystem protection: Value added of ecological risk assessment approach, Project No. 93-IRM-4(a), Water Environment Research Foundation, Alexandria, VA., 1997, 342 pp.
5. NRC, *Understanding Risk: Informing Decisions in a Democratic Society*, National Research Council, National Academy Press, Washington, D.C., 1996.
6. Funtowicz, S.O. and Ravetz, J.R., A new scientific methodology for global environmental issues, in *Ecological Economics: The Science and Management of Sustainability*, Costanza, R., Ed., 1991, Columbia University Press, NY, NY, 137.
7. Scheraga, J.D. and Furlow, J., From assessment to policy: Lessons learned from the U.S. National Assessment, *Hum. Ecol. Risk Assess.*, 7, 1227, 2002.
8. Norgaard, R., The case for methodological pluralism, *Ecol. Econ.*, 1, 37, 1989.
9. NRC, *Restoration of Aquatic Ecosystems: Science, Technology and Public Policy*, National Research Council, Commission on Geosciences, Environment and Resources, Washington, D.C., 1992.
10. NRC, *Assessing the TMDL Approach to Water Quality Management*, National Research Council, National Academy Press, Washington, D.C., 2001.
11. Holling, C.S., Bazykin, A., Bunnell, P., et al., *Adaptive Environmental Assessment and Management*, Wiley-Interscience, New York, 1978.
12. USEPA, Guidelines for ecological risk assessment, EPA/630/R-95/002F, Risk Assessment Forum, U.S. Environmental Protection Agency, Washington, D.C., 1998.
13. Serveiss, V.B., Applying ecological risk principles to watershed assessment and management, *Environ. Manage.*, 29, 145, 2002.
14. Suter, G.W., Adapting ecological risk assessment for ecosystem valuation, *Ecol. Econ.*, 14, 137, 1995.
15. Schiller, A., Hunsaker, C.T., Kane, M.A., et al., Communicating ecological indicators to decision-makers and the public, *Conserv. Ecol.*, 5, 19 [online], 2001. http://www.ecologyandsociety.org/vol5/iss1/art19/index.html. Accessed Oct. 06, 2004.

16. USDOE (U.S. Department of Energy), Use of institutional controls in a CERCLA baseline risk assessment, CERCLA Information Brief EH-231-014/1292, U.S. Department of Energy Office of Environmental Guidance, Washington, D.C., 1992.
17. PCCRARM, Framework for environmental health risk management, Presidential/Congressional Commission on Risk Assessment and Risk Management, Washington, D.C., 1997. Available at http://www.riskworld/Nreports/1997/risk-rtp/pdf/EPAJAN.PDF. Accessed Oct. 6, 2004.
18. van Leeuwen, C.J. and Hermens, J.L.M., *Risk Assessment of Chemicals: An Introduction*, Kluwer Academic Publishers, Dordrecht, 1995.
19. Rogers, C.E., Julius, S.H., and Furlow, J., Assessment as a method for informing decisions about water quality, aquatic ecosystems and global change, in *Water Resource Issues, Challenges and Opportunities: Part II: Using Science to Address Water Issues*, 2002, 10.
20. Coiner, C., Wu, J., and Polasky, S., Economic and environmental implications of alternative landscape designs in the Walnut Creek Watershed of Iowa, *Ecol. Econ.*, 38, 119, 2001.
21. Hulse, D., Eilers, J., Freemark, K., White, D., and Hummon, C., Planning alternative future landscapes in Oregon: Evaluating effects on water quality and biodiversity, *Landsc. J.*, 19, 1, 2000.
22. USACE (U.S. Army Corps of Engineers), Planning guidance notebook, ER 1105-2-100, U.S. Army Corps of Engineers, Washington, D.C., 2000.
23. USEPA, Watershed protection: A statewide approach, EPA/841/R-95/004, Office of Water, U.S. Environmental Protection Agency, Washington, D.C., 1995.
24. USEPA, Ecological restoration: A tool to manage stream quality, EPA/841/F-95/007, U.S. Environmental Protection Agency, Office of Water, Washington, D.C., 1995.
25. U.S. Water Resources Council, Economic and environmental principles and guidelines for water and related land resources implementation studies, U.S. Water Resources Council, Washington, D.C., 1983.
26. Hufschmidt, M.M., A conceptual framework for watershed management, in *Watershed Resources Management: An Integrated Framework with Studies from Asia and the Pacific*, Easter, K.W., Dixon, J.A., and Hufschmidt, M.M., Eds., Westview Press, Boulder, 1986, 2, 17.
27. Morgan, R.K., *Environmental Impact Assessment: A Methodological Perspective*, Kluwer Academic Publishers, Boston, MA, 1998.
28. Walters, C.J., *Adaptive Management of Renewable Resources*, Macmillan Publishing Company, New York, 1986.
29. USEPA, Guidelines for preparing economic analyses, EPA/240/R-00/003, U.S. Environmental Protection Agency, Washington, D.C., 2000.
30. Lindner, M., Sohngen, B., Joyce, L.A., et al., Integrated forestry assessments for climate change impacts, *For. Ecol. Manage.*, 162, 117, 2002.
31. Duraiappah, A.K., Sectoral dynamics and natural resource management, *J. Econ. Dyn. Control*, 26, 1481, 2002.
32. Geoghegan, J., Wainger, L.A., and Bockstael, N.E., Spatial landscape indices in a hedonic framework: An ecological economics analysis using GIS, *Ecol. Econ.*, 23, 251, 1997.
33. Odom, D.I.S., Cacho, O.J., Sinden, J.A., and Griffith, G.R., Policies for the management of weeds in natural ecosystems: The case of scotch broom (*Cytisus scoparius*, L.) in an Australian national park, *Ecol. Econ.*, 44, 119, 2003.
34. USEPA, Environmental monitoring and assessment program (EMAP) research strategy, EPA/620/R-98/001, U.S. Environmental Protection Agency, Washington, D.C., 1997.
35. Walmsley, J.J., Framework for measuring sustainable development in catchment systems, *Environ. Manage.*, 29, 195, 2002.
36. USEPA, A framework for the economic assessment of ecological benefits, Science Policy Council, U.S. Environmental Protection Agency, Washington, D.C., Feb. 1, 2002.

37. Stahl, R.G., Bachman, R.A., Barton, A.L., et al., *Risk Management: Ecological Risk-Based Decision-Making*, Society for Environmental Toxicology and Chemistry, Pensacola, FL, 2001.
38. World Commission on Dams, *Dams and Development: A New Framework for Decision-Making*, Earthscan Publications Ltd., London, 2002.
39. USEPA, Watershed protection: A project focus, EPA/841/R-95/003, Office of Water, U.S. Environmental Protection Agency, Washington, D.C., 1995.
40. Timmerman, J.G., Ottens, J.J., and Ward, R.C., The information cycle as a framework for defining information goals for water-quality monitoring, *Environ. Manage.*, 25, 229, 2000.
41. USFS, National forest system land and resource management planning, *Fed. Re.g.,* 65, 67513, 2000.
42. USFS, National forest system land and resource management planning; proposed rules, *Fed. Re.g.,* 67, 72769, 2002.
43.Davenport, T.E., *The Watershed Project Management Guide*, Lewis Publishers, Boca Raton, FL, 2002.
44. USEPA, Considerations in risk communication: A digest of risk communication as a risk management tool, EPA/625/R-02/004, National Risk Management Research Laboratory, U.S. Environmental Protection Agency, Cincinnati, OH, 2003.
45. USEPA, Risk communication in action: Environmental case studies, National Risk Management Research Laboratory, U.S. Environmental Protection Agency, Cincinnati, OH, 2003.
46. Norton, B.G., Improving ecological communication: The role of ecologists in environmental policy formation, *Ecol. Appl.*, 8, 350, 1998.

APPENDIX 9-A

Discussion Of Existing Frameworks That Have Been Applied To Watershed Management

CONTENTS

Table 9.1 presents a typology of frameworks that have been applied to the processes of watershed assessment and management. This appendix discusses the frameworks listed in each of the four cells of the typology, and it presents several applicable flow diagrams that serve as background for the design of the conceptual approach presented in Figure 9.1.

SITUATIONAL MONITORING OR ASSESSMENT FRAMEWORKS

Several frameworks pertain to monitoring or assessment that provide information for decision-makers but do not include the decision-making process. ERA, per the USEPA's *Guidelines*, is described in Chapter 3 and diagrammed in Figure 3.1. ERA is a situational process for decision support; it is initiated in response to past, ongoing, or potential future adverse effects to ecological resources. ERA emphasizes the natural sciences and the separation of science and policy. Stakeholder involvement may be important for development of management goals during planning and, debatably, for problem formulation, but it is considered inappropriate for analysis and risk characterization. The results of risk characterization are communicated to risk managers, but decision-making occurs outside the ERA process.[1] A *Framework for the Economic Assessment of Ecological Benefits* that explores the potential integration

1-56670-639-4/05/$0.00+$1.50

of ERA and economic valuation techniques has been described by the USEPA[2]; it has not been applied to watershed management but is included in the typology as a point of reference.

Environmental monitoring is an essential component of watershed management, and decisions about what to monitor implicitly are decisions about management. Most monitoring programs are limited to the collection of natural science data, but some include economic and institutional indicators as well. An example of the former is the USEPA's Environmental Monitoring and Assessment Program (EMAP), which estimates status and trends of selected ecological resources by monitoring indicators of ecosystem structure and function and by measuring relationships between environmental stressors and impacts.[3] An example of a broader indicators framework is one developed by the Organization for Economic Cooperation and Development (OECD).[4] The DPSIR framework (see Table 9.1) calls for indicators of the social and economic conditions that drive environmental changes, and the policy and management responses to those changes, in addition to indicators of the environmental changes themselves. Monitoring system design usually stresses input from managers but not other stakeholders. For example, EMAP's indicator development process borrows several concepts (such as ecological values, assessment questions, and conceptual models) from the ERA *Framework* but does not assume stakeholder involvement.[5,6]

REGULAR MONITORING OR ASSESSMENT FRAMEWORKS

ERA generally is not a regular process; while its steps may be reiterated as more is learned, it is not intended to be continuous. Frameworks for the set-up of monitoring systems, including indicator design, usually depict a one-time (i.e., situational) process as well. However, a cyclical (i.e., regular) redesign process can allow monitoring systems to adapt as knowledge and management needs change.[7]

SITUATIONAL PLANNING AND MANAGEMENT FRAMEWORKS

The Society for Environmental Toxicology and Chemistry has described an ecological risk management framework composed of the following steps:[8]

- Issue identification.
- Goal setting.
- Management options development.
- Data compilation and analysis.
- Option selection.
- Decision implementation.
- Tracking and evaluation.

The process is informed by stakeholders during goal setting, and effective communication with stakeholders throughout the process is considered important. It assumes that economic analysis will be involved in the decision, but processes for integrating ecological and economic aspects are not discussed.

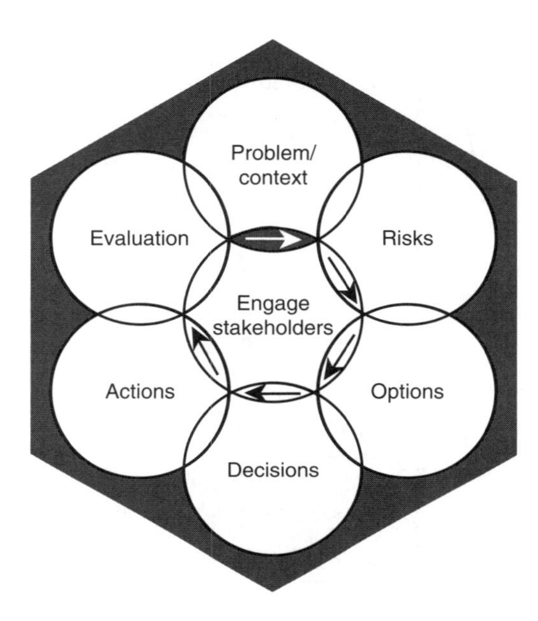

Figure 9-A.1 Framework for environmental health risk management (from PCCRARM[9]).

The *Framework for Environmental Health Risk Management* depicts a process that is similar, albeit with a slightly different ordering of steps (Figure 9-A.1).[9] Active engagement of stakeholders is encouraged throughout the process, and it is suggested that stakeholders be empowered to make decisions where allowable. While the framework is pictured as cyclical, it should be viewed as situational (responding to problems) yet amenable to adaptive management as necessary to implement effective solutions. A panel convened by the USEPA's Science Advisory Board (SAB), tasked with making recommendations on the integration of environmental decision-making, presented similar ideas[10] but depicted the process more appropriately as unidirectional, albeit with feedback loops, rather than cyclical (Figure 9-A.2).

The U.S. Army Corps of Engineers (USACE) uses a six-step planning process for civil works projects, including those related to water resources and watersheds:

- Step 1 — Identifying problems and opportunities
- Step 2 — Inventorying and forecasting conditions
- Step 3 — Formulating alternative plans
- Step 4 — Evaluating alternative plans
- Step 5 — Comparing alternative plans
- Step 6 — Selecting a plan

The process includes decision-making but in most instances does not include implementation, retrospective evaluation, or adaptive management. Stakeholder involvement is intended to play an important role in step 1, including in the selection of decision criteria; communication channels are to be maintained throughout the process; and stakeholder consultation is to occur after evaluation is completed and

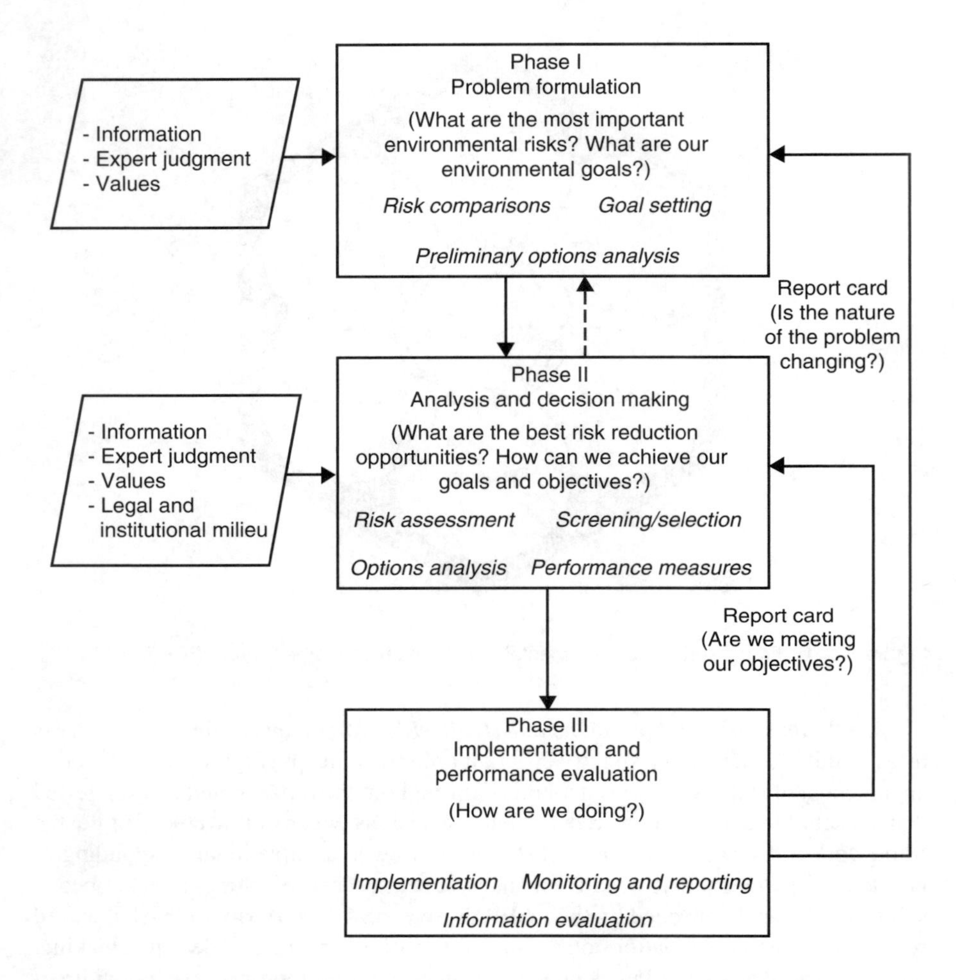

Figure 9-A.2 Framework for integrated environmental decision making (from SAB[10]).

before plan selection. Evaluation of alternative plans includes quantifiable national and regional costs and benefits as well as nonquantifiable environmental and social impacts or benefits (see Chapter 4 and USACE[11]).

By comparison, a planning and project development framework developed by the World Commission on Dams[12] provides for a more extensive stakeholder role and for adaptive management (Figure 9-A.3). Criteria ensuring, among other things, public participation, assessment of ecological risks, and consideration of a comprehensive set of alternatives are checked at the conclusion of each development phase. Analyses of alternatives include the identification of people who are affected when lands or other resources are put at risk by the project, and negotiating rights with respect to the final decision are conferred according to risk burden. The framework emphasizes compliance with negotiated agreements during and after construction. Finally, project operation is to be reviewed periodically and should adapt to changes in the project context.

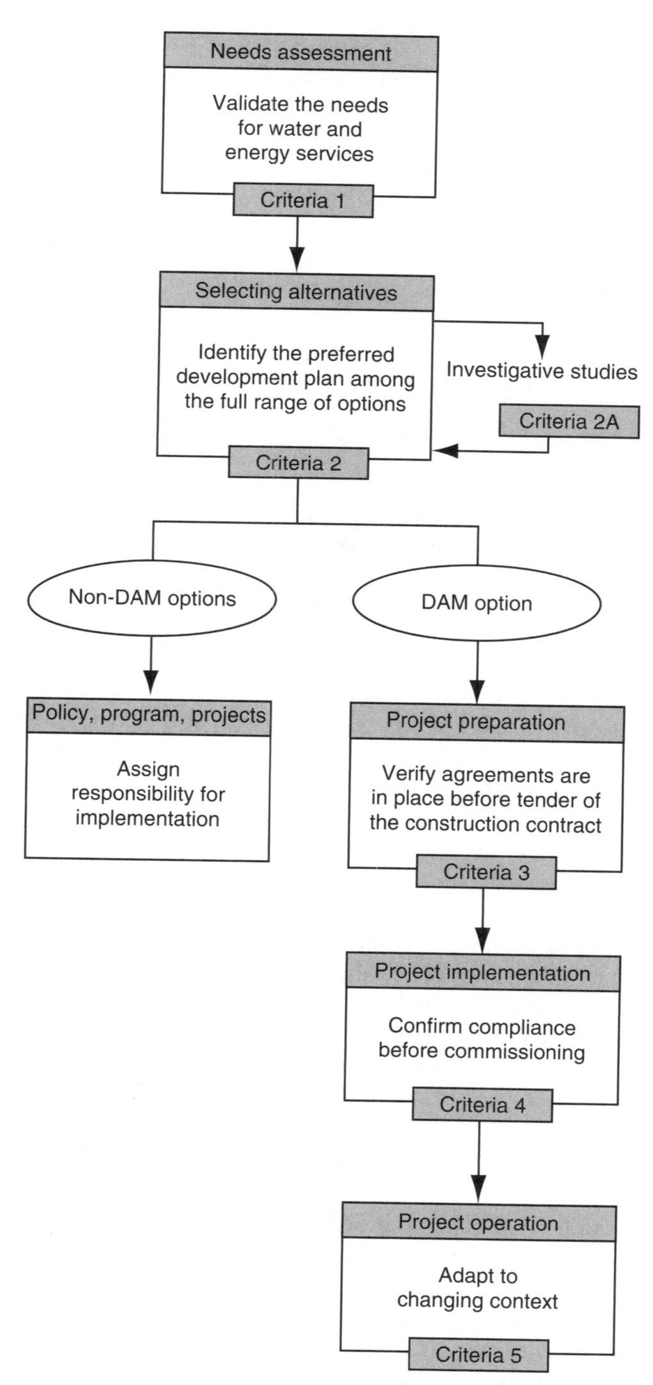

Figure 9-A.3 A framework for planning and project development of large dams, including five key decision points at which specific criteria should be evaluated (redrawn from World Commission on Dams[12]).

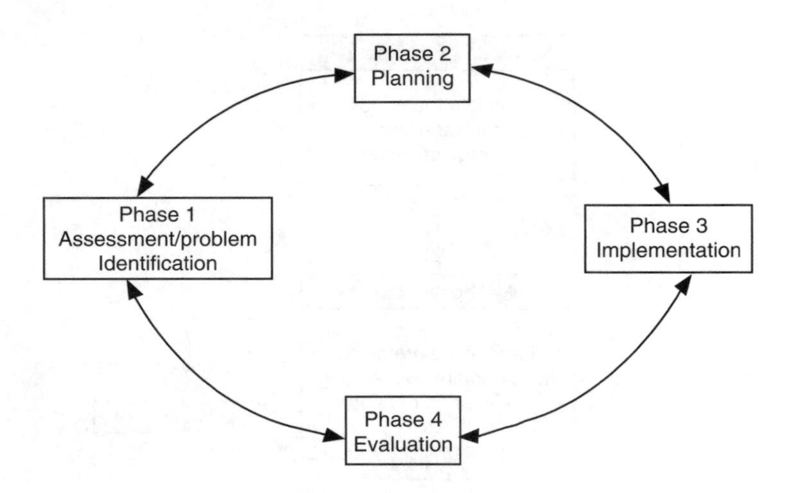

Figure 9-A.4 A watershed management model for the planning and implementation of watershed projects (redrawn from Davenport[14]).

A four-step process for the planning and implementation of watershed projects (Figure 9-A.4) was described by the USEPA[13] and more fully elaborated by Davenport.[14] The process is designed to be carried out through a partnership of government agencies and local stakeholders, and it emphasizes involvement and action. The assessment and problem identification phase consists of four parts — inventory, analysis, problem identification, and goal-setting — and is analogous to ERA. However, ERA assumes that analysis itself will require advance planning and substantial time and resources to conduct and will result in a quantitative characterization of risks, whereas the watershed project management approach emphasizes qualitative description of the most critical problems and their causes. Natural science is used to identify problems, and know-how, partnerships, and consensus-building processes are used for making and implementing decisions. Project analysis, including economic analysis, is not emphasized. Like the *Framework for Environmental Health Risk Management*, the process is pictured as circular; we have grouped it with situational methods on the assumption that efforts will conclude once conditions change. If a partnership is effective, however, an effort could be longstanding.

REGULAR PLANNING AND MANAGEMENT FRAMEWORKS

Several frameworks have been proposed for the regular and ongoing management of watershed resources. These regular processes can spawn situational analyses which may be portrayed as linked cycles.[15] For example, the U.S. Forest Service (USFS) uses a planning process (Figure 9-A.5) to guide the ongoing management of national forests and grasslands.[16,17] The spatial scale of planning ranges from national to regional to local, and it can be done at the watershed level if appropriate to the scope and scale of the issues addressed. Existing plans authorize site-specific management actions, and outcomes are monitored and evaluated according to plan

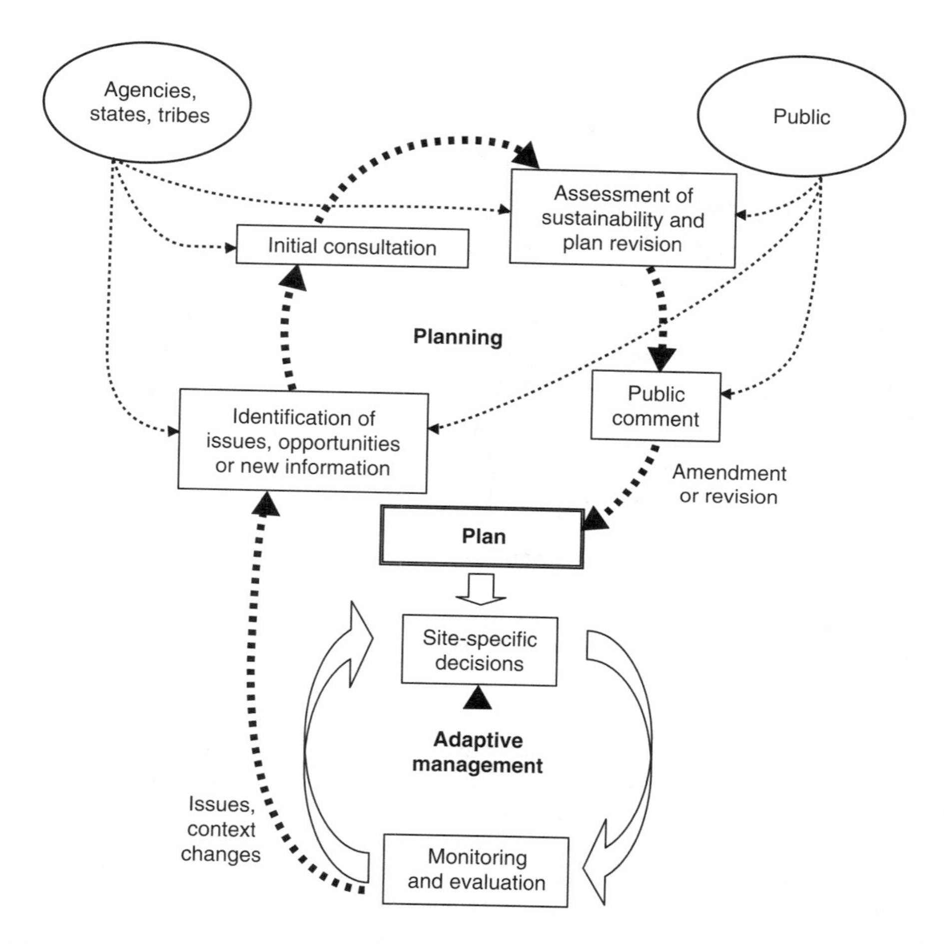

Figure 9-A.5 The USFS planning framework incorporates regular adaptive management and situational planning processes.

criteria in an adaptive cycle. New rounds of planning are undertaken after 15 years or as necessitated by issues or conditions. Stakeholders play an important role in the initial development of goals and are encouraged to participate in subsequent steps; participation opportunities are to be early, frequent, open, and meaningful, and stakeholders may lodge objections before decisions are taken. Information development includes baseline analyses of both ecological and economic sustainability of current forest or grassland management practices. Ecological analyses include the effects of current or anticipated human disturbance (as compared to natural and historical human disturbance) upon ecosystem processes and system and species diversity. Social and economic analyses examine the benefits provided by forest lands, social and economic trends, and the society–forest relationship.

Many U.S. states have adopted a watershed management cycle, sometimes referred to as a *rotating basin approach*, for implementation of the regulatory requirements and other programs of the Clean Water Act (CWA).[18,19] Whereas the

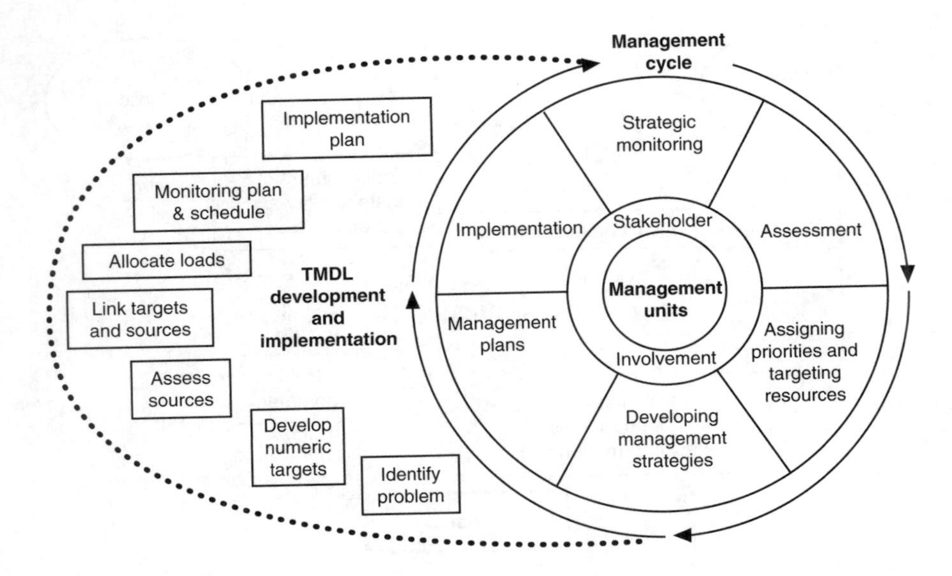

Figure 9-A.6 The watershed-based management cycle used by many states may include TMDL development and implementation (Adapted from the USEPA[18]).

approach usually is adopted to improve state agency efficiency, in most cases it has led to enhanced involvement of stakeholders as well, and the trend is toward more localized, partnership-based approaches driven by multi-stakeholder teams.[19] Typically, the state is divided into major watershed units, and CWA activities are implemented on a roughly five-year activity cycle that is staggered to begin in different years by watershed (Figure 9-A.6). The cycle begins with monitoring and assessment and continues through planning and implementation. *Assessment* as referred to here entails comparison of monitoring data and water quality standards (WQS), a process that should detect likely adverse effects from stressors for which WQS have been determined but that falls short of risk assessment per se (see Chapter 6). While economic or other social-science studies are not precluded as part of this process, natural science is emphasized. In theory, activities such as review of designated uses, listing of impaired waters, issuance or review of point-source discharge permits, and award of loans and grants for water quality improvement projects are carried out in the implementation phase of this cycle, although in practice limited resources and competing priorities make this difficult to accomplish.[19] Total maximum daily loads (TMDLs) may be developed and implemented for high-priority impaired waters; here the TMDL process is depicted as a situational cycle linked to the regular management cycle (Figure 9-A.6).

REFERENCES

1. USEPA, Guidelines for ecological risk assessment, EPA/630/R-95/002F, Risk Assessment Forum, U.S. Environmental Protection Agency, Washington, D.C., 1998.

2. USEPA, A framework for the economic assessment of ecological benefits, Science Policy Council, U.S. Environmental Protection Agency, Washington, D.C., Feb. 1, 2002.
3. USEPA, Environmental monitoring and assessment program (EMAP) research strategy, EPA/620/R-98/001, U.S. Environmental Protection Agency, Washington, D.C., 1997.
4. Walmsley, J.J., Framework for measuring sustainable development in catchment systems, *Environ. Manage.*, 29, 195, 2002.
5. Barber, M.C., Environmental monitoring and assessment program indicator development strategy, EPA/620/R-94/022, U.S. Environmental Protection Agency, Office of Research and Development, Athens, GA, 1994.
6. Jackson, L.E., Kurtz, J.C., and Fisher, W.S., Evaluation guidelines for ecological indicators, EPA/620/R-99/005, U.S. Environmental Protection Agency, Office of Research and Development, Washington, D.C., 2000.
7. Timmerman, J.G., Ottens, J.J., and Ward, R.C., The information cycle as a framework for defining information goals for water-quality monitoring, *Environ. Manage.*, 25, 229, 2000.
8. Stahl, R.G., Bachman, R.A., Barton, A.L., et al., *Risk Management: Ecological Risk-Based Decision-Making*, Society for Environmental Toxicology and Chemistry, Pensacola, FL, 2001.
9. PCCRARM, Framework for environmental health risk management, Presidential/Congressional Commission on Risk Assessment and Risk Management, Washington, D.C., 1997. Available at http://www.riskworld/Nreports/1997/risk-rtp/pdf/EPAJAN.PDF. Accessed Oct. 6, 2004.
10. SAB, Toward integrated environmental decision-making, EPA-SAB-EC-00-011, U.S. Environmental Protection Agency, Science Advisory Board, Integrated Risk Project, Washington, D.C., 2000.
11. USACE, Planning guidance notebook, ER 1105-2-100, U.S. Army Corps of Engineers, Washington, D.C., 2000.
12. World Commission on Dams, *Dams and Development: A New Framework for Decision-Making*, Earthscan Publications Ltd., London, 2002.
13. USEPA, Watershed protection: A project focus, EPA/841/R-95/003, Office of Water, Environmental Protection Agency, Washington, D.C., 1995.
14. Davenport, T.E., *The Watershed Project Management Guide*, Lewis Publishers, Boca Raton, FL, 2002.
15. Cole, R.A., Feather, T.D., and Letting, P.K., Improving watershed planning and management through integration: A critical review of Federal opportunities, IWR Report 02-R-6, U.S. Army Corps of Engineers, Institute for Water Resources, Alexandria, VA, 2002.
16. USEPA, National forest system land and resource management planning, *Fed. Reg.,* 65, 67513, 2000.
17. USEPA, National forest system land and resource management planning; proposed rules, *Fed. Re.g.,* 67, 72769, 2002.
18. USEPA, Watershed protection: A statewide approach, EPA/841/R-95/004, Office of Water, U.S. Environmental Protection Agency, Washington, D.C., 1995.
19. USEPA, A review of statewide watershed management approaches, Office of Water, U.S. Environmental Protection Agency, Washington, D.C., 2002.

APPENDIX 9-B

Sociocultural Assessment Methods

CONTENTS

The comparison of management alternatives that is called for in Chapter 9 necessitates a multidisciplinary perspective. Integrated watershed management may combine ecological risk assessment with economics, human health risk assessment, or sociocultural approaches (drawing from geography, sociology, anthropology, or political science) to assess each alternative.[1] The inclusion of sociocultural information is often vital because of the complexity and uncertainty inherent in environmental issues and policy decisions.[2,3] The public is increasingly involved in consultation or decision-making processes to value resources, inform scientific practice, and prioritize management options. Sociocultural assessment is useful for identifying root causes of pollution and appropriate mitigation strategies, for identifying impacted groups and stakeholders, and for understanding the sociocultural context that surrounds and influences decision-making and implementation. The term *sociocultural* is used here to focus on pertinent social, political, and cultural factors apart from conventional economic methods, although it does not necessarily exclude the economic context in which decisions are made. This appendix first addresses the need for sociocultural assessment and then briefly reviews methodological approaches. A case study highlights how these approaches were combined to create an integrated watershed management assessment to support decisions.

1-56670-639-4/05/$0.00+$1.50

NEED FOR SOCIOCULTURAL ASSESSMENT

The social, political, and cultural context of a locality can exacerbate or mitigate ecological risks, can determine how environmental management actions will impact people and their communities, and can help determine whether management initiatives are likely to be successful. The configuration of industry, farming, commerce, cultural traditions, social organization, political activism, and institutions of governance influence how humans impact, and in turn are affected by, their environment. For example, insufficient land-use planning or zoning may result in haphazard development patterns that increase stormwater runoff, alter riparian ecosystems, and result in increased flooding and water quality degradation. Community values with respect to cultural and natural landscapes, and associated cultural traditions, may determine whether people feel helped or harmed by an ecological restoration project. Finally, a lack of trust between a government agency or risk manager and local stakeholders can doom a management initiative before it begins. An analysis of the sociocultural context may be essential to understanding a community's level of risk and their willingness to support a management initiative.[4,5]

The importance of social analysis and public participation is illustrated by an unsuccesful management initiative in the Big Darby Creek watershed, located in central Ohio (described in Chapter 10). The unique ecological resources of the watershed are threatened by runoff from agriculture and increasing urbanization. In an attempt to protect the resources in the northwest portion of the watershed, the U.S. Fish and Wildlife Service (USFWS) proposed to create the Little Darby National Wildlife Refuge. The plan called for retiring farmland to create a 20,000-acre wildlife refuge and to manage an additional 20,000-acre conservation area for water quality protection.[6] However, the assessment of management alternatives did not adequately address the human dimensions or social impact of the watershed management strategy. Ultimately, the proposed wildlife refuge was abandoned due to strong public opposition.

The farming community of the Darby watershed supported environmental protection efforts but was skeptical of government intervention in local land management. Problems arose because the Draft Environmental Impact Statement (Draft EIS)[6] failed to adequately consider impacts on a farmer's ability to maintain an economically viable business. According to survey data collected in the area, a number of landowners indicated they were willing to sell their land or development rights for establishing the refuge. The USFWS thought they had the necessary support to proceed with the refuge. However, the majority of those willing to sell were absentee landlords. The local resident farmers relied heavily on leasing land from the absentee landlords to increase acreage under production, a necessary step to be profitable. The removal of land from circulation and the resultant price increases for remaining available farmland would threaten the economic survival of farms and the cultural cohesiveness of the farm community.[7]

Although the Draft EIS had included economic and social factors, it did not consider the impact on local farmers' livelihoods and did not place the information in the larger sociocultural context. The resident farmers were not adequately consulted in the establishment of the management alternatives, and hence became very suspicious and critical of the USFWS-proposed actions.[7] The farmers' opposition completely derailed the wildlife refuge as a strategy to protect the natural resources of the area.[8] Early interaction between the USFWS and the community, including public deliberation and negotiation on management alternatives, might have avoided the pitfalls and lack of trust that doomed the Little Darby Wildlife Refuge. The case illustrates how sociocultural information and public input into ecological assessment can be essential to successful implementation.

ROLE OF SOCIOCULTURAL ASSESSMENT

There is debate over the role of sociocultural information and public participation in analyzing environmental management alternatives. Some scientists advocate for a strict separation of ecological assessment from assessment of political or cultural factors. Conventional approaches to environmental management may place community values and public input entirely outside the scientific arena.[9,10] But scientists increasingly recognize the need for positioning analysis and technical expertise in the context of community values and social systems.[9,11,12] As outlined in Table 9.1 and discussed in Appendix 9-A, the form of public participation varies widely between assessments and can range from no public input, to consultation with educated community leaders,[13] to diversified and broad deliberation that informs decision-making,[14] to grassroots initiatives with concerned citizens sharing decision-making powers with public and private entities.[15]

What factors does sociocultural assessment examine? In the literature on social impact assessment (as defined under the National Environmental Policy Act of 1969 and also as a general approach to human dimensions of environmental management), considerable effort has been put into defining and listing elements for sociocultural analysis.[10,16,17] Vanclay[17] defines social impact assessment as (p. 190):

> . . . the process of analyzing (predicting, evaluating and reflecting) and managing the intended and unintended consequences on the human environment of planned interventions (policies, programs, plans, projects) and any social change processes invoked by those interventions so as to bring about a more sustainable and equitable biophysical and human environment.

The social impacts considered may include health and safety, demographics, quality of life, travel patterns, access to and delivery of services, neighborhood cohesion, property values, or relocations. Impacts may also address gendered use of natural resources; indigenous rights; cultural, institutional, and political dimensions of use; psychological impacts; and impacts to the community.[16,17] The social impacts of concern to a particular watershed management initiative are site and context specific, and therefore may vary widely between cases.

One critique of social assessment is that it has been too narrowly focused on quantifiable data from readily available economic and demographic information (population change, job creation, or quantitative measures of services use). The social impact assessments made with this type of information have not addressed a variety of impacts — such as on livelihood, cultural heritage, attachment to place, gender, or age factors — or considered these in the larger sociocultural context. As a result, many social impact assessments have gravely underestimated negative effects on local communities.[17]

In response to these critiques, a more formal combination of scientific analysis and public debate, called an *analytic–deliberative process*, has been advocated by the National Research Council.[18] The approach is similar to Gregory's structured deliberation approach to risk management.[14] It relies on rigorous scientific investigation in natural, social, or decision sciences combined with public or stakeholder discussion of community values, goals, analytic techniques, and assumptions used in modeling and risk characterization. It is an iterative process that allows scientific analysis and public deliberation to inform each other.[15,18] The following section uses the analytic–deliberative framework to review methods commonly employed in social analytic processes (quantitative and qualitative) and deliberative processes (public participation, focus group meetings), recognizing that these are closely intertwined rather than separate processes.

ANALYTIC METHODS: SOCIAL SCIENCE

In assessing the impacts of management alternatives, consideration of sociocultural elements may involve the use of both quantitative and qualitative analytic methods. While quantitative data use numbers and mathematical relationships to represent phenomena, qualitative data provide the "meanings, concepts, definitions, characteristics, metaphors, symbols and descriptions" of things under study.[19] Often sociocultural assessment will combine these approaches, using quantitative information to inform and guide a more in-depth qualitative analysis. In addition, socioeconomic data can help identify potentially affected populations and interested stakeholders.

Investigations of the sociocultural context will often incorporate basic information about a community such as boundaries, physical features, landscape, land cover and land use, demographics, and economic conditions. Many of the quantitative methods for social assessment use data collected from survey questionnaires or employ secondary data reported by public or private institutions (e.g., government census data or business collected data).[20]

Spatial-analytic techniques employing geographic information systems (GIS) and remote sensing technologies are powerful tools for organizing and analyzing geo-referenced information about physical and human environments.[21–23] The growing use of GIS by states, counties, and municipalities, and the availability of georeferenced data, have made demographic, housing, and economic data accessible for watershed management planning. The distributive effects of management options are more easily identified using demographic data in a GIS to identify subpopulations that may be at increased risk or bear disproportionate costs.[24,25] However, GIS is limited in identifying many sociocultural elements that may be impacted by management alternatives, such as livelihoods, community culture, goals, values, beliefs, perceptions, or ways that people use natural resources.[26] Disadvantages of GIS also include the cost of computer systems and data, and the required analytical expertise when using geo-referenced data.[23,27]

Sociocultural analysis should move beyond quantifiable factors to consider the characteristics that give a sense of place to a community including attitudes, values, perceptions, traditions, religious and spiritual practices, information flows, and decision-making.[28] Methods commonly used to elicit this information include the following: (1) interviews — open-ended questions or discussions posed to individuals or groups;[29] (2) Participation and Observation — direct observation of community features or interaction with subjects, which may take place during focus groups, organizational meetings, or community events; (3) Survey and Information Collection — various methods including written questionnaires, rank ordering or voting on key issues and impacts, counter mapping (mental maps, social networks, asset maps, conceptual diagrams), and others;[28] and (4) Content Analysis — interpretation of textual materials including built or natural landscapes, archival material, maps, literature, and visual images.[30,31]

Qualitative analysis has been criticized for being overly subjective and biased by the researcher's interpretation. However, rigorous methodological approaches exist for organizing and analyzing data and presenting results.[19,29] Some researchers employ quantitative methods (e.g., Q-method, factor analysis) to analyze qualitatively collected information.[32]

DELIBERATIVE METHODS: PUBLIC PARTICIPATION

In addition to social science analysis of people and places, there are various methods to facilitate public input into management, policy, scientific methods, and decision processes. Information collection and stakeholder input at public meetings can take the form of focus groups engaged in problem definition, voting to prioritize environmental problems, ranking of potential solutions, or consensus decision-making.[28,33] The objectives of public participation meetings or focus groups can range from educating the public on scientific or technical issues; to eliciting information from the public on their values, priorities, and perceptions; to building rapport among diverse stakeholders; to reaching a consensus decision acceptable to all participants.

Public participation meetings often use a professional facilitator to moderate dialogue, negotiation, and decision-making. Techniques for soliciting information include open brainstorming sessions or round-robin approaches to create an exhaustive list of ideas. The Delphi and Nominal techniques are particularly useful in bringing together diverse and contentious stakeholders, often for a consensus decision. These processes typically use anonymously written responses to a series of questions or issues presented. The compiled opinions and arguments for and against the issue at hand are then presented to the group. Keeping individual responses anonymous provides an opportunity for stakeholders to change their opinion without publicly admitting it and prevents one person or group from dominating the discussion or decision.[28]

Sociocultural assessment often involves multiple methods to identify and assess the relevant social, political, and cultural factors impacting management alternatives.

The multidisciplinary approach is essential to a successful process that takes seriously public values, goals, and the social context in which decisions and implementation take place.

SOCIOCULTURAL ASSESSMENT IN COLORADO SPRINGS

To highlight the contribution of sociocultural assessment to watershed management, a case study illustrates the process of combining social science analysis and public participation.[9] To meet the increasing water demands of a growing population, the city of Colorado Springs, Colorado acquired water rights in the Arkansas River Basin. It examined three proposed management plans for bringing water to the city. Several communities along the Arkansas River would be impacted, either by decreased flows or infrastructure of the dam, reservoir, pipeline, or diversion. The three management alternatives proposed by the city were:

1. Elephant Rock Dam: This alternative was the most expensive due to dam and reservoir construction, 70 miles of pipeline, and required road and railroad relocation, but it was also the most preferred by the city's water resources department professionals. It would deliver high-quality water requiring little treatment, involving low pumping costs, and utilizing existing city infrastructure for storage and distribution.
2. Mount Princeton Diversion: This option would create a low-head overflow diversion on the Arkansas River that was less costly because it avoided road and railroad relocation. It did not require a reservoir in the mainstem of the Arkansas River.
3. Pueblo Reservoir Storage: Water would be collected farther downstream, avoiding major disruption to the Arkansas River and preserving local recreation and tourism on upstream portions of the river. This option had higher costs than the second alternative due to pumping requirements and the need for new water treatment facilities.

There was strong public resistance to Elephant Rock Dam, and several of the city's public meetings turned hostile. The opposition was not only from residents of small towns along the Arkansas, but also from city residents opposing alteration of the valued river ecosystem. In trying to resolve the situation, the city of Colorado Springs realized that they needed to investigate the "social issues that often are beyond the professional concerns of the engineers who design such projects."[9] The city asked a group of social scientists to assess the socioeconomic impacts of the water development alternatives. The article by Cortese[9] outlines the details of the case study described here.

The first step of the analysis was to gain an understanding of the affected communities and their associated social histories. The analysis employed quantitative and qualitative methods to elicit the required social and economic information. First quantitative data on demographics, housing, and economics were examined. Next, content analysis on public meeting transcripts produced a list of initial stakeholders and interest groups (defined as people with an interest in the Arkansas River or those potentially impacted by the water project alternatives). Researchers then conducted

Table 9-B.1 Stakeholder Values Regarding Water Supply Development Options in Colorado Springs[a]

Primary Value	Explanation
Environmental Resources	Protection of wildlife, riparian vegetation, pristine views, and clean water outweighed recreational opportunities from the proposed reservoir.
Community Resources	A high value was placed on centers of community activity and a nature preserve that provided environmental education. It was unacceptable to lose these places (even if replaced with buildings in new locations) because it was seen as disrupting community stability.
Social Environment	There was shared concern over the expected disruptions from construction, influx of temporary workers, increased traffic, and relocation of homes and businesses.
Economic Base	Protection of the tourism industry on the Arkansas River was a priority, and surprisingly, city communities were found to be willing to suffer some financial loss and construction impacts from building a new treatment plant to protect distant communities' well-being and more pristine environments.
Secure Future	The small communities feared a loss of local control over communities and resources. The residents viewed water transfers as eroding local autonomy and threatening the loss of rural livelihoods and farms.

[a] The table summarizes information from Cortese.[9]

key informant interviews with leaders and activists engaged in the water resources arena. The information collected during the interviews helped to identify primary concerns, key players, and the relative positions of stakeholder groups.

The next step involved focus group workshops that included representatives from each of the interest groups involved, both individuals and organizations. In these workshops, investigators conducted focus group interviews, used a round-robin method to produce an exhaustive list of issues, and then facilitated participant discussion of each issue of concern. Participants were asked to identify, weigh, and rank social and economic issues. Through the process, the workshops helped identify what to study and why it was important, and they provided a forum for public participation early in the process. Finally, the analysis included a survey to see how issues were rated by the general population of the area (for a complete description of methods and results, see Cortese[9]).

The analysis revealed important information about the context of opposition and decision-making for Colorado Springs, including why residents opposed the construction of the dam and some unexpected positions supporting the publicly preferred management alternative (Table 9-B.1). The results from the analytic–deliberative process could not have been discovered without the participatory and focus group exercises to elicit information. Stakeholders in Colorado Springs strongly valued the natural functioning of the river ecosystem over recreation opportunities provided by the reservoir at Elephant Rock Dam. Differing from the usual urban-versus-rural divide in resource management issues, the social analysis revealed that city residents also favored the management alternative that minimized the ecological impacts to the river and preserved ecotourism and local uses of the river. City residents took a conservation approach (use but also protect local autonomy and economic use) over a preservation approach (minimal human use but high ecological protection).

The Colorado Springs water resources department was able to combine information from the social science analysis with engineering studies, environmental assessment, and cost analysis. The publicly preferred alternative, the Pueblo Reservoir Storage, was surprising to many stakeholders. The City Council, acting on a recommendation from the water resources department, unanimously approved this alternative with little opposition from communities along the Arkansas River.[9]

This case study features the use of an analytic–deliberative process to assess the sociocultural impacts of environmental management alternatives. The integrated assessment produced information about sociocultural factors shaping the context of decision-making and public acceptance. Although the selected management alternative differed from the initial preference of the water resource professionals, it minimized adverse sociocultural, economic and environmental impacts and had support across diverse communities. Accurate environmental and engineering analyses alone were not sufficient. Successful implementation required public participation and qualitative data gathering to fully assess the impact on local communities and to decide on the management alternative that minimized physical and social impacts.

An analytic–deliberative approach to watershed management that incorporates natural science, economics, and social assessment, and includes public participation, can better inform decision-making. The Ohio and Colorado case studies emphasize the importance of early public participation and consultation processes. In addition to the analysis of easily quantifiable economic or landscape data, sociocultural analysis must address qualitative aspects that give meaning, significance, and a sense of place to a community. The environmental assessment must be positioned in light of the larger sociocultural context. Multidisciplinary approaches to assessing management alternatives hold great promise for producing results that are biophysically sound, account for the social values of affected communities, and respond to the needs of decision-makers.

REFERENCES

1. Bellamy, J.A., Walker, D.H., McDonald, G.T., and Syme, G.J., A systems approach to the evaluation of natural resource management initiatives, *J. Environ. Manage.*, 63, 407, 2001.
2. Funtowicz, S. and Ravetz, J., The worth of a songbird: Ecological economics as a post-normal science, *Ecol. Econ.*, 197, 1994.
3. McCarty, L. and Power, M., Letter to the editor: Environmental risk assessment within a decision-making framework, *Environ. Toxicol. Chem.*, 16, 122, 1997.
4. Irwin, A., *Sociology and the Environment: A Critical Introduction to Society, Nature and Knowledge*, Polity, Cambridge, UK, 2001.
5. Slovic, P., Perceived risk, trust and democracy, in *Social Trust and the Management of Risk*, Lofstedt, R. and Cvetkovich, G., Eds., Earthscan, London, 1999.

6. USFWS, Draft environmental impact statement on the Little Darby National Wildlife Refuge, U.S. Fish & Wildlife Service, Ft. Snelling, MN, 2000.
7. Bargar, R., Open Letter to Ohio EPA on Darby Wildlife Refuge, December 20, 1999.
8. USFWS, Little Darby Creek conservation through local initiatives: A final report, U.S. Fish & Wildlife Service, Ft. Snelling, MN, 2002.
9. Cortese, C., Conflicting uses of the river: Anticipated threats to the resource, *Soc. Nat. Resour.*, 16, 1, 2003.
10. King, T., How the archeologists stole culture: A gap in American environmental impact assessment practice and how to fill it, *Environ. Impact. Assess. Rev.*, 18, 117, 1998.
11. Fight, R.D., Kruger, L.E., Hansen-Murray, C., Holden, A., and Bays, D., Understanding human uses and values in watershed analysis, PNW-GTR-489, Pacific Northwest Research Station, U.S. Department of Agriculture, Forest Station, Portland, OR, 2000.
12. Forsyth, T., *Critical Political Ecology: The Politics of Environmental Science*, Routledge, New York, NY, 2003.
13. Becker, N. and Easter, K.W., Water diversions in the Great Lakes Basin analyzed in game theory framework, *Water Resour. Manage.*, 9, 221, 1995.
14. Gregory, R., Using stakeholder values to make smarter environmental decisions, *Environment*, 42, 34, 2000.
15. Weber, E., *Bringing Society Back In: Grassroots Ecosystem Management, Accountability, and Sustainable Communities*, Massachusetts Institute of Technology, Cambridge, MA, 2003.
16. Marriott, B., *Environmental Impact Assessment*, McGraw-Hill, New York, 1997.
17. Vanclay, F., Conceptualising social impacts, *Environ. Impact. Assess. Rev.*, 22, 183, 2002.
18. NRC, Understanding risk: Informing decisions in a democratic society, National Research Council, National Academy Press, Washington, D.C., 1996.
19. Berg, B., *Qualitative Research Methods for the Social Sciences*, Allyn and Bacon, Boston, MA, 1998.
20. Sheskin, I., *Survey Research for Geographers*, Association of American Geographers, Washington, D.C., 1985.
21. Chrisman, N., Revisiting fundamental principle of GIS, in *Socio-Economic Applications of Geographic Information Science*, Kidner, D.B., Higgs, G., White, S., and Kidner, D.W., Eds., Taylor and Francis, London, 2003.
22. Kuczenski, T., Field, D., Voss, P., Radeloff, V., and Hagen, A., Integrating demographic and Landsat™ data at a watershed scale, *J. Am. Water Resour. Assoc.*, 36, 215, 2000.
23. Stillwell, J. and Clarke, G., Eds., *Applied GIS and Spatial Analysis*, John Wiley & Sons, Chichester, England, 2003.
24. Craig, W., Harris, T., and Weiner, D., *Community Participation and Geographic Information Systems*, Taylor & Francis, New York, 2002.
25. Sheppard, E., Leitner, H., McMaster, R., and Tian, H., GIS-based measures of environmental equity: Exploring their sensitivity and significance, *J. Expo. Anal. Environ. Epidemiol.*, 9, 18, 1999.
26. Rocheleau, D., Maps, numbers, text, and context: Mixing methods in feminist political ecology, *Prof. Geogr.*, 47, 458, 1995.
27. Griffith, D. and Layne, L., *A Casebook for Spatial Statistical Data Analysis*, Oxford University Press, New York, 1999.
28. USEPA, Community culture and the environment: A guide to understanding a sense of place, EPA/842/B-01/003, Office of Water, Washington, D.C., 2002.
29. Baxter, J. and Eyles, J., Evaluating qualitative research in social geography: Establishing 'rigour' in interview analysis, *Trans. Inst. Br. Geogr.*, 22, 505, 1997.

30. Johnston, R., Gregory, D., Pratt, G., and Watts, M., Eds., *The Dictionary of Human Geography*, Blackwell, Malden, MA, 2000.
31. Strauss, A. and Corbin, J., *Basics of Qualitative Research: Grounded Theory, Process and Techniques*, Sage Publications, Newbury Park, CA, 1990.
32. Addams, H. and Proops, J., *Social Discourse and Environmental Policy: An Application of Q Methodology*, Edward Elgar Publishing, Cheltenham, UK, 2000, 14.
33. Beierle, T. and Cayford, J., *Democracy in Practice: Public Participation in Environmental Decisions*, Resources For The Future, Washington, D.C., 2002.

CHAPTER 10

Evaluating Development Alternatives for a High-Quality Stream Threatened by Urbanization: Big Darby Creek Watershed

O. Homer Erekson, Orie L. Loucks, Steven R. Elliot, Donna S. McCollum, Marc Smith, and Randall J.F. Bruins

CONTENTS

1-56670-639-4/05/$0.00+$1.50

A vision for integrating ecological risk assessment (ERA), economics, and watershed decision processes has been presented in the previous chapter. The objective in this chapter is to consider a case study in which certain elements of that conceptual approach (see Figure 9.1) are implemented and field-tested with specific data. A large watershed in central Ohio, the Big Darby Creek, provides the locale and basis for the study design.

In 1993, the Big Darby Creek watershed was selected by the U.S. Environmental Protection Agency (USEPA) for one of five watershed ecological risk assessment (W-ERA) case studies for several reasons: the substantial interest by organizations at the local, state, and federal level in protecting the watershed; the outstanding character of the aquatic biological resource; the range of sources and stressors (agricultural nonpoint sources, urban nonpoint sources, permitted discharges, etc.); the existence of a large, multiple year, watershed-wide database; and a commitment by the Ohio EPA (OEPA) to co-lead the risk assessment team.

In 1999, while the W-ERA was in the later stages of completion, a USEPA-funded study was initiated by Miami University with the goal of integrating ERA and economic analysis to further inform environmental management efforts in the Big Darby Creek watershed. The methodological framework for this integrated research was rooted in a broadly based approach to sustainability that encompasses, but extends beyond, ERA. This approach views economic development as complementary with, rather than antagonistic to, the maintenance of nonrenewable resources. As such, it argues that sustainable systems require coordination between ecological, economic, and social considerations to maintain overall system resilience.[1]

Because the Miami University study was initiated well after the Big Darby Creek W-ERA (see Chapter 1), using information from the latter but carried out by a separate team, the two efforts were not integrated in an ideal sense. However, the research approach that was used illustrates some of the advantages, as well as the difficulties, of integrated study. The first section of this chapter, "Watershed Description," introduces the watershed setting, and the "Ecological Risk Assessment" section discusses the W-ERA effort and its findings. The Miami University study is presented under "Economic Analysis," and the "Discussion" section evaluates these findings in light of the larger integration problem.

WATERSHED DESCRIPTION

Big Darby Creek is a high-quality, warm water stream located in the Eastern Corn Belt Plains ecoregion of the Midwest (Figure 10.1). The watershed encompasses 1443 km^2 (557 mi^2) and is home to a diverse community of aquatic

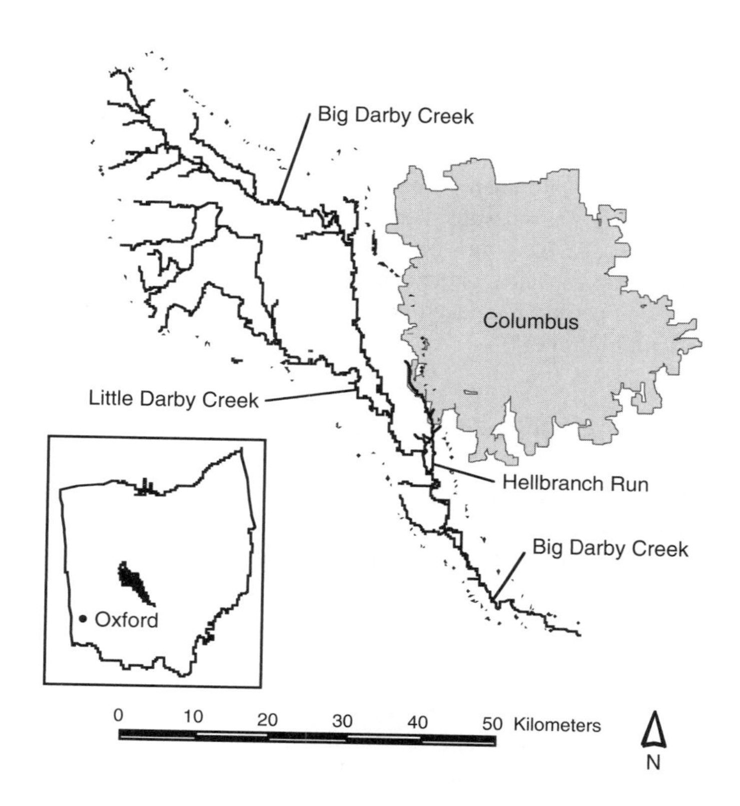

Figure 10.1 The Big Darby Creek watershed in central Ohio, USA. The Columbus metropolitan area is expanding into the easternmost area of the watershed, where Hellbranch Run is especially affected. Respondents surveyed in this study were drawn either from the watershed area, Columbus, or Oxford.

organisms including many rare and endangered fish and freshwater mussel species. The Big Darby Creek watershed was given a conservation priority by The Nature Conservancy (TNC) through its recognition as one of the "Last Great Places" in the western hemisphere.[2,3] The risks to ecological resources in the Big Darby watershed derive from ongoing changes in agriculture and suburban land use.

The watershed drains portions of six counties in rural Ohio just west of Columbus. Agriculture currently comprises 92.4% of the land use of the watershed. Cropland, most of which is actively row-cropped, is the highest use (72%), followed by livestock pasture (8.6%). However, suburban Columbus is expanding westward into the Big Darby watershed. Currently, the western tributaries drain agricultural lands almost exclusively, whereas the eastern tributaries drain areas with increasing suburban, commercial, and industrial land use. Urban development recently has quadrupled in some areas, with significant negative consequences for stream habitat. Although there have been recent improvements in fish and invertebrate indices in the Big Darby Creek mainstem, the easterly Hellbranch Run shows degradation.[3,4] A number of stream reaches in the watershed have been listed as impaired and are subject to potential regulation through development of *total maximum daily loads* (TMDLs; see Chapter 6), mostly focused on phosphorous, nitrogen, and sediment.

To the west from Hellbranch Run, the urban and industrial impacts are generally not greater than agricultural impacts, but given the present population of the region and the rapid rate of development, urban water pollution problems are a risk for a large part of the Big Darby watershed in the future. Without management, the increased frequency of damaging storm runoff and associated pollutant loads pose risks to the uncommon species, game fish, and general aquatic system functioning. These are risks that could be reduced through best management practices for both urban and agricultural runoff.[5]

ECOLOGICAL RISK ASSESSMENT

The phases of ERA as described in the USEPA's *Guidelines*,[6] that is, planning, problem formulation, analysis, and risk characterization, are summarized in Chapter 3. This section describes the work that was conducted in each phase of the W-ERA for Big Darby Creek.

Planning

The OEPA database available for this assessment included standard water quality parameters such as suspended and dissolved solids, pH, oxygen-demanding substances, nutrients, ammonia, and metals. It also included biological assemblage data describing the presence and abundance of fish species and of macroscopic sediment-dwelling invertebrates (termed *benthic macroinvertebrates* or *benthos*) collected by standard sampling procedures. Also available were a set of descriptors of stream corridor condition, including condition of substrates, instream habitat types (pools, riffles), channel stability, and riparian zone vegetation. Multimetric indices that provided a composite assessment of habitat or biological quality, based on these data, included the Qualitative Habitat Evaluation Index (QHEI) for stream corridor condition; the Index of Biotic Integrity (IBI) and the Modified Index of Well-being (MIwb), which are measures of the functional and structural organization of the fish community, respectively; and the Invertebrate Community Index (ICI), which evaluates the structural organization of the macroinvertebrate community. These indices have been used extensively by the OEPA to establish biological criteria and to evaluate stream use attainment (see Chapter 6 and Appendix 6-A for further description of these indices).[7]

Cooperators in the Big Darby Creek Watershed Ecological Risk Assessment included the W-ERA team co-chairs from the USEPA and OEPA and at various times representatives from The Ohio State University, The Nature Conservancy, the United States Geological Survey (USGS), and Operation Future, a conservation-oriented farm group. Management goals for the risk assessment were developed through review of pertinent regulations, discussions with residents and resource managers, and meetings with the Darby Partners, a loose-knit group of over 40 public agencies and private organizations united by the shared goal of watershed protection. The overarching risk reduction goal from these discussions was to "protect and maintain native stream communities of the Big Darby ecosystem." Three specific objectives were seen as necessary to meet this risk-reduction target:

1. Attaining criteria for designated uses throughout the watershed (see Chapter 6).
2. Maintaining the OEPA's exceptional warm water criteria for all stream segments having that designation between 1990 and 1995.
3. Ensuring the continued existence of all native species in the watershed.

The risk management problem was to ensure that these specific objectives could be met. The risk characterization would require understanding how various environmental factors might prevent meeting these objectives.

Problem Formulation

Ecological assessment endpoints are measurable attributes of valued ecological characteristics. Two assessment endpoints were chosen for the Big Darby risk assessment:

1. Species composition, diversity and functional organization of the fish and macroinvertebrate communities.
2. Sustainability of native fish and mussel species.

From a practical standpoint, the first of these endpoints could be evaluated utilizing three composite indices (IBI, MIwb, and ICI) and the individual measures they comprise. It was determined ultimately, however, that while some of the available data were relevant to the second endpoint, the necessary information on life history and genetic diversity of native species in the watershed was not sufficient for evaluating their sustainability.[3] Therefore, only the first endpoint was carried further.

A critical step in problem formulation is the development of a conceptual model. It articulates the risk assessors' hypotheses on the relationships among the sources of stress, stressors, effects, and endpoints. Six significant stressors were identified for this watershed as affecting the assessment endpoint: altered stream morphology, increased flow extremes, sediment, nutrients, temperature, and toxicants. A conceptual model illustrating the hypothesized relationship between land use, sources of stress, the aforementioned stressors, subsequent ecological interactions, and the stressor signatures (i.e., characteristic changes in aquatic community metrics) is presented elsewhere.[3]

Seven risk hypotheses were developed based on the relationships inherent in the conceptual model:

1. No differences exist in community structure and function among the subwatersheds.
2. No differences exist in community structure and function among time periods.
3. Community structure and function will decline downstream from identified point sources.
4. An increase in certain land uses or land-use activities will result in a change in the IBI or the ICI.

5. An increase in certain land uses or land-use activities will result in an increase in the intensity or spatial or temporal extent of in-stream stressors.
6. An increase in the intensity, or spatial or temporal extent, of in-stream stressors will result in a change in the biological community as quantified by ICI and IBI metrics and species abundances.
7. The pattern of response of the stream community can discriminate among the different type of stressors.

The first two were null hypotheses; analysis would determine whether they could be statistically rejected. The other five were maintained hypotheses, thought to be true; analysis would seek confirmatory or contradictory evidence.

Current Status of Analysis and Risk Characterization

The analysis to test these hypotheses was carried out in two phases. Hypotheses 1 and 2 were tested by analyzing historical biological assemblage data within the Big Darby Creek watershed.[4] Both hypotheses were rejected because certain spatiotemporal differences were shown in the analysis. Time series analysis, which was feasible for fish community metrics and IBI within the Big Darby Creek mainstem, indicated a general improvement over the time period 1979–1993. At the same time, spatial comparisons among the Big Darby, Little Darby, and Hellbranch Run subwatersheds revealed significant spatial differences for IBI, ICI, and several component metrics. In general, the Big Darby Creek mainstem showed superior biotic condition; however, some of this difference could be attributed simply to its comparatively larger drainage area. After correction for drainage area, many differences disappeared, but the biotic condition of the urbanized Hellbranch Run remained lower than the mainstem according to several measures.[4] These findings, while encouraging for the watershed as a whole, were consistent with concerns that suburban encroachment threatens watershed ecological resources in the eastern portion of the watershed. However, without an ability to correlate biological condition with stressors of concern or their sources, these results were of limited value for assessing risks associated with likely future changes in the watershed.

By contrast, hypotheses 3 to 7 required the analysis of point sources, land uses, and stressors in spatial relation to biological data. Relatively few point sources of pollution are present in the watershed, but most have shown negative effects on the mussel community for some distance downstream. Migration of species within the fish community making up IBI tended to remove the downstream effect. Thus, hypothesis 3 was confirmed for metrics focused on invertebrate species, but not for free swimming migratory species such as fish.

Initial attempts to analyze stressor effects derived from land-use patterns were complicated by the watershed's relatively good water quality and higher-than-average IBI. The narrow range of variability in the biotic metrics and the chemical and physical parameters seen in the Big Darby needed to be assessed in the context of the greater variability of the region as a whole. Therefore, Norton et al.[8] analyzed biological, chemical, and habitat data for the Big Darby Creek and other comparably sized watersheds within the Eastern Corn Belt Plains ecoregion in Ohio, among which a wider gradient of the stressors and subsequent responses could be observed.

Discriminant functions constructed using biological variables from this larger dataset were used to separate site groups into high-, medium-, and low-stress categories along stressor gradients. Analysis of the biological variables here did distinguish between higher- and lower-quality sites classified on the basis of six different types of stressors: degraded stream corridor structure; degree of siltation; total suspended solids, iron, and biochemical oxygen demand (BOD); chemical oxygen demand (COD) and BOD; lead and zinc; and nitrogen and phosphorus. Functions based on biological variables could also discriminate between sites having different dominant stressors.[8]

Using somewhat different methods for their data aggregation and analysis, Gordon and Majumder[9] analyzed similar data, but they also included land use (dense urban, forested, or agricultural, as a percentage of each watershed) in an effort to develop regression models that could predict the ecological effects of future land use changes. A number of models showed some ability to explain average watershed IBI. For a set of 137 watersheds, the regression model explained 39.5% of variance in the IBI when only stream corridor characteristics, land use, and stream order were included (N = 467), 47.4% when an index of chemical pollution stress was added to the model (N = 196), and 65.5% when upstream IBI was added (to correct for spatial autocorrelation, N = 177). The percentage of dense urban land use was a strongly negative predictor. For the three models described, standardized regression coefficients for the percentage of dense urban land use (which relate the variance in that factor to the variance in IBI) were –0.305, –0.258, and –0.179, respectively.

Therefore, hypotheses 4–7 were shown to hold true for the Eastern Corn Belt Plains ecoregion, and the relationships found can reasonably be applied in the Big Darby Creek watershed. These preliminary results suggested that fish and macroinvertebrate community responses to land use, stream corridor habitat, and various chemical stressors are predictable to a degree. The USEPA's efforts to apply these findings to the assessment of ecological risks in the Big Darby Creek watershed are still ongoing. Additionally, because of the identified impairments to some of its subwatersheds, the Big Darby Creek is subject to the development of a TMDL by the OEPA. Similarly, in an effort to assist planners, environmental organizations, government agencies, and concerned citizens, scientists and planners in The Ohio State University's City and Regional Planning Program, working on a USEPA-funded grant, have created an interactive, geographic information systems (GIS)–based screening tool to evaluate the biological effects of various changes within the Big Darby Creek watershed and other watersheds within the Eastern Cornbelt Plains ecoregion.[10]

ECONOMIC ANALYSIS

The overarching goal for Miami University's integrated ecological and economic analysis was to utilize the findings of ERA in an economic analysis that would be relevant to environmental management decisions in the watershed. At the time of initiating our integrated study in the Big Darby Creek watershed, the problem-formulation phase and early portion of the analysis phase of ERA had provided a clear picture of current conditions and apparent threats. Because the spatial scope of the analysis had to be expanded to all Eastern Corn Belt Plains watersheds in Ohio, a full complement of stressor-response or source-response relationships was not yet available, and the risk characterization had not been carried out. However, the following sections show that sufficient information was available for meaningful analysis.

The objective for this integrated case study was to undertake an analysis capable of informing decisions about reducing risks from suburban development. An independent modeling study sponsored by Miami University's Center for Sustainable Systems Studies[11] had quantified the range of effects on hydrology, sediment transport, and nitrogen concentrations from changes in land use. This study had considered three types of residential development in the Big Darby basin and had found that two different types of low-density development protected the stream amenities very well. The analyses by Norton et al.[8] also informed the selection of stressors considered to be key influences on stream conditions following urbanization. Thus, the goal for this case study was an integrated evaluation of ecological and socioeconomic impacts associated with several land use approaches at the peri-urban fringe.

The specific objectives of the case study, therefore, were as follows: (a) to estimate the quantitative or qualitative impacts of a set of land-use scenarios on stream ecological condition, local economic well-being, and local quality of life, (b) to communicate these impacts to the public effectively, and to measure the overall economic value (see "Economic Value" in Chapter 5) corresponding to each scenario based on individual willingness to pay (WTP), and finally (c) to better understand the particular contribution stream ecological condition makes to the value of a given scenario.

Research Approach

Based on prior work in the Big Darby watershed,[11,12] four development scenarios were used to compare outcomes for stream amenities: (1) a most-likely case of high-density, conventional subdivisions using 1/4- to 1-acre lots with water and sewer services discharging to the Big Darby; (2) a low-density ranchette development on 3- to 5-acre lots with local water and septic system disposal; (3) a low-density cluster development, with intervals between clusters to achieve the same housing density as ranchettes (e.g., as maintained through purchase or set-aside of transferable development rights); and (4) a reference case of continued agriculture, which was the predominant land use pattern actually observed in the 1990s.

A dichotomous-choice contingent valuation method (CVM) survey instrument (see Chapter 5 and Appendix 5-A for a discussion of this method) was developed that allowed presentation of technical information on how changes in stream amenities are induced or avoided during land development, followed by expression of WTP for a certain outcome. Analysis was also carried out to develop a quantitative relationship between the four land-use scenarios and stream biological integrity based on empirical relationships.

The survey approach involved in-person, multimedia presentations to noninteracting groups of 30–50 respondents who completed a questionnaire. The instrument was designed according to Arrow et al.[13] and implemented according to Dillman.[14] In addition, multiple stakeholders were brought into a pretest phase to gain insight about their viewpoints, as well as their suggestions about refining the survey instrument.

The survey and presentation was divided into three parts. The first section asked respondents their knowledge about and use of the Darby Creek before we provided information about this watershed. An example of such a question is: *"Do you believe some types of residential development lead to increased soil erosion and runoff of fertilizers and pesticides?"* In this case, 90% of respondents answered that they were aware of this issue. A follow-up question for those who answered *yes* to the previous question asked: *"If so, do you think these runoff products do significant damage to streams and water quality?"* Again, a significant portion, 85%, of the respondents answered *yes*. Finally, those respondents who answered yes to both questions were asked: *"If so, do you think these runoff products will do damage to fish and other species in the stream*?" About 85% responded that they were aware of the damage of the runoff.

The second section was designed to engage the respondents as they were presented the effects that development might have on the environmental, social, and economic characteristics of the area. These effects are discussed here, but the reader must note that as this material was presented, respondents were asked questions about it. Many questions related back to the first section. For example they were asked: *"Did you know that the food base for many fish is tiny insect larvae that live on the stream bottom?"* About 30% of the respondents were not aware of this before the presentation. In another example, respondents were asked: *"Did you know that lawn and garden chemicals could affect the fish in the stream?*" Consistent with the results in the first section, 94% indicated that *yes* they were aware of this. The other 6% were made aware by the presentation. Thus it was possible to create a uniform minimum knowledge base across respondents. The final section was the valuation and demographics questions. These results are discussed in greater detail later in this section.

The respondents were drawn from three different populations. *Residents* were defined as people who live within the study area of the Big Darby Creek watershed, both farmers and non-farmers. *Near-Residents* were people living outside the watershed but within the greater Columbus, Ohio metropolitan area. *Non-Residents* were drawn from people in the area surrounding Oxford, Ohio, a two-hour drive from Big Darby Creek. Residents and near-residents capture the value attached by people who use the area for residence or recreation (use value). These two groups also capture nonuse value if they value the watershed solely for the benefit of acknowledging its existence or are willing to contribute to its preservation for future use. Non-residents may use the area for recreation, but at a much lesser rate than those in the Columbus area. The primary value for this group was expected to be nonuse value (see "Economic Value" in Chapter 5).

Samples were drawn at random from zip codes contained within each of the targeted areas. Respondents received a payment of $30 to cover their out-of-pocket travel costs and to show appreciation for the time spent in the one-hour presentation and survey response. Each respondent later received a mailed summary of survey results. The following sections develop the sequence of topics covered in the presentation to survey respondents and provide details on the valuation question.

Communicating the Effects of Urban Development on Ecological Endpoints

Scientific understanding of the mechanisms by which residential development brings about change in streams generally can be reduced to four causal factor groups:

1. Increased nutrients (which increase algal growth and affect the kinds of fish).
2. Increased sediment (which decreases light penetration and affects the food chain).
3. Increased toxic substances (which cause mortality in the food chain and fish).
4. Increased runoff and flooding (which allow bank erosion and sedimentation).

The most difficult challenge for this project concerned the need to have the public (represented as survey respondents) understand the mechanisms inducing change in the stream well enough for them to attach value to the outcomes they prefer. This question of informing respondents about linkage mechanisms was addressed by presenting the following synthesis on watershed processes and ERA.

Increased Nutrients, Leading to Change in Fish Species

Nutrients were described as chemicals that enhance growth of plants, on land as well as in the water. In high-density subdivisions, nutrients come from lawn fertilizers, from storm water runoff and, at some "downstream" locations, from household sewage. Runoff also carries soil and fertilizers from farmland, further enriching streams. The amount of nutrients entering the stream has been shown to depend on the number of people living in an area and how they manage their fields, lawns, or gardens. Nutrient loading also has been shown to depend on the amount of hard surfaces (roads and roofs) developed in a neighborhood.

The main effects of increased nutrients on the amenities to be valued are nuisance-level growths of algae in streams. This increased growth can cause a change in water quality and the kinds of fish that live there. In enriched streams, fish species that feed on decaying stream-bottoms (many minnows, carp, and catfish) are favored over those predator fish (e.g., bass and sunfishes) that feed on small fish. If nutrient input is very high, "fish kills" can occur. It was then possible to ask the respondents what types of fish they would rather have in a stream, and they were given the choice of (1) minnows, sunfishes, carp, and catfish, or (2) bass, sunfishes, suckers, and darters.

Increased Sediment, Leading to a Decreased Insect Food Base in Streams

Survey respondents also were told and shown that the amount of sediment entering a stream from residential areas or farmland can vary widely, but poses a serious problem. During initial residential construction, erosion from bare soil can be very high during heavy rainfall. Large amounts of soil can enter the stream and remain there for years, despite being mobilized after every major rain event. After construction, less sediment enters from residential areas than from agricultural land that exhibits standard row crop cultivation, seasonally bare soils, and livestock wading in and alongside streams.

The main effect of sediment is to decrease the quality of fish and invertebrate habitat by filling small spaces between pebbles in the stream bottom that are normally home for insect larvae.[15] Such insects are the main food for many types of fish, part of the rich ecological diversity in the Big Darby Creek. Without these insects the number and kinds of fish decrease. Furthermore, very fine particles are shown to stay suspended in the flowing water, making it cloudy and also decreasing the ability of fish to find their prey.

Toxic Substances, Changing the Insect Food Base, Fish Species, and Causing Disease

In the section on effects from toxic substances in runoff, the survey respondents were given information on how storm runoff washes pesticides from cropland, lawns, or gardens into streams. Such compounds also may come from spilled oil, gasoline, and other automotive chemicals present on roads and driveways. These chemicals often cause change in the numbers and kinds of fish in streams, favoring fish that tolerate these substances. Respondents were asked whether they knew that lawn and garden chemicals could affect the fish in streams.

Changes in Runoff and Flooding Patterns, Decreasing Habitat Quality and Causing a Shift to Fewer, More-Tolerant Species

Natural streams were described as having bends, pools, and riffles, with logs and limb "dams" all the way to their headwaters, thus slowing the passage of water. Slow natural drainage from the land also allows water to seep into the ground slowly after heavy rains, replenishing ground water. However, with some residential development, streams are straightened, logjams are removed, and storm water drains quickly off the land, increasing the risk of downstream flooding.

The effects of these physical changes were described in the presentation as increasing the speed of water flow, causing further erosion from stream banks, and increasing flood heights. After the runoff, water flow can become quite low in the absence of a strong groundwater recharge. These alternating high and low flows drastically change the quality of fish habitat, reducing biological diversity. Instead of many different depths and bottom types, the channel becomes wide and shallow. During low-flow periods, water moving over or through a gravel base becomes too shallow to be inhabitable. The resulting crowded conditions lead to increased death rates for fish as they use up nearly all of the available oxygen.

Communicating the Effects of Urban Development on Economic and Social Services

While the information in the previous section sought to frame certain values attached to the Big Darby, respondents also derive other kinds of value from economic and social functions within the area. To isolate the value placed on ecological services, one must control for value related to the economic and social services. Accordingly, the economic and social dimensions included in a sustainable development framework[1] also were briefly described.

In considering the value of economic services, the dominant endpoint is increased economic well-being. Although many measures that contribute to economic well-being were considered, the presentation focused on four economic outcomes: (1) dependence upon agricultural employment; (2) distance to employment for nonagricultural workers; (3) provision of retail services; and (4) impact on the local income base. Employment opportunities for agricultural and nonagricultural workers can be expected to change significantly across the four development scenarios. As residential development increases, agricultural employment opportunities will decline, but there would be sufficient population growth to justify expansion of retail services. Dependence upon commuting for nonagricultural work not only increases travel and time costs but also has feedback effects as commuters either make purchases outside of the Big Darby area or, conversely, bring higher incomes back to the local area. This is one of the ways in which development would be expected to affect the local income base. In addition, the income profile of residents who would be expected to populate the study area would vary under the different scenarios. Questions were included to capture respondent preferences about these economic outcomes.

With respect also to social services, the ultimate endpoint is increased quality of life. Among the many factors that contribute to quality of life, the presentation focused on four social outcomes: (1) open space, (2) privacy, (3) public services, and (4) quality of education. These factors vary among the different development scenarios as well. The change in open space and privacy during the transformation from rural to suburban could be a confounding variable of importance to respondents. As residential development progresses, the availability of open space for use in recreational activities and the degree of privacy begin to decline. In addition, residential development not only brings a need for increased public services, such as police and fire services, but also a difference in access as the proximity to these services changes. Moreover, the quality of publicly provided elementary and secondary education is likely to change with increases in local income and property wealth, and as voter tastes for education change. Questions were included to capture respondent preferences about these social outcomes.

Land-Use Scenarios for Framing Expression of Preference and Value in the Stream

All the variables considered in the previous two sections vary among the land management or development options, allowing an approach that estimates stakeholder value through CVM surveys. The CVM questionnaire tries to focus on the unique amenities that could be at risk while acknowledging that other factors come into play. The information provided to survey respondents about physical stressors and ecological, economic, and social mechanisms can affect the estimate of WTP in terms of the direction and magnitude of the potential bias.[16,17] Thus, the survey instrument must have questions concerning preferences as well as values. To facilitate an understanding of the contrasts in options and outcomes, maps, data, and

Table 10.1a Relative Effect of Four Housing Development Scenarios on the Four Main Causes of Change in Big Darby Creek

	High Density Development	Low Density Ranchettes	Low Density Clusters	Agriculture
Nutrient input	Medium to high	Low to high	Low to medium	Medium to high
Sediment input	Low to high	Low to medium	Low to medium	Medium to high
Toxin input	Medium to high	Low to high	Low to medium	Medium to high
Change in flow patterns	High	Low to medium	Low to medium	Medium to high

photographs were used to frame WTP to conserve amenities described in each of the development scenarios that follow.

For easy reference, survey respondents were provided Table 10.1a and Table 10.1b, which show the levels of effect from each of the objective factors considered in the section on linking mechanisms. The tables were developed based on reviews of the relevant literature and scoring of effects by the researchers. In each case, a range of possible effects was described, categorized as low, medium, or high in both the script and color slides. These categories are intended to reflect increasing levels of risk. For instance, low nutrient input would be that input leading to nutrient concentrations in the stream that are in the range of the lowest 1/3 of the observed data on nutrient concentrations. The factors are normalized such that when the effect reaches high levels, there is risk to stream integrity.

Table 10.1b Relative Effect of Four Housing Development Scenarios on Socioeconomic Outcomes in Big Darby Creek

	High Density Development	Low Density Ranchettes	Low Density Clusters	Agriculture
Economic Outcomes				
Agricultural employment	Low to Medium	Low	Medium to High	High
Retail services	High	High	Medium	Low
Distance to employment for nonagricultural workers	Low	Medium	Medium	Medium to High
Local income base	High	Medium	Medium	Low to Medium
Social Outcomes				
Open space	Low	Medium to High	Medium to High	High
Privacy	Low	High	Medium	High
Proximity to police and fire services	High	Medium	Medium	Low
Quality of education	Medium to High	Medium	Medium	Low

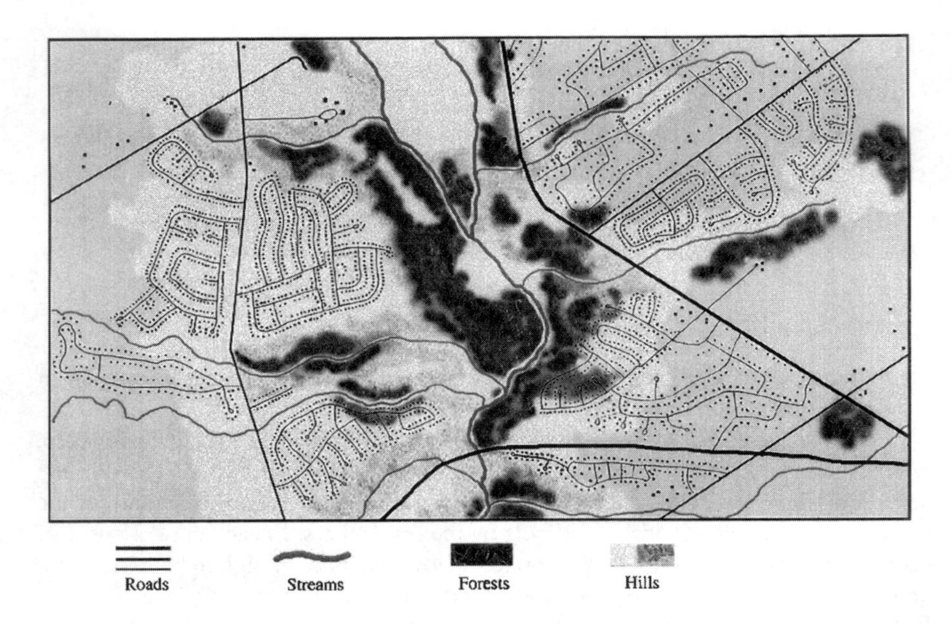

Figure 10.2a Illustration of high-density scenario (dots represent houses).

High-Density Development

The base case against which the respondents are asked to indicate preferences or WTP (to avoid) is illustrated in Figure 10.2a. It shows a 4-mi^2 area that includes both sides of the Big Darby, not far off I-70. It represents the conventional residential development that many people expect based on the patterns already being seen in the Columbus area. The characteristics defined for this high-density scenario were: 15% open or agriculture, 70% residential, 10% forest, and 5% nature preserve. The lot size is about to 1 acre and the residential density is 200 dwelling units per 100 acres of land.

Nutrient input is affected by storm water runoff that carries lawn fertilizers at certain times of the year. In this scenario, the aggregate effect is expected to be medium to high. Sediment input from this scenario will be high during construction periods, but then may be fairly low. Toxin input will be medium to high depending on lawn and garden care practices, and whether a storm water treatment system is in place to treat the chemicals scavenged from roads and driveways. The pattern of stream flow, flood frequency, and scouring is changed considerably, mainly due to the very large increase in hard surfaces. The respondents were asked several questions as to their preferences for avoiding associated enrichment, toxins, and extreme flow outcomes.

Low-Density Ranchette Development

A second scenario, shown in Figure 10.2b, illustrates the same 4-mi^2 area but with development in the form of large lots, based on patterns already observed in many

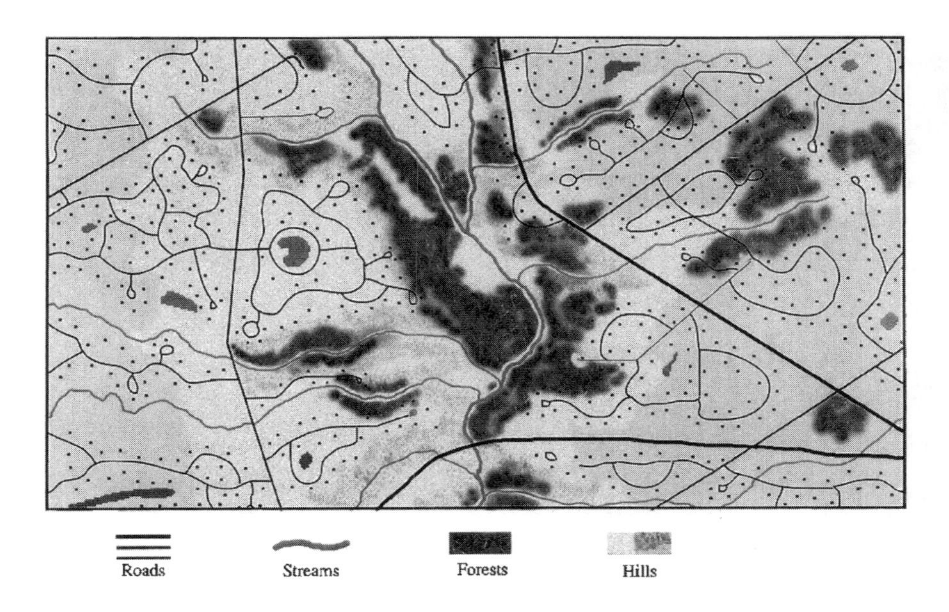

Figure 10.2b Illustration of low-density ranchette scenario (dots represent houses).

suburban areas. The characteristics defined for this "ranchette" residential development were: 10% agriculture, 70% residential, 15% forest, and 5% natural preserves. The dwelling unit density is 20 units per 100 acres, with 3- to 5-acre lots.

The inputs of nutrients and toxins can vary from low to high in this scenario depending on how much of each lot is left in natural vegetation and how the lawns are maintained. Some nutrient input to the stream from septic tank seepage also is possible. When few pesticides are used on lawns and much of the land is left in a "natural" state, then both nutrient and toxin input will be much lower than in the high-density scenario. However, when large areas are maintained as lawns using standard chemical lawn treatments, then both nutrients and toxins could be almost as high as the high-density scenario.

Sediment input also will range widely, from low to medium, with some entering the stream mainly during the construction phase, and tending to be much less over time. Changes in stream flow peaks will be low to medium, and much less than the high-density scenario. In this scenario, stream habitat will depend largely on the amount of forest and wetlands left near the stream channel. In comparison to conventional agriculture, however, the overall change in the Big Darby system from large lot development is likely to be positive. The survey respondents were asked whether it is likely that residents of this ranchette type of development will leave enough land in its natural state to protect Big Darby water quality, and whether they would be willing to pay slightly higher land and construction costs to guarantee that sediment input to the creek is minimized by erosion barriers and sediment traps. They also were asked whether taking over nearly all the farmland is a significant negative consideration for them.

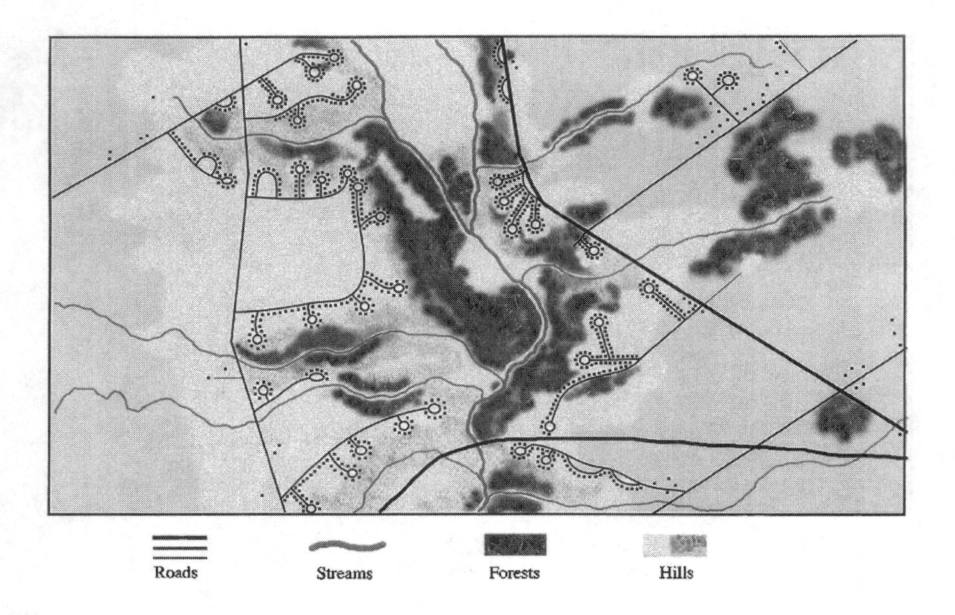

Figure 10.2c Illustration of low-density cluster scenario (dots represent houses).

Low-Density Cluster Development

A third scenario, shown in Figure 10.2c, illustrates the same 4-mi^2 area, but with a clustered development that keeps most of the land in agriculture. The characteristics defined for this type of development were: 60% agriculture, 20% residential, 15% forest, and 5% nature preserves. The dwelling unit density is 20 units per 100 acres, the same as for the ranchette development.

Nutrient input from this scenario is shown to vary from low to medium depending primarily on associated farming practices. The cluster housing developments would each include their own sewage treatment system, possibly in the form of package treatment and wetland wastewater application, with little input to the creek. Maintenance of lawn area also would contribute little because of the small lot sizes for housing. Nutrient input from farms may be insignificant, depending on fertilizer applications and the density of livestock.

Sediment input will vary here much as it does in the ranchette scenarios, with higher input during construction, decreasing with time. Because the amount of bare land in hard-surface roadways is less than in either of the other two residential scenarios, overall sediment input even during construction will be low to medium, with the input determined by the amount of land left in agriculture. Soil-conserving agricultural practices such as low-tillage could decrease the sediment load even further. Toxin input will be lower in this scenario than for either of the other residential developments because of the smaller area of lawns and hard surfaces, but the range of agricultural practices largely will determine the level of toxins

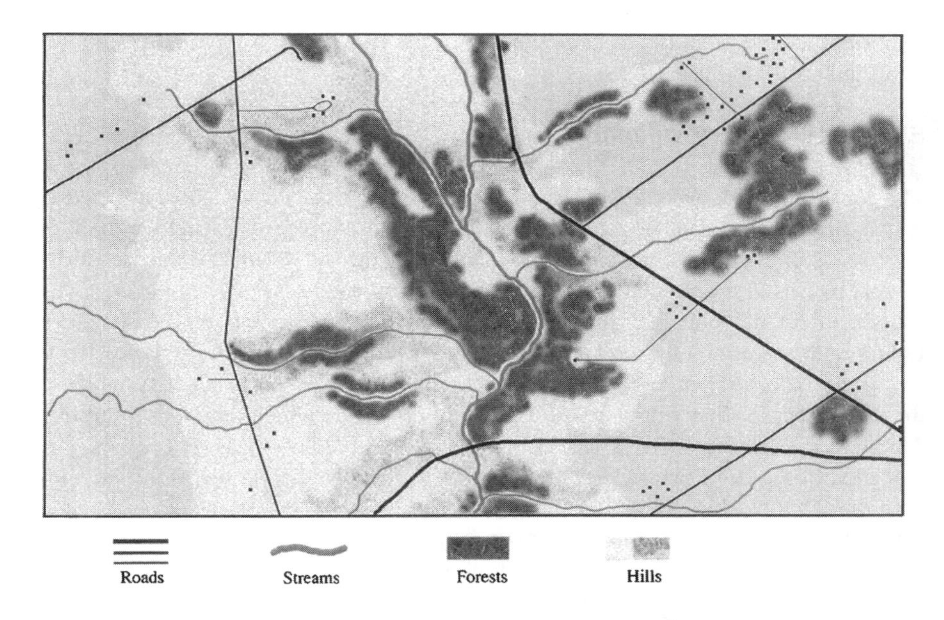

Figure 10.2d Illustration of present agriculture scenario (dots represent houses).

reaching the stream. The altering of stream flow and flooding pattern is lower here than for the agriculture or high-density scenarios.

Agriculture Land Use

The final scenario is shown in Figure 10.2d. This scenario shows the land use and residence density actually observed in the area in the early 1990's. The characteristics of this "present landscape base case" are: 75% agriculture, 10% residential (including farm lawns), and 15% forest. The dwelling unit density is 2 units per 100 acres.

The input of nutrients, sediment, and toxins in this scenario can be medium to high, depending on local agricultural practices and the amount of livestock (see Table 10.1a). The time of cultivation and the amount of fertilizer and pesticide application also influence the amount of sediment, nutrients, and toxins in runoff reaching the stream. Certain farming practices can be adopted to reduce fertilizer applications and minimize runoff after rain events. However, many farmers in the Big Darby drainage area already use conservation tillage practices to reduce nutrient, pesticide, and sediment inputs.

The altering of stream flow characteristics under this scenario is medium to high (relative to a pristine, unfarmed condition), also depending on farming practices. Because the Big Darby area is fairly flat, water does not flow to the stream quickly, and farmers are often anxious to drain the water off their fields. Tile drainage systems and straight clean waterways have been introduced locally, increasing water flow and transport of nutrients off the land. The survey respondents were asked how important it is to them that a large portion of the Big Darby watershed be retained in agricultural land use.

Eliciting Monetary Valuation

The four scenarios, and the ecological, economic, and social variables affected by residential development in the hypothetical 4-mi^2 area, were presented visually to groups of about 30 respondents, who completed the survey questionnaire during several pauses in the presentation. In the first part of each session, respondents were introduced to the potential impact of development under each of the scenarios. Photographs taken within the Darby watershed were used to illustrate these effects.

In the latter part of each session, respondents were asked to identify which of the four scenarios they felt were most likely to occur and which they most preferred. This was followed by a WTP question used in the CVM analysis. A map showing a portion of the Big Darby Creek watershed was displayed, with a 150-mi^2 area just west of Columbus highlighted as "facing likely development over the next 20 years." The Darby watershed sample was drawn from this area. Each respondent was then confronted with a choice between the high-density base case and one of the other development scenarios. This question was framed around the idea that a group of citizens, along with government officials at both the local and state levels, had developed a fund to ensure that development in the highlighted area of the Darby follows a path that would lead to a specified state. It is proposed that monies for the fund would come from a hypothetical check-off on Ohio State Income Tax forms similar to current donation opportunities for wildlife and for natural areas. The respondents were asked if such a check-off were available, asking them to contribute $____ to the fund, would they check YES or NO? The dollar amounts were filled in by a random allocation within the questionnaire of amounts ranging from $1 to $100, based upon results from focus group pretests.

A method suggested by Loomis and colleagues[18] was used to calculate mean WTP based on survey results. For a particular landscape scenario, a core logit equation was formulated as follows:

$$\text{VOTE} = f\,(\text{FUND, INC, USEFREQ, AGE, Z}) \tag{10.1}$$

where VOTE is a dummy variable indicating whether the respondent voted YES or NO on the WTP question (preferring an alternative to the high density outcome), FUND is the respondent's posed dollar value contribution, INC is household income, USEFREQ is the number of times per year the respondent or family uses the Big Darby for outdoor activities, AGE is the age of the respondent, and Z is a variable indicating special circumstances that might influence WTP.

For example, one question asked whether the respondent or a family member considered themselves to be a farmer; another asked whether the respondent was a member of an environmental group.[a] Alternate specifications of the model were estimated using different respondent variables as the basis for the core equation (i.e.,

[a] In the sample of 766 respondents, 83 stated they were members of an environmental group, 66 said they were farmers, 8 were both, and 625 were neither.

Table 10.2 Mean Willingness to Pay and Confidence Intervals for Two Model Specifications[a]

	Specification 1[b]			Specification 2[c]		
Sample	Mean WTP	90% C.I. Min	90% C.I. Max	Mean WTP	90% C.I. Min	90% C.I. Max
Entire	$37.65	$28.64	$58.18	$37.96	$28.72	$58.94
Resident	$49.82	$29.29	$156.09	$51.44	$29.15	$162.39
Near-Resident	$33.91	$23.37	$68.28	$33.38	$23.40	$67.19
Non-Resident	$25.99	$14.99	$80.57	$25.45	$15.06	$75.11
Ranchette	$25.62	$17.15	$58.91	$25.19	$17.02	$54.47
Cluster	$67.05	$30.89	$261.17	$69.73	$27.33	$291.69
Agriculture	$29.58	$20.86	$57.72	$29.24	$20.50	$57.09

[a] Residents, n = 322; Near-Residents, n = 319; Non-Residents, n = 106.

[b] Model specification includes dummy variable for "farmer."

[c] Model specification includes dummy variable for "environmental group member."

Z was a dummy, YES/NO, variable either for "farmer" or for "environmental group member"), then separately considering status of the respondent (Resident, Near-Resident, Non-Resident), and finally by scenario type (Ranchette, Cluster, Agriculture). The results can then be interpreted as the contribution of each of the variables toward an individual's probability of contributing to the fund.[a]

Mean values for all the variables are used in conjunction with the estimated regression coefficients from the logit regression to estimate a mean WTP. The resulting general estimates from two alternate model specifications are shown in Table 10.2. The upper and lower bounds for the 90% confidence intervals are estimated using a simulation model with 10,000 random draws of the estimated regression coefficients. As would be expected, the mean values are higher for residents than for near-residents, and those are higher than for non-residents. In addition, the WTP for a Cluster landscape alternative was significantly higher than for the Agriculture or Ranchette alternatives.

Linking Stream Integrity to the Development Scenarios

The approach to linking stream ecological condition with the development scenarios relied to some degree on an empirical relationship between impervious surface area (a runoff inducing condition) and IBI. Recent work by Yoder et al.[19] showed that for the lowest quartile of urbanization around Ohio stream sampling sites (with impervious surface of less than 4.3% of watershed area), modal IBI is 42. This is just above the Warm Water Habitat criterion of IBI = 40 (and well below the Exceptional Warm Water Habitat criterion of 50).[b] For the second quartile of

[a] Details of these results are available upon request.

[b] Under the OEPA's designation, exceptional warm water habitat differs from warm water habitat in having an exceptional or unusual community of species when compared to reference sites (i.e., comparable to the 75th percentile of reference sites on a statewide basis). More stringent biological criteria are established for exceptional waters (see "Water Quality Standards and Ecological Risk Assessment" in Chapter 6).

Table 10.3 Runoff-Inducing Condition and IBI per Scenario

Scenarios	Indicated Impervious Surface Assumptions (after Yoder et al.[19])	Interpolated Runoff-Inducing Conditions	Modal IBI[a]
Agriculture	3%	16.9	42
Ranchette	—	16.3	43.0
Cluster	—	17.0	41.8
High Density	20%	21.3	35

[a] Interpolated from a graph linking the results of Yoder et al.; interpolated runoff-inducing condition; and IBI. Details available from the authors.

urbanization (4.3 to 14.6% impervious), the modal IBI is 39.5. For the third quartile (14.7 to 29.3% impervious), the IBI is 35.0, while for the fourth quartile (over 29.3% impervious), the estimated mid-range IBI is 24, or highly degraded. This work suggests a likely median of 3 percent impervious surfaces for rural agricultural land, and 20 percent or more for urban areas, both reflecting a literal understanding of the term *impervious surface*: the total surface area of roads, driveways, and roofs. These results also suggest a possible threshold for serious degradation of IBI when impervious surfaces are at or above 20 percent. In addition, the majority of watersheds having more than 15% impervious surface do not meet the OEPA's Warm Water Habitat Biocriteria.[19]

However, runoff hydrologists[20,21] have over many years developed an empirical relationship between modified surface conditions (such as cultivation, or residential lawn surfaces) and the intensity of runoff induced. These papers show that intensive cultivation creates runoff-inducing conditions in agricultural areas roughly equivalent to a moderate level of impervious surfaces. Using a transformation based on the "curve numbers" adopted by the hydrologists, a measure, *runoff-inducing condition*, has been developed (as shown in Table 10.3) that captures the conditions (and IBI) associated with each of the development scenarios.

Linking Stream Integrity and Willingness to Pay

There is a great deal of interest among environmental managers in determining the dollar values that may be associated with changes in ecological condition. When respondents expressed WTP to obtain one of the development scenarios, their valuation took into account the economic, quality-of-life, and ecological ramifications of adopting that scenario in place of the expected high-density scenario. In this case study, those ecological changes were quantified as units of IBI change. A multimetric index such as IBI has the potential to respond in a complex fashion to changes in water or habitat quality. The large number of metrics it includes, however, and the functional complementarity among those metrics, apparently lend it a degree of numerical stability. In practice, IBI often has been treated as having cardinal properties

Table 10.4 Estimated WTP per Unit of IBI Improvement Over a 150-mi² Study Area for Two Model Specifications

		Specification 1		Specification 2	
	IBI Improvement	Mean WTP	Mean WTP/IBI	Mean WTP	Mean WTP/IBI
Ranchette	8	$25.62	$3.20	$25.19	$3.15
Cluster	6.8	$67.05	$9.86	$69.73	$10.25
Agriculture	7	$29.58	$4.23	$29.24	$4.18

for purposes of environmental analysis and regulation. In this section, the investigators probe the implications of their data for associating a dollar value with a unit of change in IBI.

Table 10.4 provides preliminary estimates of the relationship between WTP and IBI change in the 150-mi² area considered in the survey. For example, in the case of respondents considering the agriculture alternative, the change in runoff-inducing condition from high density to agriculture (from 21.3 to 16.9) corresponds to an IBI improvement of from 35 to 42. Respondents for the agriculture cohort had a mean WTP of $29.58, corresponding to the 7-point improvement in IBI. Thus, an estimate of the WTP per unit of IBI for this cohort would be $4.23 per unit of IBI. The corresponding estimates ($9.86 per unit of IBI) for the cluster cohort were more than double that of the agriculture cohort, and almost triple that for the ranchette cohort.

For many reasons, however, caution is necessary in interpreting these IBI-normalized WTP values, since these results do not separate changes in ecological and related risks from other environmental, economic, and social changes associated with the development scenarios. In fact, since the IBI changes associated with these three scenarios were similar in magnitude, it is likely that the expressed differences in value between the scenarios were influenced by both nonecological factors and certain perceptions about ecological factors not captured by IBI. Analyses now underway are looking more closely at the respective, marginal contributions of the ecological, economic, and social factors to WTP.

DISCUSSION

When the Big Darby Creek watershed ERA and economic analysis are considered collectively, the overall work has some of the ideal characteristics of an integrated analysis as described in "Diagramming an Integrated Management Process" in Chapter 9 and diagrammed in Figure 9.1. It also demonstrates some of the problems that result when integration is not a goal from the outset.

Assessment planning involved a wide variety of partners and stakeholder groups, resulting in clearly defined goals and objectives. The problem formulation conducted as part of the ERA identified two ecological assessment endpoints, of which one could be feasibly measured, and conceptual models were developed relating human activities in the watershed to stressors, to effects on endpoints, and to specific measures of effect. An analysis plan for the evaluation of specific risk hypotheses was developed and substantial progress was made toward the analysis and characterization of baseline risk. The ERA made use of data collected as part of the statewide watershed management cycle (see Figure 9-A.6) and had begun to provide empirical, stressor-response, and source-response relationships that will be useful in TMDL development.

The team conducting the economic analysis formulated a set of management alternatives, in this case suburban development scenarios, focused on one of the more severe concerns identified in assessment planning and problem formulation: stream degradation linked to urban encroachment in the watershed's eastern portion. The subsequent steps, analysis and characterization of alternatives and comparison of alternatives, were similar in form to the example shown in Figure 9.3 but with a number of important differences. As shown in Figure 10.3, they provided a qualitative analysis of the effects of each scenario on a set of important stressors affecting instream biota and on economic and social services to watershed residents. They did not examine the financial costs or other market-based effects of the management alternatives. In that those costs would accrue to land holders who would have to forego valuable development options, the analysis also did not address equity.

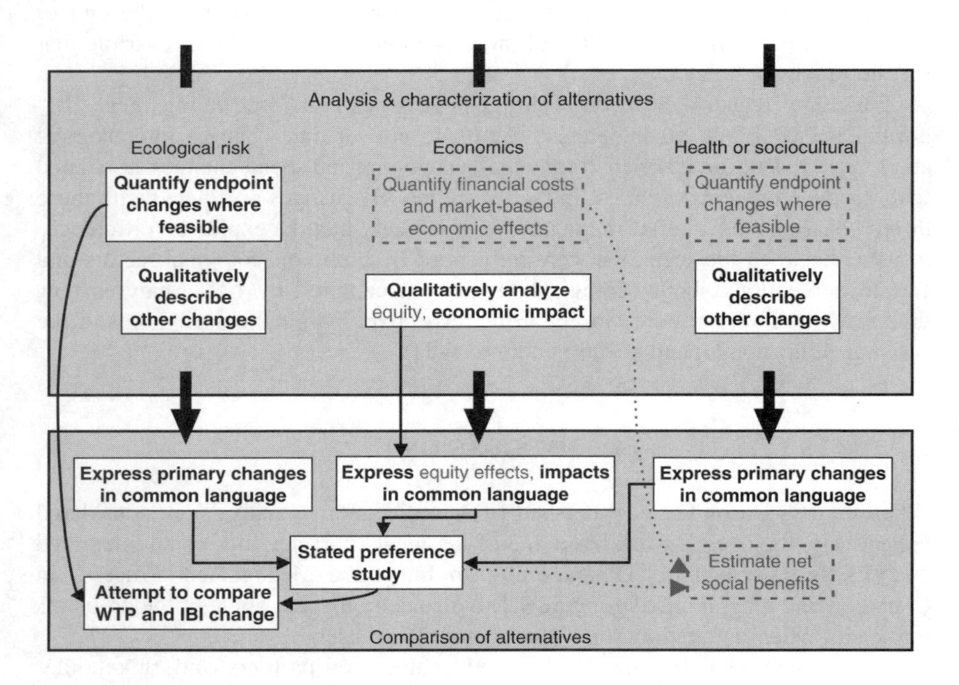

Figure 10.3 Techniques used for analysis, characterization, and comparison of management alternatives in the Big Darby Creek watershed, as compared to the example shown in Figure 9.3. White boxes and bold type show features included in this analysis.

To compare the alternatives, the ecological, economic, and social impacts of each scenario were incorporated into an integrated CVM instrument. The comparison was effected using monetary WTP associated with each scenario. That is, the economic analysis examined current WTP to avoid development changes that were expected to take place at some time during the next 20 years. Respondents were presented with a set of development alternatives and the expected ecological, economic, and social changes that would result from each.

The expected time frame for these effects was not made explicit, making interpretation of the analysis difficult. The time horizon is important both for understanding the respondents' preferences and for comparing the value of current effects to that of future effects (i.e., discounting the stream of future costs and benefits; see "Cost–Benefit Analysis" in Chapter 5).[22] Supposing, for example, that respondents assumed most of the expected, high-density development would not occur for 10–15 years in any case—and thus that any benefits of funding an alternative would be similarly delayed—they would have discounted their current WTP accordingly. If development actually were likely to occur sooner than they assumed, the WTP values measured in this study would be too small. Similarly, if they assumed that the ecological effects of high-density development would occur only much later than the other (economic and social) effects, and if this assumption was incorrect, then the ecological benefits of the other scenarios would not matter as much as the other changes, and the WTP for the more ecologically beneficial scenarios would be negatively biased.

In a subsequent step, WTP was compared to estimated IBI change. This latter step was of limited success, for reasons just discussed in the previous section, but with further analysis it could provide information that is useful in other settings. In general, this integrated assessment process provided decision support only (see Table 9.1); it did not include decisions or subsequent implementation.

In future studies of this type, if estimates of WTP for a given IBI change are sought, a more effective approach might be to elicit preferences for different fish community characteristics and preferences for different housing densities using separate CVM questions (within one survey) or representing these as separate attributes in a conjoint analysis study (see Appendix 5-A). The next step would be to use these data, along with information on the effects of the development scenarios on fish communities and the financial and market-based economic effects of the scenarios, to assess the net social benefits of the scenarios. Such an approach would be less reliant on establishing accurate respondent understanding of the ecological impacts of housing scenario, and it would also allow adjustment for new knowledge about that relationship without repeating the survey. It would also yield a more inclusive indicator (i.e., net social benefit) than WTP alone.

Nonetheless, neither WTP nor net social benefit estimates are necessarily the best endpoint for housing-related decisions in the Darby watershed. In spite of the thoroughness of the biophysical and socioeconomic framing of this CVM study, reviewers of this study at a USEPA workshop held in July 2001 were pessimistic about its likely influence on development decisions in the study area. They cited the substantial private gains to be made by developing individual tracts to the maximum allowable number of housing units, the spatial fragmentation of zoning authorities, and the tendency of zoning boards to respond to the wishes of property owners and developers. In other words, in specific zoning or development decisions there is not an effective mechanism for internalizing the negative externalities of high-density development manifested in statewide WTP. There was skepticism that the simple provision of WTP information would make an impact. Although there is some Clean Water Act authority for reducing the water-quality impacts of home construction, road construction, and imperviousness, it does not otherwise interfere with local land development.

Although the assessment planning effort that was carried out originally as part of the Big Darby Creek ERA examined a broad suite of watershed problems, the reviewers' observations suggested that this analysis did not adequately characterize the decision context (see "Planning" in Chapter 3) specific to suburban development. To better determine the applicability of WTP measured in this study to development decisions in the Big Darby Creek watershed, the assessment planning process would need to be revisited. Participants in a renewed process should include members of zoning boards, farm owners, developers, and individuals representing the residents', near-residents' and statewide interests in retaining the ecological, economic, and social amenities of the area. They should also include OEPA officials responsible for addressing local stream-reach impairments. Interactions could involve the provision of information about these amenities and the impacts of development, discussion of shared values, and an attempt to develop consensus goals for this portion of the watershed. Techniques used might include the joint development of future scenarios for the area.[23,24] Further analyses should include development of TMDLs and implementation plans that consider alternative residential (or industrial) development scenarios. Significantly, these plans should include efforts to develop compensation mechanisms whereby those who partially or completely forego development options are compensated, as is done under *transferable development rights* initiatives.

ACKNOWLEDGMENTS

The authors wish to thank the members of the Darby Partners and members of the Big Darby Creek Watershed Ecological Risk Assessment Workgroup for their participation in the development of the risk assessment on which parts of this chapter are based. We also thank attendees of a workshop held in July 2001, in Cincinnati, OH for their comments on an early draft of this work, and in particular we acknowledge John M. Gowdy, Robert V. O'Neill, Ralph Ramey, and David Szlag for their written reviews.

REFERENCES

1. Erekson, O.H., Loucks, O.L., and Strafford, N.C., The context of sustainability, in *Sustainability Perspectives for Resources and Business*, Loucks, O.L., Erekson, O.H., Bol, J.W., Gorman, R.F., Johnson, P.C., and Krehbiel, T.C. Eds., Lewis Publishers, Boca Raton, FL, 1999.
2. Zwinger, A., Darby Creek, Ohio: Back home again, in *Heart of the Land: Essays on Last Great Places*, Barbato, J. and Weineman, L. Eds., Pantheon Books, New York, 1994, 151.

3. Cormier, S.M., Smith, M., Norton, S.B., and Neiheisel, T., Assessing ecological risk in watersheds: A case study of problem formulation in the Big Darby Creek watershed, Ohio, USA, *Environ. Toxicol. Chem.*, 19, 1082, 2000.
4. Schubauer-Berigan, M.K., Smith, M., Hopkins, J., and Cormier, S.M., Using historical biological data to evaluate status and trends in the Big Darby Creek watershed (Ohio, USA), *Environ. Toxicol. Chem.*, 19, 1097, 2000.
5. USFWS, Little Darby Creek conservation through local initiatives: A final report concluding the proposal to establish a National Wildlife Refuge on the Little Darby Creek in Madison and Union Counties, Ohio, U.S. Fish & Wildlife Service, Ft. Snelling, MN, 2002.
6. USEPA, Guidelines for ecological risk assessment, EPA/630/R-95/002F, Risk Assessment Forum, U.S. Environmental Protection Agency, Washington, D.C., 1998.
7. USEPA, Biological criteria: Technical guidance for streams and small rivers. Revised edition, EPA/822/B-096/001, U.S. Environmental Protection Agency, Office of Water, Washington, D.C., 1996.
8. Norton, S.B., Cormier, S.M., Smith, M., and Jones, R.C., Can biological assessments discriminate among types of stress? A case study from the Eastern Corn Belt Plains ecoregion, *Environ. Toxicol. Chem.*, 19, 1113, 2000.
9. Gordon, S.I. and Majumder, S., Empirical stressor-response relationships for prospective risk analysis, *Environ. Toxicol. Chem.*, 19, 1106, 2000.
10. Gordon, S.I., Arya, S., and Dufour, K., Creating a screening tool for identification of the ecological risks of human activity on watershed quality, Report to the U.S. EPA on Cooperative Agreement # CR826816-01-0, City and Regional Planning Program, School of Architecture, Ohio State University, Columbus, OH, 2001.
11. Hume, H.G., Sustaining Biological Diversity and Agriculture in the Big Darby Creek Watershed MS Thesis, Institute of Environmental Sciences, Miami University, Oxford, OH, 1995.
12. Zucker, L.A. and White, D.A., Spatial modeling of aquatic biocriteria relative to riparian and upland characteristics., Alexandria, VA, June 8-12, 571.
13. Arrow, K.J., Solow, R., Portney, P.R., et al., Report of the NOAA panel on contingent valuation, *Fed. Reg.,* 58, 4601, 1993.
14. Dillman, D.A., *Mail and Internet Surveys: The Tailored Design Method*, John Wiley and Sons, New York, 2000.
15. Karr, J.R. and Chu, E.W., *Restoring Life in Running Waters: Better Biological Monitoring*, Island Press, Washington, D.C., 1999.
16. Elliott, S.R., Schulze, W.D., McClelland, G.H., et al., Reliability of the contingent valuation method, U.S. EPA Cooperative Agreement CR-812054, University of Colorado, Boulder, CO, 1989.
17. Knetsch, J.L., Environmental policy implications of disparities between willingness to pay and compensation demanded measures of value, *J. Environ. Econ. Manage.*, 18, 227, 1990.
18. Loomis, J., Kent, P., Strange, L., Fausch, K., and Covich, A., Measuring the total economic value of restoring ecosystem services in an impaired river basin: Results from a contingent valuation survey, *Ecol. Econ.*, 33, 103, 2000.
19. Yoder, C.O., Miltner, R.J., and White, D., Using biological criteria to assess and classify urban streams and develop improved landscape indicators, in *National Conference on Tools for Urban Water Resource Management and Protection, EPA/625/R-00/001*, Minameyer, S., Dye, J., and Wilson, S. Eds., U.S. Environmental Protection Agency, Office of Research and Development, Cincinnati, OH, 2000, 32.
20. Soil Conservation Service, Urban hydrology for small watersheds, Technical Release No 55, United States Department of Agriculture, Engineering Division, Washington, D.C., 1975.

21. Soil Conservation Service, Ohio supplement to urban hydrology for small watersheds: Technical release No 55, United States Department of Agriculture, Columbus, OH, 1981.
22. USEPA, A framework for the economic assessment of ecological benefits, Science Policy Council, U.S. Environmental Protection Agency, Washington, D.C., Feb. 1, 2002.
23. Hulse, D., Eilers, J., Freemark, K., White, D., and Hummon, C., Planning alternative future landscapes in Oregon: Evaluating effects on water quality and biodiversity, *Landsc. J.*, 19, 1, 2000.
24. Coiner, C., Wu, J., and Polasky, S., Economic and environmental implications of alternative landscape designs in the Walnut Creek Watershed of Iowa, *Ecol. Econ.*, 38, 119, 2001.

CHAPTER 11

Valuing Biodiversity in a Rural Valley: Clinch and Powell River Watershed

Steven Stewart, James A. Kahn, Amy Wolfe, Robert V. O'Neill, Victor B. Serveiss, Randall J.F. Bruins, and Matthew T. Heberling

CONTENTS

1-56670-639-4/05/$0.00+$1.50

WATERSHED DESCRIPTION

The Clinch and Powell Rivers originate in the mountainous terrain of southwestern Virginia and extend into northeastern Tennessee, flowing into the upper reaches of the Tennessee River (Figure 11.1). The Powell River originally was a tributary of the Clinch River, but both now flow into the upper reach of Norris Lake. The Clinch and Powell River watershed above Norris Lake, also referred to here as the upper Clinch Valley, covers 9971 km^2 and ranges between 300 and 750 meters in elevation. Historically, it contained one of the most diverse fish and mussel assemblages in North America,[1] yet most of these populations have declined dramatically or have been eliminated.[2] The mainstem Tennessee River and many of its tributaries have been dammed, resulting in the loss of habitat for many fish and mussel species, and therefore the upper Clinch and Powell Rivers represent some of the last free-flowing sections of the expansive Tennessee River system. Currently, the Clinch Valley supports more threatened and endangered aquatic species than almost any other basin in North America.[3] Despite the implementation of recovery plans for most federally protected species in this basin, there is evidence that these species are either declining or becoming extinct at an alarming rate due to impacts from mining, agriculture, urbanization, and other stressors.[4]

The Clinch Valley is a traditional rural Appalachian region. The areas are among the poorest in their respective states, with coal mining, agriculture, and scattered manufacturing as the primary industries. Although the area is very scenic, with a few exceptions tourism is poorly developed. The regional coal and tobacco industries are in decline, and the high tech economy has not found its way south of Blacksburg (Virginia Polytechnic Institute and State University) or east of Knoxville (University of Tennessee and the Oak Ridge National Laboratory). Many former miners suffer from Black Lung Disease and other problems. School districts often have trouble offering curricula that are comparable to the suburban school districts and finding qualified teachers. Children often leave the region upon completion of their university education.

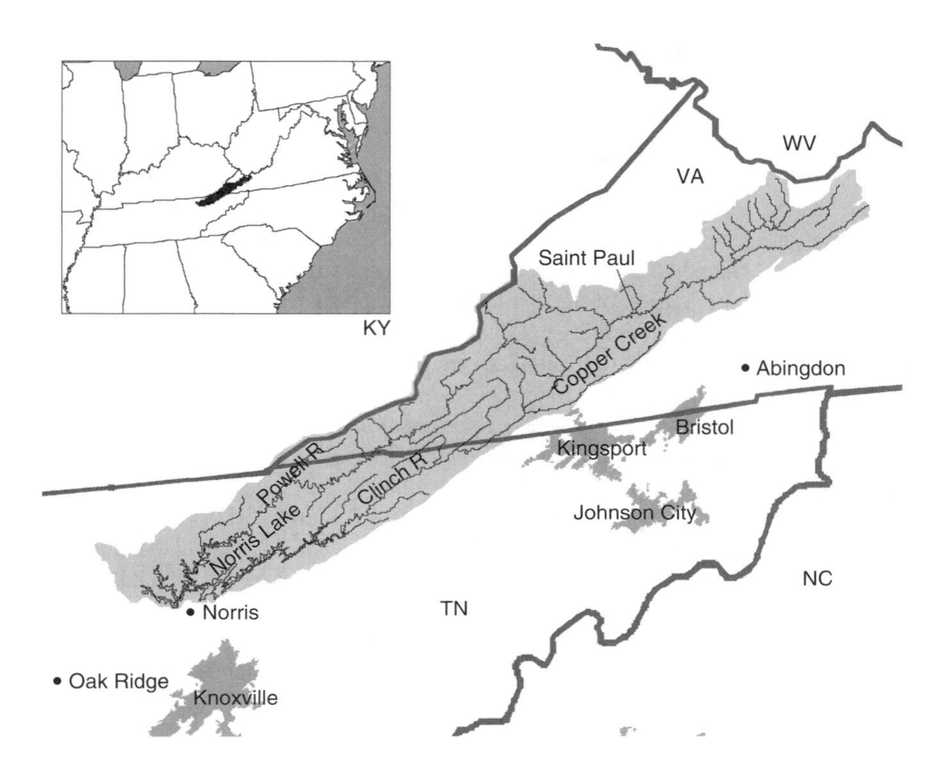

Figure 11.1 The Clinch and Powell River watershed in the Eastern USA. The study area is the portion of the watershed that is above Norris Lake. Initial ecological study focused on Copper Creek. Towns where discussions were held are shown, as are urbanized areas.

Transportation problems contribute to the area's economic isolation. Interstates I-81 and I-40 run parallel to the Clinch River, only one or two ridges east, and a quick glance at a map might indicate that transportation is not a problem; however, getting from the Clinch Valley communities to the interstate highways can be quite time consuming, often requiring more than an hour's travel on rural roads. An additional one to two hours is required to reach the Blacksburg and Roanoke area or the Knoxville area. Given the topography of the region, improving the transportation system could conflict with protecting the Clinch River and its tributaries, as the only ideal place for roads is in the flood plains of the streams.

The people of the region do appreciate its environmental resources and are very active in outdoor recreation such as hunting, fishing, and hiking. Evidence of this perspective was found in an unpublished survey. Preliminary to an ecological study of the watershed, local environmental organizations surveyed several communities in the region in 1994 to determine their attitudes and values. The results indicated strong interest in protecting local natural resources, but not at the expense of building roads, attracting industry, or creating new jobs.

A large amount of ecological information has been collected in this watershed over many years, but much of it had not been analyzed prior to this work. Entities collecting environmental data included The Nature Conservancy (TNC), Tennessee Valley Authority (TVA), U.S. Fish & Wildlife Service (USFWS), U.S. Geological Survey (USGS), Virginia Department of Game and Inland Fisheries, and Virginia Department of Conservation and Recreation. Resource managers suspected that mining, urbanization, and agricultural activities were adversely impacting the exceptional fish and mussel diversity. While several hypotheses have been advanced to explain these species' decline in other watersheds,[5] definitive answers as to their decline in this watershed have been lacking. Resource managers recognized that a comprehensive examination of the available data was needed to evaluate the relative effects of different human activities. Given the socioeconomic context of the Clinch Valley, it was also important to investigate the ways the people of the region compare environmental protection with economic development.

The following sections of this chapter describe studies carried out in the Clinch Valley by the U.S. Environmental Protection Agency (USEPA) and its partners to improve management of the area's unique ecological resources. "Ecological Risk Assessment" describes a watershed ecological risk assessment (W-ERA) initiated in 1993 and carried out by an interagency workgroup. In 1999, the USEPA awarded a grant to the University of Tennessee for an economic study that would use the results of the W-ERA and address decision-making needs; this study is described in the "Economic Analysis" section. The "Discussion" section then examines the overall work in the light of a conceptual approach for ERA–economic integration in watersheds (described in Chapter 9).

ECOLOGICAL RISK ASSESSMENT

Planning

The Clinch Valley ecological risk assessment[6–8] was one of five prototype watershed ecological risk assessments (W-ERA) sponsored by the USEPA to further develop, demonstrate, and test the use of the ecological risk assessment paradigm[9] at the watershed scale. (The reader is referred to Chapter 3 for more explanation of the procedures and terminology of ERA.) Like the other watersheds selected, the Clinch Valley was a candidate for W-ERA because it contains valued and threatened ecological resources; has been the subject of data collection efforts; is subject to multiple physical, chemical, and biological stressors, and receives attention from several organizations working to protect its resources. Federal, state, and local managers had been working with scientists from Virginia and Tennessee to study the distribution of aquatic resources in the Clinch Valley. The global significance of the faunal (especially molluscan) diversity had drawn a great number of scientists to the area.

For this risk assessment, an interdisciplinary, interagency workgroup was established in 1993 with representatives from the USFWS, TVA, TNC, Virginia Department of Game and Inland Fisheries, Virginia Cave Board, USEPA, and USGS. Unlike in the other W-ERAs, a broader stakeholder group was not convened. Information on attitudes and values from the community survey mentioned at the beginning of this chapter was taken in lieu of direct stakeholder involvement. Among six environmental concerns presented in that survey, "preserving our rare plant and animal species" was rated lowest in importance, whereas "our water quality" was rated highest.

Table 11.1 Outstanding Ecological Resources, Environmental Management Goal, and Management Objectives for the Clinch Valley Ecological Risk Assessment

Outstanding ecological resources:

- *The diversity and biological integrity of aquatic macroinvertebrates, especially the unique native freshwater mussels.*
- *The diversity and abundance of the native fish community.*

Environmental management goal and subgoals:
Establish and maintain the biological integrity of the Clinch and Powell watershed surface and subsurface aquatic ecosystem.

- Establish self-sustaining native populations of macroinvertebrates and fish.
- Improve water quality in the rivers.
- Establish and maintain functional riparian corridors of native vegetation.
- Safeguard water quality in a sustainable subsurface ecosystem.

Management objectives:

- Create and maintain vegetated riparian zones in agricultural areas to intercept sediment, nutrient, and pesticide runoff; enhance fish habitat; reduce thermal stress in smaller headwater streams; and exclude cattle from stream beds.
- Create and maintain vegetated riparian zones in urban, industrial, and developed areas to diminish sedimentation from storm water runoff and reduce instream habitat alteration.
- Implement agricultural best management practices (BMPs) such as rotational grazing to reduce sedimentation, pathogens, and nutrient enrichment instream.
- Contain and treat runoff from mining activities to reduce pollutant load and sedimentation instream.
- Install or improve sewage treatment facilities in streamside rural and urban communities to reduce inputs of toxic pollutants, pathogens, and nutrients instream.
- Adequately treat industrial discharges to reduce input of toxic pollutants instream.
- Create and maintain storm water retardation and holding facilities for highways and developed areas to reduce sedimentation runoff instream.

From Diamond et al.[8] and USEPA.[26]

This information stood in some contrast to the urgency for biodiversity protection felt by members of the interagency workgroup.

To focus the scientific information that would be analyzed in the Clinch and Powell watershed, the workgroup identified outstanding ecological resources, developed a management goal, and identified a set of management objectives considered important to achieving the management goal (Table 11.1). The workgroup agreed to focus the assessment on the unimpounded stream segment above Norris Lake since only that portion of the watershed provided suitable habitat for the fish and mussel species of concern. The assessment would use its limited funds to analyze data collected previously. Terrestrial and aquatic communities in caves associated with karst, though unique and diverse in the watershed, were not examined in this risk assessment because of insufficient information. The workgroup also recognized that there were other possible sources of stress in the watershed, including competition from exotic species (e.g., the asiatic clam *Corbicula fluminea*) and atmospheric deposition of contaminants. They opted not to consider these sources in this assessment because their impacts are relatively minor and they cannot be addressed by local managers.

Problem Formulation

During problem formulation, the broad management goal of establishing and maintaining biological integrity was more explicitly defined. Human-caused sources and stressors in

Table 11.2 Stressors and Sources Identified in the Clinch and Powell Watershed

Stressor	Sources	
	Degraded Water Quality	
Toxic chemicals	Catastrophic spills Urbanization Point-source discharges Atmospheric deposition	Agriculture Coal mining Transportation
Pathogens	Urbanization	Agriculture
Nutrients	Urbanization Atmospheric deposition	Agriculture
	Physical Habitat Alteration	
Sedimentation	Coal mining Hydrologic changes Transportation	Agriculture Urbanization
Riparian modification	Agriculture Hydrologic changes	Urbanization
Instream destruction	Agriculture Hydrologic changes	Urbanization
	Biotic Interactions	
Exotic species introductions	Accidental (Asiatic clam, zebra mussel) Recreational (brown trout, rainbow trout)	
Overexploitation	Other biota Over harvesting	Poaching

From Diamond et al.[8]

the watershed were listed (Table 11.2) and considered in detail.[8] Assessment endpoints corresponding to the outstanding biological resources were selected, and conceptual models were drawn illustrating the pathways by which the endpoints may experience adverse effects. The two endpoints selected in this assessment were: (1) reproduction and recruitment of threatened, endangered, or rare native freshwater mussels; and (2) reproduction and recruitment of native, threatened, endangered, or rare fish species.

Conceptual models developed by the workgroup traced the most important, hypothesized pathways between sources, stressors, and direct and indirect ecological effects. For example, the model for effects on mussels (Figure 11.2) shows agriculture, mining, silviculture, and urban areas to be sources of excess sediment. The resulting turbidity affects mussel survival and recruitment by interfering with filter feeding, and siltation smothers the substrates to which they attach. Siltation also smothers benthic (bottom-dwelling) macroinvertebrates, the food source of insectivorous fish, thereby reducing the availability of host species for the mussels' parasitic larval stage, or glochidia, which must attach onto the fins, epidermis, or gills of a suitable host fish. A similar model (not shown) traced the pathways for risks to fish species.

Risk hypotheses to be evaluated in the analysis phase were developed for each endpoint and eventually consolidated to three, corresponding to two categories of stressors:

Physical Habitat Alteration Hypotheses

- Greater connectivity of riparian (i.e., stream-side) vegetation, or forested riparian vegetation, is associated with greater diversity and abundance of mussels, other macroinvertebrates, and native fish.

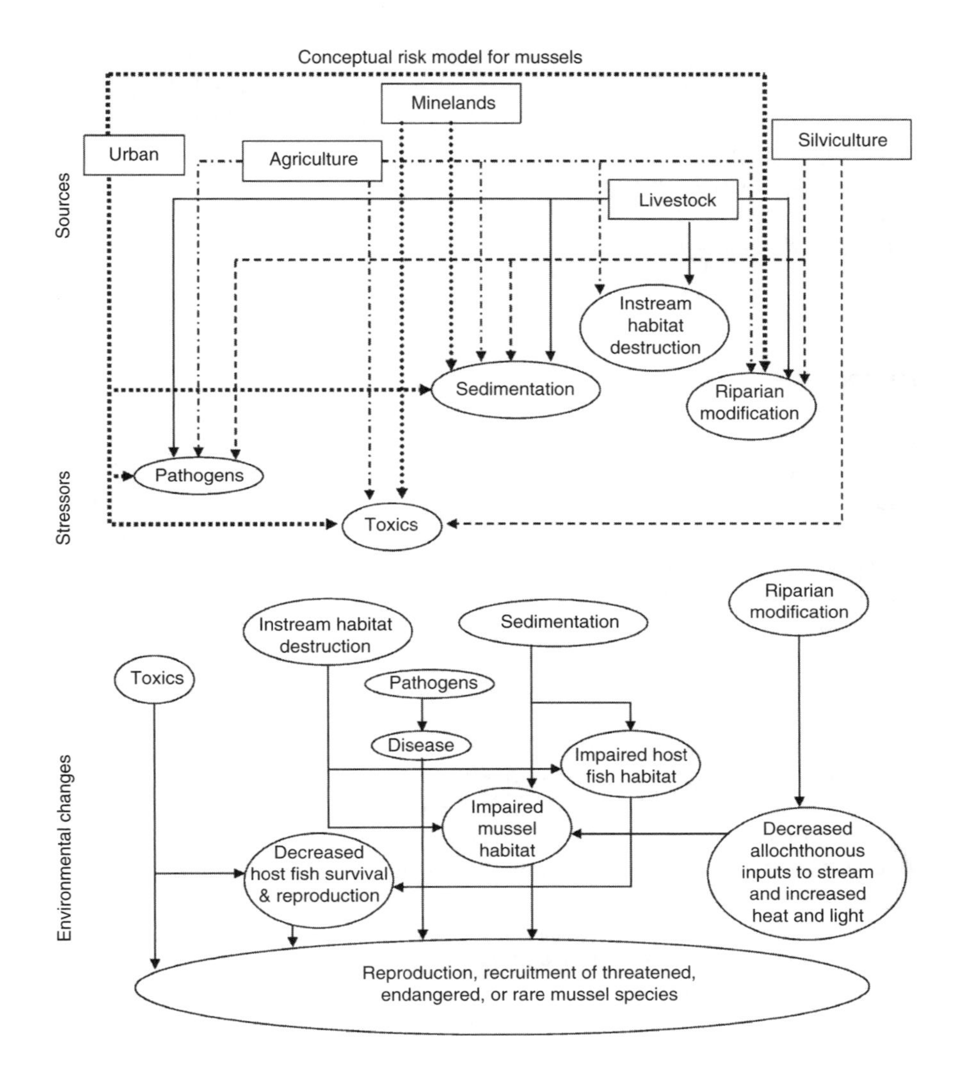

Figure 11.2 Simplified conceptual model showing major pathways between sources (land use), stressors, and effects on the assessment endpoint for native mussel species abundance and distribution and data sources available (adapted from Diamond et al.[8]).

- Watershed areas dominated by agricultural, urban, or mining land uses are associated with poorer physical habitat quality and biological diversity than are forested or naturally vegetated areas.

Water Quality Hypothesis

- Proximity to nonpoint-source runoff (from agricultural activities and urban areas) and point-source discharges (including coal mining discharges) results in detrimental structural changes to native mussel and fish populations.

Available data sets for subwatersheds of the Clinch Valley were examined, and an analysis plan was developed. Because of data limitations, it was decided to undertake a preliminary analysis in a subwatershed, Copper Creek (Figure 11.1), to determine the appropriate spatial scale for analysis of riparian vegetation and land uses, and to identify appropriate biological measures as surrogates for the assessment endpoints. It was also decided that the TVA would organize the available information in a *geographic information system* (GIS).

Risk Analysis

Methods

Analyses were based on data collected at many locations in the watershed over several years. Monitoring programs that provided key data for this risk assessment included TVA's Clinch–Powell River Action Team Survey and the Cumberlandian Mollusc Conservation Program. Land cover data used in this risk assessment were derived from LANDSAT Thematic Mapper imagery and were classified into 17 discrete categories including several different forest types, urban and developed land, pasture and cropland. All terrain data (e.g., elevation and slope) were derived from a mosaic of USGS digital elevation models (DEM) at 30-m resolution. The USEPA's River Reach File 3 provided stream network data. Locational data were also available for mines, coal preparation plants, major transportation corridors, urban centers, and biological sites in the basin. Several measures of instream habitat quality, including bottom substrate characteristics, bank stability, riparian vegetation integrity, channel morphology, and instream cover, were used to characterize habitat condition. A multimetric habitat quality index (similar to QHEI; see Appendix 6-A) was also used. However, water quality data were insufficient to allow determinations either of land-use effects on water quality or water-quality effects on the assessment endpoints. Therefore, it was necessary to directly examine the relationships between land uses, instream habitat quality, and the assessment endpoints, without reference to water quality per se.

Since data directly matching the assessment endpoints were not available, surrogate measures were used. For example, few data were available on native threatened, endangered, or rare fish species. However, the Index of Biotic Integrity (IBI), a multimetric index describing the status of the fish community, had been determined by the TVA at a number of locations throughout the watershed and was considered to be a reasonable measure for the second assessment endpoint (for more information on the IBI see Appendix 6-A). Data on mussel species richness and abundance were also limited, but preliminary study in the Copper Creek subwatershed showed a reasonable correlation between IBI score and mussel species richness, and therefore IBI values were used to supplement the mussel species data.[6] For benthic macroinvertebrates, the EPT index, consisting of the number of taxonomic families present from the orders Ephemoptera (mayfly), Plecoptera (stonefly) and Tricoptera (caddisfly) had been determined in some locations. These orders are known to be sensitive to adverse water quality and are replaced by other macroinvertebrates as water quality diminishes.

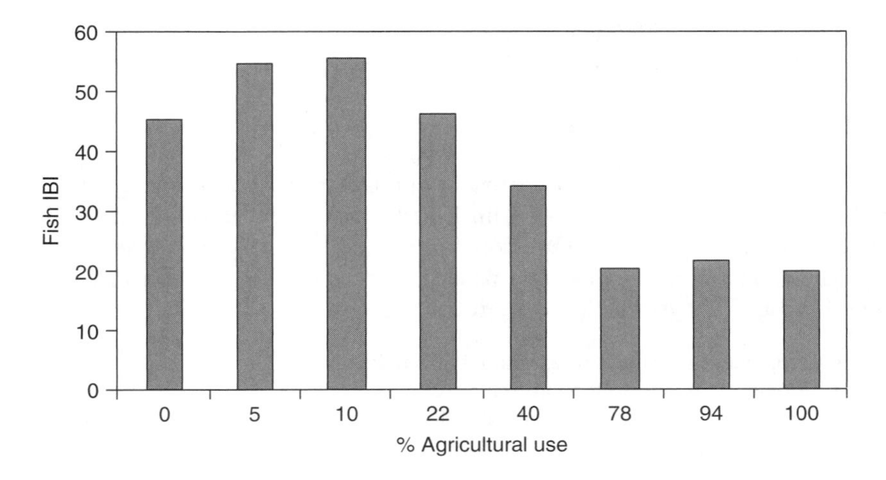

Figure 11.3 Fish community integrity as a function of agricultural land in a riparian corridor of 200 m width and 1500 m length in Copper Creek (from Diamond et al.[8]).

Forward stepwise multiple regression analyses and univariate statistical analyses of data within a GIS were used to test stressor-response associations. GIS maps were produced that examined each risk hypothesis. In many cases it was necessary to reduce the underlying variability by truncating the elevation range of sites included to detect source-response or stress-response relationships.

Copper Creek Pilot Study

Copper Creek was chosen for pilot analysis because it was a comparatively data-rich subwatershed, and it presented a simpler case in that agricultural uses were the major sources of stressors. Findings, which were used to structure the analysis of the entire Clinch Valley watershed, included the following:

- Agricultural uses in the riparian zone had more of an influence on instream habitat quality and fish community integrity (IBI) than did upland agricultural land use.
- Effects of human activity in the riparian zone could be observed in native fish and mussels as much as 1500 m downstream of the activity (see Figure 11.3).
- IBI score was correlated with mussel species richness.
- Land use in the riparian corridor had a stronger effect on IBI than did an overall index of habitat quality, although particular habitat parameters — such as instream cover score and the degree to which stream substrates were free from embedding fine sediments (clean substrate score[a]) — did correlate well to IBI and EPT.
- After analyzing riparian corridor data at widths of 50, 100, 200, and 500 m and at varying lengths, a riparian corridor zone measuring 200 m across (100 m to either

[a] TVA defines this parameter as *substrate embeddedness*. To make the directionality of the score (1 = poorest, 4 = best) more intuitive, it is here renamed *clean substrate score*.

side of the stream) and extending 500 to 1500 m upstream was found to be the appropriate spatial area in which to analyze land-use effects on fish and mussels.

Clinch Valley

The most successful analytical approaches in the Copper Creek pilot study, noted earlier, were applied to the entire Clinch Valley watershed. Because other parts of the watershed are subjected to stressors from the coal industry and urbanization, the riparian land cover analyses were expanded to include land uses other than agriculture. Land-use analyses included the following:

- Proximity to different types of mining activities.
- Proximity to urban and industrial areas.
- The percentage of land use in the area that was forested, pasture, cropland, or urban.
- Proximity to three classes of roads, including major U.S. highways, state roads, and county roads.

Effects of Land Use on Habitat Quality

Some effects of riparian-corridor land use upon instream habitat quality could be discerned when variability was reduced by limiting the analyzed sites to those occurring between 350 and 450 m elevation. Forty-two percent of among-site variability in the habitat quality index (N = 85) could be explained by riparian land use. Stream sedimentation was lower where cropland was 3% of total land use. Riparian integrity was better in areas in which pasture or herbaceous land was < 50% of the total land use. Instream cover was poor if urban use was 20% of the surrounding area upstream. Instream cover and clean substrate scores were affected by both the percentage of pasture and herbaceous cover and the percentage of urban area nearby. The relationships between land use and habitat quality suggest that instream habitat will have the highest probability of being satisfactory for aquatic life if agricultural land use is relatively low and urban influences are small.

Relationships between Land Use and Biological Measures of Effect

Among sites of 350–450 m elevation, riparian land uses explained 55% of the variability in IBI scores (N = 38) and 29% in EPT scores (N = 34). The percentage of pasture area was positively related to IBI, while proximities to mining, crops, and urban areas were negatively related. The apparently positive effect of pasture land on IBI was unexpected based on the pilot results for Copper Creek and the negative relationship between pasture area and riparian integrity observed at these sties. A likely explanation is that IBI may respond positively to moderate nutrient enrichment and that the negative effects of mining and urban development are comparatively much worse. The number of native mussel species was inversely related to several land uses including (in order of significance): percentage of urban area, proximity to mining, and percentage of cropland. In the multiple regression model these factors accounted for 26% of the observed variation in mussel species richness. Collectively, the analyses demonstrated that mining and urban areas are more detrimental than pasture areas to aquatic fauna in this watershed.

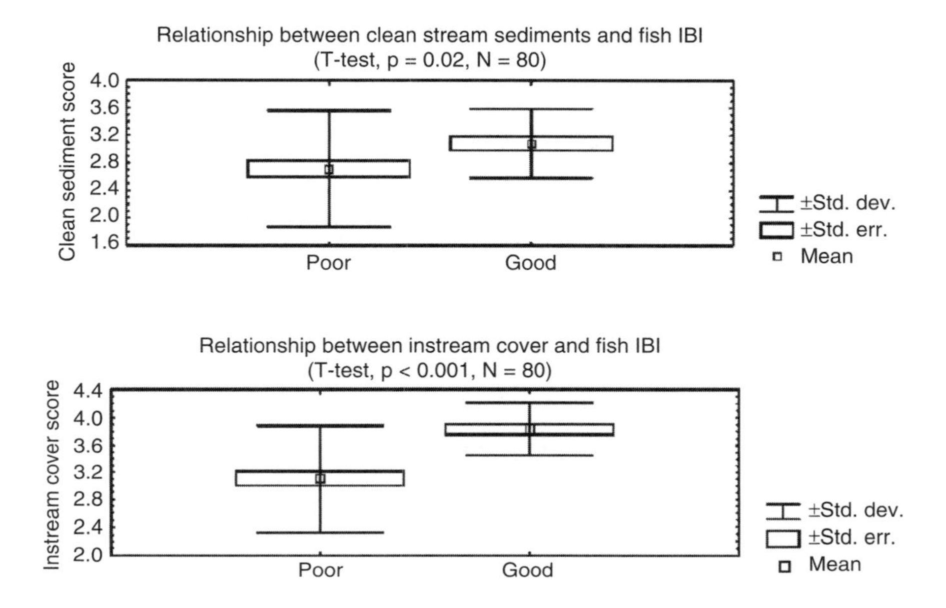

Figure 11.4 Relationship between two instream physical habitat parameters, clean sediment (substrate embeddedness) and instream cover, and IBI score, where IBI is categorized as either poor (impaired) or good (unimpaired) based on TVA's criteria; fish community impairment is associated with poorer habitat quality as measured by these two parameters (from Diamond et al.[8]).

Relationships between Habitat Quality and Biological Measures of Effect

In stepwise regression analyses of sites 350–500 m in elevation, habitat measures proved less effective than land uses at explaining variance in biological measures. Regression models explained 29% of the variance in IBI (N = 81) and 23% in EPT (N = 65). However, in univariate analyses where IBI was categorized as either poor or good based on the TVA's criteria, both instream cover and clean substrate scores were clearly related to fish IBI: sites with either low instream cover or highly embedded substrates had a >90% chance of having poor fish community integrity (Figure 11.4). The low overall explanatory power indicates either that both of these biological measures were responding primarily to non-habitat–related factors or that the habitat quality measures used were not sufficiently sensitive indicators of physical stressors in this basin.

Cumulative Source Index for Each Site

A cumulative source index for each site was computed, based on how many of four stress-causing land uses (sources of stressors) were present within 2 km upstream of the site. The four sources were active coal mining or processing; major transportation corridors; > 10% urban area; and > 10% cropland area. IBI was inversely related to the cumulative number of sources present (Figure 11.5A) and was consistently "poor" or "very poor" (TVA rating) at sites having all four sources present. In nearly all of these cases (88%), the proximal sources were urban areas and mining. Similar results were found for the

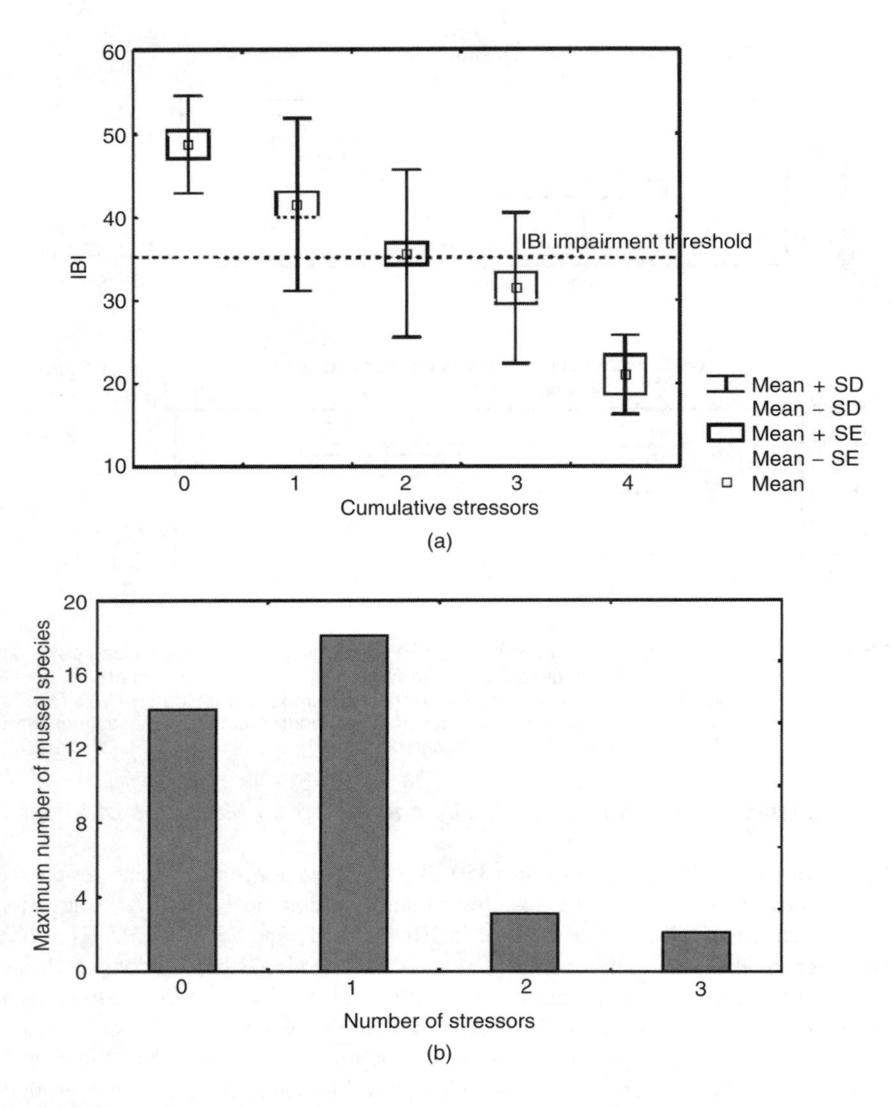

Figure 11.5 Fish IBI (a) and maximum number of mussel species (b) in the Clinch and Powell basin as a function of the number of stressors (from Diamond and Serveiss[6]).

maximum number of mussel species present at a site (Figure 11.5B). Sites having 2 or more proximal sources had a >90% probability of having fewer than 2 mussel species present. Sites with one or no sources of stress had between 4 and 18 species, which is still far less than the historical number of species reported (>35 species at many sites[10]).

Potential Effects of Toxic Chemicals

The risk analysis was hampered by the lack of water quality data sufficient for examining correlations between water quality parameters, including toxic chemical

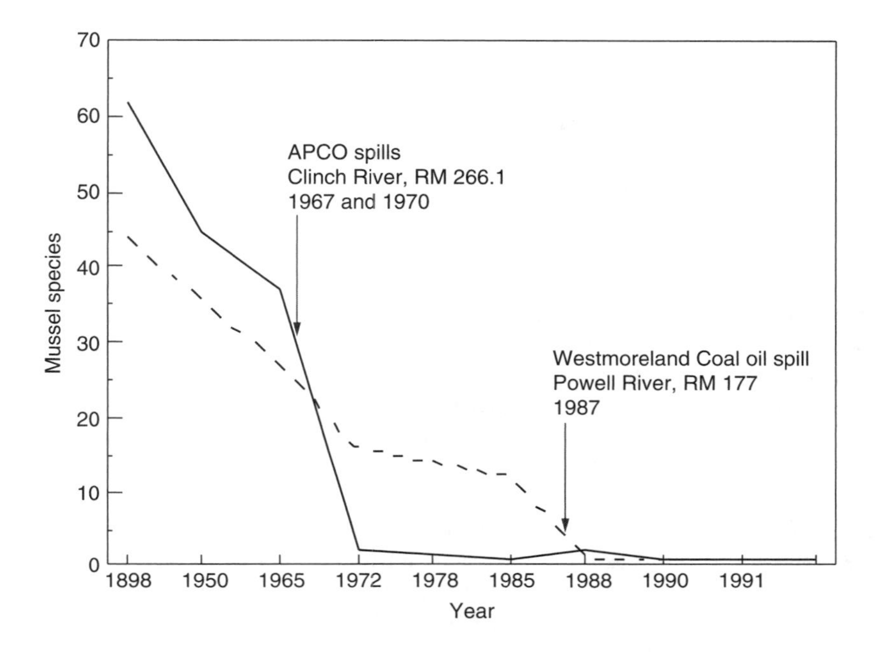

Figure 11.6 Number of mussel species recorded over time at two sites in Clinch and Powell watershed affected by large toxic point-source discharge events (from Diamond et al.[8]).

concentrations, and biological effects. The significant amount of variance in biological indices that was unexplained by land use and habitat quality data suggests that other factors were at play. Toxic chemicals may be released in municipal or industrial effluents, from coal mining or processing activities, or transportation accidents. While macroinvertebrates can recolonize an area within a relatively brief period following an episodic release, recolonization by fish and especially molluscs may require years or decades, depending on distance and barriers to other colonized areas. Figure 11.6 illustrates the effects observed after catastrophic spills at Westmoreland Coal Company and the APCO power plant on the Powell and Clinch rivers, respectively. In 1998, a large coal slurry impoundment on the upper Powell River failed, resulting in a massive fish kill and substantial mortality of native mussels for a distance of more than 20 miles downstream. A 1999 truck accident on the upper Clinch River in the Cedar Creek area resulted in a substantial loss of mussels, including more than 300 threatened and endangered mussels.[11]

Risk Characterization

Risk analysis examined the available data on land use, instream habitat parameters, and biological assemblages and produced a limited set of statistical associations. The risk characterization step interpreted these associations to suggest what the primary sources of risks were and to explain observed trends in stream faunal diversity. It also described uncertainties and presented management recommendations.

Ecological Risks

Analyses indicated that up to 55% of the variability in stream fauna could be explained by land uses, with mining and urban land uses exerting the most adverse effects. Key factors appeared to be sedimentation and other forms of habitat degradation stemming from urban and agricultural land uses and toxic chemicals from coal and urban areas. Riparian areas with more forested land cover and less cropland, urban, or mining activity tended to be associated with less sedimentation, more instream cover for aquatic fauna, cleaner substrates, and higher fish and native mussel species richness. Our results suggest that if agricultural or urban use upstream is great enough within the riparian zone, sedimentation effects and a subsequent loss of habitat will ensue for some distance downstream (1–2 km). These effects are accentuated in higher-gradient, headwater areas.

Although riparian vegetation can reduce deleterious land-use effects on water quality,[12] it is not clear that improvement of the riparian corridor alone in this watershed will necessarily result in the recovery of native mussel and fish populations. Little or no recovery of threatened or endangered mussel and fish species has been observed in this basin despite improved water quality.[1] In fact, the results of this study suggest that the risk of native species extirpation is likely to increase as more sources of potential stress co-occur. Of the 10 remaining mussel concentration sites studied, only half appeared to be reasonably isolated from major roads, urban areas, mines, and agricultural areas. This information suggests that native mussel populations are relatively vulnerable to likely sources of stress in this watershed and that further extinctions or extirpations are probable unless additional resource protection measures are taken.

Native fish and mussels have a high risk of extirpation due to endemism (i.e., restriction to a very limited geographic area) and habitat fragmentation, resulting in populations that are too inbred, smaller in size, and more susceptible to stressors. Populations are now more widely separated than they were historically,[3] which could lead to reduced recruitment success and declining populations, especially in the presence of stressors. Therefore, it may be most useful to further protect those populations that appear vulnerable due to their proximity to mining, urban areas, or transportation corridors. Protection and enhancement of the riparian corridor at these sites, as well as protection from toxic spills and discharges, is probably as important for sustaining endemic species as stocking new or historically important areas. If stream habitat as well as water quality can be maintained or improved, present mussel and fish populations might be able to expand into nearby areas, thus increasing the distribution and abundance of these species.

Uncertainties

Several uncertainties limited our ability to discern associations between causes and effects in the upper Clinch Valley. First and foremost, as has just been noted, the available biological information was only infrequently coincident in time and place with relevant instream chemical measurements. Second, physical habitat assessment data were fairly qualitative and relatively infrequent. Given the observed importance of physical stressors such as sedimentation on valued resources in this water body, resource managers should use more robust habitat assessment techniques that provide more quantitative data on impairments. Third, the macroinvertebrate measure EPT relies on family-level taxonomy, reducing its ability to discriminate changes in the benthic community; a generic- or specific-level index probably would provide better information. Fish IBI appeared to be a more sensitive index to stressors, probably because the metrics in this index have been demonstrated to be sensitive in a number of other watersheds. Fourth, the apparent relationship between fish IBI and mussel species richness or abundance, observed in the Copper Creek subwatershed, needs to be explored in more detail. IBI is composed of a number of metrics, such as native species richness, that were potentially more explanatory of mussel assemblages, but the unaggregated data were not available to this analysis. It must be noted, however, that any comparisons between native mussel and fish or macroinvertebrate data will be limited by the lack of overlap in sampling locations between the TVA's monitoring programs. Only eight sites in the entire watershed had data on mussels and either IBI or EPT. Because of the paucity of mussel species occurrence data, the risks to mussel species in the watershed could be over- or understated.

Management Recommendations

The risk assessment has helped lend further credence to what many resource managers had long conjectured were problems within the watershed, thereby providing more scientific support to take actions to address problems. Based on the assessment findings, the USFWS and TNC are considering the following types of management actions: riparian buffer protection; building spill prevention devices along transportation corridors near streams and restricting the types of materials transported over certain bridges; limited access of livestock to streams; better monitoring and control of mine discharges to streams; maintaining existing natural vegetation; BMPs for pasture and agricultural land to reduce sediment loading; and better treatment of wastewater discharges.

ECONOMIC ANALYSIS

The overarching goal for this integrated study was to utilize the findings of ERA in an economic analysis that would be relevant to environmental management decisions in the watershed. The economists' team chose to focus on values held by valley residents as important for determining how local decision-makers would act. The economic analysis therefore addressed the task of valuing potential changes in biological diversity and other ecological services at risk in the upper Clinch River Valley in Virginia and Tennessee, as expressed by valley residents. This task presented two major challenges: first, credible measures of economic value needed to be integrated with the ecological assessment endpoints such that the results would be useful in analyzing risk-relevant management and development scenarios. Second, the techniques used in the study needed to be consistent with economic principles of individual welfare maximization and to minimize biases associated with the measurement process.

Ecologists, such as those conducting the W-ERA, and Clinch Valley residents were thought to view the ecological assessment endpoints differently. Ecologists believe that biodiversity is important for a number of reasons, including its contribution to ecosystem resilience, i.e., the ability to withstand perturbations (such as from natural or human-caused stress) without shifting to a different kind of ecological state.[13] As stated earlier, however, Valley residents had rated "preserving our rare plant and animal species" lower than five other environmental concerns listed, and therefore might be unlikely to attach much value to the diversity of the Valley's mussel fauna, for example. However, mussel health is a good indicator of water quality, which residents had rated as most important. Because mussels are very sensitive to pollution, poor water quality will tend to impact mussels before other species in the river, and before human health. The economists expected that Valley residents would value the service provided by mussels as water quality indicators. Their approach was to design a survey that would interpret the results of ERA in terms most likely to be meaningful to Valley residents.

This section is organized as follows: "Methods for Valuing Biodiversity and Environmental Quality" introduces choice modeling as a potential tool for solving this difficult valuation problem, "Integrating the Choice Model with the Ecological Risk Assessment" presents a methodology for integrating a choice modeling approach with ERA in the upper Clinch Valley, and "Results of Economic Analysis" discusses the choice model results.

Methods for Valuing Biodiversity and Environmental Quality

Conjoint Analysis versus Contingent Valuation

Current approaches for assessing the value of environmental change, including changes in biodiversity, involve predicting an outcome associated with the change and then using a method such as the contingent valuation method (CVM; see Chapter 5 and Appendix 5-A) to estimate individuals' willingness to pay (WTP) for a beneficial change, or willingness to accept (WTA) a change that is detrimental.[a] For example, Rubin et al.[14] estimate the value of preserving spotted owls to determine the benefits of preserving old growth forests, and Stevens et al.[15] calculate WTP for various levels of preservation of Atlantic salmon and bald eagles. However, CVM tends to focus on losing or gaining the whole good, whereas management decisions tend to address changing characteristics of the goods.[16] For example, a typical CVM question might be worded as follows:[b]

> The upper Clinch/Powell watershed, which lies in southwestern Virginia and northeastern Tennessee, is threatened by water quality insults from agricultural operations, coal processing facilities, and urban runoff. The watershed is important habitat for many plants and animals, including eleven endangered mussels that are found only

[a] The use of WTP or WTA is a function of the perceived property right as well. See Freeman[27] for a discussion.

[b] This question was contrived for demonstration purposes only. A high-quality CVM survey would convey much more information before the valuation question was posed.

in the Clinch River. The river and adjacent areas are also used for recreational fishing, canoeing, picnicking, hunting, and to a lesser extent, commercial fishing.

> A nonprofit organization is seeking voluntary donations to purchase land and conservation easements to protect water quality in the Clinch/Powell watershed. These lands, which in total would comprise 2200 acres and would help ensure the protection of 15 miles of stream habitat, would then be managed by state land management agencies as preserved land. Would you be willing to contribute $X to aid in the purchase of the land and conservation easements?

In theory, CVM can measure both use and nonuse components of economic value (see Chapter 5); however, all these components would be lumped together in the WTP estimate. By contrast, conjoint analysis (CA) asks individuals to make choices about which state of the world they would prefer, given that different states have differing levels of certain definable attributes. The choice model, a variant of CA, elicits individuals' preferences by asking them to consider a series of trade-offs. In contrast to CVM, which asks individuals to explicitly state their WTP for a proposed change in environmental quality, choice models ask individuals to choose from a series of possible outcomes (choice sets). This allows the researcher to obtain the trade-offs that an individual is willing to make between any attributes presented in the choice sets, as well as to estimate WTP.

Choice models ask questions that may be more familiar to individuals. Individuals are asked to choose among bundles of goods according to the level of attributes of each bundle. For example, individuals routinely make choices among goods that have multiple attributes, such as among five automobiles having different colors, engines, interiors, and so on. A typical choice task might ask the subject to choose the most preferred of the five, with each having different characteristics, including price. In contrast to CVM, which would ask the individual to assign a price to each of the cars, the choice model task is more representative of the choices that individuals regularly face when making transactions. CA relies less on the information contained in the description of the scenario and more on the description of the attributes of each alternative.[17]

The family of CA models, of which the choice model is a member, is receiving increasing attention in the economics literature as well as in policy circles. Its use has been legitimized by the National Oceanic and Atmospheric Administration's (NOAA) proposed Habitat Equivalency ruling, which arose in part due to criticisms of CVM during the *Exxon Valdez* damage assessment case (60 FR 39816).[a] In particular, NOAA recommended CA as a tool to measure in-kind compensation for damaged natural assets (see Chapter 14 for a case study).

Regional development problems and multiple use management are perhaps the ideal tests of the usefulness of the choice model. With proper survey construction, the researcher can measure many characteristics including use and nonuse values, as well as indirect use values such as ecological services (see Chapter 5 for definitions of these values). Conjoint models are particularly useful for disentangling likely

[a] Habitat equivalency argues that the appropriate measure of natural resource damages due to, say, an oil spill, is provision (or augmentation) of ecological services that substitute for the services lost (e.g., improvement of wetlands in other areas). Refer to Chapter 8 for more information.

complementarities between attributes. For example, changes in water quality could be positively correlated with endangered fishes, sport fishing, and water-based recreation; with choice models, the effects of each of the attributes on welfare can be estimated independently.

Choice Modeling Framework

To explain individuals' preferences for alternative states of the Clinch River Valley, this effort used a *random utility model* (RUM) framework, which is widely used in dichotomous-choice CVM and travel-cost modeling, as well as in CA. RUMs rely on choice behavior and assume that individuals will choose the alternative that gives them the highest level of utility; that is, RUMs estimate the probability that an individual will make a selection based on the attributes and levels of each possible choice. The RUM is directly estimable from choice models (see Appendix 11-A for technical detail of the RUM framework).

Integrating the Choice Model with the Ecological Risk Assessment

The task of integrating the measurement endpoints from the upper Clinch Valley ERA (especially, IBI and mussel species richness) with indicators of social value proved a formidable challenge since they were not the type of endpoint the ordinary citizen is likely to think about in his or her day-to-day life. Meetings were held in Abington, VA and Norris, TN between the economists, ecological risk assessors, and other individuals who had shown interest in biological resource management in the Clinch Valley. The decision was made to approach the problem of lack of familiarity with the ecological endpoints in two ways. First, succinct wording was developed to express the relationship of these ecological endpoints to quality of life. After several iterations, a survey was drafted, presented to focus groups, revised, and then pilot tested.

Second, socially meaningful endpoints were included that were complementary to the ERA measurement endpoints but outside of the ERA's original scope. For example, increased forestation of the riparian corridor would not only help protect mussel and fish biodiversity but also increase the diversity and abundance of terrestrial fauna and birds and improve the quality of smallmouth bass fishing. Since these endpoints are jointly produced, it was important that they be jointly valued. Their inclusion expanded the choice sets to more fully describe the state of the Clinch Valley environment and the auxiliary benefits of management policies aimed at preserving biodiversity.

Choice Model Design

Choice model surveys are complex by nature. Each possible choice comprises bundles of attributes, with each attribute having different levels. Because the potential for miscommunication between the researcher and the survey recipient via the survey instrument is great, two formal focus groups of 6 and 11 subjects and three informal focus groups were conducted to inform our survey design. The first informal group was conducted in September 2000 using staff and students of the University of Tennessee. The second and more formal focus group was conducted by an expert facilitator in St. Paul, VA in November 2000. The third and fourth focus groups were conducted at the University of Tennessee in January and February 2001. The final focus group was conducted in Oak Ridge, TN in February 2001 using residents of Anderson County, TN, the westernmost county in our study.

The focus groups allowed the participants to home in on those attributes correlated with management changes that are likely to be important to the residents of the Clinch River Valley. Six attributes were identified, with the number of levels per attribute varying from 2 to 6 (Table 11.3); see Table 11.4 for an example choice set from the survey. The "cost to household" attribute allowed the estimation of conventional WTP measures. Interaction with the "agricultural income" attribute allowed investigation of whether individuals thought society as a whole, or farmers and ranchers alone, should bear the burden of increased environmental quality.

Choice model variables were specified based on these attribute levels, and *a priori* predictions of their signs were made (Table 11.5). The variables that represented the attributes "agriculture-free zone," "aquatic life," and "sportfish" were each decomposed into two separate, effects-coded variables to control for the three levels that each of these variables can take (see Louviere et al.[18] for a full discussion). *Effects codes* are an alternative to dummy variable codes and are useful when interpreting the coefficients of a choice model.[17,18] SMALLZONE and BIGZONE

Table 11.3 Attributes and Attribute Levels used in Survey Questionnaire

Attribute levels making up options A and B in a given choice set varied among those listed.
Attribute levels for option C were the same in all choice sets.[a]
Corresponding model variable names are given in parentheses.[b]

<table>
<tr><th>Attribute</th><th colspan="6">Attribute Levels for Options A & B</th><th>Option C: No New Action</th></tr>
<tr><td>Agriculture-free zone</td><td colspan="2">25 yards Clinch/10 yards tributaries (BIGZONE)</td><td colspan="2">10 yards Clinch/5 yards tributaries (SMALLZONE)</td><td colspan="2">none</td><td>none</td></tr>
<tr><td>Aquatic Life</td><td colspan="2">full recovery (FULLRECOV)</td><td colspan="2">partial recovery (PARTRECOV)</td><td colspan="2">continued decline</td><td>continued decline</td></tr>
<tr><td>Sportfish</td><td colspan="2">increase (SPORTINC)</td><td colspan="2">no change</td><td colspan="2">decrease (SPORTDECL)</td><td>no change</td></tr>
<tr><td>Songbirds</td><td colspan="3">increase population (SONGINC)</td><td colspan="3">no change</td><td>no change</td></tr>
<tr><td>Agricultural income</td><td colspan="3">no change</td><td colspan="3">$1 million/yr decrease (AGDECL)</td><td>no change</td></tr>
<tr><td rowspan="2">Cost to Household ($ per year)</td><td>$100</td><td>$75</td><td>$50</td><td>$25</td><td>$10</td><td>$5</td><td rowspan="2">no change</td></tr>
<tr><td colspan="6">(COST)</td></tr>
</table>

[a] The choice sets are designed to allow for the efficient estimation of the parameters of all of the attributes. While SMALLZONE and BIGZONE are our policy variables, they are varied independently of the other variables. For example, it is possible to have choice sets that include the 25yard/10yard agriculture exclusion (BIGZONE), but have SPORTDECL or have CONTINUED DECLINE for the level of aquatic life. Individuals would be expected to focus on the outcomes and not the policy attribute.

[b] See Appendix 11-B for explanatory text that was provided in the survey

Table 11.4 Sample Question and Choice Set from Survey Questionnaire

Which option for the future of agriculture and the environment in the Clinch Valley do you prefer the most, Option A, Option B, or Option C? Option C is the status quo, or what is currently happening and will continue to happen with no further environmental or agricultural policies. Note that some of these options might not seem completely realistic in real life. We ask that you do your best to assume that each option is possible and then choose your most preferred option.

	Option A	Option B	Option C: No New Action
Agriculture-free zone	10 yards Clinch/5 yards tributaries	10 yards Clinch/5 yards tributaries	none
Aquatic Life	full recovery	partial recovery	continued decline
Sportfish	no change	increase	no change
Songbirds	increase	increase	no change
Agricultural income	no change	no change	no change
Cost to Household ($ per year)	$50	$50	no change

Please check the option that you would choose:

	Option A	Option B	Option C
	❑	❑	❑

represent the size of the agriculture-free zone;[a] these are expected to be positive, albeit weakly. PARTRECOV and FULLRECOV should be positive as individuals should be more willing to choose options that lead to higher levels of recovery for aquatic life, all other factors being equal. SPORTDECL should be negative as individuals should be less likely to choose options that represent decreases in sportfish populations, whereas SPORTINC should be positive by similar reasoning. SONGINC is expected to be positive, since many people value the presence of songbirds. AGDECL is expected to be weakly negative, since income declines are detrimental to the regional economy, but not all respondents expect to be affected directly. COST is expected to be negative; individuals are less willing to choose options that have higher costs associated with them. Alternative-specific constants corresponding to options A and B (ASCA, ASCB) are included to incorporate any variation in the dependent variable that is not explained by the choice set attributes or respondent characteristics; there was no *a priori* expectation as to their signs.

Selected socioeconomic information thought to be important was also included in the choice model (Table 11.5). For example, it is common (though not universal) in the literature to see more support for measures to improve environmental quality as the level of education increases,[19] so EDUC is expected to be positive. RIVERVIS, which is equal to 1 if the subject visited the Clinch within the last year, is expected to be positive, since individuals are expected to choose outcomes that improve the quality of their visits to the river. Likewise, MOSTIMP (which equals 1 if the individual believes either that recreation is the most important use, or that environmental quality is the biggest issue in the Clinch Valley) is expected to have a positive sign. FISHLIC, which equals 1 if the individual holds a fishing license, should be positive; individuals who fish should be more likely to choose options 1 and 2, which

[a] An omitted third variable for the status quo, NOZONE, is implicit in the model; its coefficient can be determined by taking the negative of the sum of coefficients of the included variables.

Table 11.5 Choice Model Variables and Expected Sign

Variable[a]	Expected Influence of Variable
CHOICE[b]	NA
SMALLZONE[c]	+
BIGZONE[c]	+
PARTRECOV[c]	+
FULLRECOV[c]	+
SPORTDECL[c]	–
SPORTINC[c]	+
SONGINC[c]	+
AGDECL[c]	–
COST	–
EDUC	+
AGE	?
RIVERVIS	+
MOSTIMPO	+
FISHLIC	+
ENVORG	+
ASCA[d]	?
ASCB[d]	?

[a] Variable names are explained in Table 11.3 or in text.
[b] Dependent variable.
[c] Effects-coded variable.
[d] Alternative-specific constant.

generally include better environmental quality. ENVORG, which equals 1 if the individual belongs to an environmental organization, should be positive. There was no *a priori* expectation about the effect of AGE on choice.

Having defined these parameters, a RUM-based choice model (Appendix 11-A) is developed as follows:

$$\begin{aligned}\text{CHOICE} = {} & \alpha_1\text{ASCA} + \alpha_2\text{ASCB} + \beta_1\text{SMALLZONE} + \beta_2\text{BIGZONE} \\ & + \beta_3\text{PARTRECOV} + \beta_4\text{FULLRECOV} + \text{remaining} \\ & \text{attributes and socioeconomic parameters} + \varepsilon\end{aligned} \tag{11.1}$$

where the remaining attributes and socioeconomic parameters are all of the remaining terms in Table 11.5.

Survey Implementation

Final language to describe the choice attributes to respondents was developed (Appendix 11-B). Respondents were asked to answer eight choice sets,[a] similar to the example choice set in Table 11.4.

[a] A fractional factorial design was employed to develop a survey based on this choice model. A full factorial design would have required 648 (= $3^3*2^2*6^1$) different choice sets. The %MKTDES macro in SAS was used to choose 16 choice sets that are meaningful and will still allow the main and interaction effects to be estimated. These 16 choice sets were then blocked into two of eight choice comparisons. One outcome of the focus group process was that subjects indicated that the 16 choice sets that they had initially evaluated were too many.

Table 11.6 Summary Statistics

Variable	Mean	Std. Dev	Min	Max	Observations[a]
EDUC	13.409	1.426	6	16	1800
AGE	45.855	14.723	18	81	1824
RIVERVIS	0.592	0.492	0	1	1824
MOSTIMPO	0.627	0.484	0	1	1800
CHOICE	0.333	0.472	0	1	1824
SMALLZONE	–0.249	0.8239	–1	1	1824
BIGZONE	–0.236	0.836	–1	1	1824
PARTRECOV	–0.101	0.902	–1	1	1824
FULLRECOV	–0.287	0.746	–1	1	1824
SPORTDECL	–0.476	0.707	–1	1	1824
SPORTINC	–0.328	0.875	–1	1	1824
SONGINC	–0.157	0.987	–1	1	1824
AGDECL	–0.358	0.933	–1	1	1765
COST	24.391	31.810	0	100	1824
FISHLIC	0.453	0.498	0	1	1800
ENVORG	0.200	0.400	0	1	1800

[a] There are 1824 possible observations representing 3 possible choices on 8 choice occasions for each of 76 subjects. Ninety-one subjects completed this version of the choice study, but only 76 have complete responses for all eight choice sets.

Surveys were mailed to a random sample of 400 households in the Clinch River Valley, with the majority being distributed in the Virginia portion of the valley. [a,b] Principles from Dillman's Total Design Method[20] were followed. Approximately two to three weeks after the survey mailing, a reminder postcard was mailed to thank participants and encourage nonrespondents to return their surveys.

Results of Economic Analysis

Ninety-one subjects completed the choice study (response rate was 23%); 76 provided complete responses for all eight choice sets, generating 1824 acceptable observations for analysis (see Table 11.6 for summary statistics).

Results of Choice Model Estimation

The interpretation of the coefficients in conditional logit models suggests how utility or satisfaction changes given a change in the attribute. The parameters also reveal how the probability that an alternative is chosen changes as the level of the attribute changes.

[a] The delivery envelope for the survey was personalized and included a cover letter, the survey, supporting documents, and a stamped return envelope. Surveys were printed on legal size (8.5" × 14") paper folded as a 20-page booklet and stapled along the spine. The supporting documents were printed on letter-size paper.

[b] This survey was distributed as part of a larger study employing four different survey versions. The other surveys included a version that allowed the examination of the trade-offs of strictly environmental attributes such that a preference-based index could be constructed; a version where mussel protection implied trade-offs in employment in several sectors of the economy; and a version designed to test the similarities between choice and contingent valuation models. Results of the other surveys are still pending.

Table 11.7 Results for Conditional Logit with Choice as Dependent Variable

Variable	Coeff	Std. Error	T-statistic	P-value
SMALLZONE	0.697	0.155	4.497	0.000
BIGZONE	0.306	0.158	1.936	0.053
PARTRECOV	0.084	0.148	0.564	0.573
FULLRECOV	0.831	0.150	5.541	0.000
SPORTDECL	–0.727	0.179	–4.054	0.000
SPORTINC	0.593	0.127	4.679	0.000
SONGINC	0.079	0.120	0.657	0.511
AGDECL	–0.157	0.069	–2.271	0.023
COST	–0.033	0.004	–8.654	0.000
ASCA	–0.771	1.288	–0.599	0.549
ASCAxEDUC	0.010	0.088	0.119	0.905
ASCAxAGE	–0.013	0.008	–1.508	0.132
ASCAxMALE	–0.624	0.266	–2.345	0.019
ASCAXMOSTIMPO	0.790	0.264	2.993	0.003
ASCAxFISHLIC	0.256	0.250	1.024	0.306
ASCAxENVORG	0.492	0.308	1.597	0.110
ASCB	–1.505	1.465	–1.027	0.304
ASCBxEDUC	0.044	0.099	0.448	0.654
ASCBxAGE	–0.018	0.011	–1.671	0.095
ASCBxMALE	–0.696	0.313	–2.223	0.026
ASCBxMOSTIMPO	0.561	0.310	1.806	0.071
ASCBxFISHLIC	0.700	0.302	2.318	0.020
ASCBxENVORG	0.246	0.379	0.648	0.517
Number of Observations[a]		526		
Log-Likelihood		–423.759		
Log-Likelihood(0)		–577.870		
McFadden's Rho-square		0.267		

[a] There are 608 choice occasions in the data set, but only 526 observations have complete responses for the variables in the regression. A choice occasion represents a set of three alternatives: one outcome is selected as the preferred option by the individual, while the other two are not.

Parameter values obtained for the discrete choice model generally showed the expected signs, and the joint power of the model was very good, as evidenced by a McFadden's Rho-square of 0.27 (Table 11.7). The signs of the coefficients on the attribute variables were consistent with the priors. Both small and large agriculture-free zones serve to increase the probability that an alternative is chosen, but the small zone had a stronger effect than was anticipated. Full recovery for aquatic life and increases in sportfish were also positive, whereas decreases in sportfish had a negative effect on the probability of choice. AGDECL was negative and significant, indicating that individuals are less likely to choose alternatives if they know that agriculturalists have to pay part of the costs of recovery efforts. COST was negative and significant, indicating a decreased likelihood of choosing an alternative such as tax increases.

In this model, each subject generated 24 observations (i.e., 3 possible choices on 8 choice occasions) in the data set; thus, socioeconomic characteristics were invariant across choice sets. The only way to control for socioeconomic effects was through interactions with the alternative specific constants or interaction with

Table 11.8 Implicit Prices, or Implied Willingness to Pay for a Given Attribute Level as Compared with the Status Quo

Attribute	Implicit price ($)[a]
SMALLZONE	21.12
BIGZONE	9.27[b]
PARTRECOV	2.55[c]
FULLRECOV	25.18
SPORTDECL	–22.03
SPORTINC	17.97
SONGINC	2.39[c]
AGDECL	–4.76

[a] Since the payment vehicle described in the survey was a change in tax rate (see Appendix 11-B), values should be assumed to represent annual amounts.
[b] Coefficient for BIGZONE was marginally significant (see Table 11.7).
[c] Not significantly different from zero.

the attributes. The decision was made to interact education, age, gender, fishing license ownership, and membership in environmental organizations with the alternative specific constants. The interpretation of these interactions was complicated as well. For example, ASCA*MALE and ASCB*MALE both were negative and significant (Table 11.7), indicating that the probability of choosing Option A or B rather than the status quo in any of the eight choice sets was lower for men than for women.

Calculation of Part-Worths

Using the coefficients from Table 11.7, implicit prices (with respect to the COST variable) were obtained for each of the choice variables (Table 11.8). These are typically called the *part-worths* in the conjoint and choice model literature.[a] While in theory the calculation can be made in terms of any one attribute for any other, the most intuitive trade-offs are those between dollars and the other attributes. We can estimate the part-worths by dividing the coefficient on one of the attribute variables by the coefficient on the COST variable and multiplying that result by negative 1. For example, the part-worth on full recovery of aquatic life is

$$\text{Dollar value of full recovery of aquatic life} = -\left(\frac{\beta_4}{\beta_\$}\right) \tag{11.2}$$

where $\beta_\$$ is the coefficient on the variable COST. Respondents were willing to pay substantially more for a small than for a large agriculture-free zone, suggesting perhaps that (a) the idea of an agriculture-free zone is attractive in and of itself,

[a] This is the *marginal rate of substitution* concept in economics upon which indifference curves are based. Simply, it gives the trade-offs that an individual is willing to make between bundles of goods while holding utility constant.

independent of any benefits expressed in the other attributes, but that (b) such land use restrictions are most attractive when kept to a minimum. The dollar-valued part-worth for partial recovery of aquatic life was insubstantial in comparison to full recovery, and that for an increase in songbirds was similarly insubstantial in comparison to that for improved sport fishing, or to the negative part-worth associated with a decline in sport fishing.

Calculating the Value of a Biodiversity Management Program

Economists are often interested in calculating the change in welfare, or well-being (Chapter 5), due to a change in public policy. The β estimates allow the calculation of compensating surplus (CS), or total WTP, associated with any policy definable in terms of the attributes (see Chapter 5 for more information on WTP). First, the utility of the status quo is calculated by substituting the appropriate variable values defining the status quo attribute levels into Equation 11.1. Next, the utility of the policy is calculated using the values corresponding to the attribute levels that define the policy. Then, CS is given by

$$\mathrm{CS} = -\frac{1}{\beta_{\$}}(\textit{status quo utility} - \textit{utility of new policy}) \tag{11.3}$$

Following these techniques for obtaining CS[21] and using the coefficients in Table 11.7, the choice model allows valuation of the multiattribute change to be evaluated (e.g., in the case where management actions lead to simultaneous improvements [or declines] in the various facets of the ecosystem). If, for example, the status quo utility were taken as zero and a change in agricultural practices were to improve habitat for mussel populations, sportfish, and songbirds — and farmers' income were unaffected by the program — the welfare for the representative individual would increase by $54.81 (i.e., the average respondent would be willing to pay $54.81 annually to move from the status quo to the state of the world having the new agricultural practices). It is this ability to derive multiple welfare measures for complex ecosystem changes that sets choice models apart from CVM studies that allow calculation of the value for only a single policy change.

DISCUSSION

This section evaluates the cumulative outcome of the W-ERA and economic analysis conducted in the upper Clinch Valley by comparison to the generalized conceptual approach for ERA–economic integration developed in Chapter 9 (see Figure 9.1). As explained in Chapter 1, the Clinch Valley analyses were undertaken prior to the development of this conceptual approach, and the economic analysis was initiated following completion of the W-ERA. For these reasons, the studies conducted in the Clinch Valley should not be viewed as integrated in any ideal sense. However, the conceptual approach for integration can be used to examine these efforts in the larger context of watershed decision-making and management and to gain insights into ways that integration can be improved. The following discussion compares specific components of the conceptual approach with work carried out in the Clinch Valley.

Consultation with Extended Peer Community

The conceptual approach for integration has defined the *extended peer community* as consisting of interested and affected parties, decision-makers, and scientific peers and has argued, in agreement with the National Research Council[22] and others,[23–25] that these parties should be actively engaged throughout assessment processes (see Chapters 3 and 9). The ERA for the upper Clinch Valley was undertaken by a diverse, interdisciplinary, and multiagency workgroup that included both government and nongovernment representatives, and the risk characterization was conducted with scientific consultation (a workshop held by the USEPA) and formal peer review. The result was a creative, state-of-the-art analysis, the findings of which have helped to identify potential management actions by workgroup member organizations.

Decisions were made at an early stage to conduct the W-ERA without an open process for broader public involvement. Through an informal survey and long experience working in the region, analysts had indications that community residents valued biodiversity less highly than water quality, on one hand, and economic development opportunities, on the other. Therefore, the management goal on which the ERA was based, which focused on biological integrity, reflected the values of the technical specialists and environmental managers who composed the interagency workgroup, rather than a broader stakeholder consensus as in other W-ERAs. This decision undoubtedly allowed the workgroup to tackle the difficult problems of data gathering and analysis more expeditiously; arguably, it may also have limited the development of broader community awareness of biodiversity issues and mutual understanding of necessary trade-offs for environmental protection.

The economic analysis team benefited from several consultations with members of the ERA workgroup and selected stakeholder group representatives, in which ERA findings were explained and regional economic development goals were discussed. Informal and formal consultations (focus groups) with watershed residents were held to avoid miscommunication between analysts and the public. The resulting survey instrument may be thought of as a structured form of consultation with the public, in which aspects of ecological risk were presented and feedback, in the form of choices between alternative states, was elicited. Interestingly, certain results of the economic analysis ran counter to expectation about residents' values. Survey respondents appeared willing to trade-off a portion of regional agricultural income to obtain full recovery of aquatic life, and were willing to accept — even to help fund — measures that would limit agricultural use of the riparian zone to improve habitat.

Baseline Risk Assessment

The conceptual approach for integration defines baseline risk assessment as the assessment of risks currently and into the future if no new management action is taken (see Chapter 9). The upper Clinch Valley ERA used existing data to characterize the risks (and uncertainties) affecting the assessment endpoints according to current conditions and trends. It identified the impacts of multiple sources and stressors, and pointed to the future likelihood of continued extirpations of species if stressors are not more effectively managed. It provided models (in this case, empirical relationships) that could be used to assess the impacts of management policies, including spatial relationships of riparian zone land use and in-stream biological response and the impacts of multiple stressors. It did not attempt to evaluate any management alternatives, however.

Formulation, Characterization, and Comparison of Alternatives

According to the conceptual approach, economic analysis of environmental problems usually requires the evaluation of some action or policy to determine who would be affected, how they would be affected, and to what extent. Therefore, it includes the steps in which alternatives are formulated ("Formulation of Alternatives" in Chapter 9), analyzed and characterized ("Analysis and Characterization of Alternatives" in Chapter 9) and then compared to one another ("Comparison of Alternatives" in Chapter 9). In the Clinch Valley case study, the economic analysis had to examine management alternatives, even though the ERA had not done so. The economic analysis specified two hypothetical agricultural policies (in addition to a status quo alternative) for use in choice model construction. The apparent coherency of the choice model results suggests that respondents understood the proposed policies and choice sets and that the model is valid. However, it should be understood that the model does not characterize a specific alternative per se. Rather, it is a flexible, albeit semiquantitative, tool that could be useful for *comparison* of specific policies *after* they had been analyzed and characterized, as Figure 11.7 illustrates.

Figure 11.7 compares the analytic processes used in two steps, analysis and characterization of alternatives and comparison of alternatives, with those of a hypothetical example that was presented in Figure 9.3. In the hypothetical example, the ecological risks, economic effects, and health or other (sociocultural) effects of the management alternatives were analyzed quantitatively to the extent feasible. Endpoint changes that could not be quantified were expressed qualitatively. A stated preference study was used to value the nonmarket welfare effects of the alternatives and improve the estimation of their net social benefits (see Figure 9.3).

Methods used in this case study comprise a subset of those described in the example. Although the Clinch Valley W-ERA quantified relationships between land uses and ecological endpoints, the endpoint changes expected to result from the two riparian management policies introduced in the economic study were not quantified. Similarly, the financial costs and other economic effects of implementing the policies were not analyzed. Equity issues were not examined, and human health or other effects were not considered relevant to this case study. The stated preference survey used qualitative language to describe expected ecological improvements, whereas both the cost attribute and the attribute describing potential regional impacts on agriculture were numerical (Table 11.4 and Appendix 11-B).

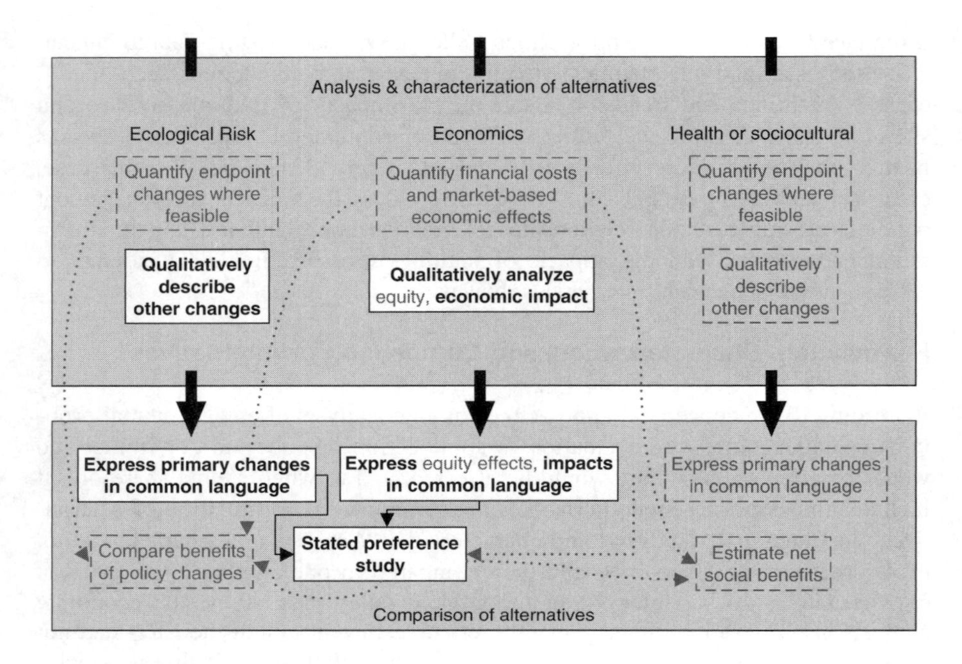

Figure 11.7 Techniques used for analysis, characterization, and comparison of management alternatives in the Clinch Valley watershed, as compared to the example shown in Figure 9.3. White boxes and bold type show features included in this analysis.

As a result, the choice model derived from the stated preference study would be capable of comparing the benefits of these or other policies only after additional work was done. The analysis and characterization of real alternatives would require the following additional steps:

- Determination of the decision context, including who could make the decision to implement a given alternative, how they would decide, and who would stand to gain or lose as a result (as part of planning, see "Assessment Planning" in Chapter 9).
- Detailed formulation of the alternatives, including design of structural (e.g., fencing) and nonstructural (e.g., institutional) implementation measures (see "Formulation of Alternatives" in Chapter 9) determination of the ecological outcomes (efficacy, in terms of instream biological response), economic outcomes (costs, including opportunity costs) and uncertainties of the policy (see "Analysis and Characterization of Alternatives" in Chapter 9).

Using the choice model as a comparison tool would present several additional challenges. Since the actual efficacy of a given exclusion zone for enhancing aquatic life can be estimated only with substantial uncertainty, it would be difficult to determine how a given best estimate of increase in IBI should be evaluated in the choice model if the available choices are partial and full recovery. Respondents ascribed statistically significant value only to full recovery. Yet even a substantial, predicted increase in IBI would not necessarily signal a recovery of extirpated species (and certainly not of extinct species), and implementation of an exclusion zone would not reduce the very substantial risks from transportation spills, for example; therefore it would be hard to rate *any* agricultural policy as leading to "full" recovery. Similar problems would be encountered in coding the effects of an actual policy on sportfish and songbirds. Ultimately there would be heavy reliance on expert judgment to interpret the ecological data and to apply the choice model.

Nonetheless, the apparently successful development of this choice model suggests that models of this type can be used for comparative welfare analysis of watershed management policies. What remains unanswered, however, is the important question of whether welfare estimates are useful to decision-makers in a given case. Whereas large water resource development projects may require welfare estimates, other kinds of decisions may not. For example, if biodiversity protection in the upper Clinch Valley will continue to depend largely on success by organizations such as The Nature Conservancy at acquiring federal grants for voluntary riparian protection programs, and private funds for land acquisition, as is presently the case, it is not clear that welfare estimates are needed. For any other protection mechanism under consideration, the decision context specific to that mechanism would need to be examined to determine what information is needed for decision support.

Adaptive Implementation

The conceptual approach for integration suggests that when uncertainties are great, management decisions should be implemented in an adaptive fashion, with continual reevaluation of effectiveness and, as necessary, redesign (see Chapter 9). The nature and magnitude of biological response that may result from any program of riparian zone protection are uncertain. However, programs can be designed in such a way that early stages of implementation will yield the information needed to resolve specific questions and improve the effectiveness of later stages. Riparian dimensional analysis indicated that the instream impacts of riparian land use were most observable over a downstream distance of 500–1500 m (see "Copper Creek Pilot Study" in this chapter). This suggests that stream reaches of appropriate lengths in different subdrainages could be preselected as treated and untreated replicates, with protection efforts targeted accordingly. Such an approach could yield valuable information on the amount of investment required to meet voluntary or regulatory goals for stream quality improvement in the upper Clinch Valley and other, similar watersheds.

ACKNOWLEDGMENTS

The authors wish to thank the members of the Clinch and Powell Watershed Ecological Risk Assessment Workgroup for their participation in developing the USEPA assessment report, upon which this manuscript is based. Dennis Yankee and Jeff White provided GIS support and database management. We also thank attendees of a workshop held in July 2001 in Cincinnati, OH for their comments on an early draft of this work, and in particular we acknowledge Leonard Shabman, Charles Menzie, Glenn Skinner, and James E. Smith for their written reviews.

REFERENCES

1. Neves, RJ., Mollusks, in *Virginia's Endangered Species*, Terwilliger, K., Ed., McDonald and Woodward Publishing Company, Blacksburg, VA, 1991, 251.
2. Neves, R.J., Pardue, G.B., Benfield, E.F., and Dennis, S.D., An evaluation of endangered mollusks in Virginia, Virginia Commission of Game and Inland Fisheries, Richmond, VA, 1980, 149.
3. Stein, B., Kutner, L., and Adams, J., *The Status of Biodiversity in the United States*, The Nature Conservancy, Oxford University Press, New York, 2000.
4. Jones, J., Patterson, M., Good, C., DiVittorio, A., and Neves, R., Survey to evaluate the status of freshwater mussel populations in the upper Clinch River, VA, Final Report, U.S. Fish and Wildlife Service, Abingdon, VA, 2000.
5. Watters, T., Small dams as barriers to freshwater mussels (Bivalvia, Unionidae) and their hosts, *Biol. Conserv.*, 75, 79, 1996.
6. Diamond, J.M. and Serveiss, V.B., Identifying sources of stress to native aquatic fauna using a watershed ecological risk assessment framework, *Environ. Sci. Technol.*, 35, 4711, 2001.
7. Serveiss, V.B., Applying ecological risk principles to watershed assessment and management, *Environ. Manage.*, 29, 145, 2002.
8. Diamond, J.M., Serveiss, V.B., Gowan, D.W., and Hylton, R.E., Clinch and Powell Valley watershed ecological risk assessment, EPA/600/R-01/050, U.S. Environmental Protection Agency, Office of Research and Development, National Center for Environmental Assessment, Washington, D.C., 2002.
9. USEPA, Guidelines for ecological risk assessment, EPA/630/R-95/002F, Risk Assessment Forum, U.S. Environmental Protection Agency, Washington, D.C., 1998.
10. Ortmann, A.E., The nayades (freshwater mussels) of the upper Tennessee drainage with notes on synonomy and distribution, *Proceedings of the American Philosophical Society*, 52, 1918.
11. Hylton, R., Setback hinders endangered mussel recovery, *Triannual Unionid Rep.*, 16, 25, 2002.
12. Allen, J.D., *Stream Ecology, Structure and Function of Running Waters*, Chapman & Hall, New York, 1995.
13. Peterson, G.D., Allen, C.R., and Holling, C.S., Ecological resilience, biodiversity and scale, *Ecosystems*, 1, 6, 1998.
14. Rubin, J., Helfand, G., and Loomis, J., A benefit–cost analysis of the northern spotted owl: Results from a contingent valuation survey, *J. For.*, 89, 25, 1991.
15. Stevens, T.H., Echeverria, J., Glass, R., Hager, T., and More, T.A., Measuring the existence value of wildlife: What do CVM estimates really show?, *Land Econ.*, 67, 390, 1991.
16. Hanley, N., Wright, R.E., and Adamowicz, V., Using choice experiments to value the environment, *Environ. Resour. Econ.*, 11(3-4), 413, 1998.
17. Boxall, P.C., Adamowicz, W.L., Swait, J., Williams, M., and Louviere, J., A comparison of stated preference methods for environmental valuation, *Ecol. Econ.*, 18, 243, 1996.
18. Louviere, J.J., Hensher, D.A., and Swait, J.D., *Stated Choice Methods: Analysis and Application*, Cambridge University Press, Cambridge, UK, 2000.
19. Sanders, L., Walsh, R., and Loomis, J., Toward empirical estimation of the total value of protecting rivers, *Water Resour. Res.*, 26, 1345, 1990.
20. Dillman, D.A., *Mail and Telephone Surveys, the Total Design Method*, Wiley and Sons, New York, 1978.
21. Cameron, T.A., A new paradigm for valuing non-market goods using referendum data: Maximum likelihood estimation by censored logistic regression, *J. Environ. Manage.*, 15, 355, 1988.

22. NRC, *Understanding Risk: Informing Decisions in a Democratic Society*, Washington, D.C., 1996.
23. Funtowicz, S.O. and Ravetz, J.R., A new scientific methodology for global environmental issues, in *Ecological Economics: The Science and Management of Sustainability*, Costanza, R. Ed., 1991, 10, 137.
24. Scheraga, J.D. and Furlow, J., From assessment to policy: Lessons learned from the U.S. National Assessment, *Hum. Ecol. Risk Assess.*, 7, 1227, 2002.
25. PCCRARM, Framework for environmental health risk management, Presidential/Congressional Commission on Risk Assessment and Risk Management, Washington, D.C., 1997.
26. USEPA, Clinch Valley Watershed ecological risk assessment planning and problem formulation - draft, EPA/630/R-96/005A, U.S. Environmental Protection Agency, Risk Assessment Forum, Washington, D.C., 1996.
27. Freeman, A.M., *The Measurement of Environmental and Resource Values: Theories and Methods*, Resources for the Future, Washington, D.C., 1993.

APPENDIX 11-A

Random Utility Model

Using random utility theory, one can model discrete choices by assuming that individuals make choices that maximize their *utility*, or well-being.[1] If the utility of alternative i is greater than the utility of alternative j, the individual will choose i. Utility is composed of both deterministic (environmental quality, income, etc.) components and random, individual-specific components that are unobservable to the researcher. The random utility model (RUM) framework is directly estimable from conjoint rankings and choice models.

Following Roe et al.[2] and Stevens et al.[3], the utility of a management program i is given by

$$U^i (q^i, z) \tag{11-A.1}$$

where the utility (U) of program i for the individual is a function of the attributes (q) of i and where z represents individual characteristics. While utility is an interesting measure of preferences, it is not particularly valuable because it does not reflect the trade-offs, financial or otherwise, that individuals must make to consume a bundle of goods. Thus one typically considers the indirect utility function, which expresses utility as a function of income and prices:

$$U^i = v^i (p^i, q^i, m, z) + \varepsilon^i \tag{11-A.2}$$

where v is indirect utility and p and m represent price of the state of the world i and income of the individual, respectively. Then the standard RUM can be estimated from the discrete choice conjoint data using conditional logit:

$$\Pr(i) = \Pr\{v^i(p^i, q^i, m, z) + \varepsilon^i > v^0 (p^0, q^0, m, z) + \varepsilon^0\} \tag{11-A.3}$$

The probability that the program having attributes i is chosen is the probability that the indirect utility of program i plus a random, unobservable error is greater than the indirect utility of program *0* and its error term.

Then v is estimated using a linear functional form of the indirect utility function, by means of the conditional logit model specified generally as:

$$v = const + \beta_1\ Attributes + \beta_2\ Socioeconomic \tag{11-A.4}$$

The stylized model in Equation 11-A.4 generates the probability of choosing a particular option given the levels of attributes of the option and the individual's (socioeconomic) characteristics. The β's generated from the above equation are the coefficients associated with each of the attributes in the choice model.

To estimate the welfare impacts, or willingness to pay, for a change from the status quo state of the world to the chosen state, one calculates:

$$v^i\ (p^i, q^i, m - CS, z) + \varepsilon^i = v^0(p^0, q^0, m, z) + \varepsilon^0 \tag{11-A.5}$$

where CS (compensating surplus) is the income adjustment necessary to leave the individual as well off with bundle i as they were with bundle 0. Additionally, the β's from Equation 11-A.4 can be used to calculate implicit prices, or part-worths, for each variable with respect to all of the other variables in the model (see Section 11.3.3.2).[a]

REFERENCES

1. Boxall, P.C., Adamowicz, W.L., Swait, J., Williams, M., and Louviere, J., A comparison of stated preference methods for environmental valuation, *Ecol. Econ.*, 18, 243, 1996.
2. Roe, B., Boyle, K.J., and Teisl, M.F., Using conjoint analysis to derive estimates of compensating variation, *J. Environ. Econ. Manage.*, 31, 145, 1996.
3. Stevens, T.H., Barret, C., and Willis, C., Conjoint analysis of groundwater protection programs, *Agric. Res. Econ. Rev.*, October, 229, 1997.

[a] This is the *marginal rate of substitution* concept in economics upon which indifference curves are based. Simply, it gives the trade-offs that an individual is willing to make between bundles of goods while holding utility constant.

APPENDIX **11-B**

Excerpt from Survey Administered by the University of Tennessee: Explanation of Hypothetical Agricultural Policies and their Potential Impacts

CONTENTS

BACKGROUND INFORMATION ON THE CLINCH RIVER VALLEY

The upper Clinch and Powell Rivers represent some of the last free-flowing river segments in the Tennessee River system. Together, they drain approximately 3800 square miles of land area. The Clinch and Powell Valley has one of the most diverse concentrations of freshwater mussels and fish species of any river in North America. Many of the valley's mussel and fish species are on the decline. Twenty-two mussels and eleven fish species are listed as endangered or threatened. Moreover, the Clinch River Valley has many species that are found nowhere else. Of the 50 mussel species that are listed by the U.S. Fish and Wildlife Service as "Threatened" or "Endangered," 16 are found in the Clinch River Valley.

Ecologists believe that biodiversity is important for a number of reasons, including its contribution to the health of the ecosystem (diverse ecosystems can better withstand and recover from stressors such as drought). Mussel species are good indicators of the health of the ecosystem. Because mussels are very sensitive to pollution, poor water quality will often affect mussels before it has an impact on other species in the river and before it has a direct impact on human health.

Although employment in the region is increasingly migrating to the manufacturing, service, and tourism sectors, the economy of the valley has historically been based on coal mining and agriculture. More than 40% of coal production in Virginia

1-56670-639-4/05/$0.00+$1.50

occurs within the Clinch/Powell Valley, and much of the discharge of pollutants in the region is not regulated.

The combined effects of raising livestock, pesticide runoff, and soil erosion from farming, forest clearing for development, coal mining and processing, discharge from sewage treatment facilities and septic tanks, chemical spills, runoff from roads, parking lots, and chemically treated lawns decrease water quality and reduce mussel and fish abundance and diversity.

EVALUATING CHANGES IN AGRICULTURE TO PROTECT THE ENVIRONMENT

One cause of reduced water quality in the river is that livestock get into the river, crushing mussels, eroding river banks, and muddying the water. Intensive cultivation of crops near the river allows fertilizers, pesticides, soil, and other substances to contaminate the river as well.

These problems could be lessened by the development of an "agricultural free zone" in the immediate proximity of the river. This zone, where crop planting and grazing would be restricted, could be of different widths. In our study, we ask you to compare the present case of no agriculture free zone with two alternative zone sizes: a zone 10 yards wide on the Clinch and 5 yards wide on tributaries or a zone that is 25 yards wide on the Clinch and 10 yards wide on tributaries.

Farmers who keep cattle would need to construct fences to keep the livestock out of the exclusion zones. Fences would keep the cattle from trampling the mussels, reduce erosion and sedimentation of the river. Trees would shade the river water, reducing its summertime temperature and increasing the dissolved oxygen level, which would benefit aquatic life. As the pastures revert to more naturally occurring types of vegetation, songbird and wildlife populations could increase. The construction of fences and substitute watering facilities for the cattle and the loss of the use of the land are costly for farmers. Farmers who grow crops would not be able to plant in the zones, which may be among their most fertile (and flattest) land holdings.

However, the farmers need not bear the full cost of the policy. A pilot project has been underway where nonprofit organizations such as The Nature Conservancy have been compensating farmers who construct fences and take lands near the river out of production. This type of project could be expanded and funded through a small increase in taxes for everyone in the Clinch Valley. The questions below ask respondents to compare possible alternative policies. One primary difference among the policies is the extent to which the farmers or the taxpayers bear the costs. Farmers could be fully or partially compensated for their losses. Another set of differences involve the levels of the environmental characteristics. These changes in agricultural practices may have effects on aquatic life, sportfish, and songbirds. The ranges of these effects that we would like you to consider are as follows:

Aquatic life: Includes all nongame fish and mussels. Changes are in terms of diversity, abundance, and distribution throughout the watershed.

Continued Decline = continued decreases in diversity, abundance and distribution in the Clinch River and its tributaries
Partial Recovery = some improvement in the Clinch River, but no improvement in tributaries
Full Recovery = improvement in the Clinch River and its tributaries

Sportfish: Includes smallmouth bass, trout, etc. Changes are in terms of number and average size.

No change = current numbers and distribution of sizes
Increase = 20% increase in Clinch and tributaries
Decrease = 20% decrease in Clinch and tributaries

Songbirds: Changes are in terms of variety of species and number of birds found in the Clinch River Valley.

No change = current numbers of birds and varieties of species in the valley
Increase = 20% increase in numbers of birds in the valley

Agricultural income: Changes are in terms of lost income in the agricultural sector of the Clinch River Valley economy. These losses would accrue to farmers in the 21 counties that are part of the valley as a result of decreased production.

No change = no change in agricultural income
Small decrease = $1 million/year total decrease in production; this represents less than 1 percent of total farm income for the valley

Cost to household: One way of financing improvements to the quality of the Clinch River is to ask residents of the valley to share in the costs of protection. If you live in the *Virginia portion of the valley,* this could be implemented through small *changes in state income taxes*. If you live in the *Tennessee portion of the valley,* this protection could be paid for through small *changes in local property taxes.*

CHAPTER 12

Seeking Solutions for Interstate Conflict over Water and Endangered Species: Platte River Watershed

Raymond J. Supalla, Osei Yeboah, Bettina Klaus, John C. Allen, Dennis E. Jelinski, Victor B. Serveiss, and Randall J.F. Bruins

CONTENTS

1-56670-639-4/05/$0.00+$1.50

WATERSHED DESCRIPTION

Watershed Resources and Impacts of Development

The central Platte River floodplain in Nebraska, which includes the 130 km of river known as the "Big Bend Reach," is rich in biodiversity and ecologically complex. The reach extends from near Lexington, NE on the west to immediately below Grand Island on the east. Nested within the Platte River watershed (Figure 12.1), which encompasses 223,000 km^2 (86,000 mi^2) in Colorado, Wyoming, and Nebraska, the central floodplain occupies 13,280 km^2 (5130 mi^2) and hosts a diverse assemblage of ecosystems, plants, and animals. Approximately 50 species of mammals and several hundred species of terrestrial birds use the cottonwood-willow forests and wet meadow grasslands near the river for breeding or stopover habitat during migration.[1] Nearly one-half million sandhill cranes (*Grus canadensis*) and several million ducks and geese use the Platte River during their annual migration.[2] In addition, the central Platte River floodplain supports nine species of plants and animals that are listed as threatened or endangered, including the interior least tern (*Sterna antillarum athalassos*), the piping plover

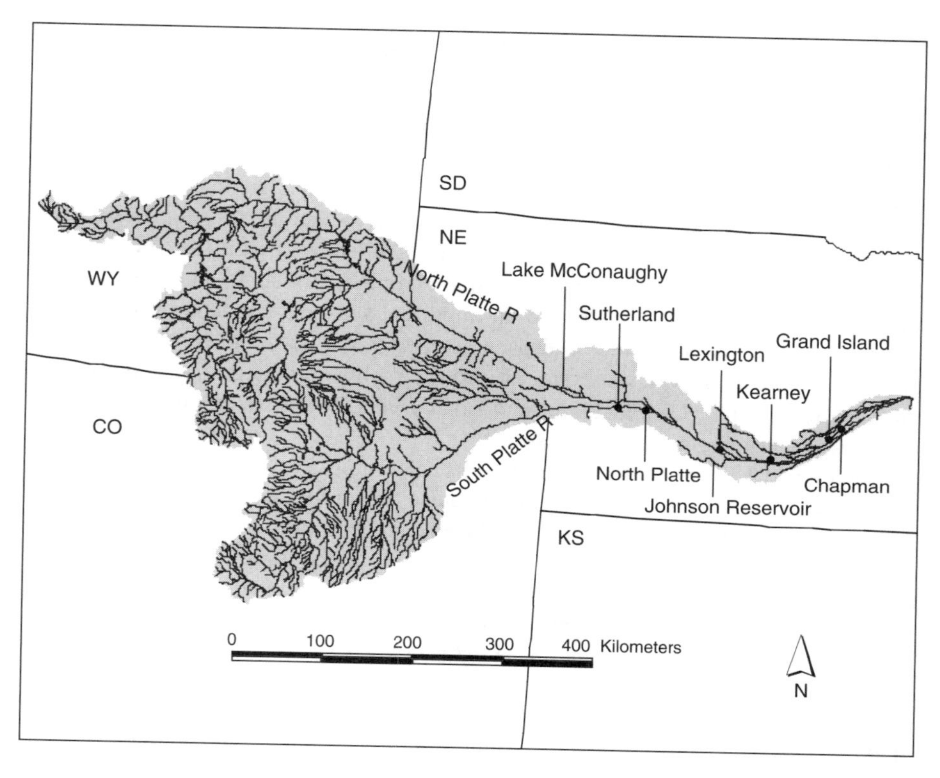

Figure 12.1 The watershed of the North Platte, South Platte, and Big Bend Reach of the Platte River in the Great Plains of the USA. Towns and reservoirs mentioned in the text are indicated.

(*Charadrius melodus*), and the whooping crane (*Grus americana*), and another 12 species that are candidates for federal listing.[3] The high levels of biodiversity found in this reach are at risk, however, due to the cascading effects of reduced water flows and development on ecosystem structure and function.

Irrigation water from the Platte River and adjacent aquifers has made the Platte Valley a highly productive agricultural region, providing irrigation water to over one million acres. Water storage reservoirs such as Lake McConaughy and Johnson Reservoir (see Figure 12.1) have provided increased recreational and sportfishing opportunities, contributing to the more than two million recreational visitor days per year provided by the river. Platte River hydropower stations help meet regional energy demand by supplying 300 MW of hydroelectric power. As a result, the natural hydrologic regime has been influenced by more than 200 upstream diversions as well as by 15 dams and reservoirs on the North and South Platte Rivers, all but one of which are in Colorado and Wyoming.[4] This elaborate network of dams, diversions, and irrigation canals has resulted in a 70% decline in peak discharge.[5]

From a hydrogeomorphologic perspective, the Platte River is a braided stream whereby the main channel contains a network of smaller channels separated by small islands called *braid bars*. Braided rivers are also characterized by highly erodible banks and an abundance of sediment. In a braided system that is unregulated, the number and location of the channels and braid bars may change quickly as a function of stream discharge and sediment load. In turn, the dynamic nature of braided rivers creates a mosaic of habitats such as shifting sandbars, side-arm channels, backwaters, and temporally inundated floodplains. Combined, this rich array of habitats supports high levels of floral and faunal biodiversity. Critically, however, flood-pulsed hydrology[6] is needed to sustain this diversity of habitats and species. These flood pulses typically occur in spring as a function of snow melting in the stream's headwaters with river disturbance scouring established habitats and creating new ones. The flood pulse also maintains an important seasonal connection of the river channel to the floodplain, which distributes energy and nutrients between the river and the land, and supports ecosystem functions such as production, decomposition, and consumption.[6–8] On the Platte and other rivers, these water fluctuations also drive patterns of vegetation succession.[9–11]

In contrast to unregulated river systems, damming and other alterations to the natural flow regime alter the nature of the pulse transmitted to the Platte floodplain. As a result, the Platte has experienced reduced channel movement and environmental heterogeneity. In addition, in regulated rivers such as the Platte, sediments become trapped behind dams, so downcutting and erosion occur in the downstream channel, further isolating the channel from the floodplain.[12–14]

Channel width on the Platte has been reduced 85–90% over the last century or so.[15] Establishment of *Populus* (cottonwood, poplar)–dominated forests has followed narrowing of the main channel and stabilization of river braids. Approximately half of the active channel present in the middle Platte in the 1930s had succeeded to woodlands by the 1960s due to the combined effects of irrigation, streamflow regulation, and drought.[11] In total, some 9500 ha of *Populus* woodland are established in the Big Bend Reach.

The significant alteration of the natural flow regime notwithstanding, high levels of faunal biodiversity are associated with the present channel structure. Two species of particular concern are Platte River populations of the least tern and piping plover — listed as endangered and threatened, respectively, by the U.S. Fish and Wildlife Service (USFWS). Terns and plovers nest on large, high-elevation, barren sandbars. Historically, spring flooding during ice pack breakup would scour vegetation off of midstream sandbars, leaving the necessary open nesting substrate. Establishment of riparian forest has significantly reduced available habitat. Sandhill cranes, perhaps the flagship species of the Platte, are also highly dependent upon open channel habitat. Approximately 80% of the continental population of cranes spend about six weeks in spring staging on the central Platte River. Sandhill cranes roost in open channels and forage for invertebrates in nearby wet meadows and for waste corn in nearby farm fields.[16,17] Much has been written about the preferences of roosting cranes for open channel habitat. In general, cranes prefer roosting in shallow water and with channel widths of 500 feet or more and rarely inhabit those that are less than 150 feet. They may roost in concentrations of 20,000 per mile. Roosting on the river protects them from their predators. The issue is complex, however, because many factors are involved in the selection of roosting sites, including availability of and distance to off-channel food (wet meadows and corn fields), weather, water depth, stream flow, and distance of roosts to tall vegetation.[2,18–23] Crane use has declined in the upper Platte River coincident with dramatic channel narrowing between 1930 and 1957, and has since increased farther downstream where channels have narrowed less.[24] However, large populations of cranes roost in the relatively narrow channels of the North Plate River or roost away from the river in wet meadows.[22] The effects, if any, of such displacement are unknown.

The channels are also important to a wider variety of migratory water-birds including whooping cranes and a variety of ducks and geese.[16,17,25–27] Waterfowl population estimates during migration range from 5 to 9 million individuals in spring.[28,29] Most of the migration population consists of snow geese, Canada geese, greater white-fronted geese, mallard, and northern pintail.

Wet meadows that flank the Platte River support a rich assemblage of migratory and breeding grassland birds.[30] Of principal concern to this avian community are the effects of lower water tables on habitat structure and forage and particularly habitat fragmentation.[30,31] An important conservation objective is the maintenance of sufficiently large habitat patches for core-grassland (no-edge) species including upland sandpiper, bobolink, grasshopper sparrow, dickscissel, and meadowlark.[30]

While alteration of the natural hydrological regime poses significant risk to many species, the establishment and evolution of the riparian *Populus* forest has created significant ecological opportunity for other species, principally those that use riparian forests. For example, based on a two-year study of 72 woodland patches, Colt[32] showed that these forests support some 50 species of breeding birds, including 32 neotropical migrant species, a guild of birds that includes several species with populations at risk. Further, Colt and Jelinski (unpublished data) have preliminary findings that suggest nest success is high, and that some species are not rendered as vulnerable to the deleterious edge effects (e.g., predation and nest site parasitism) found elsewhere on the Great Plains. The resulting increase in avian biodiversity as a result of altered flows broadens the number of stakeholders to include those concerned about off-channel species.

The seeming bonanza of forest bird species may substantially change, however. In less than a century, and barring a catastrophic major disturbance, the *Populus* dominated forests will be almost completely replaced via succession by equilibrium forests dominated by *Fraxinus* (ash), as Johnson[10] has predicted for the Missouri River floodplain forests. A profound biodiversity decline may result because a large proportion of flora and fauna is restricted to, or strongly associated with, *Populus* communities (Jelinski and Colt, unpublished paper). It is well established that maximum diversity of trees, birds, and small mammals occurs in older *Populus* forests midway along the series of successional stages (*sere*).[9,33,34]

In summary, the flood-pulse system[6,8] that is characteristic of the central Platte River floodplain links hydrology with biological communities and ecosystem processes in complex ways.[9,11,31] Alteration of the natural flow regime for hydropower, food production, and recreation has changed the dynamic nature of the river and places some species and habitats at risk. At the same time, hydrologic alteration of the Platte has created ecological windows of opportunity for a number of other species.

The effects of altered flows, habitat fragmentation, and agrochemical runoff on riparian vegetation in the central Platte River floodplain have been extensively studied,[35] whereas effects on some avian communities have barely been investigated,[2] and science is only in the early stages of predicting impacts on fish and other wildlife communities.[31,36]

Watershed Management Efforts

A long history of efforts to protect the resources of the central Platte River floodplain forms the backdrop for the ecological and economic analyses discussed in this chapter. Conservation organizations and governmental agencies have worked to improve avian habitat along the Big Bend Reach, while federal and state agencies and various stakeholders have sought ways to resolve enmeshed conflicts between economic demands for water withdrawal and environmental needs for increased, and seasonally varying, instream flows as determined under the Endangered Species Act (ESA). Over the past 25 years, a number of management initiatives, often backed by technical analyses, have been tried.

To improve habitat suitability for cranes, waterfowl, and native grassland birds, the National Audubon Society, Platte River Whooping Crane Maintenance Trust (PRWCMT), and The Nature Conservancy have acquired tracts of wet meadow and river channel. They have eliminated roads, fences and buildings and have consolidated land units to reduce disturbance and habitat fragmentation. The Natural Resource Conservation Service (NRCS) of the U.S. Department of Agriculture and the USFWS have cooperated to restore wet meadow and open-channel roost habitat for cranes by removing woody vegetation from sandbars in the river channel. These actions have not been without controversy, however, as the mechanical removal of some tracts of late seral vegetation to recreate early-successional habitats has favored the requirements of certain wildlife species while destroying established habitat for others. There is also scientific disagreement over the extent to which riparian land management can effectively substitute over the long term for restoration of stream flow.[23,37]

Concerning required flows, some scientists contend that high stream flows are needed periodically to prevent vegetative growth on sandbars and sustain the wide and shallow riverine habitat preferred by whooping and sandhill cranes,[38,39] whereas others contend that such scouring flows are of little value and may actually be harmful in the case of fish, because scouring flows lead to lower reservoir levels and higher water temperatures.[40] The terms of the legal debate over stream flow are defined by ESA provisions that prohibit any federal action jeopardizing the continued existence of a species designated as threatened or endangered, and provide that the USFWS determine species' requirements based on the best available scientific information. The USFWS has determined that an additional 417,000 acre-feet (514 hm^3) per year of water is needed to meet endangered species needs for the Big Bend Reach in a wet-to-average year.[a,3] Absent any agreement as to how to make up that deficit, this determination is sufficient to preclude any major water consuming action that constitutes a federal nexus. In other words, the U.S. Forest Service (USFS) water leases in Colorado cannot be easily renewed; Wyoming cannot pursue additional,

[a] This annual volume does not include less frequent flow recommendations such as a 5-year peak flow of 16,000 cfs for channel maintenance.

federally permitted upstream water storage projects that would increase consumptive use; and the public power districts in Nebraska cannot be assured of getting a long-term hydropower license from Federal Energy Regulatory Commission (FERC) unless some accommodation of the competing demands can be made.

Stakeholder groups have been actively involved in management discussions that have occurred in the context of water right litigation, power plant licensing hearings, legislative debates, and other venues.[4] Environmental interests in all political jurisdictions (Colorado, Wyoming, Nebraska, and the USFWS) tend to agree on the need for increased and re-regulated stream flow and management of riparian lands for endangered species protection. Irrigation interests are much more parochial both between and within states. Upstream surface water irrigators have sought the right to continue irrigating and, in some instances, the right to develop additional acreage. Downstream surface water irrigators want their water supply protected against additional depletions from upstream irrigation or environmental demands. Groundwater irrigators in all locations have sought the right to pump at will, irrespective of stream flow considerations. Hydropower interests want high reservoirs to maximize feet of head and would like to make reservoir releases during the summer months when electricity is worth the most. Coal-fired electric utilities want assured cooling water supplies and expansion opportunities. Finally, recreation interests have mixed demands, including moderate reservoir storage levels, stream flows that sustain fishing and waterfowl hunting, and easy access to the river and to bird-watching opportunities.

Since 1976 the Nebraska Department of Water Resources (DWR) has held over 400 days of public hearings to address proposed diversions or requested instream uses of Platte River water. From 1983–1997, the public power districts in Nebraska were in negotiations with the FERC over the relicensing of Lake McConaughy. In addition, from 1986–2001 the states of Wyoming and Nebraska were in litigation over the interstate allocation of Platte River water.

The struggle to manage the Platte system has led to several attempts to facilitate resource management decisions, including empirical modeling with and without stakeholder input, several negotiation formats, multistate litigation and, most recently, a tristate–federal Cooperative Agreement[41] that takes an interim, adaptive management approach to the problem. One of the first organized attempts to reach a compromise solution was an adaptive environmental assessment process that began in 1983. Called the Platte River Forum, this approach involved identifying a group of experts and stakeholders and assembling them in a single location for one week. This group first identified the relevant impact variables and policy options. Then, with the help of experts, the associated technical relationships were described in mathematical terms and computerized. The idea was that stakeholder participation and input would lead to a widely supported simulation model and agreement regarding the consequences of management options.[42] This expectation proved to be invalid. Not only did participants fail to agree on all the facts, but even when there was general agreement on how the natural system worked, differing value judgments and varying objectives prevented completing a model that was very useful for determining how the water should be used.[43]

The Platte River Forum was responsible in part for the formation of a small research group to develop a multi-objective model of the Platte. This model was built by a group of university professors without stakeholder involvement.[44] Whereas the Platte River Forum focused on the physical aspects of the river system and considered only a small set of alternatives, the multi-objective model focused on the delineation of trade-off curves for numerous alternatives. The intent was to improve on the Platte River Forum by producing additional information for decision-making and to do so without the inefficiencies and biases of a committee of 30, many of whom represented stakeholder interests rather than areas of expertise. The outcome of the multi-objective modeling approach can best be characterized as good science that was unused and ineffective. The scientists involved, operating independent of political pressure, were able to produce a credible operational model, but the results were not embraced by any interest group or decision-maker.

A third attempt to resolve the water management problem involved relicensing of hydropower plants. From 1986 until a provisional hydropower license was issued in 1997, the Central Nebraska Public Power and Irrigation District and the Nebraska Public Power District were involved in an intensive effort to get the FERC to relicense their Platte River hydropower facilities. The central issue was protection of threatened and endangered species, but National Environmental Policy Act requirements associated with licensing a public resource also meant that broader fish and wildlife issues, including sandhill crane habitat, had to be addressed. The major hydropower facility involved is part of the Kingsley Dam, which creates Lake McConaughy.

Lake McConaughy (see Figure 12.1) is the largest reservoir on the Platte River and the closest one to the endangered species habitat. Historically, Lake McConaughy has been used to directly irrigate over 200,000 acres (77,000 ha) and to enhance the groundwater supply for an additional 300,000 acres (112,000 ha).[45] It has also been managed as a fishery in cooperation with the Nebraska Game and Parks Commission and is a significant recreational resource, drawing over 600,000 annual visitors per year. For nearly 50 years, however, the water entering Lake McConaughy was managed in a serial dictatorship with irrigation receiving first priority for the water, followed by hydropower and recreation. Endangered species were not considered. This all changed when the original hydropower license expired in 1987. The FERC required the Districts to address wildlife habitat maintenance and enhancement, which led to extensive study by the Districts and by environmental interest groups, and eventually to intensive negotiations among the Districts, environmental interests, and FERC. However, the parties were unable to agree on how to balance endangered species with other needs. Licenses were nevertheless issued provisionally, with a requirement that the districts' operations be coordinated with the proposed Cooperative Agreement.

As the pressures for reallocating water to meet endangered species needs mounted, Nebraska interests sought to broaden the responsibility for meeting these needs to include Colorado and Wyoming. Of the two million acres irrigated with surface water within the Platte Basin, Colorado has 56 percent, Wyoming 12 percent, and Nebraska 32 percent. It seemed unfair to Nebraska water interests that they should have to meet endangered species needs without appropriate contributions from Colorado and Wyoming.[45] At the same time Colorado was facing endangered species problems with USFS water rights and with potential irrigation projects, while the threat of subjecting U.S. Bureau of Reclamation projects to consultations under the ESA had eastern Wyoming and western Nebraska irrigators nervous.[45] All three states found that cooperation was in their mutual interest and negotiated the Cooperative Agreement, initiated in 1994 and signed on July 1, 1997.

The Cooperative Agreement constituted a multistate–federal effort to protect Platte River endangered species without unduly constraining the availability of water for other uses. It established a preliminary agreement to increase instream flow by an average of 130,000–150,000 acre-feet (160–185 hm^3) and to acquire an initial 10,000 acres (3900 ha) of an eventual 29,000 acres (11,200 ha) of riparian habitat, but it did not set forth where all of the water would come from or what land would be acquired. The participants originally had three years to study alternatives and to agree on sources of water and land, including a distribution of the costs. Progress has been slow, however, and the period for reaching agreement has been extended repeatedly; as of this writing it stands at June 2005 and may be extended further.[46] If agreement is reached, the plan is to be put in place and monitored for 10–13 years to determine how well the program is meeting endangered species needs. If an agreement is not reached, the public power districts in Nebraska may lose their provisional hydropower licenses, holders of water right leases on USFS lands will find renewal very difficult, new surface water development in all states will be difficult (if not impossible), and actions to protect endangered species will be further delayed.

Whether the Cooperative Agreement is successful remains to be seen, but thus far none of the management approaches used have led to a comprehensive resource management plan that addresses the conflicting demands of competing interest groups.

ECOLOGICAL RISK ASSESSMENT

Planning

Concern over threats to the valued biodiversity of the central Platte River floodplain, coupled with evidence that various agencies and stakeholders would be willing participants (Table 12.1), motivated the U.S. Environmental Protection Agency

Table 12.1 Participants in Planning for the Central Platte River Floodplain W-ERA

Participants
Central Nebraska Public Power and Irrigation District
Nebraska Public Power District
Nebraska Department of Environmental Quality
Nebraska Natural Resources Commission
Central Platte Natural Resources Districts
Nebraska Game and Parks Commission
Tri-Basin Natural Resources Districts
Nebraska Department of Agriculture
The Nature Conservancy
Prairie Plains Resources Institute
Platte River Whooping Crane Maintenance Trust (PRWCMT)
University of Nebraska — Lincoln and Kearney
U.S. Fish and Wildlife Service
U.S. Geological Survey
U.S. Department of Agriculture

(USEPA) in 1993 to establish an interdisciplinary workgroup to begin a watershed ecological risk assessment (W-ERA). The goal was to obtain a better understanding of how the central Platte River landscape and associated flora and fauna are being impacted by water withdrawal and other stressors. The workgroup was composed of individuals with disparate interests and responsibilities and many years of experience working in the central Platte River watershed. The planning process included face-to-face dialogue between assessors and resource managers, a group tour of the watershed, symposia, public meetings, focus group meetings, and teleconferences.

Recognizing that any protective management actions would have to be weighed against the need for human uses, the workgroup developed the following management goal for the watershed: protect, maintain, and where feasible, restore biodiversity and ecological processes in the central Platte River floodplain to sustain and balance ecological resources with human uses. The management goal is a qualitative statement that addresses concerns expressed by various agencies and management organizations as well as the floodplain residents and other stakeholders.

Problem Formulation

This section summarizes the problem formulation exercise conducted for the central Platte. The intricacies of that process, and the limitations of the resulting analyses, presented in the following section, illustrate the difficulty of narrowing a broad management goal for a large and complex system to a tractable set of risk assessment problems.

The management goal was interpreted by representatives from the USEPA's Region VII and Office of Water, the USFWS, the U.S. Geological Survey, and Nebraska officials (listed in Table 12.1) into potentially implementable environmental management objectives (Table 12.2). A more detailed description of the watershed than that presented earlier in this chapter was developed, along with a description of the environmental problems in the watershed. The environmental problems emanate from a combination of physical and chemical stressors. Of the many human-caused stressors thought to be interfering with attainment of the goal, eight principal stressors were selected by the workgroup (Table 12.3) using a Delphi ranking technique[47] that documents iterative group input and helps groups reach a consensus (see also Appendix 9-B). Nine ecological assessment endpoints, representing three spatial scales, were selected (Table 12.4) that met the criteria of (a) relevance to environmental management objectives, (b) ecological relevance, and (c) susceptibility to stressors (see "Problem Formation" in Chapter 3).

As in other risk assessments discussed previously, detailed conceptual models, developed for each endpoint, were used to hypothetically attribute stressors to their sources and to explain their impact on the assessment endpoints. Three of the nine assessment endpoints, or representative elements of them, subsequently were selected as priorities for detailed quantitative analysis. Those endpoints, and the corresponding risk hypotheses that were derived from the conceptual models, are presented in Table 12.5. These three were selected because they capture the predominant concerns regarding birds and unique habitat in the floodplain and because they crystallize water and riparian management conflicts. All three are linked to the fact that lower

Table 12.2 Eleven Environmental Management Objectives that are Implicit In and Required to Achieve the Management Goal

Affected Area	Environmental Management Objective
Channel	1. Restore and maintain stream channel dynamic equilibrium
	2. Maintain sufficient flows to prevent high temperatures detrimental to native fish populations
Riparian Forest	3. Maintain range of successional stages of forest vegetation
Backwaters	4. Maintain and reestablish backwater ecosystems
	5. Maintain and restore hydrologic connectivity between river channels through surface flows
Floodplain	6. Maintain hydrologic connectivity between river channels and wet meadow ecosystems
	7. Maintain and reestablish natural diversity in wet meadow systems
	8. Maintain and reestablish natural diversity in native upland systems
Landscape	9. Protect and where feasible reestablish the mosaic of habitats in the central Platte River floodplain to support key ecological functions and native biodiversity
	10. Maintain diversity of water-dependent wildlife including migratory and nesting birds, mammals, amphibians, reptiles, and invertebrates
	11. Prevent toxic levels of contamination in water consistent with state water quality standards

Table 12.3 Principal Stressors (and their Primary Sources) in the Central Platte River Floodplain

Altered surface water regime (dams and diversions)
Truncated sediment supply (dams and diversions)
Altered ground water regime (dams, diversions, groundwater withdrawal, and irrigation)
Physical alteration of habitat (land conversion to agriculture, including drainage of wet meadows, and clearing of vegetation for wildlife management)
Nutrients (fertilizer use)
Toxic chemicals (agricultural biocide use)
Harvest pressure (fishing, seining, waterfowl hunting)
Direct disturbance (roads, off-road vehicles, bird watching)

Table 12.4 Ecological Assessment Endpoints for the Central Platte River Floodplain W-ERA

Landscape scale	Floodplain landscape mosaic structure, function, and change
Habitat scale	Open channel configuration and distribution for migratory birds
	Side channel and backwater area and connectivity to main channels
	Riparian vegetation successional stage, areal extent, and dispersion
	Wet meadow composition and abundance
Organism/Population level	Sandhill crane and waterfowl diversity, abundance, and dispersion
	Core grassland breeding bird diversity and abundance
	Amphibian survival and reproduction
	Riverine and backwater fish and invertebrate survival and reproduction

rates of flow reduce channel habitat for species such as sandhill cranes, piping plovers, and least terns[2,17,18,48] and reduce shallow groundwater levels, thereby desiccating wet meadows and reducing habitat diversity.[49] However, lower flows promote the establishment of riparian forests favored by other avian species.

The embattled nature of the Platte River management problem was evident during the problem formulation process. An initial draft of the planning and problem formulation report was presented to, and amended by, the stakeholder group in February of 1996. Subsequently, the draft was further revised by the risk assessment team in accordance with the USEPA's concurrently developing ERA guidance. Upon release of the revised draft,[50] some of the stakeholders considered the revised draft overly environmentalist in tone and a breach of group process, and they formally complained to the USEPA by way of their Congressional representatives. To some extent, this disagreement reflects a divergence in values and objectives between the larger environmental community and those who live in the region. As such, it is characteristic of the problems encountered when the benefits of environmental improvements accrue to a broad community, while most of the costs are incurred locally.

Analysis

Because of reassignments and shifting priorities, only a portion of the quantitative exposure and stress-response analyses that were contemplated could be completed, even for the reduced list of three assessment endpoints. This section presents those partial analyses.

Riparian Vegetation Successional Stage, Areal Extent, and Dispersion

The risk hypotheses attributed fragmentation and loss of heterogeneity of riparian vegetation to reductions in instream flow and sediment supply, as well as to riparian habitat management measures, including mowing to create crane roosting habitat. It was also hypothesized that agricultural herbicide use may pose additional stress. Reductions in mean annual flow, peak flow, and sediments in the central Platte River during the period of regulation are well documented, as are reductions of active (unvegetated) channel area, increases in wooded area, and decreases in wet meadow area since the onset of regulation.[3,11,29,35,51] Therefore, the veracity of hypotheses 1 and 2 (Table 12.5) is not much questioned, but efforts to develop quantitative relationships between these variables to enable estimates of risk were not completed. Analysis of herbicide impacts on riparian vegetation was not undertaken, nor was there an analysis of riparian management effects on patch dimensions.

Core Grassland Breeding Bird Diversity and Abundance

Risk hypotheses postulated that lowered ground water levels and habitat destruction and fragmentation reduced habitat suitability for, and survival of, several grassland

Table 12.5 Selected Assessment Endpoints and Stressors and the Associated Risk Hypotheses Developed During Problem Formulation for the Central Platte River Floodplain W-ERA

Priority Assessment Endpoints	Principal Stressors	Risk Hypotheses
Riparian vegetation successional stage, areal extent, and dispersion	Altered surface water regime	1. Lower flows have led to reduced reworking of channels, greater cottonwood regeneration, less heterogeneity of riparian vegetation
	Truncated sediment supply	2. Reductions in sediment may alter development of river braids by lowering riverbed elevation, decreasing sediment deposition on floodplain, increasing stability, and reducing riparian heterogeneity
	Physical alteration of habitat	3. Removal of riparian woodland vegetation by mowing and cutting reduces patch size and diversity of riparian vegetation
	Toxic chemicals	4. Herbicide drift and runoff from agricultural fields have caused physiological stress and perhaps increased mortality in riparian vegetation
Core grassland breeding bird diversity and abundance	Altered ground water regime	5. Lowered water table reduces diversity of wet meadow vegetation and renders adults, eggs, and young more susceptible to predation
	Physical alteration of habitat	6. Loss of habitat, reduction of patch size, and fragmentation of habitat may lead to decline of species requiring large wet meadows
Sandhill crane abundance and distribution	Altered surface water regime	7. Lower flows lead to additional woody plant establishment, channel narrowing and deepening, and roosting habitat fragmentation; these changes reduce roost suitability, increase crowding, and may increase susceptibility to disease or other catastrophic events
	Truncated sediment supply	8. Reductions in sediment supply reduce channel braiding and thus open-channel roosting habitat
	Physical alteration of habitat	9. Wet meadow conversion to crops has fragmented crane foraging, loafing, and resting habitat; channelization has reduced roosting habitat suitability
	Direct disturbance	10. Auto and rail traffic and crane-based tourism disturb migrating cranes
	Altered ground water regime	11. Lowered water tables reduce the production of wetland invertebrates, tubers, and seeds that provide forage for migrating cranes

Source: Jelinski.[31]

nesting species. Therefore, an analysis of habitat use data was performed. Helzer and Jelinski[30] surveyed 45 and 52 grassland patches, in 1995 and 1996 respectively, in the central Platte River valley. Patch size ranged from 0.12 to 347 ha; roughly half of these meadows were used for grazing, the others for haying. In each patch, four randomly selected, 100-m transects (4 ha total area) were surveyed twice between May 17 and July 5, and species that are exclusively grassland nesters were censused. Where intended sampling area exceeded patch size, patches of similar size characteristics were combined. Patch area and perimeter were determined using aerial photographs and digital planimeter. Thirteen wet-meadow breeding species were found during the two field seasons; the six most common were used in species occurrence models, and all 13 were used in species richness analysis.

Occurrences of all six common species and species richness were most strongly (and inversely) correlated to perimeter-area ratio, indicating that habitat use by wet-meadow nesting species is maximized in patches that provide the most abundant interior area, free from edge effects. These findings directly supported hypothesis 6 (Table 12.5). Since wetness or vegetational diversity within these patches was not measured, hypothesis 5 was not evaluated.

An analysis of the diversity and abundance of 50 woodland breeding bird species was also carried out[32] but was not completed (Colt and Jelinski, unpublished data). During the 1995 and 1996 breeding seasons, birds were censused in 72 woodland habitat patches ranging in size from 0.02–44 ha and were analyzed in relation to five spatial variables (related to patch size and shape) and 15 habitat structural variables (e.g., tree species richness, average tree basal area, canopy height, tree density, percent area flooded). In preliminary findings,[a] both richness models and occurrence models (the latter were significant for 24 species) tended to indicate that although structural variables (including canopy cover, shrub stem density, and percent area flooded) were significant for some species, spatial variables related to patch size were more important in general. These findings suggest that a statement similar to hypothesis 6 can be made for woodland avifauna.

Sandhill Crane Abundance and Distribution

Over a six-week period during spring migration, approximately 500,000 sandhill cranes stage in the central Platte River floodplain, with an individual staying about 2–4 weeks to rest and accumulate fat reserves. Cranes roost in the evening in broad, shallow segments of the river channel. They prefer channels at least 150 m wide and 10–15 cm deep, with unobstructed views. Though they will roost in channels less than 150 m wide, they avoid those less than 50 m in width.[23,52] Faanes and LeValley[24] evaluated population changes among four staging areas and found that a west-to-east shift had occurred. This shift was attributed to loss of roost habitat in some of the western river segments and to scouring river flows and human removal of woody vegetation, providing more desirable roost sites in some eastern segments. Controversy exists, however, as to whether the river channel is now in a state of equilibrium with respect to suitability for crane roost habitat, or in a state of decline.[11,39]

[a] Communication by D. Jelinski to V. Serveiss and R. Fenmore (May, 2000).

Risk hypotheses attributed reductions in roost suitability to reduced river flows, reduced sediment supply, reduced acreage (and wetness) of wet meadows, channelization, and direct disturbance (Table 12.5). The Cadmus Group[53] attempted to evaluate relationships between sandhill crane distribution and habitat and to develop a model capable of predicting future changes in crane use of staging habitat in the central Platte River valley. Using habitat data determined in 1982,[29] coupled with USFWS annual, one-day crane census data for the flanking years 1980–1984, evaluations were performed by bridge-to-bridge river segment (N = 15), by river reach (N = 10), and by crane staging area (N = 4). Associations by bridge segment were weak, most likely because bridge segments are not ecologically meaningful. On the river reach scale, mean unobstructed channel width showed the best relationship to crane density ($r^2 = 0.45$; $p < 0.05$), while the density of wet meadows (ha of wet meadows per river kilometer) showed a rather weak relationship to crane density. When data are aggregated by staging area, the relationships improve, and crane density is a function of both mean channel width and the density of wet meadows, in a two-step relationship. First, if mean channel width is less than about 50 m, cranes will not be present. For staging areas with mean channel widths greater than 50 m (i.e., Kearney to Chapman, Lexington to Kearney, Sutherland to North Platte), the following best-fit regression model was obtained:

$$\text{ABUND} = 318 + 3.74\ \text{MEADOW} - 1.39\ \text{ALFALFA} \tag{12.1}$$

where ABUND is crane density (numbers/km of river) and MEADOW and ALFALFA are density (ha/km of river) of wet meadows and alfalfa fields, respectively. For this model, the adjusted r^2 was 0.754, and p was 0.0002; the standard errors of the intercept, MEADOW, and ALFALFA were 147, 1.28, and 0.67, respectively. The regression of crane abundance versus density of wet meadows alone was also significant ($p = 0.0002$; adjusted r^2 of 0.665); the best fit equation for this model was:

$$\text{ABUND} = 39.9 + 5.49\ \text{MEADOW} \tag{12.2}$$

in which the standard errors for the intercept and MEADOW were 69 and 1.08, respectively. These findings are generally consistent with aspects of hypotheses 7–9 and 11 (Table 12.5). They demonstrate that there is an apparent threshold for acceptable channel width, above which the availability of forage habitat (especially wet meadows, and to a more limited extent alfalfa) is most important. However, data on channel widths and areas of wet meadows and alfalfa fields more recent than 1982 were unavailable to test the model, limiting its confidence and reliability.[a] Furthermore, the relationship between the primary stressors — that is, reductions of flow, sediment, and ground water level—and either the habitat variables or crane abundance could not be investigated by this approach, and thus the analysis was not

[a] The PRWCMT has collected additional data on crane use between 1998 and 2002, but as of this writing not all of it has been converted to a useable form.

directly applicable to decisions related to water management. Data on direct disturbance (hypothesis 10) were not available for analysis.

Risk Characterization

As mentioned above, risk analyses for the central Platte River floodplain were not completed, and therefore risk characterization, or the translation of exposure and response analyses into meaningful — and, where possible, quantitative — statements about risk, could not be carried out. Nonetheless, the W-ERA served to summarize existing knowledge about risks to a set of valued ecological endpoints in the region, to focus information needs on a set of risk hypotheses, and to provide new data and quantitative relationships for several of these endpoints. These findings are potentially valuable because factual disagreements underlie some of the ongoing resource management disagreements, as discussed earlier. Whereas the questions currently driving policy are specific to the water and habitat needs of federally threatened and endangered species, the ecological risk problem was formulated more broadly to examine the ecological integrity of the region as a whole. These results do not directly address the question of target flows in the Big Bend Reach, but they do speak directly to the importance of maintaining broad, active river channels and a diverse riparian landscape mosaic — that is, one that includes wet meadow patches with large interior dimensions and forested patches of varying seral stage — as the means to protect regional biodiversity, particularly of birds.

ECONOMIC ANALYSIS

Environmental economics often approaches environmental management problems as budget-constrained, social-utility-maximization problems, in which a key role for analysis is the quantification of policy-relevant costs and benefits, including those related to nonmarket goods (see Chapter 5), so that a socially optimal policy can be found. Ecological economics often takes a similar approach, while adding a sustainability or other biophysical constraint. Experience with Platte River decision-making, however, suggests that technical analysis alone does not lead to a resource management equilibrium, either optimal or suboptimal. First, information asymmetries create principal–agent problems (see "Welfare Economics" and "Game Theory" in Chapter 5). For example, states have an incentive to overstate their political compensation costs for providing environmental water (see "Data Sources" later in this chapter). Second, the presence of multiple objectives and stakeholder groups means that the optimal management plan is different for each stakeholder group and that a global social optimum cannot be achieved without weighting the relative importance of each. Such weights are never explicitly assigned, but instead are implied by the decisions that are taken. A resource management equilibrium is reached only when each stakeholder group believes that the cost of further negotiations or political action exceeds the value of the expected change in outcome, a condition that closely approximates the classic Nash equilibrium in economic game theory.[54]

All participants in the dispute over environmental management in the central Platte River floodplain have a strong incentive to reach a solution. Without a negotiated solution, the federal government will have greater difficulty meeting its ESA obligations; agriculturalists could face federal imposition of very high instream flow requirements; environmentalists may encounter further delays before instream flows are increased; and the states face continued uncertainty, hampering their individual water management and economic development programs and threatening higher costs if a settlement is imposed. In spite of these incentives, the parties have been unable to reach an agreement. A case in point is the need to follow-up the general agreement reached in the Cooperative Agreement — to increase Platte River flows by 130,000–150,000 acre-feet for 10 years and to monitor the results — with a specific agreement as to how the state and federal parties will provide and pay for the water. All stakeholder groups continue to argue over technical issues and to take strategic positions designed to improve the resource management outcome from their point of view.

Recent developments suggest that selected game theory techniques may be useful in resolving this conflict. Game theory occasionally has been applied to water resource management problems during the last decade. Becker and Easter[55] used game theory to analyze the dependency among eight states and two provinces concerning water diversions from the Great Lakes. Diversion decisions were modeled under different scenarios, with different restrictions on the lakes where diversions could occur. The results suggested that states do not necessarily divert water because they stand to gain relative to the status quo, but because they may lose more if they follow an alternative future strategy. In a case similar to the central Platte, Adams et al.[56] proposed game theoretic models in the form of computer simulations to investigate the likely outcome of negotiations among agricultural water users, environmental groups, and municipal water users in California. Their results indicate that the outcome of the negotiation process depends crucially on the institutional structure of the game, the input each group has in the decision-making process, the coalitions of groups that can implement proposals, the scope of negotiations, and the consequences if parties fail to reach agreement.

The principal appeal of game theory to the central Platte bargaining problem is that it offers the potential of inverting the problem from a case where stakeholder representatives propose solutions to each other to one where stakeholders respond to solutions suggested by game models. This increases the possibility that an equilibrium solution will be found, because all bargaining strategies are simultaneously considered and because mathematical manipulation is likely to reveal solutions that may not emerge in a round-table bargaining process. Although a realistic game model for this situation is unlikely to have a solution that meets all constraints, and it will certainly not have a unique solution, the game theory approach may still have considerable merit. It forces the participants to consider the role of incentives and strategic behavior in bargaining and, if nothing else, increases the likelihood that individual stakeholder groups will pursue policy options that are attractive enough to all participants to have a reasonable potential for successful implementation.

The decision to focus the economic analysis on the Cooperative Agreement process, and to use game theory, was made by the economic research team of the

University of Nebraska – Lincoln (UN-L) in their application for a USEPA grant. Some team members had a longstanding involvement with the instream-flow negotiations. After the grant was awarded, and prior to the start of work, an informational meeting was held in 1999 involving USEPA, the UN-L research team, a representative of the Nebraska Department of Natural Resources (familiar with stakeholder concerns and the Cooperative Agreement), the Platte Watershed Program Coordinator of the UN-L Cooperative Extension Service (familiar with habitat management efforts), and the lead researcher for the W-ERA. Participants were informed regarding the status of the W-ERA, the status of the Cooperative Agreement, and the proposed economic research approach.

For this analysis the central Platte management problem was defined in terms of two game models: Model I, which addresses who should provide and pay for environmental water (i.e., water reallocated to instream flow for purposes of maintaining or enhancing biodiversity), and Model II, which addresses how much water should be so reallocated. Data for Model I were obtained from available reports, whereas Model II required a survey of households in Colorado, Nebraska, and Wyoming. The next sections present the methods and results of each model.

Model I: Determining Who Should Provide and Pay for Environmental Water

The parties to the Cooperative Agreement (initiated in 1994 and signed in 1997 by Colorado, Wyoming, and Nebraska) have agreed as to an incremental amount of instream water (i.e., 140,000 acre-feet) that would constitute a first step in the adaptive implementation of measures to protect threatened and endangered species in the central Platte River floodplain. However, they have not been able to fully agree on the source of the water or who would pay for it (as well as a number of other administrative details). This study hypothesized that an auction approach capable of addressing information asymmetries would lead to an agreement in circumstances where other negotiating strategies may break down. After examining auction techniques (see Klemperer[57] for a comprehensive review), the approach selected was a second-price, sealed-bid sequential procurement auction with descending bidding and predetermined cost shares. In a sequential procurement auction, one unit (in this case, a given quantity of water) is auctioned at a time, and a single buyer receives bids from several sellers. In a descending-bid (or English) procurement auction, price falls incrementally until only one seller remains. If the auction is of the second-price (or Vickrey) variety, the winning seller receives the second-lowest bid, which eliminates the incentive for a seller to bid higher than his minimum price. Most of the auction literature deals with auctions where a single unit is sold at a time. Sequential versions of each standard auction type exist, although their use is not well researched.[57]

The only players in this game are the three states. Environmental or agricultural interest groups are not players because their primary concern is assumed to be the amount of reallocation, not who pays and at what price. The federal government's only role is to commit to a given cost share at the beginning of the game. The predetermined cost shares define how much each state and the U.S. Department of

Interior (DOI) contribute to the cash pool for purchasing environmental water. The state with the winning bid incurs an obligation to supply the water in return for a payment from the cash pool. Although the use of predetermined cost shares may be unusual or unexpected, it is consistent with the terms of the Cooperative Agreement mentioned earlier.

It is well known that for a sealed-bid, second-price auction, it is a dominant strategy for each player to announce costs truthfully.[57] The descending English auction design does not necessarily result in truthful revelation of all costs, but it does result in a dominant strategy equilibrium that minimizes welfare costs.[a] All players bid until only the two lowest-cost players remain; then the agent with the second-lowest cost stops at his cost, and the lowest-cost player wins the auction with a bid equal to (or slightly below) the second-lowest cost. Mathematical details and proof that the strategies result in a Nash equilibrium have been reported elsewhere.[58]

Data Sources

The data needs for this model consisted of acquisition costs, third party costs, and political compensation costs. Acquisition costs represent what each state will need to spend to acquire the water for reallocation to environmental uses, such as for acquiring water rights, for providing additional storage, or other costs depending on the water source. Acquisition costs were compiled from a recent report[59] prepared for use by the states and the DOI in resolving the central Platte management problem. Third-party costs were assumed to be 10 percent of the acquisition costs based on historical levels of unemployment and underemployment and on regional input–output model[b] results for the central Platte region[60] and the states of Nebraska, Colorado, and Wyoming.

Political compensation costs are the payments above expected opportunity costs (i.e., foregone economic benefits) that the states may demand as compensation for the political turmoil and economic uncertainties associated with agreeing to supply a given quantity of water. These values can be inferred from game results if the game is actually played, rather than simulated as in this study. For purposes of this analysis, three different levels of political compensation were defined that, based on the investigators' observation of the Cooperative Agreement Governance Committee's discussions on this issue, were expected to bound the problem: no compensation, moderate, and high. Political compensation for the moderate case, expressed as a multiple of the real cost, started at near zero for the first blocks of water supplied by a single state and increased exponentially to 20 percent of real cost at 50,000 acre-feet and to 57 percent at 140,000 acre-feet of water supplied. Corresponding points on the political compensation function for the high compensation case were 40 percent of real costs at 50,000 acre feet and 113 percent at 140,000 acre feet.

[a] Welfare costs represent the real cost of the water to all parties combined. Net welfare is equal to the budget cost less that part of the budget cost that represents transfer payments. Both second-price gain and political compensation payments affect the distribution of welfare among the parties but not total welfare, because the loss to the paying party equals the gain to the receiving party.
[b] See "Socioeconomic Submodel: Geo-Referenced Social Accounting Matrix" in Chapter 13 for further explanation of input–output modeling.

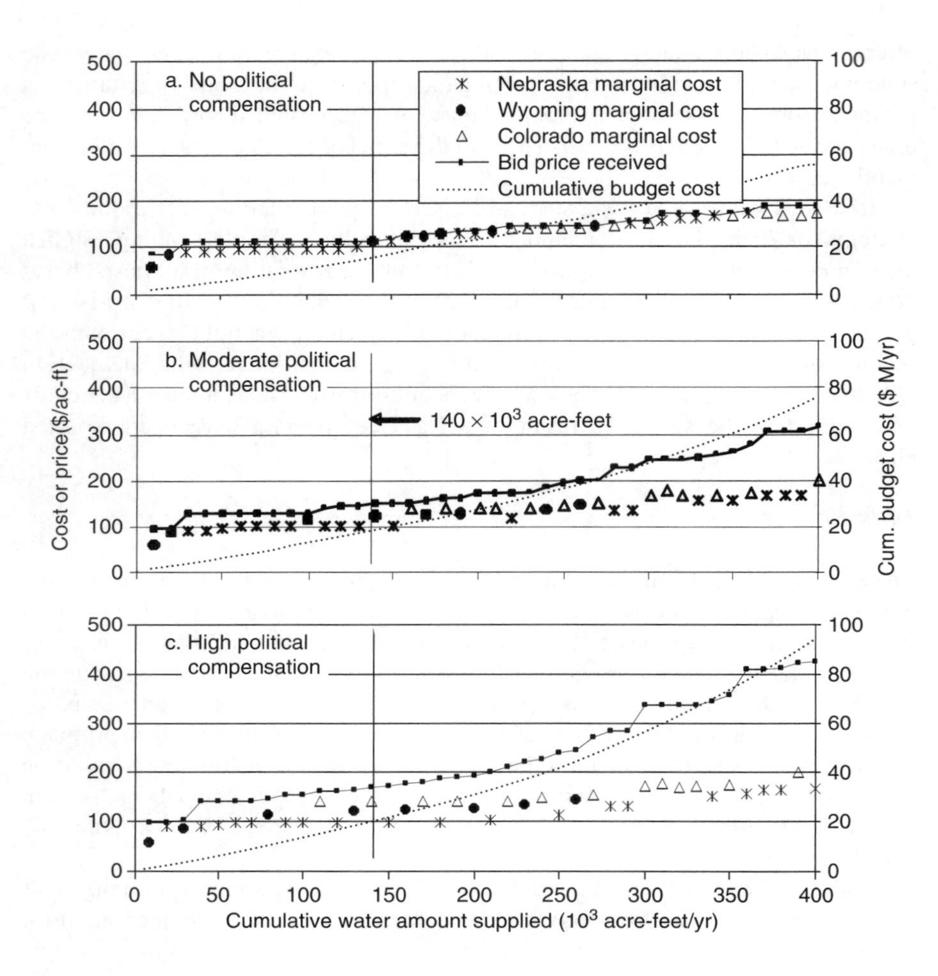

Figure 12.2 Price of 10,000-acre-foot increments of environmental water, and cumulative cost, assuming different levels of political compensation.

Simulations assumed a cost-share policy consisting of Colorado 0.2, Nebraska 0.2, Wyoming 0.1, and the DOI 0.5. These shares are based on the initial cost allocations that were incorporated in the 1997 Cooperative Agreement between the states and the DOI. Water was procured in blocks of 10,000 acre-feet, with minimum bid increments of $0.50 per acre-foot. Results were computed for water supply quantities ranging from 10,000 to 420,000 acre-feet per year (i.e., the total increment recommended by the USFWS), but all welfare comparisons were calculated for a quantity of 140,000 acre-feet, the target quantity adopted under the Cooperative Agreement.

Model I Results

Water supply costs under three different political compensation policies are depicted in Figures 12.2a to 12.2c. In Figure 12.2a, the observed difference between marginal cost and bid price is the second-price gain, whereas in Figures 12.2b and 12.2c that difference includes political compensation costs as well. Under a no-political-compensation policy (Figure 12.2a), the costs are lowest, but Nebraska would need to supply 110,000 out of 140,000 acre-feet, or 79 percent of the water. This finding reflects the fact that most of the low-cost water is in Nebraska,[61] but results of preliminary multistate negotiations to develop a water supply plan suggest that a cost minimization approach is not likely to be politically acceptable. Under these circumstances one would expect Nebraska to bid high to either get adequate political compensation or to induce another player to supply the water, whichever comes first. Under the simulated effect of political compensation (Figures 12.2b and 12.2c), exponential increases in Nebraska's bid price, and a corresponding increase in cumulative budget costs, provide incentives that cause supply by Wyoming and Colorado to increase. However, net welfare costs (Table 12.6) increase less than budget costs, because the budget increase is largely in the form of political compensation transfers among the parties. The second price effect increases as

Table 12.6 Welfare Effects from Supplying 140,000 Acre-Feet (AF) of Environmental Water

Level of Political Compensation	Welfare Costs[a]				
	Colorado	Nebraska	Wyoming	Federal	Total
None					
Water Supplied (AF/yr)	0	110,000	30,000	0	140,000
Budget Cost ($/yr)	–3,057,823	–3,057,803	–1,528,912	–7,644,560	–15,289,120
Second Price Gain	0	+1,772,443	+360,652	0	+2,133,095
Political Compensation	0	0	0	0	0
Net Welfare[a]	–3,057,823	–1,285,360	–1,168,260	–7,644,560	–13,156,003
Moderate					
Water Supplied (AF/yr)	0	100,000	40,000	0	140,000
Budget Cost ($/yr)	–3,552,900	–3,552,900	–1,776,450	–8,882,250	–17,764,500
Second Price Gain	0	+1,362,207	+404,612	0	+1,776,819
Political Compensation	0	+2,201,000	+442,700	0	+2,643,700
Net Welfare	–3,552,900	+10,307	–929,138	–8,882,250	–13,353,981
High					
Water Supplied (AF/yr)	20,000	80,000	40,000	0	140,000
Budget Cost ($/yr)	–3,960,740	–3,960,740	–1,980,370	–9,901,850	–19,803,700
Second Price Gain	+111,191	+867,424	+469,561	0	+1,448,176
Political Compensation	+372,900	+2,881,100	+889,500	0	+4,143,500
Net Welfare	–3,476,649	–212,216	–621,309	–9,901,850	–14,212,024

[a] Welfare costs are equal to budget costs net of transfers between the parties (i.e., second price gain and political compensation costs).

political compensation increases, with second price gains going to those who supply the water. Most of the increased welfare costs accrue to the federal share because they supply no water and therefore receive no second price gains or political compensation transfers.

In summary, these findings present a scenario in which a mutual supply agreement, unachievable up to this point, could be reached for a modest increase in total welfare costs (when compared to a least-cost scenario). The auction approach would resolve principal–agent problems (see "Game Theory" in Chapter 5) by creating incentives for each state to incrementally reveal its true political compensation costs. The resulting agreement is likely to benefit all the parties because each can choose between supplying the water at an acceptable minimum price or paying someone else to supply it.

Model II: Determining How Much Water to Allocate to Environmental Use

Whereas Model I examined only the provision of water and constrained the problem to a negotiation among three states, Model II casts the negotiation problem more broadly. Questions (policy attributes) examined in this model were the following:

1. What method or approach should be used for meeting endangered species' needs in the central Platte River floodplain (*Method* attribute)?
2. What is the appropriate level of investment in meeting species' needs (*Cost* attribute)?
3. Who should make that investment (*Who pays* attribute)?

The players included the federal government and environmental and agricultural interest groups, as well as the states. Because all parties stand to gain if agreement is reached, the decision process was modeled as a cooperative multilateral bargaining game. Policy options were defined as a combination of the three policy attributes. For each attribute there were five choices or levels, i.e., five methods, five cost alternatives, and five payment policies, which produced a potential for 125 different policies ($5^3 = 125$). Policy evaluation criteria were based on the utility of (i.e., relative preference for) each policy on the part of each of the game participants. Utility was also expected to vary not only by group but also according to the level of knowledge about ecological risks and the likely regional impacts of environmental policies. To develop this game, it was necessary to conduct a survey of preferences in Nebraska, Colorado, and Wyoming. The following subsections will discuss, respectively, the survey approach, the mathematical definition of Model II, and the results of Model II simulations.

Household Survey of Environmental Preferences

In November 2000, a total of 4150 households in Colorado, Nebraska, and Wyoming were randomly selected from lists compiled by Experian (Costa Mesa, CA), a private company specializing in the compilation of mailing lists. Survey procedures consisted of a first mailing, followed by a reminder postcard about 10 days later; those who had not responded within 10 days following the postcard were sent a second copy of the survey.

The survey consisted of four parts. In Parts 1–3, respondents were posed a series of statements and asked to indicate whether they agreed or disagreed (Parts 1 and 2) or opposed or supported (Part 3) each statement, on a five-point scale.[62] Part 1 assessed general attitudes regarding water and threatened and endangered species policy in the three Platte River states, but because these responses did not figure directly in model construction, they are not discussed in detail here. Part 2 examined technical beliefs, and the responses were used to assess the effect of respondents' levels of knowledge on policy preferences. Part 3 examined policy attributes and options, and these responses were used to compute respondent and interest group preferences for various policy attributes. Part 4 asked questions about demographics, and this information was used to identify respondents with particular bargaining groups (state of residence, and agricultural or environmental interest groups) to be represented in the model.

Level of Knowledge

The 10 statements posed to respondents in Part 2 (Table 12.7) were similar in form to risk hypotheses, which postulate a causal relationship between a source or a stressor and an endpoint. Whereas risk hypotheses generally refer to existing relationships, these statements tended to be in the form of inferences about the future to emphasize their relevance to policy. Six of the 10 statements are regarded as having correct answers; the other four were of interest because they are often claimed, but their veracity is uncertain. Seven of the 10 statements pertained to ecological endpoints, including the shallow water table, wet meadows, cranes, and other wildlife. These statements roughly corresponded to several of the risk hypotheses (Tables 12.5 and 12.7).[a] Three of the seven ecological statements dealing with the habitat needs of cranes were based on the expert opinion of the researchers. A simple sum of responses to the six verifiable statements constituted the knowledge level, KL, after appropriate transformations so that a higher value meant more knowledge in all cases.

Utility of Policy Attributes

In Part 3, each of the three policy attributes, *Method*, *Cost*, and *Who pays*, was described, and five different levels were defined for each (Table 12.8). Respondents were asked to rate their support of each of these 15 attribute levels individually. Next, seven policy options (each consisting of one *Method* level, one *Cost* level, and one *Who pays* level) were selected out of the total of 125 possible combinations that would capture the range of potential responses over each attribute. Utility ratings for these options were used to derive attribute weights in Model II, as further described later. These attribute weights were multiplied by utility scores for each

[a] The fact that many of those hypotheses were not evaluated in the W-ERA does not mean that these statements are not scientifically supported; in many cases the hypothesis is regarded as supported, but the underlying relationship needs to be better quantified.

Table 12.7 Statements Used in the Household Preferences Survey to Assess Respondent Level of Knowledge; Answers Regarded by Researchers as Correct; and Basis. Respondents Were Asked to Rate Agreement/Disagreement on a Five-Point Scale

Technical Statement Appearing in Part 2 of Household Preference survey	Correct Answer	Basis for Statement/Answer (and Relationship to Risk Hypotheses as Numbered in Table 6.1)
a. Maintaining a wider Platte River channel is not necessary for sustaining a large and healthy Sandhill Crane population	False	Cadmus Group;[53] Currier & Ziewitz[29] (Risk Hypotheses 1 & 7)
b. Increased stream flow will help maintain a wide Platte River channel for use by cranes and other wildlife	True	Sidle et al.;[15] McDonald & Sidle,[65] (Risk Hypotheses 1 & 7)
c. Increased wet meadow acreage is needed to meet the food needs of cranes and other wildlife in the Central Platte Valley	True	Cadmus Group;[53] Faanes & LeValley;[24] Currier & Ziewitz[29] (Risk Hypotheses 9 & 11)
d. Increased instream flows would significantly increase the quantity and quality of wet meadows	True	Hurr;[66] *The Groundwater Atlas of Nebraska*[67] (Risk Hypotheses 5 & 11)
e. The changes in regional income and employment that result from reallocating up to 420,000 acre-feet of water from agriculture to endangered species are likely to be so small that they will go unnoticed by most of the people living in the Platte Valley region	unknown	
f. Policies to maintain or increase the current flows in the Platte River will lead to increased water costs for people living in communities located near the river	unknown	
g. Ground water irrigation has lowered the water table in some parts of the Central Platte Valley	True	*The Groundwater Atlas of Nebraska*[67]
h. Ground water irrigation has adversely affected wet meadows in some parts of the Central Platte Valley	True	*The Groundwater Atlas of Nebraska*[67]
i. Improved habitat will result in an increased number of Sandhill Cranes using the Platte River	unknown	
j. An increased number of Sandhill Cranes will result in increased tourism in the Central Platte region	unknown	

attribute and summed across the three attributes that define a policy to determine the utility scores for all 125 policy options.

The utility of a given environmental policy, within a particular interest group, was defined as the adjusted sum of preference scores for the attributes of that policy, as follows:

$$U_{ij} = W_{i1} M_{ij} + W_{i2} C_{ij} + W_{i3} P_{ij} + KAF_{ij} \tag{12.3}$$

Table 12.8 Descriptions of the Three Policy Attributes and their Respective Levels, a–e, that were Evaluated in Part 3 of the Household Preferences Survey

Method:	Five different methods for meeting threatened and endangered species needs on the central Platte are described here: a. Meet all endangered species needs using least-cost methods of water conservation, water reallocation, and riparian land management, even if this means purchasing or leasing substantial quantities of water from agriculture. b. Meet all endangered species needs using a combination of water conservation, water reallocation, and riparian land management programs, but minimize the purchase or leasing of water from agriculture, even if this increases the cost of meeting these needs. c. Meet as many endangered species needs as possible using riparian land management and water conservation programs to provide for endangered species, but do not purchase or lease any additional water from agriculture, even if this means that the continued existence of the species involved may be at risk. d. Use a combination of water conservation, water reallocation, and riparian land management implemented on a trial basis over several years to make certain that the program is necessary and effective before making large public investments, even if this means there is a potential for continued risk to threatened and endangered species. e. Invest in all endangered species protection methods as long as the economic benefits from such investments are greater than the costs, even if this means continued risk to threatened or endangered species.
Cost:	To provide for threatened and endangered species on the Platte River, the cost to federal taxpayers throughout the U.S. and state taxpayers in Colorado, Nebraska, and Wyoming could range from zero to $40,000,000 per year. The amount will depend on what priority we choose to attach to species protection; on the level of risk to species extinction that we choose to accept; and on the species protection methods that we choose to use. Five different investment policies for meeting threatened and endangered species needs on the central Platte are described here: a. Invest nothing to protect Whooping Cranes, Least Terns, and Piping Plovers. b. Invest whatever the U.S. Fish and Wildlife Service (USFWS) says is needed for the species to return to nonthreatened status (currently estimated to cost as much as $40,000,000 per year). c. Invest about 25 percent of what the USFWS says is needed, or $10,000,000 per year. d. Invest about 50 percent of what the USFWS says is needed, or $20,000,000 per year. e. Invest about 75 percent of what the USFWS says is needed, or $30,000,000 per year.
Who pays:	Another important policy dimension concerns the question of who should pay for species protection. Should it be the federal government, the states involved in using the resources, private environmental interests, or some combination? The following five potential policies reflect these choices: a. All costs paid by the federal government. b. Federal government pays 50 percent and private environmental interests pay the remaining 50 percent. c. Federal government pays 50 percent and the states of Colorado, Nebraska, and Wyoming pay equal shares of the remaining 50 percent. d. Federal government pays 50 percent and the states of Colorado, Nebraska, and Wyoming pay the remaining 50 percent in proportion to the amount of Platte River water consumed in each state (Colorado 20%, Nebraska 20% and Wyoming 10%). e. Federal government pays one-third, private environmental interests pay one-third, and the states of Colorado, Nebraska and Wyoming split the remaining one-third in proportion to the amount of Platte River water consumed in each state (Colorado 13%, Nebraska 13% and Wyoming 7%).

where:

U_{ij} = utility or preference score for interest group *i*, policy option *j*
M_{ij} = attribute score by interest group *i* for *Method*, policy *j*, on a 1 to 5 scale
C_{ij} = attribute score by interest group *i* for *Cost*, policy *j*, on a 1 to 5 scale
P_{ij} = attribute score by interest group *i* for *Who pays*, policy *j*, on a 1 to 5 scale
KAF_{ij} = knowledge adjustment factor for interest group *i*, policy *j*

and W_{i1}, W_{i2}, and W_{i3} are attribute weights. The knowledge adjustment factor (KAF) was defined as the difference between the mean U_{ij} for those in interest group *i* whose knowledge level KL_i, as defined earlier under "Level of Knowledge," was one standard deviation or more above the mean and the mean U_{ij} for the entire interest group *i*. However, KAF was set to zero unless the participants in the game chose to invest in education as one method of reaching agreement, or chose to ignore the preferences of those in each interest group who were not technically knowledgeable.

The attribute weights W_{i1}, W_{i2}, and W_{i3} would be unnecessary if *Method*, *Cost*, and *Who pays* were of equal importance to respondents within a given interest group. If this were the case, then the overall utility U_{ij} of a policy option (after adjusting to equivalent scales) would be similar whether it was derived by summing a group's mean utility scores for the individual attribute levels that composed the policy or using that group's utility scores for the policy evaluated as a whole. Because this was not the case, raw attribute weights B_1, B_2, and B_3 were determined for each interest group *i* by regressing raw utilities RU_i for the seven whole policies over the scores of the three individual attributes to obtain the following equation for each group:

$$RU_i = B_{i0} + B_{i1}M_i + B_{i2}C_i + B_{i3}P_i + \varepsilon_i \tag{12.4}$$

where M, C, and P are the 1 to 5 scores for the three policy attributes and ε is an error term. The regression coefficients were then normalized across attributes to get a total value of 1.0 by dividing each nonnormalized "B_i" value by the quantity ($B_{i1} + B_{i2} + B_{i3}$), such that for each group the normalized weights become:

$$W_{i1} + W_{i2} + W_{i3} = 1.0 \tag{12.5}$$

These normalized weights were then used to adjust the individual attribute scores for all 125 policy alternatives as shown in Equation 12.3.[a]

Bargaining Theory and Model Solutions

The previous subsection defined utility for each policy by bargaining group. Here the problem of combining those utilities to identify the most globally preferred policies

[a] The concept of utility as used here is simply a preference rating. It depends on how important the consequences of a policy choice are to the respondent and also on what he or she believes the consequences will be. Knowledge can influence utility by changing the respondents' beliefs regarding consequences.

is addressed. The primary objective of the bargaining process is to find the policy option, defined as a combination of policy attributes, that maximizes total utility and is acceptable to all groups. In the bargaining literature and the broader literature of social and public choice, certain solution concepts seem to prevail. This section will introduce three of the most commonly used solution concepts for the bargaining model at hand: the utilitarian, Nash, and egalitarian solutions. Each of these solutions will later be applied to the data obtained from the survey to determine whether there are policy options that emerge repeatedly. An option chosen by different bargaining processes, which represent different social judgments, is most likely to be the policy option that would emerge from a real bargaining game. If, on the other hand, the policy options chosen by different bargaining solutions are very different, then one has to investigate the conditions of the bargaining process and the background for the social judgment much more carefully. If there is no attribute-level combination that is minimally acceptable to all groups, the players have four options: (1) negotiate a lower level of minimally acceptable utility; (2) change the water supply costs by negotiating a reduction in the political compensation factor in Model I; (3) change the preference functions of participants by providing improved biological and economic information; or (4) declare an infeasible solution.

Let X denote the set of available alternatives. In our case, X equals the set of 125 policy options that could be chosen. Let N denote the set of agents. Later on, three different sets of agents will be considered:

N = {Colorado, Nebraska, Wyoming} = {CO, NE, WY}
N = {Agricultural Interest, Environmental Interest} = {Ag,En}
N = {AgCO, AgNE, AgWY, EnCO, EnNE, EnWY}

To model the theory applied to these agents, a generic set of agents N={1, . . ., n} and a generic agent *i* are denoted. Similarly, there are generic alternatives x and y and a generic set of alternatives X. Next, it is assumed that each agent associates a cardinal utility u_i (x) with each policy option x, estimated as u_{ij} in Equation 12.3. (Alternatively, the ordinal ranks of alternatives are taken as utility information, ignoring intensities of utility across alternatives and across agents.) Since each policy option x now induces a vector (u_1 (x), . . ., u_n (x)), the decision of choosing a policy option boils down to deciding which vector of utilities is acceptable to all agents. To preserve efficiency of bargaining outcomes, only bargaining solutions that are Pareto efficient are considered; that is, for any policy option chosen by the bargaining solution, there does not exist another policy option such that all agents are weakly better off and at least one agent is strictly better off (see "Welfare Economics" in Chapter 5).

Utilitarian Solution

The utilitarian solution is the policy option that maximizes the sum of all agents' utilities and can be depicted as

$$\max_{x \in X} \sum_{i=1}^{n} u_i(x) \tag{12.6}$$

where u_i is the cardinal or ordinal utility for agent *i*, for some vector of policy options x.

Nash Solution

The Nash solution is the policy option that maximizes the product of all agents' utilities and can be depicted as

$$\max_{x \in X} \prod_{i=1}^{n} u_i(x) \tag{12.7}$$

with all terms as defined previously.

Egalitarian Solution

The egalitarian solution is the Pareto-efficient policy option that minimizes the sum of the differences between all agents' utilities. Mathematically this solution can be defined as

$$\min_{x \in X} \sqrt{\sum_{i=1}^{n} \left[\frac{\sum_{j=1}^{n} u_j(x)}{n} - u_i(x) \right]^2} \tag{12.8}$$

where u_i is the cardinal or ordinal utility for agent *i*, and u_j is that for all other agents, for some vector of policy attributes x.

In terms of social policy, the utilitarian solution represents that set of decision rules where there is no concern for the relative utility of agents. Any gain in total utility is considered an improvement irrespective of how the total is distributed across agents. The Nash solution essentially incorporates the concept of diminishing marginal utility, while the egalitarian solution takes the potential concern for equity or fairness one step further. Let us demonstrate with a simple example. Suppose there are two agents and three policy options. Option A produces 1 unit of utility for agent 1 and 10 units for agent 2; option B produces 4 units for each agent; and option C produces 6 units for agent 1 and 3 units for agent 2. In this case, the utilitarian solution would favor option A ($1 + 10 > 6 + 3 > 4 + 4$), whereas the Nash solution would favor option C ($6 \times 3 > 4 \times 4 > 1 \times 10$), and the egalitarian solution would favor option B ($4 - 4 < 6 - 3 < 10 - 1$). The respective solutions can also be referred to as the sum, product, and equity solutions.

Survey Results

This section summarizes the survey findings with an emphasis on their application to model calculations; tabularized responses to survey questions are presented in

Supalla et al.[63] A total of 1187 useable surveys were returned, for an overall response rate of 26 percent. The response rate for Nebraska residents was highest at 32 percent, followed by Wyoming at 24, percent and Colorado at 22 percent. These relatively low response rates suggest a likelihood of response bias, although there were no particular indications of response biases within or between interest groups. One would generally expect, however, that those who were better educated and most interested in the problem would be the most likely to respond.

Demographics

Demographic responses showed that respondents were in fact somewhat older and better educated than the general population. The average age of respondents was 53, over 38 percent had a Bachelor's degree or better education, and less than 5 percent had not graduated from high school. The age distribution was essentially the same for each state, but the Colorado respondents were significantly better educated than those from Nebraska or Wyoming. Approximately 14 percent of respondents were farmers or ranchers, over 18 percent were self-employed in other ways, about 13 percent worked for state or local government, and the remainder were either employed by other types of organizations or retired. The employment distribution was very similar for each state, except for agriculture. Very few of the Colorado respondents were farmers or ranchers (7 percent), compared to 12 percent for Wyoming, and 19 percent for Nebraska. Differences in the proportion of state respondents who were farmers or ranchers reflect, in part, actual differences in the proportion of each state's population that is engaged in agriculture, but these differences may also reflect a self-selection bias. Farmers in Nebraska, especially central Platte irrigators, are more likely to be directly impacted by central Platte programs and, thus, more likely to take the time to respond to the survey.

A relatively large number of respondents was affiliated with agricultural or environmental interest groups. In total, about 17 percent of respondents were affiliated with agricultural groups and 31 percent with environmental groups. The three states were quite similar, except that only 8 percent of Colorado respondents were affiliated with agriculture, and only 19 percent of Nebraska respondents were affiliated with environmental groups, compared to 48 percent in Colorado. This suggests that interest groups may be a major source of information on central Platte issues for that part of the population that was interested enough in the issues to respond to the survey.

Attitudes Regarding Environmental Policy

About one-third of respondents agreed that society should ensure species protection regardless of cost. There was very strong support for having the federal government and private environmental organizations pay for species protection rather than the states. Two-thirds of respondents agreed that the federal government rather than the states should pay for most of the cost, and 80 percent agreed that private environmental organizations should also contribute. There was also strong support for the idea that the economic base provided by irrigated agriculture should be protected.

Over 70 percent would be willing to pay more for species protection to protect the economic base, and over 50 percent were willing to protect the economic base even if it meant increased risk to endangered species. Surprisingly, 55 percent would support paying twice as much for environmental water as an alternative to reducing irrigation.

There were few significant attitudinal differences among the states. Colorado residents were much more likely than Wyoming or Nebraska residents to agree that society should ensure environmental integrity regardless of the cost. Wyoming respondents were not supportive of each state supplying one-third of the environmental water, while Nebraska respondents supported this alternative. This probably reflects a concern among Nebraska residents that the state may be asked to provide more than a one-third share and a belief by Wyoming residents that their equitable share is less than one-third.

Beliefs Regarding Central Platte River Environmental Problems

There was considerable disagreement and lack of knowledge concerning physical environmental attributes. Only 24 percent of Colorado residents, 29 percent of Wyoming residents, and 41 percent of Nebraska residents were aware, defined as agreed or strongly agreed, that a wide river channel is important to cranes. Less than 50 percent of the respondents in all states recognized that increased stream flow would help maintain a wide river channel. There was greater recognition of the environmental importance of wet meadows and of the link between groundwater irrigation and wet meadow production, but the number of correct responses was still below 50 percent in nearly all cases. Respondents in all states also expressed considerable uncertainty with respect to the economic effects from management alternatives. Nearly an equal number of people agreed as disagreed with statements concerning the effects of changes in the amount of irrigation or tourism on the regional economy.

Differences between the states may suggest some reasons for the technical beliefs that are held. Over 21 percent of Nebraska respondents disagreed with the statement that groundwater irrigation adversely affects wet meadows, compared to 11 percent for Colorado and 12 percent for Wyoming. Similarly, 22 percent of Nebraska respondents disagreed with the contention that improved habitat will increase the number of cranes, compared to 10 and 13 percent for Colorado and Wyoming, respectively. These differences suggest that there may be an inclination on the part of some respondents to deny recognition of technical relationships that do not support their policy position or that imply some responsibility for an adverse impact. The Nebraska sample contains a relatively large proportion of irrigators, many of whom may be reluctant to accept scientific claims about how their activities may affect the Middle Platte ecosystem.

Level of Support for Policy Attributes

Data on the level of public support for each of five different levels of each of three policy attributes were used in game models to find bargained policy solutions.

Table 12.9 Respondent Classification into Bargaining Groups, by State

Bargaining[a] Group	Colorado	Nebraska	Wyoming	All States
	Numbers of Respondents			
Agriculture	24	105	55	184
Environmental	143	86	110	339
Other	132	257	166	555
Total	299	448	331	1078

[a] Based on type of employment, interest-group affiliation, and attitude regarding endangered species, a respondent could be classified as either agriculture, environmental, both, or neither.

Preferences were analyzed by state and for each of two interest-related bargaining groups, agricultural and environmental (Table 12.9). Respondents were classified as agricultural if they indicated that they were self-employed as a farmer or rancher; employed by an agricultural interest group; or affiliated with the Farm Bureau, the Farmers Union, or an irrigation district. Respondents were classified as environmental if they indicated that they were employed by an environmental interest group; affiliated with the Sierra Club, The Nature Conservancy, or the Audubon Society; or agreed or strongly agreed with the statement "Society should ensure that the needs of threatened and endangered species are met regardless of economic cost." Respondents who qualified as agricultural based on employment or interest group affiliation, but who also agreed that society should meet the needs of endangered species irrespective of economic cost, were considered as both agricultural and environmental. Those respondents who either could not be classified as exclusively agricultural or exclusively environmental were classed as "other" and were included in state totals but were not analyzed as a separate bargaining group.

For the *Method* attribute, the level receiving the strongest support from all states as measured by the average score for all residents was adaptive management (Appendix 12-A). Colorado's second-best choice was to meet all needs while minimizing reallocation of irrigation water, but the second-best option preferred by Nebraska and Wyoming respondents was to do the best possible job of meeting endangered species needs with no reallocation of water. Agricultural interests in all states strongly preferred either an adaptive management approach or a program that produced as much endangered species protection as possible without reallocating any water from agriculture (Appendix 12-A). They were most strongly opposed to the idea of meeting all needs irrespective of the costs. Environmental interests preferred to meet all needs, although they also expressed considerable support for an adaptive management approach.

Expressed support for different levels of investment (*Cost* attribute) was somewhat mixed, but the strongest support in all states was for a $10M annual investment, which is about 25 percent of what many observers believe it would take to fully implement USFWS recommendations. However, 32 percent of all Colorado respondents expressed strong support for investing whatever it took to meet USFWS recommendations. Agricultural interests preferred to invest nothing, or perhaps $10M per year, but there was very little support among agriculturalists in all states for spending more than $10M per year.

The payment policy results (*Who pays* attribute) were especially interesting. All states preferred that private environmental groups pay a significant part of the cost, which is contrary to current proposals to address the problem. The reasons for preferring private contributions are unknown, but the leading hypothesis is that respondents believe those who get the most utility from environmental improvements should also pay the most. The first choice of all states was a payment policy consisting of one-third federal, one-third private, and one-third state funds, with the state one-third being distributed between the three states in proportion to current water use. Wyoming respondents objected strongly to each state paying an equal share of the aggregate state share, but there were no other significant differences between the states. The strongest support for some private contribution to the cost of meeting endangered species needs came from agricultural interests, but surprisingly there was also substantial support from environmental interests for requiring some private cost sharing. This may reflect a belief that the benefits from endangered species protection accrue disproportionately to environmental interests and, thus, the entire burden should not fall to general taxpayers.

Model II Results

Weights for Policy Attributes

Responses to a sampling of 7 of the 125 policies were used as described in Equations 12.4 and 12.5 to derive attribute weights for each of the three states and for agricultural and environmental bargaining groups within each state (Appendix 12-A). Except for the environmental interest group in Wyoming, the most heavily weighted policy attribute was payment policy and the least important was the method of meeting endangered species needs. Environmental interests generally placed more weight on method and less on payment policy, compared to agricultural interest groups.

Policy Preferences

Weighted cardinal utility scores were computed for all 125 policy options for each bargaining group using Equation 12.3 (not presented). To facilitate comparisons, the cardinal utility scores for each group were ranked from 1 to 125, where the best option is ranked 125, and the poorest has a ranked score of one; this yielded ordinal utility scores. The full array of 125 policy options was then reduced to 17 by eliminating those that were not Pareto-efficient (Tables 12.10, 12.11, and 12.12). An option was considered Pareto-inefficient if it was possible to improve the level of total utility across groups without making one or more groups worse off. The level of support for the more efficient options was considered in more detail.

Surprisingly, the highest-ranked option in each state was the same, option N, which consists of an adaptive management program using both riparian land management and improved stream flow to protect endangered species, at an investment level of $10M per year, with the federal government paying one-third, the states one-third, and private environmental groups one-third (Table 12.11). Under this option the states' share is split proportionally between the states according to historical

Table 12.10 Definition of Pareto-Efficient Policy Options: Attribute Levels Corresponding to Each Policy

Policy Option	Attribute Level[a]		
	Method	Cost	Who pays
A	d. Adaptive Management	c. Invest $10M, 25% of Need	a. All Costs Paid by Feds
B	d. Adaptive Management	a. Invest Nothing	b. Feds 50%, Private 50%
C	d. Adaptive Management	b. Invest $40M, per USFWS	b. Feds 50%, Private 50%
D	a. All Needs, Least Cost	c. Invest $10M, 25% of Need	b. Feds 50%, Private 50%
E	d. Adaptive Management	c. Invest $10M, 25% of Need	b. Feds 50%, Private 50%
F	a. All Needs, Least Cost	d. Invest $20M, 50% of Need	b. Feds 50%, Private 50%
G	e. Benefit–Cost Approach	c. Invest $10M, 25% of Need	d. Feds 50%, States 50% Proportional to Use
H	d. Adaptive Management	a. Invest Nothing	e. Feds 1/3, Pvt.1/3, States 1/3 Proportional to Use
I	a. All Needs, Least Cost	b. Invest $40M, per USFWS	e. Feds 1/3, Pvt.1/3, States 1/3 Proportional to Use
J	b. All Needs, Minimum Water	b. Invest $40M, per USFWS	e. Feds 1/3, Pvt.1/3, States 1/3 Proportional to Use
K	d. Adaptive Management	b. Invest $40M, per USFWS	e. Feds 1/3, Pvt.1/3, States 1/3 Proportional to Use
L	a. All Needs, Least Cost	c. Invest $10M, 25% of Need	e. Feds 1/3, Pvt.1/3, States 1/3 Proportional to Use
M	b. All Needs, Minimum Water	c. Invest $10M, 25% of Need	e. Feds 1/3, Pvt.1/3, States 1/3 Proportional to Use
N	d. Adaptive Management	c. Invest $10M, 25% of Need	e. Feds 1/3, Pvt.1/3, States 1/3 Proportional to Use
P	a. All Needs, Least Cost	d. Invest $20M, 50% of Need	e. Feds 1/3, Pvt.1/3, States 1/3 Proportional to Use
Q	b. All Needs, Minimum Water	d. Invest $20M, 50% of Need	e. Feds 1/3, Pvt.1/3, States 1/3 Proportional to Use
R	b. All Needs, Minimum Water	e. Invest $30M, 75% of Need	e. Feds 1/3, Pvt.1/3, States 1/3 Proportional to Use

[a] A full description of each policy attribute and level is found in Table 12.8.

water use. The lowest-ranked options in all states were generally those that called for investing nothing.

Policy preferences of interest groups within a state were much more varied (Table 12.12). The first choice of agricultural interests in both Nebraska and Wyoming was option B, which consists of adaptive management at a very low level of investment,

Table 12.11 Pareto-Efficient Policy Preferences,[a] by State

Policy Option[b]	Ranked Utility Scores		
	Colorado	Nebraska	Wyoming
A	78	104	89
B	80	89	110
C	93	84	103
D	70	94	91
E	98	117	119
F	65	69	77
G	69	82	60
H	117	119	123
I	113	92	92
J	119	99	104
K	123	114	121
L	116	120	114
M	121	122	120
N	125	125	125
P	115	105	96
Q	120	113	113
R	118	102	101

[a] Policy options are ranked from 1 to 125, with 125 being the highest or best option.
[b] See Table 12.10 for a description of each policy option.

with all costs paid by the federal government and private environmental interests. Agricultural interests in Colorado preferred option N, which is surprisingly consistent with the preferences of all citizens in each of the three states. Environmental interests in Colorado and Wyoming preferred meeting all endangered species needs, while reallocating as little water as possible, with expenditures of up to $40M per year and costs shared equally by the federal government, the states, and private interests.

Bargaining Solutions

The bargaining challenge, therefore, lies in finding a solution to differences of opinion within, rather than between, states. The magnitude of this challenge can be seen by analyzing how acceptable a given group's preferred option is to competing bargaining groups (Table 12.13). For example, examining the seventh row of Table 12.13, note that all agriculture interests prefer an adaptive management plan with minimal water reallocation and minimal investment, with 50 percent of the costs paid by private environmental groups and 50 percent by the federal government (option B). Moving to the end of the seventh row, environmental interests aggregated across states rank option B as their ninth-poorest option, which places it in the bottom 10 percent of the 125 choices being considered. Environmental interests (last row) prefer option J, which would meet all endangered species needs at a cost of up to $40M per year, with costs shared equally between the federal government, the states, and private environmental interests. Agricultural interests rank option J as their third-poorest option. These comparisons suggest that a bargaining process is needed to find an acceptable middle ground that lies somewhere between, at one extreme, a program

Table 12.12 Pareto-Efficient Policy Preferences,[a] by Bargaining Group and State

	Colorado		Nebraska		Wyoming			
Policy Option[b]	Ag Utility Rank	Envl. Utility Rank	Ag Utility Rank	Envl. Utility Rank	Ag Utility Rank	Envl. Utility Rank	All Ag Utility Rank	All Envl. Utility Rank
A	115	42	94	76	111	46	105	42
B	116	8	125	28	125	27	125	9
C	53	82	106	55	106	73	98	71
D	113	46	114	39	113	68	114	47
E	124	43	122	59	123	55	123	45
F	63	62	103	30	93	69	92	53
G	90	58	47	83	46	85	50	79
H	117	54	112	113	107	94	115	91
I	33	123	51	115	42	125	45	124
J	44	125	60	120	47	124	52	125
K	52	122	72	122	65	123	72	123
L	114	109	84	117	73	121	84	115
M	120	117	93	124	82	119	94	120
N	125	105	108	125	97	115	110	114
P	64	115	71	114	51	122	65	117
Q	75	121	77	118	60	120	74	121
R	58	119	61	121	49	117	56	122

[a]Policy options are ranked from 1 to 125, with 125 being the highest or best option.
[b]See Table 12.10 for a description of each policy option.

that meets all endangered species needs (as determined by the USFWS), involves a major reallocation of water from agriculture, and costs up to $40M per year; and at the other extreme, a program that reduces the reallocation of water to an absolute minimum, costs much less, and exposes endangered species to significant risk.

Three solutions to a multilateral bargaining game were computed in a search for the policy options most likely to be acceptable to all of the principal interest groups

Table 12.13 Comparison of Preferred Policy Options Between Competing Interest Groups

Group	Preferred Option	CO	NE	WY	CO Ag	NE Ag	WY Ag	All Ag	CO Envl	NE Envl	WY Envl	All Envl
					Rank of Preferred Option[a]							
CO	N	**125**	125	125	125	108	97	110	105	125	115	114
NE	N	125	**125**	125	125	108	97	110	105	125	115	114
WY	N	125	125	**125**	125	108	97	110	105	125	115	114
CO Ag	N	125	125	125	**125**	108	97	110	105	125	115	114
NE Ag	B	80	89	110	116	**125**	125	125	8	28	27	9
WY Ag	B	80	89	110	116	125	**125**	125	8	28	27	9
All Ag	B	80	89	110	116	125	125	**125**	8	28	27	9
CO Envl	J	119	99	104	44	60	47	52	**125**	120	124	125
NE Envl	N	125	125	125	125	108	97	110	105	**125**	115	114
WY Envl	I	113	92	92	33	51	42	45	123	115	**125**	124
All Envl	J	123	40	108	12	49	104	3	125	125	92	**125**

[a]Policy options are ranked from 1 to 125, with 125 being the highest or best option.

Table 12.14 Results of bargaining Models, All Bargaining Groups

Pareto-Efficient Options	Cardinal Utility			Ordinal Utility		
	Utilitarian	Nash	Egalitarian	Utilitarian	Nash	Egalitarian
	Rank of Policy Option[a]					
A	101	104	123	98	95	121
B	102	103	87	85	56	106
C	87	87	73	92	98	105
D	105	109	125	102	96	120
E	116	117	118	110	105	94
F	81	82	90	82	80	40
G	72	69	63	76	81	88
H	122	122	95	121	119	25
I	103	91	8	99	90	2
J	111	105	9	107	103	98
K	115	112	20	115	114	11
L	123	123	49	123	123	44
M	124	124	32	124	124	1
N	125	125	70	125	125	91
P	110	108	24	112	111	4
Q	117	115	29	117	117	78
R	109	106	19	109	109	101

[a] Options are marked from 1 to 125, with 125 being the highest or best option.

(Table 12.14). Policy N is both the utilitarian and Nash solution (Equations 12.6 and 12.7), whether using cardinal or ordinal utility. However, the egalitarian solution (Equation 12.8) is policy option D when using cardinal utility and option A when using ordinal utility.

These results suggest that if the bargaining agents were not concerned about equity between groups they would adopt policy N, which is an adaptive management approach meeting only some of the endangered species needs, spending $10M per year, with the costs split evenly between the federal government, the states, and private environmental groups. However, if equity was more of a concern, the solution would involve a similar approach with about the same level of investment, but with no state contribution to program costs.

If policy option N is selected, environmental groups are likely to be reasonably satisfied because a reasonable amount of endangered species protection will be provided and the costs will be widely shared. However, at least part of the agricultural community is likely to be uncomfortable with a program that reallocates water away from agriculture in ways they believe may not be justified on a cost–benefit basis, especially when the states are paying a significant share of the cost.

Potential Impact of Education on Policy Preferences

An important policy issue concerns the extent to which education might reduce the level of disagreement between bargaining groups. Two questions would need to be answered. First, does the tendency for groups to disagree appear to be related to the level of technical knowledge within the groups? If the answer is yes, then would education improve the level of technical knowledge and the level of agreement? While the second question was beyond the scope of the current project, the first question was analyzed by comparing the policy preferences of more and less knowledgeable survey respondents.[a]

Knowledgeable respondents were defined as those whose knowledge level score, as defined earlier under "Level of Knowledge," was at least one standard deviation above the mean in each state. Average utility scores for the knowledgeable and nonknowledgeable classes were computed and compared for the 17 Pareto-efficient policy options. An aggressive education program was arbitrarily assumed to be able to change the level of support for the Pareto-efficient policies by nonknowledgeable citizens by an amount equal to one-half the average difference between the knowledgeable and nonknowledgeable classes. Hence, the appropriate adjustments were made to the nonknowledgeable scores, and a new interest group average was calculated for each Pareto-efficient policy option. Rank orderings of the 17 options with and without the assumed education effect were then compared to determine whether there was any appreciable effect on what option was most preferred by each interest group and, most importantly, to determine whether the knowledge effect brought the interest groups closer to an agreement on the best policy option.[63]

In all states the effect of improved knowledge was to bring the agricultural and environmental interest groups closer to agreement. In Nebraska, the effect was primarily on the agricultural interest group. Nebraska agriculture's first choice went from option B, which calls for investing nothing in endangered species protection, to option N, which was the first choice of Nebraska environmental interests before the effect of improved knowledge. With improved knowledge the first choice of Nebraska environmental interests became option J, which is similar to option N but calls for a higher level of investment. For Wyoming, the effect of improved knowledge was also to make environmentally strong options more acceptable to agricultural interests. Both Wyoming agricultural and Wyoming environmental interests preferred option I after the knowledge effect was imposed, whereas previously, Wyoming agricultural interests preferred a much lower level of investment in endangered species protection. For Colorado, there was no significant knowledge effect on environmental interests, but agricultural preferences changed from preferring adaptive management option N to preferring to meet all needs, option L.

Policy Implications of Model II

The results from Model II suggest that the most important differences of opinion regarding central Platte management policies exist between agricultural and environmental interest groups within each state, rather than between states. At the aggregate level, all three states preferred a policy that called for an adaptive management approach that minimized the reallocation of water from agriculture and involved a modest level of investment, with the costs shared equally between the

[a] The effect of knowledge on policy preferences was also addressed with a logit model that analyzed the effect of knowledge on the probability that an individual would support environmentally intense policies. This analysis found a strong statistical relationship between knowledge and level of policy support.

federal government, the states, and private environmental interests. Within Nebraska and Colorado, however, agricultural interests preferred to invest nothing, with everything paid for by the states and private environmental interests, while environmental interests preferred a much more aggressive program to ensure endangered species protection, with costs split evenly between the federal government, the states, and private environmental interests. Colorado agricultural interests were more supportive of environmental objectives, but still preferred less endangered species protection than did Colorado environmental interests.

An analysis of policy attributes found that the dominant attribute in nearly all cases was payment policy (i.e., *Who pays*; see Appendix 12-A, Table 12-A.5). Private environmental interests showed a surprising willingness to support private contribution to the costs of central Platte management programs, and agricultural interests were much more willing to endorse a significant endangered species protection program if the state cost share was minimized and there was a substantial private contribution. All interest groups were quite receptive to an adaptive management approach that is quite similar to the programs now being pursued by the states and the DOI under the terms of the Cooperative Agreement.

Application of three different sets of bargaining rules all resulted in solutions that called for an adaptive management approach that minimized the reallocation of water, with an equal sharing of the costs between federal, state, and private entities. The egalitarian solution, however, suggested that if the agents were more concerned about equity, they should pursue a somewhat more aggressive program of endangered species protection with less of a state contribution to the total cost.

An analysis of the impact of technical knowledge on policy preferences found that well-informed people had much stronger environmental preferences compared to those who were less well informed. It was found that much of the disagreement between agricultural and environmental interest groups would cease to exist if both groups had technical beliefs that were similar to those held by well-informed individuals. This finding suggests that ecological risk information might have a role in changing public opinion, leading to reduced conflict and perhaps improved resource management. However, there is also a possibility that some respondents knowingly answered technical questions incorrectly in cases where an incorrect answer supported their strongly held values and policy positions. It is also possible that individuals may reject as biased any new information that does not support such values. Before definitive conclusions can be drawn, further research is needed regarding the effectiveness of ecological-risk education in changing technical beliefs and policy preferences.

DISCUSSION

Chapter 9 put forward a conceptual approach for the integration of ERA and economic analysis for watershed management (Figure 9.1). In that ideal approach, integration occurs in all stages of assessment. Because economists' involvement began late in the assessment process for the central Platte River floodplain, the process depicted in Chapter 9 was not followed in several respects. That ideal process nonetheless provides a useful framework for evaluating the methods used and degree of ecological–economic integration achieved in this case study.

Assessment Planning and Problem Formulation

The conceptual approach calls for the interrelated steps of assessment planning and problem formulation to be carried out in advance of analysis (Figure 9.1). In this case, a formal planning process that included stakeholders was conducted at the outset of the W-ERA. Planners discussed watershed values and challenges and crafted a very broad management goal — "to protect, maintain, and where feasible restore biodiversity and ecological processes . . ." — and a list of eleven management objectives (see "Planning" earlier in this chapter). The W-ERA assessment team then worked to distill those objectives and existing knowledge of the watershed into assessment endpoints, conceptual models, and risk hypotheses (see "Problem Formulation").

The economic research effort was not yet conceived at this stage, and economists were not involved in this process. The economic study was initiated later (see Chapter 1), with a coordination meeting that had minimal stakeholder representation and occurred after most of the W-ERA work had already been completed. Therefore, ecological risk assessors did not have the benefit of considering economic concepts, research approaches, or management insights, and while economists heard a brief report of the W-ERA approach, they did not benefit from a close collaboration with that effort, nor did they engage a broad range of stakeholder groups in their work. This limited degree of coordination resulted in a divergence of analytic objectives and perspectives. The ecological analysis studied the habitat requirements of dozens of riparian-dependent avian species, whereas the economic analysis addressed only the needs of threatened and endangered species.

Formulating Alternatives, and Baseline Ecological Risk Assessment

Whereas ERA alone does not necessarily require the formulation of management alternatives, economic analysis usually is concerned with alternatives, so their formulation usually is a condition for integrated study (Figure 9.1). The Platte River W-ERA sought only to characterize baseline risks, that is, risks that exist now or are likely to occur if no new management is undertaken. The risk models that were developed (i.e., models describing floodplain segment use by sandhill cranes, and meadow or woodland patch use by nesting birds) dealt with a subset of the ecological assessment endpoints. They are potentially applicable to management questions but were not developed with a specific decision context or set of alternatives in mind.

The economic analysis, on the other hand, formulated two sets of management alternatives. Model II focused on finding a compromise solution from among 125 different options (later narrowed to 17 Pareto-efficient policies) for floodplain management, especially dealing with instream flow amount and payment. Model I provided a tool, an auction market, for use by stakeholders in deciding who would provide alternative levels of environmental water. The economic analysis thus focused directly on resource-management choices that were linked to the dominant issue in the basin, rather than addressing a broader, yet less pragmatically focused, array of baseline ecological risks. Had the economists been part of the W-ERA planning discussions, there would have been an opportunity to discuss these alternatives and thus better harmonize the ecological and economic analyses. Discussion of management alternatives during assessment planning might also have narrowed the scope of the W-ERA, limiting the number of management objectives and risk hypotheses, and sharpening its analytic focus.

Analysis and Characterization of Alternatives, and Comparison of Alternatives

The analysis and characterization of management alternatives and the comparison of the alternatives are two closely related steps in the conceptual approach (Figure 9.1). Each management alternative is to be examined in the light of both ecological risks and economic outcomes and, as applicable, other analyses (e.g., health or quality of life). Diagrammatic examples of a variety of approaches to these two steps were given in Figures 9.2, 9.3, and 9.4.

The approach employed in this case study is illustrated in Figure 12.3, which is a modification of Figure 9.3. The likely ecological and economic outcomes of various watershed management policy attributes were described in a survey of preferences, and survey results were used to evaluate specific policies (i.e., attribute combinations).

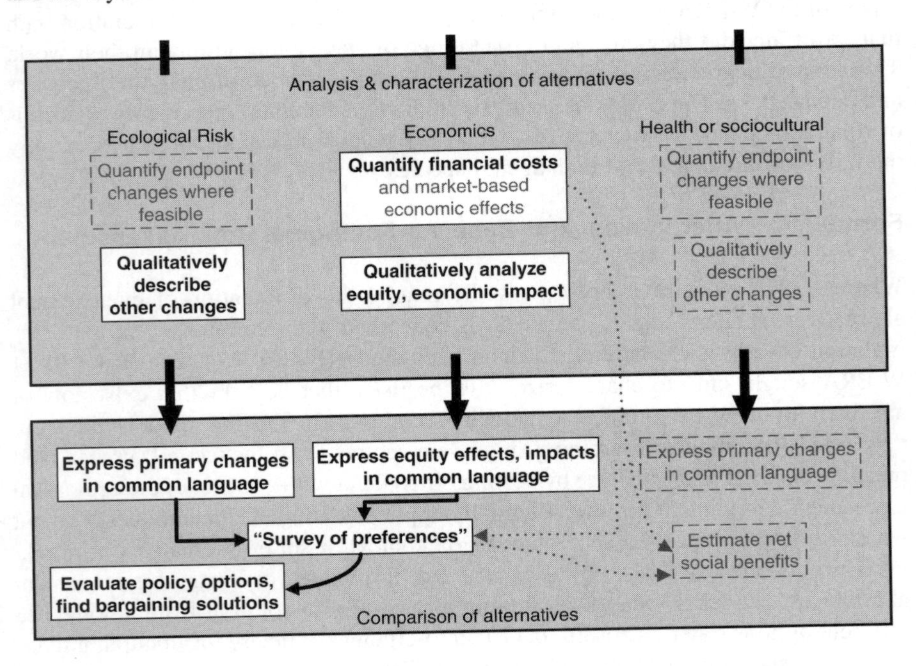

Figure 12.3 Techniques used for analysis, characterization, and comparison of management alternatives in the central Platte River floodplain, as compared to the example shown in Figure 9.3. White boxes and bold type show features included in this analysis.

The ecological point of departure for the economic study was a determination by the USFWS that a given increment of instream flow and restoration of wet meadow acreage are needed to ensure protection of endangered species.[3] This level of provision was described qualitatively in the survey as "meeting the needs" of endangered species; lesser levels of protection were described as placing the species "at risk" (Table 12.8). Annual costs to fund the USFWS program were described in dollar terms. The market-based economic effects of the program (such as the impacts of foregoing water diversion or pumping, or removing land from production) were not estimated or described. However, equity and economic impact concerns were implicit in the wording of policy options that minimized the purchase of water from agriculture or that discussed different cost-sharing options.

In Figure 12.3 the term *survey of preferences* is substituted for *stated preference survey*, because the latter usually refers to methods that ask individuals to place a value on specific changes to the environment, whereas in this case the results will not provide estimates of value either directly or indirectly. Analysis of survey results yielded policy-specific estimates of utility for each of several bargaining groups. A subsequent step used the utilitarian, Nash, or egalitarian approaches to rank-order the policies. Estimates of the net social benefit of policies could not be derived, in part because market-based economic effects of the policy options were not determined, but also because the survey of preferences did not estimate willingness to pay.

Ecological economics stresses that economic analyses should account for the biophysical constraints that exist in the ecological systems that support all human activity (see "Ecological Economics" in Chapter 5). The W-ERA for the Platte River did not formulate or evaluate any management alternatives. It is important, therefore, to examine the degree to which the economic models were informed or constrained by information on ecological risks. In general, the economic analysis regarded ecological risk as technical information that could influence the preferences of stakeholders. Ecological risk was constraining only to the extent that stakeholders regarded risk reduction an important objective relative to the trade-offs involved. With this approach no answer is regarded as scientifically correct; all that science does is provide trade-off and preference information to facilitate public decision-making. Model I, the auction model, did several things: (1) it provided a tool for efficiently "negotiating" who will supply a given quantity of water and at what price; (2) it provided a method of estimating the budgetary supply costs associated with different quantities of environmental water; and (3) it provided an indication of the price that stakeholders would pay in the form of welfare and budget costs for using the negotiating efficiencies inherent in a second-price auction instead of a direct negotiation first price approach. Providing these functions required no ecological risk information.

Model II used preference information for policy options that ranged from providing "whatever the USFWS says is needed . . ." to providing nothing. The Model II bargaining solutions were based on utility and were not constrained by conditions ensuring species' survival beyond respondents' preference for doing so. If respondents preferred policies that were lower cost or involved less reallocation of water, it is not clear whether they were accepting as valid the biological opinion of the USFWS and voting against full support for maintaining the species, whether they did not believe that water reallocation would be helpful to the species, or whether they were uncertain about key technical relationships and therefore preferred an incremental try-it-and-see approach. An analysis of the impact of technical information on policy preferences suggested that facts were a very important determinant of policy preferences. Policy preferences changed markedly and the differences between interest groups narrowed substantially if one assumed that with education the less well-informed stakeholders would develop preferences similar to those of their better-informed colleagues. If this assumption were substantiated, it would raise the possibility that an effective program of educational outreach, carried out in conjunction with a bargaining process, could provide an effective biophysical constraint. However, this study did not investigate the actual effectiveness of education in a situation of longstanding conflict, and therefore it cannot be concluded that the bargaining approach, per se, is effectively constrained.

It is possible that the process of adaptive implementation, such as that envisioned by the Cooperative Agreement, would afford constraints ensuring species survival, but much depends on the view one takes of adaptive implementation as a management and political strategy. If it serves as a reliable feedback mechanism, whereby stakeholders' preferences are updated by new information, then biophysical constraints may be effective, even when not explicit in a preference-based model. An adaptive management approach that is politically feasible may reach desired ecological goals at a slower pace than some would prefer, but it may still be the most effective approach if full and immediate implementation is not politically feasible.

Consultation with Extended Peer Community

USEPA's *Ecological Risk Assessment Guidelines* recommend fully involving stakeholders in planning but maintaining strict separation of science from policy in subsequent steps, whereas others have emphasized the limitations of science and the importance of ongoing consultation, throughout the analysis, with an extended peer community (see "Critiques of Ecological Risk Assessment" in Chapter 3 and "Consultation with Extended Peer Community" in Chapter 9). The W-ERA formally established a stakeholder panel for participation in planning, but problem formulation was conducted by a more limited technical team. Near the end of problem formulation, consultations with stakeholders were held, and a draft was reviewed, but subsequent changes made by the technical team alienated at least one stakeholder group. The economic analysis was not constrained by a formal requirement for stakeholder involvement and used more limited and informal mechanisms. Lacking their strong involvement, however, it is not yet clear whether the parties to the Cooperative Agreement will make use of the game theory results.

Decisions and Adaptive Implementation

Even if ecological risk, economics, and other information are well integrated and well tuned to the decision context, it is normal for any high-stakes decision to require negotiation after the analyses are completed. The game theory models developed here may be well suited to the support of an ongoing negotiation because they can respond quickly to changes in negotiating position and suggest new solutions. The approach may also be useful over a longer period of adaptive implementation, in which system modification and feedback result in new learning, and a new set of policy solutions is sought.

Adaptive implementation is important not only for its merit as a management approach but also as an aid to difficult negotiations. When disagreements about the true behavior of the system prevent the parties from agreeing on costly remedies, an adaptive approach can present an attractive compromise in that it holds out the promise of improved knowledge about the system. But care must be taken to distinguish between a policy that is truly adaptive and one that is simply incremental. Walters[64] argues that *incrementalism* (making small improvements without taking large risks) is not effective as an information-generating strategy. "Such policies result in strongly correlated inputs, and in state variables being correlated with inputs . . . so the effects of each cannot be distinguished." An ideal strategy from an informational standpoint would consist of repetitive sequences moving from one extreme to the other, each of sufficient duration to allow observation of responses of key variables. Managers tend to be risk averse, however, and under substantial pressure to avoid extremes. An actively adaptive policy, therefore, must somehow establish a balance between learning (via policies designed to maximize probative value) and short-term performance (maintaining the system nearest its status quo).[64]

A key question, therefore, about the value of the Cooperative Agreement as an informative policy is whether the initial increment of 140,000 acre-feet and an evaluation period of 10–13 years will be sufficient, in light of natural hydrologic variability and the slowness of successional processes, to induce unambiguous changes in key variables such as area of active channel. Since only an unambiguous response would be likely to promote agreement about subsequent actions, the prospects for reducing conflict over the long term through this game theoretic approach are closely tied to adaptive implementation's effectiveness.

ACKNOWLEDGMENTS

The authors wish to thank the members of the Middle Platte Watershed Ecological Risk Assessment Workgroup for their effort in performing the activities upon which this report is based, and Nancy Pritchett for technical support. We also thank attendees of a workshop held in July 2001, in Cincinnati, OH for their comments on an early draft of this work, and in particular we acknowledge Glenn Suter, Haynes Goddard, and Ann Bleed for their written reviews.

REFERENCES

1. Johnsgard, P.A., *Birds of the Great Plains: Breeding Species and Their Distribution*, University of Nebraska Press, Lincoln, NE, 1979.
2. Sidle, J.G., Nagel, H.G., Clark, R., et al., Aerial thermal infrared imaging of Sandhill Cranes on the Platte River, Nebraska, *Remote Sens. Environ.*, 43, 333, 1993.
3. Sidle, J. G. and Faanes, C.A., Platte River ecosystem resources and management, with emphasis on the Big Bend Reach in Nebraska, U.S. Fish and Wildlife Service, Grand Island, NE and Northern Prairies Wildlife Research Center, 1997. Accessed Jan. 29, 2004 at http://www.npsc.nbs.gov/resource/othrdata/platte2/platte2.htm#contents.

4. FERC, Draft environmental impact statement: Kingsley Dam and North Platte/Keystone diversion dam projects, FERC/DEIS-0063, Federal Energy Regulatory Commission, Office of Hydropower Licensing, Washington, D.C., 1992.
5. Eschner, T.R., Hadley, R.F., and Crowley, K.D., Hydrologic and morphologic changes in channels of the Platte River Basin in Colorado, Wyoming and Nebraska: A historical perspective, U.S. Geological Survey Professional Paper 1277-A, U.S. Government Printing Office, Washington, DC., 1983.
6. Junk, W.J., Bayley, P.B., and Sparks, R.E., The flood pulse concept in river-floodplain systems, in *Proceedings of the International Large River Symposium (LARS)*, Dodge, D.P., Ed., Canadian Special Publication of Fisheries and Aquatic Sciences, Ottawa, Canada, 1989, 110.
7. Sparks, R.E., Bayley, P.B., Kohler, S.L., and Osborne, L.L., Disturbance and recovery of large floodplain rivers, *Environ. Manage.*, 14, 699, 1990.
8. Sparks, R.E., Need for ecosystem management of large floodplain rivers and their floodplains, *BioScience*, 45, 168, 1995.
9. Currier, P.J., The Floodplain Vegetation of the Platte River: Phytosociological Forest Development and Seedling Establishment, PhD Dissertation, Iowa State University, Ames, IA, 1982.
10. Johnson, W.C., Dams and riparian forests: Case study from the upper Missouri River, *Rivers*, 3, 229, 1992.
11. Johnson, W.C., Woodland expansion in the Platte River, Nebraska: Patterns and causes, *Ecol. Monogr.*, 64, 45, 1994.
12. Petts, G.E. and Lewin, J., Physical effects of reservoirs on river systems, in *Man's Impact on the Hydrological Cycle in the United Kingdom*, Hollis, G.E., Ed., Geo Abstracts Ltd., Norwich, U.K., 1979, 79.
13. Hickin, E.J., River channel changes: Retrospect and prospect, in *Modern and Ancient Fluvial Systems*, Collinson, J.D. and Lewin, J., Eds., Blackwell Scientific Publications, Oxford, U.K., 1983, 61.
14. Petts, G.E., *Impounded Rivers: Perspectives for Ecological Management*, John Wiley & Sons, Chichester, U.K., 1984.
15. Sidle, J.G., Currier, P.J., and Miller, E.D., Changing habitats in the Platte River Valley of Nebraska, *Prairie Nat.*, 21, 91, 1989.
16. Krapu, G.L., Reineche, K.J., and Frith, C.R., Sandhill cranes and the Platte River, Transactions of the 47th North American Wildlife and Natural Resources Conference, 542.
17. Faanes, C.A., Aspects of the nesting ecology of least terns and piping plovers in central Nebraska, *Prairie Nat.*, 15, 145, 1983.
18. Krapu, G.L., Facey, D.E., Fritzell, E.K., and Johnson, D.H., Habitat use by migrant Sandhill Cranes in Nebraska, *J. Wildl. Manage.*, 48, 407, 1984.
19. Lingle, G.R., Strom, K.J., and Ziewitz, J.W., Whooping crane roost site characteristics on the Platte River, Buffalo County, Nebraska, *Neb. Bird Rev.*, 54, 36, 1986.
20. Iverson, G.C., Vohs, P.A., and Tacha, T.C., Habitat use by mid-continent sandhill cranes during spring migration, *J. Wildl. Manage.*, 51, 8, 1987.
21. Norling, B.S., Anderson, S.H., and Hubert, W.A., Nocturnal behaviour of Sandhill Cranes roosting in the Platte River, Nebraska, *Naturalist*, 23, 17, 1991.
22. Folk, M.J. and Tacha, T.C., Sandhill crane roost site characteristics in the North Platte River Valley, *J. Wildl. Manage.*, 54, 480, 1990.
23. Davis, C.A., Sandhill crane migration through the central Great Plains: A contemporary perspective, Proc. Great Plains Migration Symposium, Lincoln, NE, Mar. 7, 2002.
24. Faanes, C.A. and LeValley, M.J., Is the distribution of Sandhill Cranes on the Platte River changing?, *Great Plains Res.*, 3, 297, 1993.
25. Sharpe, R.S., The origins of spring migratory staging by sandhill cranes and white-fronted geese., *Trans. Nebr. Acad. Sci.*, 6, 141, 1978.

26. Ducey, J., Breeding of the least tern and piping plover on the lower Platte River, Nebraska, *Neb. Bird Rev.*, 49, 45, 1981.
27. Jorde, D.G.H., Krapa, G.L., Crawford, R.D., and Hay, M.A., Effects of weather on habitat selection and behavior of mallards wintering in Nebraska, *Condor*, 86, 258, 1984.
28. USFWS, The Platte River ecology study, Special Research Report, Northern Prairie Wildlife Research Center, Jamestown, ND, 1981, 187.
29. Currier, P.J. and Ziewitz, J.W., Application of a sandhill crane model to the management of habitat along the Platte River, Proceedings of the 1985 Crane Workshop, Platte River Whooping Crane Habitat Management Trust and U.S. Fish and Wildlife Service, Grand Island, NE, 315.
30. Helzer, C.J. and Jelinski, D.E., The relative importance of patch area and perimeter-area ratio to grassland breeding birds, *Ecol. Appl.*, 9, 1448, 1999.
31. Jelinski, D.E., Middle Platte River floodplain ecological risk assessment planning and problem formulation, Completed under EPA Assistance Agreement CR 826077, School of Environmental Studies, Queens University, Kingston, Ontario, 1999.
32. Colt, C.J., Breeding Bird Use of Riparian Forests Along the Central Platte River: A Spatial Analysis, MS Thesis, University of Nebraska, Lincoln, NE, 1997.
33. Keammerer, W.R., Johnson, W.C., and Burgess, R.L., Floristic analysis of the Missouri River bottomland forests in North Dakota, *Can. Field Nat.*, 89, 5, 1975.
34. Hibbard, E.A., Vertebrate Ecology and Zoogeography of the Missouri River Valley in North Dakota, PhD Dissertation, North Dakota State University, Fargo, ND, 1972.
35. Johnson, W.C., Equilibrium response of riparian vegetation to flow regulation in the Platte River, Nebraska. *Regulated Rivers*: *Res. Manage.,* 13, 403, 1997.
36. Strange, E.M., Fausch, K.D., and Covich, A.P., Sustaining ecosystem services in human-dominated watershed: Biohydrology and ecosystem processes in the South Platte River Basin, *Environ. Manage.*, 24, 39, 1999.
37. Habi Tech, Inc., Hydrologic components influencing the conditions of wet meadows along the Central Platte River, Nebraska, Lincoln, NE, 1-31-1993.
38. Johnson, W.C., Channel equilibrium in the Platte River, 1986–1995, Department of Horticulture, Forestry, Landscape, and Parks. South Dakota State University, Brookings, SD, 1996.
39. Currier, P.J., *Woody Vegetation Expansion and Continuing Declines in Open Channel and Habitat on the Platte River in Nebraska*, The Platte River Whooping Crane Critical Habitat Maintenance Trust, Grand Island, NE, 1995.
40. Chadwick and Associates, Forage fish monitoring study, Central Platte River, Nebraska, 1993, Chadwick and Associates, Littleton, CO, 1994.
41. PRESP, Cooperative agreement for the Platte River Research and other efforts relating to endangered species habitat along the Central Platte River, Nebraska, Platte River Endangered Species Partnership, 1997. Accessed Jan. 29, 2004 at http://www.platteriver.org/library/CooperativeAgreement/index.htm.
42. Gilliland, M.W., Becker, L., Cady, R., et al., Simulation and decision making: the Platte River Basin in Nebraska, *Water Res. Bull.*, 21, 1985.
43. Bleed, A., Nachtnebel, H.P., Bogardi, I., and Supalla, R.J., Decision making on the Danube and the Platte, *Water Resour. Bull.*, 26, 1990.
44. Razavian, D., Bleed, A.S., Suppal, R.J., and Gollehon, N.R., Multistage screening process for River Basin planning, *J. Water Res. Plan. Manage.*, 116, 323, 1990.
45. Aiken, J.D., Balancing endangered species protection and irrigation water: The Platte River Cooperative Agreement, *Great Plains Nat. Resour. J.*, 3, 119, 1999.
46. PRESP, Platte River Endangered Species Partnership: Implementing the Cooperative Agreement among the States of Nebraska, Wyoming, Colorado and the United States Department of Interior, Platte River Endangered Species Partnership, May 20, 2004. Accessed June 2, 2004 at http://www.platteriver.org/.

47. Mitchell, B., *Resource and Environmental Management*, Longman, London, 1997.
48. Kirsch, E. M., Habitat selection and productivity of least terns on the lower Platte River, Nebraska, Wildlife Monographs, 132, 1996, 48.
49. Wesche, T.A., Skinner, Q.D., and Henzey, R.J., Platte River wetland hydrology study, University of Wyoming, Laramie, WY, 1994.
50. USEPA, Middle Platte River floodplain ecological risk assessment planning and problem formulation, draft, EPA/630/R-96/007a, Risk Assessment Forum, U.S. Environmental Protection Agency, Washington, D.C., 1996.
51. Johnson, W.C., Adjustment of riparian vegetation to river regulation in the Great Plains, USA, *Wetlands*, 18, 608, 1998.
52. Armbruster, M.J. and Farmer, A.H., Draft Sandhill Crane habitat suitability model, Proceedings from the 1981 Crane Workshop, National Audubon Society, Tavernier, FL, 136.
53. Cadmus Group, Ecological risk assessment for watersheds: Data analysis for the Middle Platte River, EPA Contract 68-C7-002, Work Assignment B-02, Cadmus Group, Laramie, WY, 1998.
54. Gibbons, R., *Game Theory for Applied Economists*, Princeton University Press, Princeton, NJ, 1992.
55. Becker, N. and Easter, K.W., Water diversions in the Great Lakes Basin analyzed in game theory framework, *Water Resour. Manage.*, 9, 221, 1995.
56. Adams, G., Rausser, G., and Simon, L., Modeling multilateral negotiations: An application to California water policy, *J. Econ. Behav. Organ.*, 97, 1996.
57. Klemperer, P., Auction theory: A guide to the literature, *J. Econ. Surv.*, 13, 1999.
58. Supalla, R., Klaus, B., Yeboah, O., and Bruins, R., A game theory approach to deciding who will supply instream flow water, *J. Am. Water Resour. Assoc.*, 38, 959, 2002.
59. Boyle Engineering Corp., Platte River water conservation/supply reconnaissance study, Lakewood, CO, 1999.
60. Jenkins, A. and Konecny, R., The Middle Platte socioeconomic baseline, Platte River Studies, Inst. of Agric. and Natural Resour., Platte Watershed Program, University of Nebraska, Lincoln, NE, 1999.
61. Boyle Engineering Corp, Reconnaissance-level water action plan, Prepared for Governance Committee of the Cooperative Agreement for Platte River Research, Boyle Engineering Corp, Lakewood, CO, Sept. 14, 2000.
62. Babbie, E.R., Index and scale construction, in *The Practice of Social Research*, Wadsworth Publishing Company, Belmont, CA, 1979, 15.
63. Supalla, R., Klaus, B., Yeboah, O., and Allen, J.C., Game theory approach as a watershed management tool: A case study of the Middle Platte ecosystem, Project Completion Report for U.S. EPA Assistance Agreement R 82698701, Department of Agricultural Economics, University of Nebraska, Lincoln, NE, 2002.
64. Walters, C.J., *Adaptive Management of Renewable Resources*, Macmillan Publishing Company, New York, 1986.
65. McDonald, P.M. and Sidle, J.G., Habitat changes above and below water projects on the North Platte and South Platte Rivers in Nebraska, *Prairie Nat.*, 24, 149, 1992.
66. Hurr, R.T., Groundwater hydrology of the Mormon Island Crane Meadows Wildlife Area near Grand Island, Hall County, Nebraska, U.S. Geological Survey Professional Paper 1277, U.S. Government Printing Office, Washington, DC, 1983.
67. Anonymous, The Groundwater Atlas of Nebraska, Conservation and Survey Division, Institute of Agriculture and Natural Resources, University of Nebraska, Lincoln, NE, 1998.

APPENDIX 12-A

Summary of Survey Response Information Used to Calculate Utility of Environmental Policy Options for the Central Platte River Floodplain

Table 12.8 describes three environmental policy attributes (*Method*, *Cost*, and *Who pays*), each having five levels, by which 125 policy options (i.e., 5^3 attribute level combinations) for addressing the central Platte River environmental management problem are defined. Bargaining groups with respect to that environmental problem are determined as a combination of state residency and interest group membership, as defined in "Level of Support for Policy Attributes" in Chapter 12 and Table 12.9. Equations 12.3, 12.4, and 12.5 define the methods by which survey response data for several bargaining groups are used to derive each group's utility scores for each policy option. This Appendix summarizes certain information used in the calculation of utility. First, the degree of support for individual policy attribute levels is presented by state (Table 12-A.1) and interest group (Tables 12-A.2, 12-A.3, and 12-A.4). Next, the results of regression analyses conducted to establish the relative weights of the attributes are presented (Table 12-A.5).

Table 12-A.1 Degree of Support for Policy Attributes, by State

Policy Attribute and Level[a]	Do Support[b]			Don't Support[c]		
	CO	NE	WY	CO	NE	WY
	Percent of all Respondents					
Method						
a. All Needs, Least Cost	41	24.8	29.2	39.7	49.5	52.7
b. All Needs, Minimum Water	52.6	37.6	37.9	25.8	33.6	35.7
c. Best Possible, No Ag Water	36.7	43.2	45.1	46.7	31.6	34.4
d. Adaptive Management	64.1	63	63.9	21.6	17	17.1
e. Benefit–Cost Approach	26.9	38.7	37.9	47.6	31	36.6
Cost						
a. Invest Nothing	15.7	19.6	23.3	73.1	62.2	59.7
b. Invest $40M, per USFWS	31.8	16.6	19.7	51.7	63.2	63.9
c. Invest $10M, 25% of Need	36.2	39	30.6	39.4	37.6	43.8
d. Invest $20M, 50% of Need	33.6	22.4	18.9	42.4	51.5	54.3
e. Invest $30M, 75% of Need	23.8	13.2	11.7	48.4	57.6	60.3
Who pays						
a. All Costs Paid by Feds	32.2	34.9	33	57	48.7	51.7
b. Feds 50%, Private 50%	39.9	39.6	49.2	44.5	39.8	35
c. Feds 50%, States 50% Equal	27.8	26.2	17.5	53.9	54.7	64.3
d. Feds 50%, States 50% Prop.	43.5	29.4	34.4	37.9	46.6	45.1
e. Feds 1/3, Pvt.1/3, States 1/3 Proportional to Use	61.6	51.3	53.2	25	29.6	31.1

[a] A full description of each policy attribute and level is found in Table 12.8.
[b] Includes responses of "strongly support" and "support."
[c] Includes responses of "strongly oppose" and "oppose."

Table 12-A.2 Degree of Support for Policy Attribute Levels in Colorado, by Interest Group

Policy Attribute and Level[a]	Do Support[b]		No Opinion		Don't Support[c]	
	Ag	Envl.	Ag	Envl	Ag	Envl.
	Percent of Classified Respondents					
Method						
a. All Needs, Least Cost	26.1	56.7	4.3	22.7	69.6	20.6
b. All Needs, Minimum Water	60.9	69.3	13	22.1	26.1	8.6
c. Best Possible, No Ag Water	73.9	22.9	8.7	15.7	17.4	61.4
d. Adaptive Management	87	55.3	8.7	12.8	4.3	31.9
e. Benefit–Cost Approach	43.5	17.7	26.1	27	30.4	55.3
Cost						
a. Invest Nothing	49.9	4.4	4.5	7.3	54.5	88.3
b. Invest $40M, per USFWS	0	58	9.1	13.8	90.9	28.3
c. Invest $10M, 25% of Need	56.5	30.8	8.7	27.1	34.8	42.1
d. Invest $20M, 50% of Need	18.2	40	9.1	24.4	72.7	35.6
e. Invest $30M, 75% of Need	4.5	38.1	13.6	28.4	81.8	33.6
Who pays						
a. All Costs Paid by Feds	29.4	37.7	4.3	9.4	66.5	52.9
b. Feds 50%, Private 50%	58.8	39.7	8.7	17.6	39.1	42.6
c. Feds 50%, States 50% Equal	5.9	43.8	0	22.6	87	33.6
d. Feds 50%, States 50% Prop.	5.9	58.7	0	23.2	87	18.1
e. Feds 1/3, Pvt.1/3, States 1/3 Proportional to Use	47.1	18.4	4.2	18.4	45.8	13.2

[a] A full description of each policy attribute and level is found in Table 12.8.
[b] Includes responses of "strongly support" and "support."
[c] Includes responses of "strongly oppose" and "oppose."

Table 12-A.3 Degree of Support for Policy Attribute Levels in Nebraska, by Interest Group

Policy Attribute and Level[a]	Do Support[b]		No Opinion		Don't Support[c]	
	Ag	Envl.	Ag	Envl.	Ag	Envl.
	Percent of Classified Respondents					
Method						
a. All Needs, Least Cost	19.6	38.8	14.7	24.7	65.7	36.5
b. All Needs, Minimum Water	35.6	52.9	19.8	28.2	44.6	18.8
c. Best Possible, No Ag Water	57.3	27.1	17.5	21.2	25.2	51.8
d. Adaptive Management	69.9	56.5	12.6	21.2	17.5	22.4
e. Benefit–Cost Approach	47.1	22.4	29.4	23.5	23.5	54.1
Cost						
a. Invest Nothing	35.6	7.1	16.8	8.2	47.5	84.7
b. Invest $40M, per USFWS	9.8	32.1	13.7	23.8	76.5	44
c. Invest $10M, 25% of Need	35	44.4	16	18.5	49	37
d. Invest $20M, 50% of Need	21.6	30.5	17.6	22	60.8	47.6
e. Invest $30M, 75% of Need	5.9	33.3	18.8	25	75.2	41.7
Who pays						
a. All Costs Paid by Feds	38	37.3	12	20.5	50	42.2
b. Feds 50%, Private 50%	50	36.1	15	20.5	35	42.4
c. Feds 50%, States 50% Equal	19	34.5	15	20.2	66	45.2
d. Feds 50%, States 50% Prop.	24.8	44.6	15.8	24.1	59.4	31.3
e. Feds 1/3, Pvt.1/3, States 1/3 Proportional to Use	41	54.8	16	23.8	43	21.4

[a]A full description of each policy attribute and level is found in Table 12.8.
[b]Includes responses of "strongly support" and "support."
[c]Includes responses of "strongly oppose" and "oppose."

Table 12-A.4 Degree of Support for Policy Attribute Levels in Wyoming, by Interest Group

Policy Attribute and Level[a]	Do Support[b]		No Opinion		Don't Support[c]	
	Ag	Envl.	Ag	Envl.	Ag	Envl.
	Percent of Classified Respondents					
Method						
a. All Needs, Least Cost	13.2	61.1	11.3	13.9	75.5	25
b. All Needs, Minimum Water	34.6	52.8	13.5	28.7	51.9	18.5
c. Best Possible, No Ag Water	71.2	18.7	15.4	15.9	13.5	65.4
d. Adaptive Management	83.3	48.1	9.3	19.4	7.4	32.4
e. Benefit–Cost Approach	50.9	23.6	22.6	19.8	26.4	56.6
Cost						
a. Invest Nothing	42.3	6.5	19.2	5.6	38.5	88
b. Invest $40M, per USFWS	3.8	50	13.5	10.2	86.7	39.8
c. Invest $10M, 25% of Need	37.7	27.8	18.9	28.7	43.4	43.5
d. Invest $20M, 50% of Need	13.7	27.5	19.6	26.6	66.7	45.9
e. Invest $30M, 75% of Need	0	25.9	19.6	28.7	80.4	45.4
Who pays						
a. All Costs Paid by Feds	37	34.9	7.4	12.8	55.6	52.3
b. Feds 50%, Private 50%	60.4	41.1	7.5	14	32.1	44.9
c. Feds 50%, States 50% Equal	5.9	36.2	5.9	15.2	88.2	48.6
d. Feds 50%, States 50% Prop.	13.5	61.1	15.4	14.8	71.2	24.1
e. Feds 1/3, Pvt.1/3, States 1/3 Proportional to Use	47.1	57.5	3.9	12.3	49	30.2

[a]A full description of each policy attribute and level is found in Table 12.8.
[b]Includes responses of "strongly support" and "support."
[c]Includes responses of "strongly oppose" and "oppose."

Table 12-A.5 Policy Attribute Weights by Bargaining Group[a]

Interest Group	Intercept	*Method,* M	*Cost,* C	*Who pays,* P
Colorado, State, N = 994				
Reg. Coefficients, B	0.772	0.211	0.119	0.413
Standard Error		0.021	0.020	0.020
Normalized Weights, W		0.28	0.16	0.56
Colorado Agricultural, N = 154				
Reg. Coefficients, B	0.855	0.068	0.382	0.242
Standard Error		0.070	0.081	0.087
Normalized Weights, W		0.10	0.55	0.35
Colorado Environmental, N = 840				
Reg. Coefficients, B	0.873	0.191	0.192	0.321
Standard Error		0.030	0.030	0.030
Normalized Weights, W		0.27	0.27	0.46
Nebraska State, N = 1,179				
Reg. Coefficients, B	0.900	0.093	0.204	0.387
Standard Error		0.017	0.018	0.017
Normalized Weights, W		0.14	0.30	0.57
Nebraska Agricultural, N = 674				
Reg. Coefficients, B	0.628	0.056	0.198	0.524
Standard Error		0.031	0.036	0.035
Normalized Weights, W		0.07	0.25	0.67
Nebraska Environmental, N = 505				
Reg. Coefficients, B	1.729	0.055	0.026	0.332
Standard Error		0.043	0.043	0.043
Normalized Weights, W		0.13	0.06	0.81
Wyoming State, N = 999				
Reg. Coefficients, B	0.663	0.129	0.198	0.420
Standard Error		0.018	0.019	0.018
Normalized Weights, W		0.17	0.26	0.56
Wyoming Agricultural, N = 646				
Reg. Coefficients, B	0.840	0.078	0.177	0.396
Standard Error		0.043	0.054	0.050
Normalized Weights, W		0.12	0.27	0.61
Wyoming Environmental, N = 353				
Reg. Coefficients, B	1.035	0.154	0.117	0.370
Standard Error		0.031	0.032	0.030
Normalized Weights, W		0.24	0.18	0.58

[a]See Equations 12.4 and 12.5 for explanation of variables and attribute weights.

CHAPTER 13

An Ecological Economic Model for Integrated Scenario Analysis: Anticipating Change in the Hudson River Watershed

Jon D. Erickson, Caroline Hermans, John Gowdy, Karin Limburg, Audra Nowosielski, John Polimeni, and Karen Stainbrook

CONTENTS

THE TYRANNY OF SMALL DECISIONS

Many communities across the nation and world have succumbed to what Alfred Kahn[1] referred to as "the tyranny of small decisions." The tyranny describes the long-run, often unanticipated, consequences of a system of decision-making based on marginal, near-term evaluation. Land-use decisions made one property, one home, and one business at a time in the name of economic growth have accumulated without

1-56670-639-4/05/$0.00+$1.50

regard to social and environmental values. The tyranny results when the accumulation of these singular decisions creates a scale of change, or a conversion from one system dynamic to another, that would be disagreeable to the original individual decision-makers. In fact, if given the opportunity to vote on a future that required a redirection of near-term decisions, a community of these same individuals may have decided on a different path.

Incremental decisions made by weighing marginal benefits against marginal costs by an individual isolated in a point in time are the hallmark of traditional economics. But maximizing the well-being of both society and the individual requires an exercise in identifying and pursuing a collective will, quite different from assuming that community-held goals will result simply from individual pursuits of well-being.

At the watershed scale, the tyranny of small decisions has emerged in the form of urban sprawl — a dispersed, automobile-dependent, land-intensive pattern of development. One house, one subdivision, one strip mall at a time, the once-hard edge between city and country throughout the United States has incrementally dissolved. This pattern of development has costs and benefits to both the individual and society. To the individual, the choice to purchase or build a home in the suburbs over the city may initially carry the benefits of a larger home, more green space, and better schools, all at a more affordable price. Initially, individual costs may be related only to transportation to employment and services. However, by structuring the land-use decision problem as a series of individual choices, a tyranny can result in the loss of community services provided by watersheds, such as water supply, purification, and habitat provision – so-called *natural capital* depreciation. Associated social capital depreciation can include decline in city school quality and loss of social networks. The city often subsidizes the suburbs on services such as fire and police protection, water and sewer lines, road construction and maintenance, and health and emergency care. The total social costs (for city and suburbs) of sprawl may surpass the private benefits — an outcome that a democracy may not have chosen if given the chance, yet individuals often can not appreciate in their own land-use decisions. The point is not that this will always be the case for suburbanization, but rather that the accumulation of small decisions should be considered in the calculus of the small decisions themselves.

To emerge from the tyranny, the challenge is not to predict but to anticipate the future. *Prediction* in integrated social, economic, and ecological systems often requires a simplification of multiple scales and time dimensions into one set of assumptions. It implies a defense against alternative predictions, rather than an exploration of possible futures. Quantitative assessment and model building is often limited to one system, with others treated as exogenous corollaries.

In contrast, *anticipation* implies a process of envisioning future scenarios and embracing the complexity that is inherent among and within the spheres of social, economic, and ecological change. As a process-oriented approach to decision-making, anticipation focuses on the drivers of change and the connections between spheres of expertise, and relies on local knowledge and goal setting. Through scenario analysis, decision-makers can vary the assumptions within degrees of current knowledge, foresee the accumulation of small decisions, and decide upon group strategies that decrease the likelihood of undesirable consequences.

The following case study describes a project in Dutchess County, New York that has developed in this spirit. The next section introduces Dutchess County and its own version of the "tyranny of small decisions." "Economic Analysis, Land Use, and Ecosystem Integrity: An Integrated Assessment" describes an integrated approach to model development in Dutchess County, including economic, land-use, and ecological submodels that provide both the detail within and connectivity among their spheres of analysis. The "Scenario Analysis" section incorporates the scenario of an expanding semiconductor industry in Dutchess County to illustrate the connectivity and chain of causality between economic, land-use, and ecological submodels. "Multicriteria Decision Aid" then introduces a multicriteria decision framework to aid watershed planning efforts in the context of multiple decision criteria, social values, and stakeholder positions. The "Discussion" section concludes with a discussion of the strengths and weaknesses of this approach, and places this case in the context of other book chapters.

WATERSHED COMMUNITIES AND THE DUTCHESS COUNTY DEVELOPMENT GRADIENT

Watershed communities include the physical, ecological, and human components of a topographically delineated water catchment. Our study area is part of the larger Hudson River watershed of eastern New York State, which draws water from over 34,000 square kilometers of land (mostly in New York, but also reaching into Massachusetts, Connecticut, New Jersey, and Vermont) on its journey from the southern slopes of the High Peaks of the Adirondack mountains to the Atlantic Ocean.[2] Dutchess County (2077 km^2) is located in the lower Hudson watershed, midway between the state capital of Albany and New York City. Figure 13.1 highlights the county's two principal Hudson tributary watersheds of Wappingers (546.5 km^2) and Fishkill (521 km^2) Creeks, which together drain over half of the county landscape. The full county includes approximately 970 km of named streams that provide public water, irrigation, recreation, and waste disposal. This study incorporates models of the county's economy, land-use patterns, and the general health of the Wappingers and Fishkill systems into the design of a decision aide to support county and state land-use planners, ongoing intermunicipal efforts to improve watershed health, and local citizens' groups working to improve the quality of life of county residents.

The Dutchess economy through the mid-twentieth century was principally agrarian, specifically mixed row-crop, dairy, and fruit agriculture. While today's county economy is characterized by 203 distinct sectors, with a total employment of over 132,000, much of the recent economic history has reflected the rapid growth and then cyclical behavior of the International Business Machine Corporation (IBM). In 2000, IBM was the second largest employer (>11,000) in the county, preceded only by local government institutions (13,800), and followed by state government (7600).[3] Other major economic themes cutting across the county — identified at an early stakeholder meeting of this project — included the influence of seasonal home ownership and commuting patterns (particularly in relation to New York City wealth

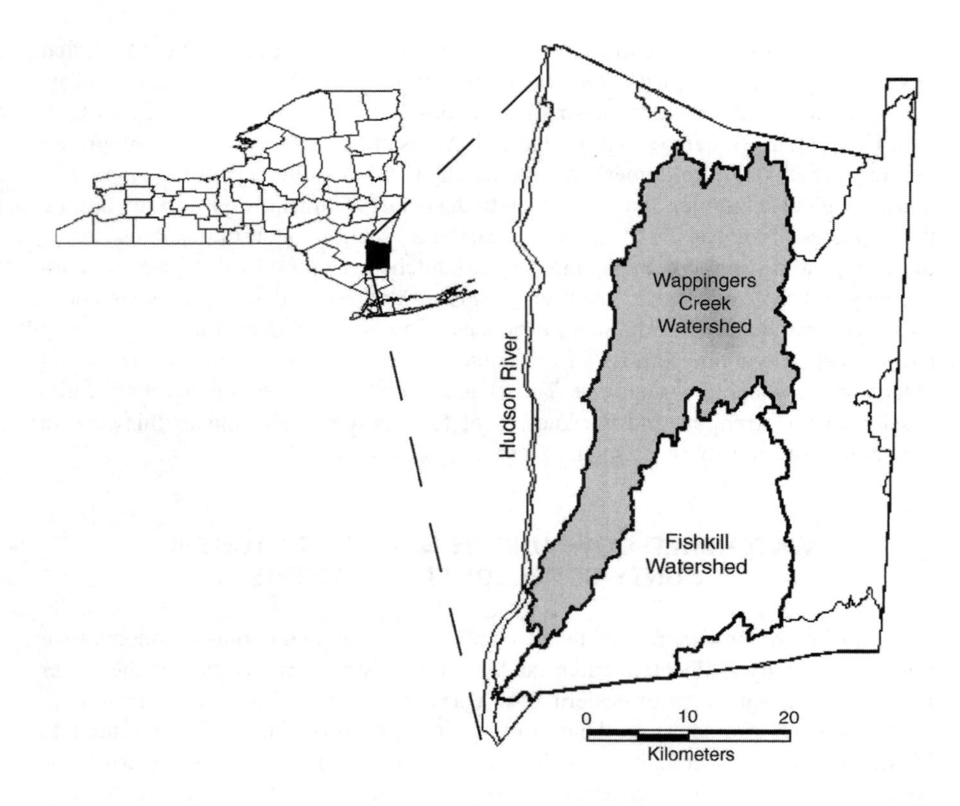

Figure 13.1 Dutchess County, New York, and its main Hudson tributary watersheds.

and employment), the decline of traditional agriculture in favor of agro-tourism activities, and the aging of the population and growth in retirement homes and services.

County land-use intensity follows a development gradient from the rural northeast to urban southwest. The Wappingers Creek watershed mirrors this gradient, beginning in mostly forested headwaters, continuing through a predominantly agricultural landscape, flowing through mixed suburban use, and discharging into the Hudson in the urban areas of Wappingers Falls and Poughkeepsie. The Fishkill Creek follows a similar northeast-southwest development gradient with generally higher population densities, and enters the Hudson through the city of Beacon. The geology of both watersheds is primarily a mix of limestone, dolostone, and shale, and annual precipitation is approximately 1040 mm.[4]

These rural-to-suburban-to-urban development gradients provide a unique opportunity to model the impact of economic change on land-use intensity and watershed health. In particular, a pattern of urban sprawl that stretches up each watershed creates a gradient of increasing impervious surfaces and corresponding impacts on aquatic health. Land use is changing most rapidly in the south-central portion of the county as a consequence of high-tech industrial growth and a general push of suburban expansion radiating out from the New York City greater metropolitan area. Residential development, in particular, is rapidly converting forest and field to roads and housing. According to county planners, about 75% of the houses in Dutchess are located in the southern half, but new building is spreading north and east. Since 1980, the average annual number of building permits for single-family dwellings was 877.[5] However, this average is significantly skewed by the 1983–1989 and 1998–2000 building booms, with each year surpassing 1000 permits, compared with an off-peak annual average closer to 500 permits. The slowdown in the early 1990s can be attributed to IBM's downsizing. These layoffs "glutted the housing market, depressing prices and making houses more affordable to people looking to move out of New York City."[6]

With new households comes new income that cascades across the county economy, creating further business and household growth and consequent land-use change. With the waxing and waning of the housing market (tied in part to the ups and downs of the IBM labor force), nonresidential building permits averaged 744 between 1980 and 1995, without much annual variation. Average per capita income in Dutchess County is the seventh highest of 62 New York counties. Dutchess households have had a median buying power of $47,380, much higher than the New York State ($38,873) and U.S. ($35,056) medians.[7] Dutchess County's effective buying income (EBI) ranks 15th in the United States, with over 46 percent of county households having an EBI of over $50,000. This household income creates multipliers that are a cause for concern for some of the more rural municipalities. A planning report from the small town of Red Hook[8] in the northwest of the county states, "These factors will continue to bring commercial development pressures on any significant highway corridors, as businesses seek to exploit the growing pool of disposable income in Red Hook and Rhinebeck." Growth is viewed as both an opportunity for business and a challenge for municipalities that struggle to preserve their rural landscape and level of community and ecosystem services.

Many of these ecosystem services, including the provision of aesthetic qualities and opportunities for recreation, depend on the ecological attributes of the watershed. Ecological risks associated with current and changing land use include the loss of water quality, hydrological function, physical habitat structure (e.g., alterations of riparian zone), and biodiversity. To anticipate and perhaps avoid irreversible loss in these attributes, the challenge is to link ecological change to land-use change and its economic drivers. The next section outlines an approach to integrated modeling, combining synoptic ecological surveys with economic and land-use models in a framework capable of stakeholder-informed scenario analysis and multicriteria decision making.

ECONOMIC ANALYSIS, LAND USE, AND ECOSYSTEM INTEGRITY: AN INTEGRATED ASSESSMENT

The analytic building blocks for the integrated watershed model include a social accounting matrix (SAM) describing economic activity in Dutchess County; a geographical information system (GIS) of land-use, socioeconomic, and biophysical attributes; and an assessment of aquatic ecosystem health based on indices of biotic

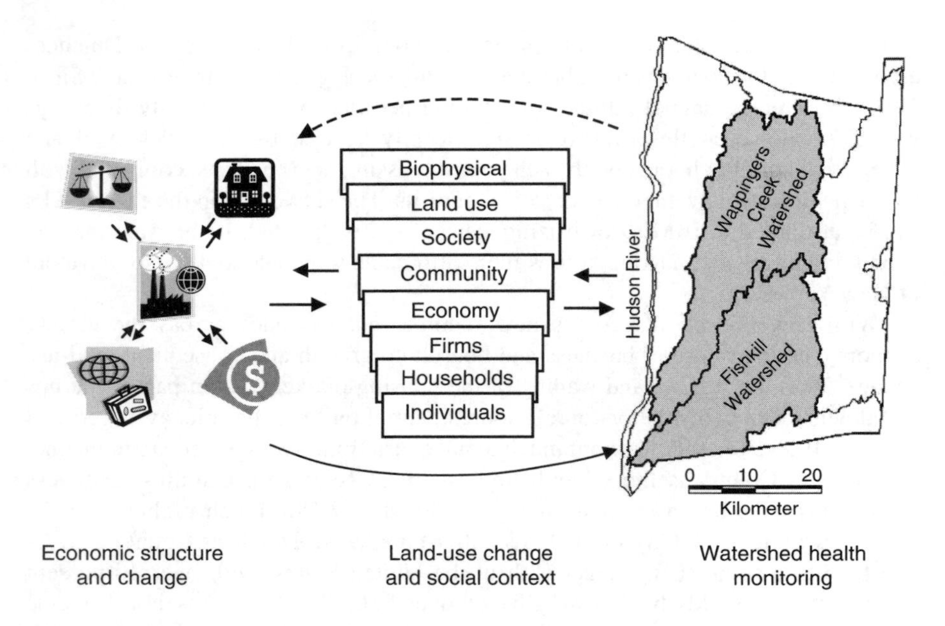

Figure 13.2 Conceptual model components and linkages.

integrity (IBI). Figure 13.2 illustrates these sequential model components, with system drivers and feedback loops denoted in solid and dashed arrows, respectively.

Starting with the left side of the diagram, regional economic activity is characterized as dollar flows between industry (in the center), households (top right), government (top left), capital markets (bottom right), and the outside economy (bottom left). The middle panel illustrates the multiple layers of biophysical and social context within which land-use decisions are made. The right panel highlights the watershed as the scale of ecosystem impact from economic and land-use change. Total economic activity has a direct effect on watershed health through material input and waste output, and an indirect effect through land use change. Land use change and ecosystem health[a] can similarly impact economic activity through feedback loops. For example, soil erosion impacts agricultural industries, water quality impacts water-based tourism, and environmental amenities influence real estate values. Drivers or feedbacks can be either marginal or episodic, accounting for system surprises.

Socioeconomic Submodel: Geo-Referenced Social Accounting Matrix

A widely used tool in national and regional economic analysis is the input–output model (IO) developed in the 1930s by Nobel laureate Wassily Leontief. As a system of accounting that specifies interdependencies between industries, IO has been used to understand how changes in final demand (household consumption, government

[a] The concept of ecosystem health is controversial (see "Planning" in Chapter 3). In this paper, the term is operationally defined using indices of biotic integrity (see also Appendix 6-A).

	Inputs ↓ / Outputs →	Processing Sectors						Final Demand Sectors			
		Agriculture	Manufacturing	Transportation	Wholesale/retail	Services	Households	Exports	Investment	Government	Total sales
Processing Sectors	Agriculture	34	290	0	0	0	7	137	0	1	469
	Manufacturing	25	1134	5	13	188	607	12303	27	10	14312
	Transportation	6	304	54	25	80	22	111	5	3	610
	Wholesale/retail	13	490	18	45	156	1171	723	29	11	2656
	Services	35	472	53	258	418	1387	816	573	229	4241
	Households	208	3242	252	881	1816	869	1203	0	244	8715
Payments Sectors	Imports	77	5712	83	456	892	2539				
	Depreciation	24	2157	129	805	446	489				
	Government	47	511	16	173	245	1624				
	Total purchases	469	14312	610	2656	4249	8715				

Figure 13.3 Hypothetical transaction table in input–output analysis.

expenditure, business investment, and exports) are allocated across an economy. To meet new demand requires industrial production, which in turn requires industrial and value-added inputs, which in turn requires more production, and so on. Each addition in the production chain sums to an output multiplier that accounts for the original demand and all intermediate production generated to meet this demand. Value-added inputs include income contributions from labor as wages, capital as profits, land as rents, and government as net taxes, and can be related to output to capture various income (wage, profit, rent, and tax) and employment multipliers.

Figure 13.3 illustrates a simplified, hypothetical example of an IO transactions table. Numerical values represent real dollar flows between the processing, final demand, and payment sectors of a regional economy (perhaps in millions of dollars). For instance, reading across the manufacturing row, firms in the manufacturing industry sell their output to firms in the agriculture (25), manufacturing (1134), transportation (5), wholesale and retail trade (13), and service (188) industries in the form of intermediate inputs; and to households (607), exports (12303), business investment (27), and government (10) in the form of final outputs.[a] Manufacturing itself requires inputs, read down the manufacturing column, including labor from households paid as wages (3242), imported goods and services from outside the region (5712), depreciation of capital assets (2157), and government services (511). The payment sectors are often captured as payments to labor (wages), capital (interest), entrepreneurship (profits), and land (rent), and collectively are called

[a] Households in this example are treated as a processing sector (or industry), even though they are also counted as a final demand sector. The distinction is based on a decision of what is exogenous and what is endogenous to the model. Exogenous sectors stimulate growth only in the model economy, but cannot themselves be stimulated in subsequent rounds of buying and selling. Assuming households are endogenous in an IO model implies that as industrial output expands it will generate new household income that will "induce" more household spending, which will create subsequent rounds of industrial expansion and labor income generation.

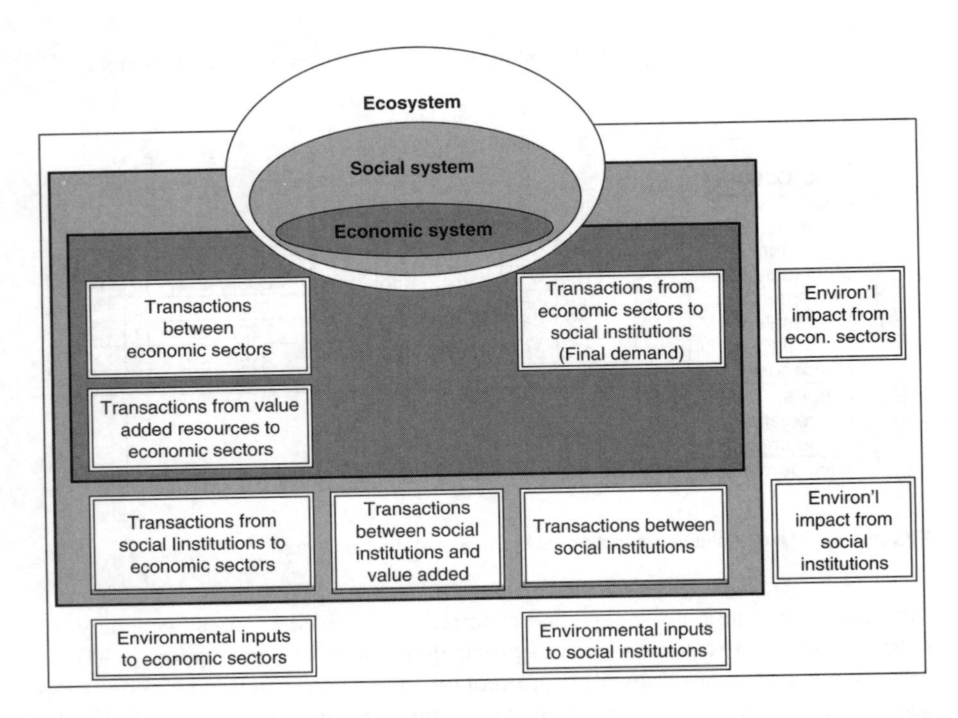

Figure 13.4 Integrated system of accounts, including economic sectors, social institutions, and ecosystem resources.

value-added inputs. The total economic production of a regional economy can be measured as either the sum of final demand or value-added inputs.

An IO system such as this forms the basis for the economic sphere in Figure 13.4. The three boxes of the economic sphere symbolize the main systems of accounts — final demand, industry production, and value-added inputs — in a traditional IO system. These accounts are specified as matrices, as in Figure 13.3, with rows read across as outputs and columns read down as inputs. For instance, reading down the column of the semiconductor industry for the Dutchess County IO table, the top ten sector inputs include other firms within the semiconductor industry, wholesale trade, maintenance and repair, computer and data processing, electric services, legal services, real estate, electronic computers, personal supply services, and banking. The sum of all these regional inputs, value-added, and any imports required from outside the region equals total inputs to the industry. Similarly, the sum of the semiconductor industry's outputs generated for other industries to use in intermediate production and final products to demand equals its total output. To balance the accounts within a particular time period, inputs must equal outputs.

By itself, the economic sphere misses key dependencies between the economic and social systems. Traditional IO has focused on the structure of production, the matrix in the upper left corner of Figure 13.4, with industry disaggregated into over 500 sectors, each with its own input–output relations specified. In contrast, the structure and detail of final demand has typically been highly aggregated, most often specified only as its four major components of household, government, business investment, and foreign consumption (as in the example of Figure 13.3). This restricted treatment of households in particular — the major driving force in economies as both consumers and suppliers of labor and capital — limits the ability of the IO model to specify income distribution, investigate the effect of welfare and tax policies, and model the impacts of changing patterns of household spending. The need for a more detailed treatment of households led researchers, beginning with the work of Nobel laureate Richard Stone in the 1960s, to expand the IO system into a *social accounting matrix* (SAM).[9,10]

In the SAM, components of final demand and value-added are called *institutions*. The interdependencies between and among industry and institutions are illustrated by the three boxes linked to the social sphere of Figure 13.4. For instance, households specified as an institution (not just as a supplier of labor) can reveal their nonlabor inputs to industry in the left box, such as land, capital, energy, and anything else besides labor that a household might supply to firms as an input. The distribution of labor income is captured in the center box. The interdependencies with other institutions are captured in the right box, for instance earnings by corporations redistributed back to households as dividends, or taxes paid to government redistributed back to households as welfare payments. Households — as consumers in final demand and labor supply in value-added — can be disaggregated into columns and rows according to criteria (and data) relevant to the policy question at hand. For instance, households have been disaggregated by income category, wage group, and skill or occupation class.

Figure 13.5 is a schematic of the Dutchess County SAM, with the six major transaction tables denoted by blocks not containing a zero.[11] The full SAM specifies 203 industry sectors, 11 occupation and skill classes, and 9 household categories as endogenous components. Exogenous changes to final demand come from government institutions, capital expenditures, trade flows (both domestic and international), and inventory adjustments.

The creation of a SAM for this study is based on a regional database called IMPLAN (IMpact analysis for PLANning).[12] IMPLAN tables are available for any collection of states, counties, or zip codes in the U.S. based on federal and state databases, which can then be modified using the best available local data. The main modification for the Dutchess County SAM was the disaggregation of IMPLAN's single labor income row into 11 occupation categories (in Figure 13.5, the *Employee Compensation and Profits* Matrix). Using Bureau of Labor Statistics data from the 2000 census, and following a procedure outlined by Rose et al.,[13] each occupation row shows the input relation to each industry column, and each occupation column shows the distribution of labor income to nine household institutions categorized by income ranges (*Industry Sales to Households* Matrix).

Finally, to complete the image of a nested system of accounts within Figure 13.4, economic activity and its distribution is linked to the ecosystem. To explore these linkages, the basic IO–SAM framework has been expanded to incorporate environmental and natural resource accounts.[14–16] In Figure 13.4, inputs from the environment to industry and institutions are tallied in the bottom two boxes, and outputs from industry and institutions to the environment are tallied in the far right boxes. Environmental inputs include energy, minerals, water, land, and numerous ecosystem

<table>
<tr><th colspan="3" rowspan="3">Social Accounting Matrix</th><th colspan="4">Processing Sectors</th><th colspan="27">Final Demand Sectors</th><th rowspan="3"></th></tr>
<tr><th colspan="4">Industries</th><th colspan="11">Value-Added by Occupation Category</th><th colspan="9">Households by Income Category</th><th colspan="7">Institutions</th></tr>
<tr><th>Agric.</th><th>Maunf.</th><th>Services</th><th>Other</th><th>Exec.</th><th>Prof.</th><th>Tech.</th><th>Sales</th><th>Admin</th><th>Service</th><th>Prec. Prod.</th><th>Mach. Op.</th><th>Transp.</th><th>Laoborers</th><th>Farm, etc.</th><th>< 5K</th><th>5 — 10K</th><th>10 — 15K</th><th>15 — 20K</th><th>20 — 30 K</th><th>30 — 40K</th><th>40 — 50K</th><th>50 — 70K</th><th>> 70K</th><th>Fed. Govt.</th><th>State Govt.</th><th>Local Govt.</th><th>Enterprises</th><th>Capital</th><th>Inv. & Tr.</th><th>Other</th></tr>
<tr><td rowspan="4">Process Sectors</td><td rowspan="4">Industries</td><td>Agriculture</td><td colspan="4" rowspan="4">Inter-Industry Transaction</td><td colspan="11" rowspan="4">0</td><td colspan="9" rowspan="4">Industry Sales to Households</td><td colspan="7" rowspan="24">Exogeneous Accounts</td><td rowspan="29">Total Sales</td></tr>
<tr><td>Manufacturing</td></tr>
<tr><td>Services</td></tr>
<tr><td>Other</td></tr>
<tr><td rowspan="25">Payments Sectors</td><td rowspan="11">Value-added Occupations</td><td>Executive</td><td colspan="4" rowspan="11">Employee Compens. and Profits</td><td colspan="11" rowspan="11">0</td><td colspan="9" rowspan="11">0</td></tr>
<tr><td>Professional — Specialty</td></tr>
<tr><td>Technicians / Support</td></tr>
<tr><td>Sales</td></tr>
<tr><td>Admin. Supp./Clerical</td></tr>
<tr><td>Service</td></tr>
<tr><td>Prec. Production/Craft/Repair</td></tr>
<tr><td>Machine Op./Assemblers</td></tr>
<tr><td>Transportation</td></tr>
<tr><td>Laborers/Handlers</td></tr>
<tr><td>Farm/Forest/Fish</td></tr>
<tr><td rowspan="9">Households</td><td>< 5K</td><td colspan="4" rowspan="9">Industry Transfers to Households</td><td colspan="11" rowspan="9">Income and Profit Distribution</td><td colspan="9" rowspan="9">Household to Household Transactions</td></tr>
<tr><td>5 — 10K</td></tr>
<tr><td>10 — 15K</td></tr>
<tr><td>15 — 20K</td></tr>
<tr><td>20 — 30K</td></tr>
<tr><td>30 — 40K</td></tr>
<tr><td>40 — 50K</td></tr>
<tr><td>50 — 70K</td></tr>
<tr><td>> 70K</td></tr>
<tr><td rowspan="5">Institutions</td><td>All Government</td><td colspan="31" rowspan="5">Exogenous Accounts</td></tr>
<tr><td>Enterprises</td></tr>
<tr><td>Capital</td></tr>
<tr><td>Inventory and Trade</td></tr>
<tr><td>Other</td></tr>
<tr><td colspan="3"></td><td colspan="31">Total Purchases</td><td></td></tr>
</table>

Figure 13.5 Major SAM accounts in Dutchess County model.

services. Outputs discarded into the environment include the gamut of solid, liquid, and gaseous wastes.

For the current study, the main consideration is the use of land as an input to the socioeconomic system. Of particular interest is how scenarios of industrial sector change and growth lead to changes in household institutions that ultimately drive new residential land development. To link economic change and social distribution to spatial patterns of land use, the Dutchess County SAM was referenced to a geographical information system (GIS). For example, the geo-referenced SAM (GR-SAM) can place household institutions (disaggregated by both occupation class and income range) within the spatial context of race, education, age, commuting patterns, wealth, income, and numerous other census-defined household characteristics. Spatial patterns and concentrations of industry sectors can be viewed with business point data and linked to information on business size, year of establishment, and income range. The spatial dimensions of the entire economy (both institutions and industry) can be further referenced to tax parcel data with information on acreage, taxable use, zoning, infrastructure, and various ownership characteristics. These ownership units can then be linked to biophysical characteristics such as soil, slopes, wetlands, and location within watersheds.

The main advantage to this integrated system of economic, social, and environmental accounts is to visualize the interconnectivity of these system components. This can then serve as the basis to conduct scenario analysis within the confines of this snapshot in time. The main weaknesses of this approach are the linear structure of input–output relationships, the lack of any time dimensions in the analysis of multiplier effects, and the inability of the model parameters to adjust to changes in relative scarcity (for instance, price signals). The static nature of IO models has been addressed to some degree with the advent of dynamic IO models and general equilibrium models, although the data limitations are severe.[17] However, the fixed coefficient assumption implicit in most IO models is in many cases a more realistic representation of technology than traditional production functions that assume away the problems of complementarity (when certain input combinations are required for production) and sunk costs (when specific investments in capital stock are required for production). A more serious problem is the difficulty of finding and modeling the critical interfaces between economic and environmental systems. Ecosystems, even more than economic systems, are characterized by nonlinearity, threshold effects, synergistic relationships, and pure uncertainty. Economic models require that these effects and interactions be drastically simplified.

Land Use Submodel: Probabilistic Geographical Information System

Moving from the first submodel to the second outlined in Figure 13.2, scenarios generated by the GR-SAM then inform a model of land-use change. The GR-SAM is a static tool that helps to identify the source of new land demands, but not necessarily how these demands could play out on the landscape. Most economic models do not include spatial variation of activity; however, location is critical to estimating environmental loading.[18]

Of particular interest to Dutchess County is growth in residential land use. Land currently characterized on the tax rolls as vacant-residential, agriculture, and private forest provides an inventory of total vacant land potentially available for conversion to residential use. By this characterization, in the Wappingers Creek watershed in 2001 there were 19,024 parcels in residential use and 4507 vacant. The conversion from vacant to residential was modeled with a binomial logit regression procedure to estimate the probability of land conversion of individual tax parcels throughout the Wappingers Creek watershed.[19] Data were cross sectional for the year 2001 due to the limited availability of digital tax maps over time. These probabilities were assumed to depend on both tax parcel characteristics and neighborhood characteristics (defined by census blocks). Tax parcel independent variables included 2001 land assessment value per acre and distance to the nearest central business district. Neighborhood independent variables included household income and population growth (between 1990 and 2000 census years) and the 2001 density of residential land-use classes in each census neighborhood.

Polimeni provides a detailed discussion of model calibration, results, and diagnostics (including tests for spatial autocorrelation).[19] The final model provides the basis for simulating residential development patterns given changes in independent variables. For instance, if incomes or population continue to grow according to intercensus year rates (1990–2000), a Monte Carlo procedure can demonstrate where conversion to residential use would likely occur. Figure 13.6 plots the outcome of a single status quo Monte Carlo run, assuming a continuation of the 1990s decadal growth rates of 53% in per capita income and 8% in population. The average of 100 runs provides a point estimate of 1120 parcels converted to residential use. Development favors the upper (rural) watershed, with 677 new residential parcels averaging 18.39 acres each. The lower (urban) and middle (suburban) watershed includes 228 and 215 new residential parcels, with an average size of 3.3 and 8.61 acres, respectively.

The model only simulates conversion of land use class, not the percentage or acreage of parcels that become physical homes. To estimate the maximum number of new homes on new residential parcels, tax parcels were screened according to both biophysical and zoning layers. Biophysical GIS layers included slope, hydric soils, wetland vegetation, riparian river corridors, and agricultural land. Acreage can be removed from the inventory of developable land according to rules imposed by these layers. Zoning maps further limit the number of principal buildings allowed per acre. To account for development infrastructure requirements, particularly new roads, various percentages of buildable land can also be assumed. Following a biophysical screening of wetlands, hydric soils and >10% slopes, a town-specific zoning overlay, and assuming 80% buildable land on remaining acreage, the status quo scenario (highlighted by Figure 13.6) can accommodate a maximum of 10,370 new homes.

Given economic scenarios from the GR-SAM sub-model, the binomial logit model can simulate residential land conversion for the Wappingers Creek watershed. These scenario-derived land-use profiles are then used to hypothesize changing land-use intensity within each of the 16 subcatchments of the Wappingers Creek watershed. This provides the empirical link to the ecosystem health assessment.

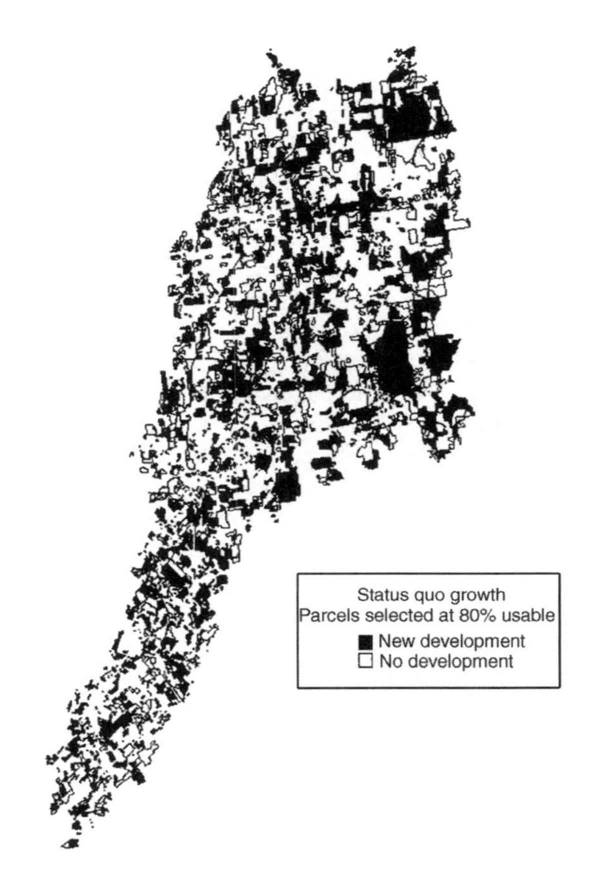

Figure 13.6 New residential land use in the Wappingers Creek Watershed under intercensus year trend in population and household income growth.

Ecosystem Health Submodel: Spatially Correlated Indices of Biotic Integrity

The third component of the Figure 13.2 model overview is an estimate of ecosystem health impact based on land development scenarios. The well-worn concept of ecosystem health, while controversial (see "Planning" in Chapter 3), is enjoying resurgence as a useful means of assessing the impacts of human activities and for protecting and restoring ecosystems,[20–22] while considering societal goals.[23] Ecosystem health may be defined as the maintenance of biotic integrity, resistance and resilience to change in the face of anthropogenic disturbance, and the absence of factors that degrade population, community, and ecosystem structure and function. Ecologists have spent the past 25 or more years exploring various indicators that best reflect ecosystem responses to anthropogenic stress and have found them to vary with the particulars of the ecosystem of interest. Nevertheless, some indicator methods have emerged as robust, when adjusted to local or regional biogeographic and geomorphological constraints.

Among these is the *index of biotic integrity* or IBI method.[24,25] Karr worked out a set of criteria for assessing the health of midwestern streams and argued that fish were a good end-point for observing ecological effects. As the "downstream receiving end" of numerous complex ecological processes, fish can serve as integrating indicators of the quality of the system. A stream IBI combines a number of different metrics that reflect fish biodiversity, community structure, and health of populations. For example, a water body that has high species richness (number of species present), a high proportion of which are endemic, including a mix of species occupying different trophic positions and showing very few indications of disease or starvation would be scored with a high IBI. Conversely, an ecosystem with only a few pollution-tolerant species, low biomass, or containing only stocked or exotic species would be scored with a low IBI. The basic methods and caveats to the use of IBIs are included in Appendix 6-A.

This study uses the metrics developed for a recently published New England fish IBI[26] and a benthic macroinvertebrate index developed by the New York State Department of Environmental Conservation.[27] In addition, several other parameters are being examined, including whole-ecosystem metabolism,[28,29] nonpoint source enrichment of ^{15}N as indicated by the $\delta^{15}N$ isotopic ratio of standardized ecosystem components, and water quality parameters — including their variability, as this varies with degree of urbanization.[30] These system-level metrics and water quality parameters are known to vary with land use.[31–34]

A series of synoptic surveys were conducted in 2001 and 2002 at a total of 33 stream sites in the two watersheds. The surveys included physical habitat assessments, using a modification of the U.S. Environmental Protection Agency standard protocols,[35] fish surveys, macroinvertebrate surveys, and water chemistry surveys. Surveys also included collection of materials for stable isotope analysis, focusing on the simplified food chain of seston (suspended organic particles), macrophytes (rooted plants), and a cosmopolitan fish species (blacknose dace, *Rhinichthys atratulus*). A short-term study of diurnal oxygen variation was also carried out simultaneously in six subcatchments with different land use. Diurnal oxygen variation generally followed expectations from first principles, with the least variation (and hence, least ecosystem metabolic activity) in the forested catchment and the highest in the suburbanized catchment, which not only had less canopy cover but also had high nutrient concentrations (Figure 13.7).

Although much of the data analysis is still in progress, patterns are beginning to emerge. As expected from visual impressions, variation is high and patterns of some parameters would likely not be visible without geographic presentation of the entire dataset. For example, Figure 13.8 shows an index of anthropogenic nitrogen (the percent of total N in the inorganic forms of nitrate and ammonia) over the watersheds. The Fishkill, the more developed watershed, has significantly higher inorganic N to total N ratio than does the Wappingers Creek. In general, nutrient and conductivity values follow degree of urbanization.

The next step is to correlate such metrics with land-use intensity within each watershed. We have also observed that many indicators of human activity in the watersheds are strongly correlated with distance away from the tributary confluence

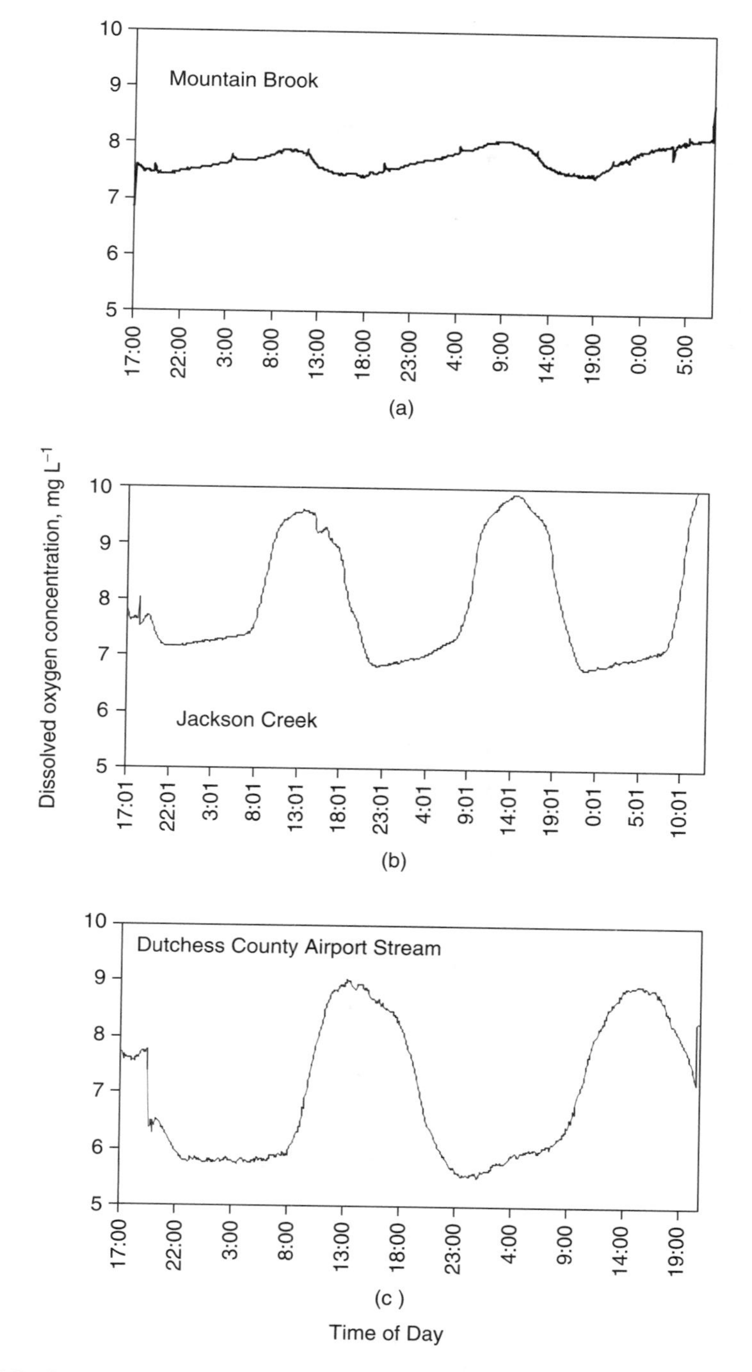

Figure 13.7 Dissolved oxygen profiles collected simultaneously, August 13–15, 2001 in (a) forested, (b) agricultural, and (c) suburbanized watersheds within the Wappingers and Fishkill drainages.

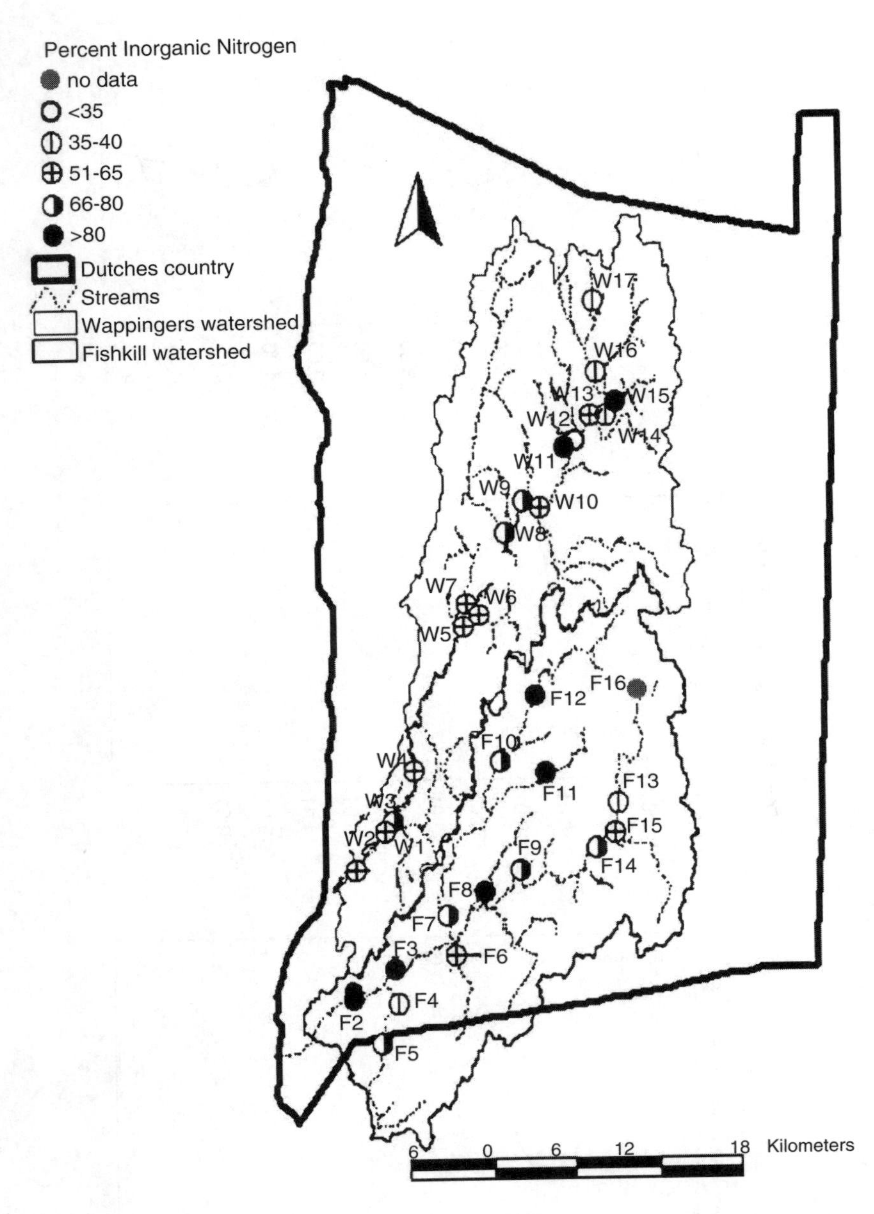

Figure 13.8 Mean percent inorganic nitrogen (May–August) measured at sites in the Wappingers and Fishkill watersheds, Dutchess County, New York.

with the Hudson River (Figure 13.9). This also corresponds to an elevation gradient, with the least-developed portions of the watersheds being the most remote headwater areas. We suspect that this will often be the case in urbanization studies; analysis of environmental metrics in relation to anthropogenic disturbance will have to be corrected for this strong geographic influence.

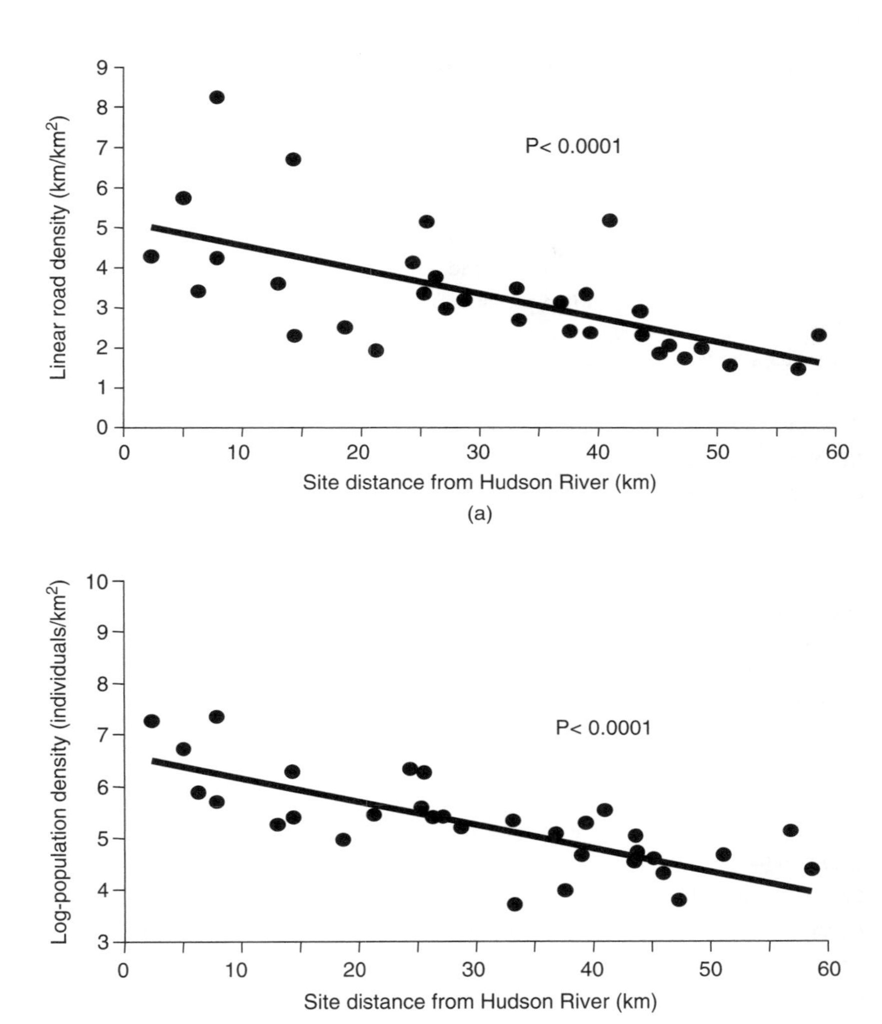

Figure 13.9 Plots of (a) road and (b) population densities versus distance from Hudson River for sites within the Wappingers and Fishkill watersheds, Dutchess County, New York.

In the analysis to date, only one fish community in the study area was severely depressed, and the rest, by the IBI index, vary widely. The reasons for this are complex but include factors such as degree of shading by riparian cover, benthic habitat type, and the presence of physical structures that might serve as attractive habitat.

One indicator of health that is comparable to past studies has been species richness. The New York State Conservation Department (NYSCD) undertook a comprehensive biological survey of watersheds across the state in the 1920s and 1930s, with the lower Hudson and its watershed being assessed last.[36] Schmidt and Kiviat also surveyed the Fishkill system in the 1980s and made comparisons of their findings with unpublished NYSCD data from the 1950s, 60s, and 70s where possible.[37] Tables 13.1 and 13.2

Table 13.1 Changes in Fish Species Present in the Fishkill Mainstem Over Time

Species	Common name	1930s	1950s	1980s	2001
Anguilla rostrata	American eel			x	x
Esox americanus americanus	redfin pickerel	x	x	x	
Ameiurus natalis	yellow bullhead				x
Ameiurus nebulosus	brown bullhead	x	x		x
Catostomus commersoni	white sucker	x	x	x	x
Carassius auratus	goldfish	x			
Cyprinus carpio	carp	x		x	
C. auratus X C. carpio	goldfish-carp hybrid	x			
Cyprinella spilopterus	spotfin shiner			x	x
Erimyzon oblongus	creek chubsucker	x			
Exoglossum maxillingua	cutlips minnow	x	x	x	x
Luxilus cornutus	common shiner	x	x	x	x
Notemigonus crysoleucas	golden shiner	x			x
Rhinichthys atratulus	blacknose dace	x	x	x	x
Rhinichthys cataractae	longnose dace	x		x	x
Semotilus atromaculatus	creek chub	x	x	x	x
Semotilus corporalis	fallfish	x		x	
Oncorhynchus mykiss	rainbow trout				
Salmo trutta	brown trout	x	x	x	x
Salvelinus fontinalis	brook trout			x	
Fundulus diaphanus	banded killifish			x	
Ambloplites rupestris	rock bass	x	x	x	x
Lepomis auritus	redbreast sunfish	x		x	x
Lepomis gibbosus	pumpkinseed	x	x	x	
Lepomis macrochirus	bluegill	x		x	x
L. gibbosus X L. auritus	pumpkinseed X redbreast	x			x
Micropterus dolomieu	smallmouth bass	x	x	x	
Micropterus salmoides	largemouth bass	x	x	x	x
Pomoxis nigromaculatus	black crappie	x			
Etheostoma olmstedi	tessellated darter	x	x	x	x
Percina peltata	shield darter	x			
Perca flavescens	yellow perch	x		x	x
Total number of species		**26**	**13**	**22**	**19**

Current study compared with NYSCD[36] and Schmidt and Kiviat.[37]

Table 13.2 Changes in Fish Species Present in Sprout Creek Over Time

Species	Common name	1930s	1960s	1980s	2001
Anguilla rostrata	American eel			x	x
Esox americanus americanus	redfin pickerel	x	x	x	x
Ameiurus nebulosus	brown bullhead	x			
Catostomus commersoni	white sucker	x	x	x	x
Erimyzon oblongus	creek chubsucker	x			
Exoglossum maxillingua	cutlips minnow	x	x	x	x
Luxilus cornutus	common shiner	x	x	x	x
Notemigonus crysoleucas	golden shiner	x			x
Rhinichthys atratulus	blacknose dace	x	x	x	x
Rhinichthys cataractae	longnose dace	x	x	x	x
Semotilus atromaculatus	creek chub	x	x	x	
Semotilus corporalis	fallfish	x	x		x
Oncorhynchus mykiss	rainbow trout				x
Salmo trutta	brown trout	x	x	x	x
Salvelinus fontinalis	brook trout	x			
Micropterus dolomieu	smallmouth bass	x			
Micropterus salmoides	largemouth bass	x	x		
Ambloplites rupestris	rock bass	x	x	x	
Lepomis auritus	redbreast sunfish	x	x	x	x
Lepomis macrochirus	bluegill	x	x	x	
Lepomis gibbosus	pumpkinseed	x	x	x	
Etheostoma olmstedi	tessellated darter	x	x	x	x
Perca flavescens	yellow perch	x			
Cottus cognatus	slimy sculpin	x	x		x
Total number of species		**22**	**16**	**14**	**14**

Current study compared with NYSCD[36] and Schmidt and Kiviat.[37]

compare these surveys with work completed in 2001 on the Fishkill mainstem and Sprout Creek, one of the largest tributaries in the Fishkill system. This characterization of fish community structure and change offers two main observations. First, the mix of species in any given time frame differs, reflecting natural ecological processes (e.g., competition, predation) as well as anthropogenic drivers (e.g., water quality alteration, species introductions). Second, there has been a general decline in species richness at all sites, with the greatest percent decline (36%) in the Sprout Creek. Overall, most of the change has occurred since the 1936 survey. The 1950s species minimum in the Fishkill Creek mainstem may have been associated with pollution from small industry, and the rebound in species numbers is likely associated with water quality improvements.

Further analysis and synthesis are in progress. The ecological data sets will be used to develop statistical relationships between ecological variables and land use, indicators of urbanization (e.g., catchment imperviousness)[38,39] and associated economic activities. These relationships can be displayed as graphs, and as suggested by Karr[40] become "ecological dose-response curves" that "show a measured biological response to the cumulative ecological exposure, or dose, of all events and human activities within a watershed." These statistical relationships will then form the basis of ecological indicators of economic and land-use scenarios. Early analysis has hinted at the complexity of the relationship between anthropogenic drivers and metrics of ecosystem health. Indicators will not likely point in the same direction, and that information may be the key to understanding how the upstream inputs are responding to human activities on the landscape.

SCENARIO ANALYSIS

To follow the progression within and between the three submodels described above, this section illustrates a scenario analysis based on growth in the Dutchess County semiconductor industry. During the first project workshop — with state and county planners, representatives of local and regional nongovernmental organizations, and technical advisors from academia and local research institutes — growth in the semiconductor industry was identified as a priority scenario. In particular, IBM was building a semiconductor plant in East Fishkill that would employ approximately 1000 people when operational in 2003. An existing building at IBM's Fishkill site was to be adapted and remodeled to house the chip plant, an activity expected to create another 1400 temporary construction jobs.[41] IBM was expected to invest $2.5 billion in the plant,[42] and the county was promoting the project by offering IBM tax exemptions valued at $475 million, as well as grants and loans worth $28.75 million.[43]

The first step in the scenario analysis is to identify the semiconductor industry within the GR-SAM and estimate its input–output and spatial linkages with other components of the regional economy. The semiconductor industry of Dutchess County was identified as a key sector based on both its relative size and its connectivity to regional industry and households.[11] In relative terms, it has the highest location quotient[44] in the county economy (a relative measure of importance of the industry locally as compared to the national average), followed closely by the related industries of computer peripheral equipment and electronic computers. By further combining location quotients with data on input–output linkages, keystone sectors are identified as those with both high location quotients and strong linkages to other high location quotient industries.[45] This method further highlights the semiconductor industry's relative importance to the Dutchess County economy. A final step in the keystone sector description delineates those sectors with strong forward linkages (i.e., selling proportionately larger amounts of inputs to other sectors within the region), as well as those with strong backward linkages (i.e., purchasing larger amounts of inputs from other sectors in the region's economy).[46] The semiconductor industry demonstrates above-average forward and backward linkage to other regional industries.

Within this context, the next step is to understand how an increase in employment in the semiconductor industry translates into countywide income and employment change, distribution of income amongst households and wage classifications, and generation of new households. An increase of 1000 new employees in the semiconductor industry multiplies into nearly 2300 economy-wide jobs in the model economy. The industry has the eleventh-highest earnings per employee ($79,604) in the county, contributing to a relatively large countywide income increase of over $238 million. About 70% of the impact is direct (due to semiconductor wages, profits, etc.), with 17% indirect (due to inputs from other sectors) and 13% induced (due to new household income expenditures). The top five industries affected are semiconductors (by a large margin), maintenance and repair facilities, eating and drinking (due primarily to induced household expenditures), wholesale trade, and personnel supply services.

Growth in employment by wage and household income category is next translated into new households. The cases of both local full employment and unemployment within the occupation classes were considered. The percentage of new commuters from outside the county economy was also considered when estimating new within-county household generation. Depending on the degree of local production assumed to fill new input demands from the semiconductor industry, new households are estimated to be between 1777 and 2051. The full employment scenario was assumed more likely as unemployment in the county is generally low, and IBM currently employs many high-skill workers at their other facilities. However, commuting may play a factor. The IBM facility in Poughkeepsie employs several people who live in neighboring Ulster County.

To locate new households, the binomial logit model can be employed to estimate the most likely tax parcels to convert to residential. The degree of new home construction will depend on the amount of real estate on the market, the distance workers are willing to drive to a new IBM plant, and other locational variables considered in the logit model. Using the GIS system, the high-probability vacant parcels can be identified within various radii from direct, indirect, and induced business locations in the GR-SAM. Various biophysical and zoning constraints can also define subscenarios that would shape development patterns.

To complete the scenario, new residential land use identified in the GIS is translated into land-use indices by use class and impervious surfaces in each subcatchment impacted. The correlation with human disturbance and ecosystem health is then drawn according to estimated statistical relationships. An ecological risk assessment can then be viewed in the full light of its primary determinants, spatial relationships, and potential for amelioration by economic and land-use policy.

MULTICRITERIA DECISION AID

A major challenge for effective watershed management is to direct and control economic growth while balancing the often conflicting goals of diverse stakeholders. Scenario analysis is illustrative of the potential economic, land-use, and ecological consequences of a particular development impact, and thus is a form of impact analysis (see "Complementary Analyses" in Chapter 5). By visualizing the accumulation of small decisions at the watershed scale, the approach portrayed in Figure 13.2, and illustrated with the semi-conductor scenario, is capable of informing a decision-making process. The next step is to frame the multiple attributes of scenarios — including descriptors of economic, land-use, and ecological change — into a decision-making framework capable of evaluating trade-offs and compromises in the formulation of management and policy alternatives.

To place scenario analysis in this decision-making context, a multi-criteria decision aid (MCDA) is under development to assist in identifying and prioritizing land use plans in the Wappingers and Fishkill watersheds. MCDA is a framework that is transparent to decision-makers, adaptable to many situations across multiple metrics and scales, and amenable to both expert and local stakeholder pools of knowledge. An explicit consideration of multiple objectives and viewpoints is in contrast to decision frameworks that seek to maximize one objective or reach an optimal solution from the perspective of a "representative" decision-maker. MCDA attempts to structure this complexity, as opposed to conventional economic tools such as cost–benefit analysis that seek to reduce complexity to a single dimension, unit, and value system.[47]

Figure 13.10 illustrates the typical hierarchy of multicriteria decision problems including goal, decision alternatives, general criteria, and specific indicators. In Dutchess County, one of the chief architects of goal formation at the watershed level has been

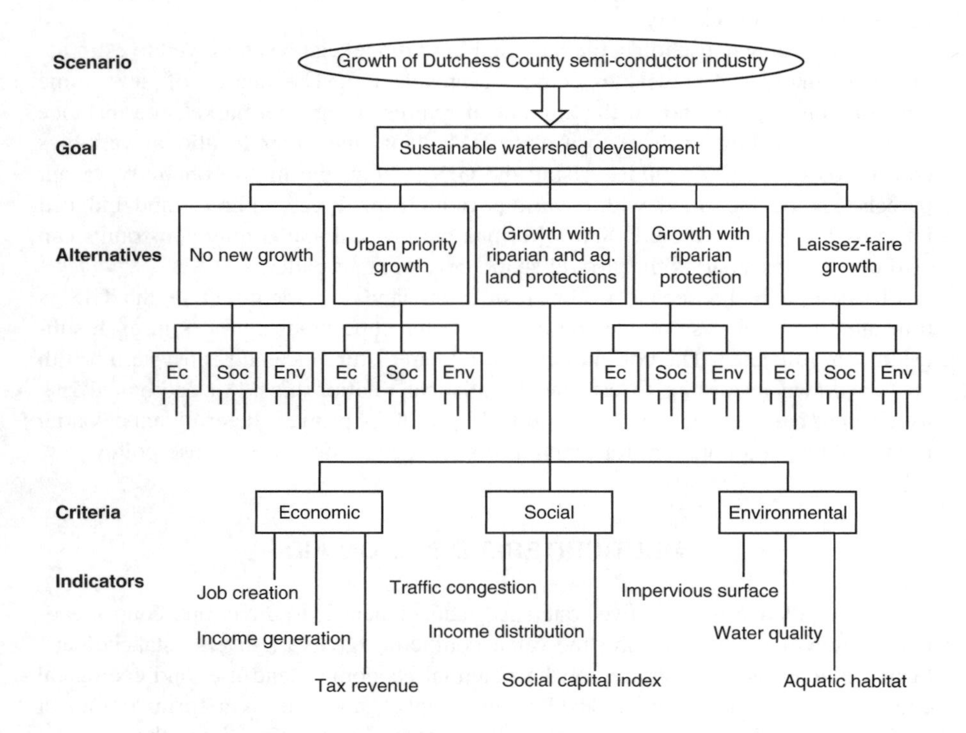

Figure 13.10 Decision hierarchy in watershed management problem.

the Dutchess County Environmental Management Council (EMC), a not-for-profit organization focused on providing research based, nonadvocacy educational resources to the community. Funded through the County and with third-party grants, the EMC works with volunteers — including members of 21 town Conservation Advisory Commissions and Conservation Boards, 11 at-large members appointed by the Dutchess County Legislature, and other interested community members — to identify, research, and prioritize environmental goals. One of the functions of the EMC is to coordinate watershed management between municipalities. A key planning body within the Wappingers watershed is the Wappingers Creek Intermunicipal Council (WIC), formed in 2000 by the umbrella Wappingers Creek Watershed Planning Council with the express goal of fostering intermunicipal cooperation in land-use decision-making.

With this charge the WIC, composed of representatives from all 13 municipalities in the watershed, has been meeting throughout 2003 to establish planning goals, craft a shared vision for their joint future, and make specific policy recommendations to their constituents. Figure 13.10 is an example of a decision hierarchy under development, which the integrated modeling effort can inform. Each constituency of the watershed has a different vision of how something like the semiconductor growth scenario might play out on the landscape to meet a goal of sustainable watershed management. The alternatives range from a rejection of the residential housing growth stimulated by an IBM expansion (an unrealistic option at this point) to letting the growth occur where it may (completely undirected according to estimated development priorities). In between these two extremes are a series of directed growth alternatives, including prioritizing urban in-fill, protecting riparian buffers and current agricultural land, or pursuing riparian buffers alone. Specific policy instruments to achieve these alternatives — for instance zoning, tax incentives, or purchase of development rights — might become part of a secondary MCDA, informed by the goals set by the first.

The future outcome of each of these alternatives is then characterized by a suite of indicators, broadly captured in Figure 13.10 as economic, social, and ecological criteria. For example, economic criteria may include job creation, income generation, and tax revenue estimated by the input–output model. Social criteria may include traffic congestion, income distribution, and a social capital index estimated by the social accounts and the probit model of development. Ecological criteria may include impervious surface, water quality, and aquatic habitat indices estimated by the probit model scenarios and spatially correlated biotic and chemical data. Each criterion can be measured in its own units (both quantitative and qualitative) and dimensions (both spatial and temporal), each evaluated by a particular stakeholder position.

These economic, social, and ecological criteria are all at the societal level, and try to anticipate the accumulation of separate multicriteria problems solved by each individual in the system. For example, the costs to the individual of longer commute times due to traffic congestion may be less than the benefits of a larger, more-inexpensive home in the suburbs. However, the watershed-wide MCDA problem tries to anticipate the accumulation of hundreds of these individual decisions and to evaluate the tradeoffs at the societal level. Additionally, a criterion such as impervious surface may itself capture criteria such as green space, stormwater run-off, and flood potential that are set aside to help simplify the comparisons. However, water quality, which also relates to impervious surfaces, may be evaluated on its own merits depending on the priorities of stakeholders.

To quantify these trade-offs, our research team is conducting two workshops in the spring of 2004 to help structure the MCDA problem: the first with municipal representatives from the WIC, and the second with representatives from various stakeholder groups in the county (as a follow-up to our first project workshop). Once the MCDA problem is structured, the next step is to elicit the preferences of the stakeholders using one of several methods within the family of MCDA frameworks. We have selected PROMETHEE (Preference Ranking Organization METHod of Enrichment Evaluation)[48] and the associated Decision Lab 2000 software package[49] after reviewing current models for flexibility to handle indifference and uncertainty, ease of use and understanding in a workshop setting, and the ability to visualize a group-based process of goal-setting and compromise.[47]

PROMETHEE requires criteria-specific and stakeholder-identified information, including: (1) choice of maximizing or minimizing, (2) weight of importance to the overall decision, (3) preference function that translates quantitative or qualitative metrics to consistent rankings, and (4) various decision threshold parameters for each function (for example, indifference thresholds identify ranges where a decision-maker cannot clearly distinguish his or her preferences). This exercise is carried out by each stakeholder in a decision problem. During sensitivity analysis, criteria weights, preference functions, and decision thresholds can all be varied to estimate stability intervals for the rankings of alternatives and to evaluate both imprecision of criterion measurement and uncertainty of preference. The outcome of PROMETHEE includes both complete and partial rankings (depending on the incomparability of decision alternatives) and both pairwise and global comparisons of decision alternatives. Global comparisons can be illustrated with GAIA (Graphic Analysis for Interactive Assistance) plane diagrams that represent a complete view of the conflicts between the criteria, characteristics of the actions, and weighing of the criteria.

With multiple stakeholders, MCDA analyses can be used to visualize conflict between stakeholder positions and opportunities for compromise, alliances, and group consensus, or to revisit and redefine the goal, alternatives, and criteria themselves.[50] In a group context, the entire MCDA process has been described as a group decision support system (GDSS), and examples of its use can be found in resource planning and management, forest management, watershed planning, public policy planning, pollution cleanup, transportation planning, and the siting of industrial and power facilities.[51–54]

The advent of spatial decision-support systems (SDSS) — the family of MCDA models that our study most closely contributes — provides an important new opportunity for the evolution of MCDA methods and applications.[55,56] Examples where SDSS and GDSS have been used together include an examination of riparian revegetation options in North Queensland, Australia,[57] land-use conflict resolution involving fragile ecosystems in Kenya,[58] watershed management in Taiwan,[59] housing suitability in Switzerland,[60] and water quality issues in Quebec.[55]

DISCUSSION

The complexities involved in economic and watershed systems are enormous. Both are evolutionary systems characterized by nonlinearities, historical contingencies, and pure uncertainty. The task of analyzing either of these systems alone would be daunting. Economic analysis is particularly hamstrung by a long history of reliance on static equilibrium models that have proved to be of limited value in modeling evolutionary change. Models of land-use change typically ignore any connection to the economic system. Similarly, ecological studies focus on point estimates of current conditions, which are divorced from landscape and economic change. Granted, there are many key gaps in knowledge and data preventing accurate forecasts of the ecological response to land-use change.[61] However, the goal of this study is to begin to integrate disciplinary expertise for the pieces of the puzzle to help visualize and inform current stakeholder decision-making processes at the watershed scale.

These pieces have been extensively used elsewhere, and thus the modeling framework doesn't avoid the limitations of any one piece. Social accounting is most limited by the static nature of characterizing the economy, lack of regional data, and lag in national structural data. The probit analysis captures only broad relationships between residential land use and assessment and census data, and is also lacking in time series data and any supply and demand dynamics in the real estate market. The chemical and biological assessment is similarly a shapshot in time and depends on the interpolation of satellite imagery to tease out relations between aquatic variables and landscape characteristics. Each component also comes with its own degree of uncertainty and variability, which is compounded when the economic, land-use, and ecological spheres of analysis are integrated. In separate publications, we deal extensively with varying model assumptions, results of Monte Carlo experiments, and variability analysis.[11,19,62]

Our integrated modeling and evaluation approach is in step with the conceptual approach outlined in Figure 9.1 and Figure 9.4. Assessment planning and problem formulation led to the integrated conceptual model of Figure 13.1 and the generation of particular development scenarios (illustrated by the semiconductor industry scenario). As shown in Figure 13.11, analysis and characterization of alternatives via economic, social and ecological criteria is the explicit outcome of the three submodels comprised by our approach (Figure 13.2). Comparison of decision alternatives under various economic change scenarios is conducted through an MCDA outranking procedure. Although not described in detail here, the MCDA can potentially lead to a negotiated decision on intermunicipal watershed cooperation. Consultation with both expert and stakeholder peer communities has occurred throughout model development, scenario characterization, alternative development, and criteria measurement with the aim to develop a decision tool amenable to adaptive management, evaluating changing conditions, priorities, and goals.

The social accounting approach proved to be flexible enough to capture the major economic drivers of Dutchess County and was amenable to the scenario approach crucial to this study. Collecting and analyzing land-use data was straightforward and proved to be a reliable way to link economic and ecosystem changes. Ecological metrics are proving complex, but some have shown promise in capturing the relationship

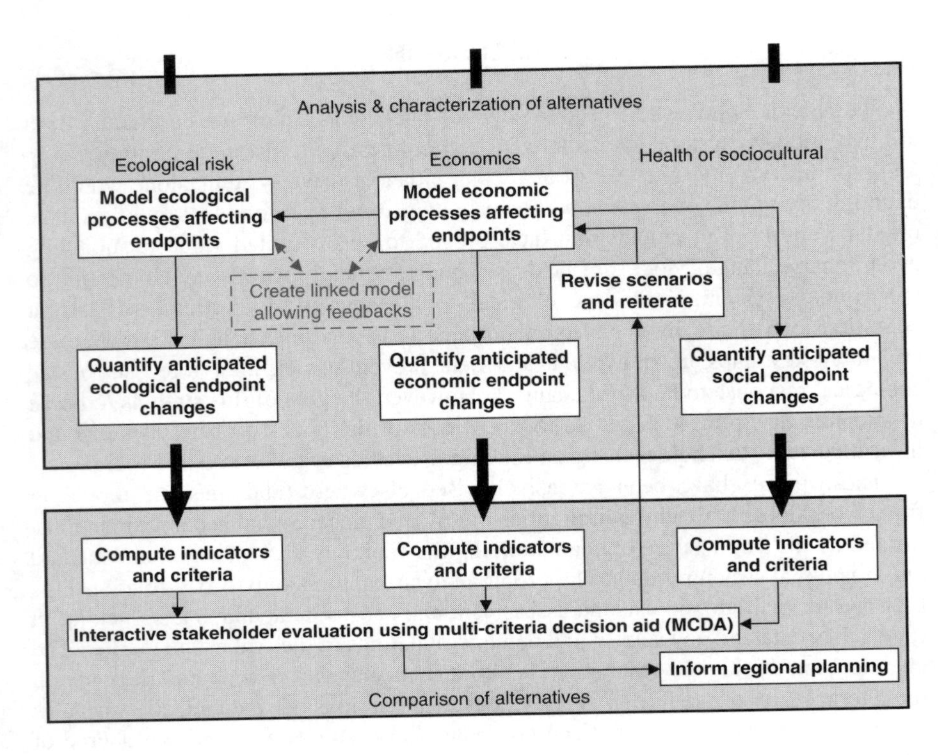

Figure 13.11 Techniques used for analysis, characterization, and comparison of management alternatives in the Hudson River watershed, as compared to the example shown in Figure 9.4. White boxes and bold type show features included in this analysis.

between land-use patterns and the biological health of the streams studied. Where this study falls short of the process outlined in Figure 9.4 is estimating any potential feedbacks from the ecological impacts to the social and economic drivers. For example, water quantity or quality could eventually become a limiting factor to locating new industry and constructing new homes.

Within the field of sustainability studies, few support traditional economic analysis as the sole guidepost for societal planning. Rather, a pluralistic view that espouses different perspectives, analytic frameworks, and metrics is seen as a more robust means to anticipate the future.[63] In addition, anticipating the future means that one should also anticipate surprise, with the practical implication of building some buffering capacity into the system. On the ground, this may translate into decisions such as not to build out completely, but rather to preserve some areas in anticipation of unspecified change, for example due to climate. This study provides a framework upon which to build a transparent model that can illuminate interconnections among economy, society, and ecosystems, and provide a basis for planning decisions. To emerge from under the tyranny of small decisions will require such tools that envision long-run change, help to shape shared community goals, and encourage dialogue between local and credentialed expertise.

ACKNOWLEDGMENTS

This research was made possible by a grant from the Hudson River Foundation entitled "Modeling and Measuring the Process and Consequences of Land Use Change: Case Studies in the Hudson River Watershed."

REFERENCES

1. Kahn, A., The tyranny of small decisions: Market failures, imperfections, and the limits of economics, *Kyklos*, 19, 23, 1966.
2. Stanne, S., Panetta, R.G., and Forist, B.E., *The Hudson: An Illustrated Guide to the Living River*, Rutgers University Press, New Brunswick, NJ, 1996.
3. Dutchess County Department of Planning and Development, Dutchess County major employers, 1997–2000, Dutchess County Department of Planning and Development, 2000. Accessed June, 2001 at http:///web.archive.org/web/20010628232214/http://dutchessny.gov/mjr-lst.html.
4. Phillips, P.J. and Handchar, D.W., Water-quality assessment of the Hudson River Basin in New York and adjacent states: Analysis of available nutrient, pesticide, volatile organic compound, and suspended-sediment data, 1970–1990, Water-Resources Investigations Report 96-4065, U.S. Geological Survey, Troy, NY, 1996.
5. Real Estate Center, Dutchess County, NY single-family building permits, Real Estate Center, 2000. Accessed July, 2001 at http://web.archive.org/web/20010708001306/http://recenter.tamu.edu/Data/bpm/sfm2281a.htm.
6. Lynch, E., Merchants cheer, but some residents wary of growth, Poughkeepsie Journal, Oct. 11, 2000. Accessed June, 2001 at http://web.archive.org/web/20010419010123/http://www.poughkeepsiejournal.com/projects/ibm/bu101100s3.htm.
7. Dutchess County Department of Planning and Development, Income and retail, Dutchess County Department of Planning and Development, 1997. Accessed July, 2001 at http://web.archive.org/web/20020103160106/www.dutchessny.gov/i-r.html.
8. Town of Red Hook, Southern gateway small area plan, Town of Red Hook, 2002. Accessed May, 2002 at http://web.archive.org/web/20020610132940/www.redhook.org/gateway/GatewayIntro.html.
9. Stone, R., Demographic input–output: An extension of social accounting, in *Contributions to Input-Output Analysis, Volume 1*, Carter, A.P. and Brody, A., Eds., North-Holland Publishing, Amsterdam, 1970.
10. Pyatt, G. and Round, J., *Social Accounting Matrices: A Basis for Planning*, World Bank, Washington, D.C., 1985.
11. Nowosielski, A., Geo-Referenced Social Accounting With Application to Integrated Watershed Planning the Hudson River Valley, Ph.D. Dissertation, Department of Economics, Rensselaer Polytechnic Institute, Troy, NY, 2002.
12. IMPLAN, Home page, Minnesota Implan Group, Inc., 2003. Accessed Feb. 4, 2004 at http://www.implan.com.
13. Rose, A., Stevens, B., and Davis, G., *Natural Resource Policy and Income Distribution*, Johns Hopkins University Press, Baltimore, MD, 1988.
14. Victor, P., *Pollution: Economy and the Environment*, Allen and Unwin, London, UK, 1972.
15. United Nations, *The System of Integrated Environmental and Economic Accounts*, United Nations, New York, 1993.

16. Lange, G., From data to analysis: The example of natural resource accounts linked with input–output information, *Econ. Syst. Res.*, 10, 113, 1998.
17. Duchin, F., *Structural Economics*, Island Press, Washington, D.C., 1998.
18. Bockstael, N.E., Modeling economics and ecology: The importance of a spatial perspective, *Am. J. Agric. Econ.*, 78, 1168, 1996.
19. Polimeni, J., A Dynamic Spatial Simulation of Residential Development in the Hudson River Valley, New York State, Ph.D. Dissertation, Department of Economics, Rensselaer Polytechnic Institute, Troy, NY, 2002.
20. Rapport, D.J., Evaluating ecosystem health, *J. Aquat. Anim. Health*, 1, 15, 1992.
21. Shrader-Frechette, K.S., Ecosystem health: A new paradigm for ecological assessment?, *Trends. Ecol. Evol.*, 9, 456, 1994.
22. Rapport, D.J., Gaudet, C.L., Calow, P., and Eds., *Evaluating and Monitoring the Health of Large-Scale Ecosystems*, Springer-Verlag, Berlin, 1995.
23. Rapport, D.J., Gaudet, C.L., Karr, J.R., et al., Evaluating landscape health: Integrating societal goals and biophysical processes, *J. Environ. Manage.*, 53, 1, 1998.
24. Karr, J.R., Assessment of biotic integrity using fish communities, *Fisheries*, 6, 21, 1981.
25. Karr, J.R., Biological integrity: A long-neglected aspect of water resource management, *Ecol. Appl.*, 1, 66, 1991.
26. Daniels, R.A., Riva-Murray, R., Halliwell, D.B., Vana-Miller, D.L., and Bilger, M.D., An index of biological integrity for northern Mid-Atlantic slope drainages, *Trans. Amer. Fish. Soc.*, 131, 1044, 2002.
27. Bode, R.W., Novak, M.A., and Abele, L.E., Twenty-year trends in water quality of rivers and streams in New York State based on macroinvertebrate data, 1972–1992, Technical Report, New York State Department of Environmental Conservation, Albany, NY, 1993.
28. Odum, H.T., Primary production in flowing waters, *Limnol. Oceanogr.*, 1, 102, 1956.
29. Bott, T.L., Primary productivity and community respiration, in *Methods in Stream Ecology*, Hauer, F.R. and Lamberti, G.A., Eds., Academic Press, San Diego, 1996, 533.
30. Limburg, K.E. and Schmidt, R.E., Patterns of fish spawning in the Hudson River watershed: Biological response to an urban gradient?, *Ecology*, 71, 1238, 1990.
31. Parsons, T.L. and Lovett, G.M., Land use effects on Hudson River tributaries, Ch. 9 in Tibor T. Fellowship Program 1991, Final Reports, Hudson River Foundation, New York, 1992.
32. Wahl, M.H., McKellar, H.N., and Williams, T.M., Patterns of nutrient loading in forested and urbanized coastal streams, *J. Exp. Mar. Biol. Ecol.*, 213, 111, 1997.
33. Wall, G.R., Riva-Murray, K., and Phillips, P.J., Water quality in the Hudson River Basin, New York and adjacent states, 1992–95, *U.S. Geol. Surv. Circ.*, 1165, 1998.
34. Bunn, S.E., Davies, P.M., and Mosisch, T.D., Ecosystem measures of river health and their response to riparian and catchment degradation, *Freshw. Biol.*, 41, 333, 1999.
35. Fitzpatrick, F.A., Waite, I.R., D'Arconte, P.J., et al., Revised methods for characterizing stream habitat in the national water-quality assessment program, Water Resources Investigations Report 98-4052, U.S. Geological Survey, 1998.
36. NYSCD, A biological survey of the lower Hudson watershed, Supplement to twenty-sixth annual report, Biological Survey No. X1, New York State Conservation Department, 1936.
37. Schmidt, R.E. and Kiviat, E., Environmental quality of the Fishkill Creek drainage, a Hudson River tributary, Submitted to the Hudson River Fishermen's Association and the Open Space Institute, Hudsonia Limited, Bard College, Annandale, NY, 1986.
38. Klein, R.D., Urbanization and stream quality impairment, *Water Res. Bull.*, 15, 948, 1979.
39. Wang, L., Lyons, J., Kanehl, P., and Bannerman, R., Impacts of urbanization on stream habitat and fish across multiple spatial scales, *Environ. Manage.*, 28, 255, 2001.

40. Karr, J.R., Defining and measuring river health, *Freshw. Biol.*, 41, 221, 1999.
41. Dutchess County Economic Development Corporation, Dutchess County continues to capture attention of corporate and business site-selection pros, Press Release, Dutchess County Economic Department Corporation, May 21, 2001. Accessed Sept., 2001 at http://web.archive.org/web/20011021101006/www.dcedc.com/preleases/News.html.
42. Schantz-Feld, M. R., Destinations: New York, area development online, Area Development Site and Facility Planning Online, 2001. Accessed June, 2001 at http://web.archive.org/web/20010624192316/http://www.area-development.com/destination/newyork.html.
43. Lyne, J., IBM's cutting-edge $2.5 billion fab reaps $500 million in NY incentives, Interactive Publishing, 2000. Accessed Feb. 4, 2004 at http://www.conway.com/ssinsider/incentive/ti0011.htm.
44. Watkins, A.J., *The Practice of Urban Economics*, Sage Publications, Beverly Hills, CA, 1980.
45. Kilkenny, M. and Nalbarte, L., Keystone sector identification: A graph theory-social network analysis approach, The Web Book of Regional Science, Regional Research Institute, West Virginia University, 1999. Accessed Feb. 4, 2004 at http://www.rri.wvu.edu/WebBook/Kilkenny/editedkeystone.htm.
46. Sonis, M., Hewings, G.J.D., and Guo, J., A new image of classical key sector analysis: Minimum information decomposition of the Leontief Inverse, *Econ. Syst. Res.*, 12, 401, 2000.
47. Hermans, C. and Erickson, J.D., Multicriteria decision-making tools and analysis: Implications for environmental management, Working Paper, unpublished, Rubenstein School of Environment and Natural Resources, University of Vermont, Burlington, VT, Available from authors.
48. Brans, J.P., Vincke, P., and Mareschal, B., How to select and how to rank projects: The PROMETHEE method, *Eur. J. Oper. Res.*, 28, 228, 1986.
49. visual Decision, Home page, Visual Decision, May 1, 2001. Accessed Feb. 4, 2004 at http://www.visualdecision.com.
50. Macharis, C., Brans, J.P., and Mareschal, B., The GDSS PROMETHEE procedure, *J. Decision Syst.*, 7, 283, 1998.
51. Hokkanen, J., Lahdelma, R., and Salminen, P., Multicriteria decision support in a technology competition for cleaning polluted soil in Helsinki, *J. Environ. Manage.*, 60, 339, 2000.
52. Iz, P.H. and Gardiner, L.R., A survey of integrated group decision support systems involving multiple criteria, *Group Decis. Negot.*, 2, 61, 1993.
53. Malczewski, J. and Moreno-Sanchez, R., Multicriteria group decision making model for environmental conflict in the Cape region, Mexico, *J. Environ. Plann. Manage.*, 40, 349, 1997.
54. Van Groenendaal, W.J.H., Group decision support for public policy planning, *Infor. Manage.*, In press, 2004.
55. St-Onge, M.N. and Waaub, J.P., Geographic tools for decision making in watershed management, Report, Department of Geography, University of Quebec, Montreal, Canada, 1998.
56. Malczewski, J., *GIS and Multicriteria Decision Analysis*, John Wiley & Sons, New York, 1999.
57. Qureshi, M.E. and Harrison, S.R., A decision support process to compare riparian revegetation options in Scheu Creek Catchment in North Queensland, *J. Environ. Manage.*, 62, 101, 2001.
58. Mwasi, B., Land use conflict resolution in a fragile ecosystem using multi-criteria evaluation and a GIS-based Decision Support System (GDSS), Nairobi, Kenya, International Conference on Spatial Information for Sustainable Development, Oct. 2, 2001.

59. Yeh, C.H. and Lai, J.H., The study of integrated model of geographical information systems and multiple criteria decision making for watershed management, Ann Arbor, ME, Association of Chinese Professionals in GIS, Proceedings of Geoinformatics' 99 Conference, Mar. 9, 1999.
60. Joerin, F. and Musy, A., Land management with GIS and multicriteria analysis, *Int. Trans. Oper. Res.*, 7, 67, 2000.
61. Nilsson, C., Pizzuto, J.E., Moglen, J.E., et al., Ecological forecasting and the urbanization of stream ecosystems: Challenges for economists, hydrologists, geomorphologists, and ecologists, *Ecosystems*, 6, 659, 2003.
62. Stainbrook, K., Using Ecological Indicators to Detect Environmental Change in Urbanizing Watersheds: A Case Study in Dutchess County, NY, Masters Thesis, College of Environmental Science and Forestry, State University of New York, Syracuse, NY, 2004.
63. Limburg, K.E., O'Neill, R.V., Costanza, R., and Farber, S., Complex systems and valuation, *Ecol. Econ.*, 41, 409, 2002.

CHAPTER 14

Determining Economic Trade-Offs among Ecological Services: Planning for Ecological Restoration in the Lower Fox River and Green Bay

Jeffrey K. Lazo, P. David Allen II, Richard C. Bishop, Douglas Beltman, and Robert D. Rowe

CONTENTS

1-56670-639-4/05/$0.00+$1.50

BACKGROUND

For decades, governmental agencies, academics, environmental groups, and industries have struggled with the dilemma presented by widespread pollution in the Lower Fox River and Green Bay in Wisconsin and Michigan. Pollution problems in Green Bay are mirrored at sites throughout the Great Lakes and around the country. The U.S. Fish and Wildlife Service (USFWS) led an effort in Green Bay to use natural resource damage assessment (NRDA) authorities under the Superfund Law (the Comprehensive Environmental Response, Compensation, and Liability Act, or CERCLA) to address the problem of polychlorinated biphenyl (PCB) contamination and to compensate the public through natural resource restoration[a] for the harm the PCBs caused to natural resources. This effort was designed to crystallize decision-making with a transparent public process. It relied on millions of dollars worth of scientific and economic research conducted over many decades, and was directly relevant to injury determination or ecological risk assessment[b] and economic benefits analysis. The results of this effort ultimately will be decided through complex negotiations or litigation between (1) the parties responsible for pollution, (2) the governmental agencies charged with cleaning up and restoring the environment and (3) the public represented by the agencies.

The USFWS undertook the Green Bay NRDA to determine the extent that Lower Fox River PCBs were causing environmental harm in Green Bay, the extent that practical action could prevent further harm, and the extent that practical restoration could compensate for losses that cannot be realistically prevented. This chapter focuses on a total value equivalency (TVE) study conducted on behalf of the USFWS as part of the NRDA to elicit public preferences and values regarding PCB cleanup, other kinds of environmental restoration that would benefit the natural resources of Green Bay, and the trade-off between them. The USFWS used the results to determine

[a] *Restoration* is a term used in NRDAs referring to "restoration, rehabilitation, replacement, or acquisition of the equivalent." Under the statute and regulations, restoration is distinct from *cleanup* or *remediation*; however, restoration can include actions similar to cleanup to break the pathway between releases and natural resource injuries (42 U.S.C. 9601 *et seq* and 43 CFR Part 11).

[b] *Injury determination* is a term used in NRDAs for the process of determining what (and to what extent) adverse effects result from the release of hazardous substances. *Ecological risk assessment* is a term used in Superfund cleanups for the process of determining the likelihood (and uncertainty) of adverse ecological effects occurring. Though treated separately under the statutes, these processes are conceptually similar. Therefore, we use *injury determination* and *ecological risk assessment* interchangeably throughout this chapter, and we use *injury* synonymously with *adverse effects.*

the appropriate magnitude of environmental restoration required to make the public whole[a] for the PCB problem in Green Bay.

The Lower Fox River and Green Bay Watershed

The Lower Fox River basin is located in northeastern Wisconsin and encompasses a 638 square mile drainage basin (Figure 14.1). The Lower Fox River, which empties a larger drainage basin of over 6000 square miles (including drainage from the Wolf River and Upper Fox River basins), empties into Green Bay. The Bay of Green Bay includes waters of Michigan and Wisconsin, and receives waters from major tributaries in both states, including the Lower Fox River, the Peshtigo River, the Menominee River, the Escanaba River, and the Whitefish River. There is also significant interchange of water with Lake Michigan between Wisconsin's Door Peninsula and Michigan's Garden Peninsula. The Lower Fox River and Green Bay form a unique and important part of the larger Lake Michigan and Great Lakes ecoregion. The various habitats of the Lower Fox River and Green Bay environment support a wide diversity of fish and wildlife, including a number of rare, threatened, and endangered species.

River habitats are found in the Lower Fox River and in tributaries to Green Bay. Sandbars and estuaries characterize the western and southern shores of Green Bay, rocky steep shorelines are typical of the eastern shore, and cold, deep waters characterize the open waters of outer Green Bay. This diversity of habitats supports many fish species at different trophic levels, including trout, salmon, walleye, perch, pike, bass, and chub. The fishery resource, one of the most productive in the Great Lakes, is critical to the Green Bay food web because it provides food for the region's many piscivorous (i.e., fish-eating) fish and wildlife. Birds and mammals that depend on the fishery resource for food include eagles, terns, herons, gulls, waterfowl, cormorants, otter, and mink.

Situated on one of the major bird migration routes in North America, the Mississippi Flyway, the Lower Fox River and Green Bay environment provides essential habitat for large populations of breeding and migratory birds. At least 16 species listed by Wisconsin, Michigan, or the federal government as threatened or endangered are found in the area, including the bald eagle, peregrine falcon, great egret, Caspian tern, and Forster's tern.[1] Human uses of the Green Bay and Lower Fox River resources include waterfowl hunting; recreational, commercial, and subsistence fishing; bird watching; boating; tourism; and tribal cultural uses.

The PCB Problem in Green Bay

From 1954 until 1971, the Monsanto Corporation sold tens of thousands of tons of PCBs, in the form of the commercial mixture Aroclor 1242, to the National Cash Register Company (NCR). NCR used Aroclor 1242 at facilities on the Lower Fox

[a] The phrase *make the public whole* is a term of art in common law and the statutes that authorize NRDA. It refers to sufficient relief to the plaintiff for damages caused by the defendants. The primary goal of the Green Bay NRDA was to make the public (plaintiff) whole through restoration (relief) for PCB-caused losses (damages).

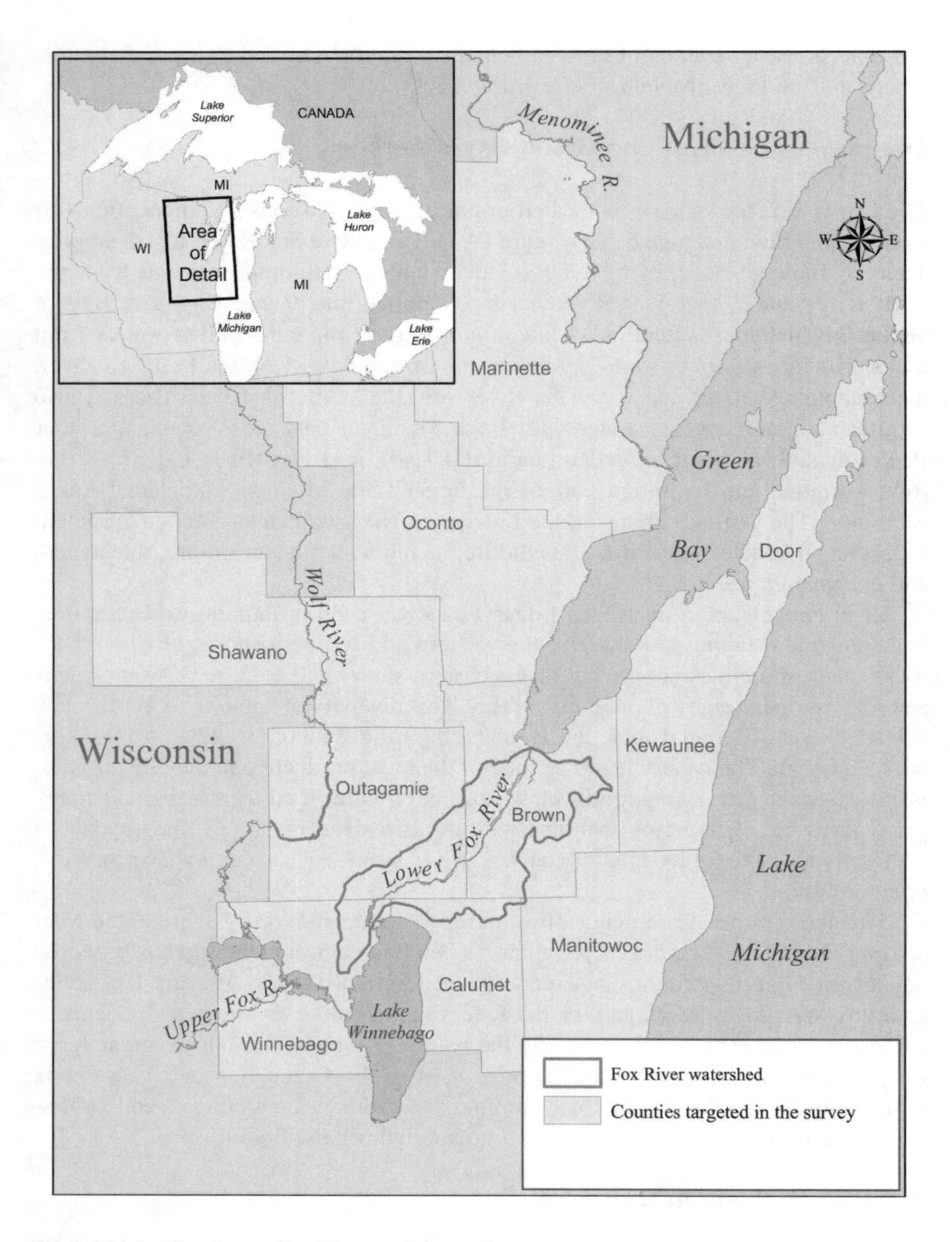

Figure 14.1 The Lower Fox River and Green Bay.

River and elsewhere to manufacture carbonless copy paper, which was sold and widely distributed in the United States.[1] The greatest concentration of paper mills in the world is located along the Lower Fox River. Several of these mills pioneered techniques to de-ink and recycle carbonless copy paper for use in low-grade paper products, such as tissue. Until the era of the Clean Water Act in the 1970s and 1980s, both the manufacture and recycling of carbonless copy paper caused hundreds of thousands of pounds of PCBs to be released into the Lower Fox River, most of which settled into sediments. Enhanced wastewater treatment in the late 1970s and 1980s eliminated paper mill pipes as a major source of pollution, but it did nothing to address the highly persistent PCBs that had already settled into millions of cubic yards of sediment.

The discovery of high levels of organochlorines in the biota of Green Bay in the 1960s, of PCBs in Lower Fox River sediments in the 1970s, and of adverse effects in Green Bay biota in the 1980s led to efforts in the late 1980s and 1990s to track the movement and effects of PCBs within and between the Lower Fox River, Green Bay, and Lake Michigan. Because of these efforts, it is now known that most of the PCBs released to the Lower Fox River have already moved into Green Bay, Lake Michigan, or beyond; hundreds of additional pounds of PCBs are scoured from Lower Fox River sediments each year.

Natural Resource Damage Assessment

CERCLA and the Oil Pollution Act (OPA) give natural resource trustees the authority to assess and restore natural resources that have been injured[a] by hazardous substances — in other words, to conduct an NRDA. Under both CERCLA and OPA, trusteeship is delegated to federal, state, and tribal agencies.[b]

The Department of the Interior, acting through the USFWS; the Department of Commerce, acting through the National Oceanic and Atmospheric Administration (NOAA); the Menominee Indian Tribe of Wisconsin; the Oneida Tribe of Indians of Wisconsin; and the states of Michigan and Wisconsin are currently working together as cotrustees on the Lower Fox River and Green Bay NRDA. These entities are also working together with the U.S. Environmental Protection Agency (USEPA) as the Intergovernmental Partnership to address cleanup and restoration of the site and resolution of liability at once.

The purpose of the NRDA is to restore natural resources and services and to provide compensation to the public for their loss. Restoration includes the return of natural resources or the services they provide to the baseline condition that would have existed absent the release of the hazardous substances in question, and compensation for losses that occur in the interim. Determining needed restoration and compensation requires assessing natural resource injuries, assessing realistic restoration opportunities, and scaling the latter to the former. *Scaling* thus refers to determining the appropriate combination, and "scale," of restoration and compensation programs — including their location, magnitude, and timing — relative to characteristics of the injuries.

Eliminating the injuries is the primary goal of both cleanup and restoration. However, in some cases this is not practical because of high environmental cleanup costs or practical engineering limits. Instead, compensation for the injured resources

[a] Injuries are defined in federal regulations at 43 CFR §11.14(v) as "a measurable adverse change, either long- or short-term, in the chemical or physical quality or the viability of a natural resource resulting either directly or indirectly from exposure to a discharge of oil or release of a hazardous substance, or exposure to a product of reactions resulting from the discharge of oil or release of a hazardous substance."
[b] Federal regulations at 40 CFR §300, 600.

Table 14.1 Overview of the Green Bay NRDA Process

Step 1: Determining Injuries
Releases and liability: PCBs from Lower Fox River paper mills (pathway determination)
Pathways: Lower Fox River to Green Bay and Lake Michigan (pathway determination)
Injuries: surface waters, fishery resources, and avian resources (injury determinations)
Step 2: Determining Restoration Opportunities
Restoration planning criteria: initial Restoration and Compensation Determination Plan (iRCDP)
Universe of restoration projects and ideas: Restoration and Compensation Determination Plan (RCDP)
Categories of acceptable restoration projects under the criteria: RCDP
Step 3: Scaling Restoration to Injuries
Amount of future restoration (e.g., wetland restoration) equivalent in value to different timelines of PCB-caused injuries (e.g., 20 more years of continuing injuries): TVE study
Step 4: Producing a Comprehensive Damage Claim
Past: value of recreational angling and enjoyment lost because of fish consumption advisories, from enactment of CERCLA to present: recreational fishing valuation study
Future: cost of restoration (scaled by the TVE study) to offset future losses: (RCDP)
Assessment costs: the reasonable costs incurred by the trustees to conduct the natural resource damage assessment

or services may be required. Additionally, in some cases, such as Green Bay, it may not be possible to restore exactly the same services as those lost to PCBs, so equally valued alternative natural resource services may need to be restored. Under the statute and regulations, these various possibilities are recognized by the phrase "restoration, rehabilitation, replacement, or acquisition of the equivalent."[a]

The measure of damages includes the cost of restoration to baseline plus the value of interim losses until baseline is achieved plus the reasonable costs of assessment. Therefore, trustees can weigh the increased costs of accelerated or more complete baseline restoration, which is preferred under the statute, against the increased amount of interim lost value should baseline restoration be incomplete or slower, perhaps to avoid restorations that are unrealistic, cost ineffective, or grossly disproportionate in cost compared to value. The TVE approach is designed to handle complicated, but not atypical, situations where natural resources cannot simply be fixed through elimination of hazardous substances, or increased in abundance through replacement.

Table 14.1 gives an overview of the Green Bay NRDA process. The Department of the Interior (DOI) has drafted regulations, which have not yet been formally proposed, that form the basis for this process. The current regulations[b] use a

[a] 43 CFR § 11.82 (a).
[b] 43 CFR Part 11.

compatible but harder to understand paradigm of *injury determination*, *quantification*, and *damage determination*. The Green Bay NRDA followed the current regulations, but we present the NRDA process in Table 14.1 using the more transparent paradigm.

Determining injuries is the first step in conducting an NRDA because injuries provide the motivation for taking action. Determining realistic restoration options is the second step because restoration provides the means for action. The "Injury Determination and Restoration Options" section describes how these steps were accomplished in the Green Bay NRDA. Scaling techniques such as TVE were then used to balance injury losses with restoration gains, as described in "Scaling Restoration Gains to PCB-Caused Losses." "Incorporating the Results of the TVE into a Comprehensive Position: The Restoration and Compensation Determination Plan" describes Step 4, producing a damage claim.

The TVE Approach

The TVE approach is built on the same simple concept as habitat equivalency analysis (HEA) and resource equivalency analysis (REA). In all three approaches, the amount of losses caused by hazardous substance releases is weighed against the amount of gains generated by natural resource restoration. However, the REA weighs natural resources (e.g., number of bald eagles), the HEA weighs habitat (e.g., acres of wetlands), and the TVE weighs utility or values associated with injury and restoration, where values need not be expressed directly in monetary terms (see Chapter 8 for more information on HEA, REA, and TVE). For example, the utility to individuals of eliminating PCB-caused increases in cancer rates, deformity rates, and reproductive failure rates among fish and wildlife resources might be compared to the utility of nonpoint-source runoff control to improve fish and wildlife habitat.

As described in Chapter 8, HEA can be used when injuries result in measurable loss of habitat, and REA can be used when injuries result in measurable losses of organisms.[2] A key advantage of HEA and REA is that they are relatively easy to perform when injuries and restoration are focused on the same habitats or resources. This is often the case in oil spills, where acres of a habitat are completely lost to the spill but very similar acres can be restored nearby. A key advantage of TVE is that it can scale different kinds of natural resources and services when (1) injuries cannot be easily quantified in terms of lost habitat or organisms or (2) no practical restoration alternatives exist that address the exact natural resources and services injured or lost. This is often the case at industrial sites where large quantities of highly persistent chemicals have been widely distributed and resultant injuries are both subtle and long-lived.

For instance, in Green Bay, even with cleanup some injuries are expected to continue for decades, and many of the injuries cannot be prevented through additional restoration. Fish and wildlife population levels may not have been affected by consumption advisories applicable to fish and wildlife, increased cancer rates in walleye, increased deformity rates in terns, and decreased reproduction in bald eagles, terns, and cormorants, but these injuries may be important to the public even if population effects are not occurring or not measurable. Increasing these same species or their habitats could actually increase the amount of injury because exposure to PCBs will continue for decades. In addition, Green Bay is a unique resource that has been impacted by PCBs throughout, and neither fixing nor replacing Green Bay is feasible; yet natural resource restoration opportunities abound in and around Green Bay. Therefore, a method was needed to trade off the gains from realistic restoration opportunities in Green Bay against the actual losses that are occurring there because of PCBs. TVE value-to-value scaling is such a method.

The TVE study used a survey in which respondents were asked to indicate their preference among scenarios comprising different levels of certain attributes.[a] For Green Bay, each of the following attributes was varied between scenarios: (1) the timeline for eliminating PCB-caused injuries; (2) the amount of wetland habitat to be restored in the basin above status quo levels; (3) the amount of nonpoint-source runoff control to be implemented in the Green Bay watershed above the status quo to improve water quality — for example, clarity — in the bay;[b] (4) the amount of improvements at area parks to be implemented above the status quo; and (5) and the amount of additional taxes to be borne by the respondent.

The TVE was designed to measure the marginal utility of these restoration programs directly so that various programs could be scaled against realistic timelines of PCB-caused injuries. Focus groups during the design of the survey instrument indicated that respondents believed that taxes would be needed for part of cleanup and restoration of the Fox River and Green Bay. Furthermore, focus groups and pretests indicated that a monetary measure made trading between programs seem more realistic. Therefore, taxes were included in the scenarios, which also allowed calculation of willingness to pay (WTP) in dollars for the restoration programs, as well as elimination of PCB-caused injuries under various timelines.[c] Therefore, the TVE has both important similarities to and important differences from the contingent valuation method (see Appendix 5-A).[3,4]

DETERMINING INJURIES

In the early 1960s, the USFWS discovered very high levels of organochlorines in herring gull eggs on the Sister Islands in outer Green Bay. This discovery preceded the ability of chemical laboratories to differentiate between various organochlorines such as PCBs, DDT, dioxins, furans, and others. One of the consequences of the discovery was that Green Bay fish and wildlife became the object of many studies to determine the exact nature, origins, and effects of the chemical exposure. The story

[a] The TVE falls into the *comparison of alternatives* phase of the conceptual approach described in Chapter 9. It is thus a tool for use in decision-making and, given the flexibility of the TVE approach, is amenable to use in adaptive implementation.

[b] In the survey instrument, improved water clarity and reduced days of excess algae are presented as results of nonpoint-source runoff control. Data on these are presented in perfect correlation, and thus we focus on water clarity as the measure of runoff control benefits.

[c] WTP was elicited rather than willingness to accept (WTA) as a more conservative approach.

that began to unfold over the decades was very complicated. Therefore, one of the first tasks for the USFWS in conducting the Green Bay NRDA was to determine what injuries were and are caused by exposure of natural resources to PCBs released by identifiable, potentially responsible parties.

The USFWS published a series of formal pathway[a] and injury determinations,[1,5–9] which can be downloaded from the USFWS Web site. These determinations are significant beyond their explanatory information because they represent formal decisions by the officials delegated by law to make NRDA-related decisions on behalf of the public, pursuant to CERCLA and federal regulations.[b] As such, these formal determinations established the factual and legal underpinnings for the government's cause of action, as well as for the economic instruments that were used to scale restoration and damages.

For release and pathway, USFWS concluded that paper companies had released PCBs into the Lower Fox River as a result of processes involving carbonless copy paper, which contained PCBs as a carrier solvent. Releases began in the 1950s, increased dramatically in the 1960s, and dropped sharply after 1971, amounting to approximately 660,000 pounds of PCBs released into the river. The river is the dominant source of PCBs to Green Bay; water circulation patterns transport PCBs throughout the bay, and PCBs are transported out of the river and bay system via water currents into Lake Michigan and air currents. The lines of evidence used to make these conclusions included analyses of currents in Green Bay; spatial patterns of PCBs in sediment, surface water, and fish; temporal patterns of PCB concentrations in various media; and PCB congener[c] profiles in bay sediments.

The USFWS also concluded that surface waters, fishery resources, and avian resources throughout the river and bay are injured by PCBs released from the paper mills, consistent with release and pathway evidence. Surface water concentrations of PCBs exceed by orders of magnitude federal criteria and state standards to protect aquatic life and wildlife. PCB concentrations in fish and waterfowl exceed Food and Drug Administration tolerance levels in 14 Lower Fox River species, 29 Green Bay species, and 6 northern Lake Michigan species, and have triggered state consumption advisories for 16 Lower Fox River species and 21 Green Bay and Lake Michigan species. Compared to reference areas, Green Bay walleye have higher PCB concentrations and a higher prevalence of liver tumors and pretumors known to be caused by PCB exposure. PCBs also cause many adverse effects in birds, including death in adults and juveniles, death in embryos, altered reproductive behavior, reduced fertility and egg production, reduced or delayed chick growth, subtle neurological effects such as impaired avoidance behavior, hormonal disruption, and deformities. PCBs in eggs cause toxicity at

[a] *Pathways* are defined in federal regulations at 43 CFR §11.14(dd) as "the route or medium through which oil or a hazardous substance is or was transported from the source of the discharge or release to the injured resource."

[b] See 43 CFR Part 11.

[c] The term *PCB* refers to a class of 209 individual compounds, called *congeners*, that share a similar chemical structure but vary in the number and position of chlorine atoms attached to the basic structure. PCBs occur in the environment as a mixture of congeners, and the relative abundance of the different congeners can be used to characterize and identify environmental PCB mixtures.

low parts-per-million total PCB concentration and low or sub parts-per-billion concentration as dioxin equivalents.[a] Numerous bird species in the river and bay are exposed to PCBs at higher concentrations than these, and PCB concentrations in the assessment area are statistically significantly higher than reference areas in all 10 species measured. Further, evidence from field studies shows that fish-eating birds in the assessment area are injured because of PCB exposure. In particular, Forster's and common terns have suffered low reproductive success, behavioral abnormalities, and physical deformations; and double-crested cormorants and bald eagles have suffered reduced hatching success.

DETERMINING RESTORATION OPPORTUNITIES

In February 1986, committees of local experts and stakeholders formed to create the Green Bay Remedial Action Plan (RAP), under the auspices of the International Joint Commission and the Great Lakes Water Quality Agreement. Those committees began documenting impaired uses in the Lower Fox River and Green Bay, and began exploring restoration opportunities to address the various impairments. The Green Bay RAP made quick progress and was featured as an example for the other 42 Great Lakes RAPs that existed at that time. As with injury determination, the USFWS relied on existing information about restoration opportunities in Green Bay to a great extent and transformed this information into formal determinations that carry the weight of legal authority.[5,10]

First, the USFWS provided an overview of the restoration planning and damage determination process that included coordination with ongoing remedial planning activities; criteria based on statutory and regulatory requirements and agency mandates for determining restoration project acceptability, focus, implementation, and benefits; and a process for ranking and scaling projects. Next, the USFWS created a universe of projects and ideas applicable to Green Bay by meeting with diverse experts and stakeholders, including Green Bay RAP committees, The Nature Conservancy, local land trusts, multiple divisions of each of the trustee agencies, environmental groups, local academics with restoration expertise, representatives of commercial and recreational anglers, local governments, and the potentially responsible parties. Projects that did not meet minimum NRDA acceptability criteria were removed from further consideration, which resulted in a database of over 600 projects and ideas grouped into four primary categories of highly ranked projects: (1) cleanup of PCBs from the environment, (2) wetland and related habitat restoration, (3) nonpoint source control to improve water quality, and (4) natural resource–based enhancement of local parks. These four broad programs, described later, were then used in the scenarios of the TVE study.

[a] Some PCB congeners and *dioxin* (2,3,7,8-TCDD) have similar chemical structures that can cause certain types of toxicity through similar modes of action. The toxicity of each PCB congener can be expressed as a percentage of the toxicity of dioxin. A toxic equivalency factor for each PCB congener can be multiplied by the concentration of each congener in a sample, and the toxicity of all congeners can be added together to determine the total dioxin-equivalent toxicity.

Cleanup of PCBs from the Environment

When the USFWS launched the Green Bay NRDA in May 1994, no response actions were planned.[a] Therefore, the USFWS included activities to break the pathway between PCBs in river sediments and natural resource injuries in Green Bay as part of assessment planning.[b] The USFWS also made it clear that any sediment cleanup activities, whether voluntarily by the Fox River Coalition or pursuant to the remedial authorities of CERCLA, would necessarily lessen the amount of primary sediment restoration that would be required for the NRDA. In June 1997, the USEPA announced that it would invoke Superfund authorities and pursue remediation of the Lower Fox River and Green Bay.[c] By October 2000, when the USFWS published the TVE study, Wisconsin and the USEPA had made significant progress on the remedial investigation and feasibility study, including projections about likely cleanup scenarios and timelines to remove various injuries and adverse ecological effects.[11–15] Consequently, the USFWS was able to use the TVE timelines associated with various realistic remedial scenarios published by the response agencies. Briefly, each remedial scenario corresponded to the removal (probably by hydraulic dredging) or immobilization (probably by capping) of different numbers of deposits (out of 38 total) in the upper 32 miles, and sediment management units (out of 96 total) in a continuous deposit in the lower 7 miles, of the Lower Fox River. The USEPA and the Wisconsin Department of Natural Resources then used fate and transport models to determine the amount of time until most adverse ecological effects would be eliminated under the different scenarios, and these corresponded to 20, 40, 70, and 100 years.[d]

The TVE summarized the injuries presented in the various NRDA determinations, described in "Determining Injuries," and used 20-year (intensive remediation), 40-year (intermediate remediation), 70-year (little remediation), and 100-year (no remediation) scenarios for the elimination of most PCB-caused injuries. These timelines were then traded against various levels of enhancement of the other three restoration categories. In addition, the USFWS decided that natural resource–based restoration projects (corresponding to the three categories of restoration other than PCB cleanup) were the preferred alternative compared to two others: (1) the no-action alternative and (2) PCB cleanup beyond the Superfund remedy. The decision against pursuing additional PCB cleanup was based on the low likelihood that such projects could pass NRDA feasibility and cost-effectiveness criteria. In fact, the estimated cost of dredging Green Bay to the same cleanup levels being considered

[a] The Wisconsin Department of Natural Resources took the position that activities of the Fox River Coalition, formed in June 1992 to pursue voluntary cleanup, were equivalent to remedial planning. The USFWS did not concur, citing that the Fox River Coalition had no charter, no enforcement mechanisms, and no standing under CERCLA.

[b] *Breaking the pathway* can involve sediment restoration or other activities that are similar to cleanup activities when no response agency is pursuing cleanup of a site.

[c] *Remediation* and *remedial action* are terms used in CERCLA to describe cleanup activities undertaken by response agencies such as the USEPA. Remediation often refers to long-term cleanup of a site, as distinct from *removal actions*, which refer to short-term emergency actions.

[d] Since that time, the Wisconsin Department of Natural Resources and the USEPA issued records of decision (ROD) for all five operable units of the Lower Fox River and Green Bay. The RODs mostly call for a combination of dredging and natural attenuation, which probably corresponds most closely with the 40-year injury elimination scenario.

for the Lower Fox River was over $100 billion, and Green Bay dredging could cause as much ecological harm as the PCBs there. Therefore, the TVE was used to determine the amount of restoration in the other three restoration categories that corresponded to the various timelines to eliminate PCB-caused injuries.

Wetland and Related Habitat Restoration

One of the natural resource–based restoration programs that was traded against timelines to eliminate PCB-caused injuries was wetland and related habitat restoration. Wetlands, an integral part of the Green Bay ecosystem, provide valuable habitat for many plants, birds, fish, and other wildlife that are dependent on wetlands for their survival. Three different types of wetlands were identified: 32,600 acres of disturbed and undisturbed coastal wetlands around Green Bay, 58,600 acres of wetlands near large human populations that could be preserved from continuing losses, and wetlands with particularly high ecological value or that support rare species.

Nonpoint-Source Runoff Control

A second natural resource–based restoration program traded against timelines to eliminate PCB-caused injuries was nonpoint-source runoff controls to improve water quality in Green Bay. Nonpoint-source runoff of sediments and nutrients into Green Bay impacts the ecosystem and human uses, including phosphorous-induced algal blooms and die-offs that eliminate or displace native fish, irritate swimmers' eyes and skin, and produce foul odors; and excess algal growth and total suspended solids (TSS) that reduce water clarity, eliminate submerged aquatic vegetation needed by many fish and waterfowl species, and affect sight-feeding strategies employed by fish such as walleye and northern pike. Nonpoint-source loadings of sediment and nutrients into Green Bay originate largely from agricultural fields and can be dramatically reduced through conservation tillage and the creation of riparian buffer strips.

Enhancement of Parks

A third restoration program traded against timelines to eliminate PCB-caused injuries was the enhancement of parks in the area by adding facilities such as picnic grounds, boat ramps, and biking and hiking trails. However, the purpose of the NRDA provisions of CERCLA is restoration of *natural* resources. Therefore, the USFWS expressed a preference for natural enhancements at existing or new parks. Nevertheless, the TVE included recreation enhancements ranging from no improvements up to a 10% increase in facilities at existing parks and a 10% increase in new park acreage.

SCALING RESTORATION GAINS TO PCB-CAUSED LOSSES

Restoration scaling (Step 3 from Table 14.1) determines the amount of restoration required to compensate the public for PCB-caused losses. The TVE study was undertaken to support restoration planning by (1) eliciting public preferences for types and mixes of restoration alternatives, and (2) providing methods to scale restoration projects to provide ecosystem services of societal value that are equivalent to the total value of all PCB-caused service losses.

For a large share of the PCB-caused ecosystem service losses, particularly within Green Bay where most of the PCBs have settled, providing restoration with the same or similar services may be technically infeasible (i.e., sufficient natural resource restoration opportunities may not be available), undesirable (e.g., increasing the population of fish or birds that may continue to experience injuries from PCB exposure), or too expensive. Therefore, trustees may prefer restoration actions that provide resources and services of a similar type or quality as those injured. In these cases, value-based scaling methods provide a basis for selecting and scaling restoration activities.

The TVE study used *value-to-value scaling* to determine how much restoration provides services similar to, but not necessarily the same as, those lost. Scaling is computed such that the value of the services gained through restoration equals the lost value of PCB-caused injuries. Value is measured by the utility (benefits or satisfaction) that people derive from all uses of the resources. Dollar measures of value are not required for value-to-value scaling.[a]

The TVE study is a total value assessment because it addresses most or all PCB-caused service losses, including but not limited to recreational fishing and other recreational losses such as waterfowl hunting and wildlife viewing; casual or indirect losses such as reduced enjoyment while driving or walking by or working near a site, and when hearing about, reading about, or seeing photographs of a site; and option and bequest losses tied to preserving resource services for future use for oneself or for others.[16]

Economics of Trade-offs of Resource Services

While the value-to-value scaling of the TVE approach does not require putting dollar figures on damages from PCBs or benefits from restoration, it is based on accepted economic theory (i.e., utility theory). A basic assumption of economics is that individuals maximize their utility from consumption, where consumption is broadly defined to include "consuming" such things as the knowledge that wetlands exist or endangered species are protected. Individuals' choices are assumed to be constrained by their income and wealth, and by limited environmental, natural, and ecological resources. Economic theory asserts that individuals make choices among different goods, services, and money to maximize their utility subject to these constraints. Choices that individuals make thus provide information about the utility or value they gain from the goods and services they select.

Given prices that exist in the economy and the initial endowment people hold of income and natural resource services, how much extra utility an individual gets from an increase in natural resource services is the *benefit* from that change. To measure that benefit, the USFWS implicitly determined how much income an individual

[a] This approach has similarities to the study of biodiversity values in the Clinch and Powell watershed (see Chapter 11). A footnote to "Survey Implementation" in Chapter 11 states that the design "allowed the examination of the trade-offs of strictly environmental attributes" (i.e., non-monetary measures).

could give up, while increasing the level of natural resource services, and leave her just as well off as she was before. The maximum amount of income that an individual is willing to pay to obtain that increase in services is called the compensating surplus (CS). CS is simply a measure of the welfare effect of giving an individual the increase in natural resource services.[16]

Freeman[16] defines the concept of CS as follows:[a] Let $v(P, Y, d)$ represent the consumer's indirect utility function,[b] where P is the vector of market prices, Y is income, and d a measure of natural resource services. Let the subscript 0 represent quality before a policy change and the subscript 1 represent quality after a policy change. For a policy that would improve natural resource quality (e.g., $d_1 > d_0$), compensating surplus is defined implicitly as the WTP value that solves the equation:

$$v(P_0, Y_0 - WTP, d_1) = v(P_0, Y_0, d_0) = u^0 \tag{14.1}$$

where u^0 is utility before a policy change, and it is assumed that there is no change in prices or income as a result of the policy change, only an increase in natural resource service flows. WTP is thus the dollar measure of enhanced individual welfare generated by increasing natural resource services from d_0 to d_1.

This can be measured without using dollars by looking at trade-offs between levels of natural resource services. Suppose that residents of northeastern Wisconsin have three types of resource services, wetlands, W, water quality, Q (measured as clarity), recreation, R, and a level of PCB contamination, PCB (measured as years until clean). The subscript 0 now represents quality at baseline, and the subscript 1 represents quality with PCB contamination. For PCBs, baseline is the level that would be present in the environment over time if the releases from Fox River paper mills had not occurred. For wetlands, water quality, and recreational opportunities, baseline consists of the levels that would exist over time without any steps being taken to improve or increase those services to compensate for PCB losses. ΔW, ΔQ, and ΔR represent changes from the baseline level of wetlands, water clarity, and recreation, respectively, as a result of steps taken to compensate the public for PCB losses. This framework takes baseline as a world "without contamination" and thus looks at how much individuals should be compensated for in terms of ΔW, ΔQ, and ΔR for PCB injuries that lower utility.

$$\begin{aligned} &v(P_0, Y_0, W_0 + \Delta W, Q_0 + \Delta Q, R_0 + \Delta R, PCB_1) \\ &\quad = v(P_0, Y_0, W_0, Q_0, R_0, PCB_0) = u^0 \end{aligned} \tag{14.2}$$

If $PCB_1 > PCB_0$, meaning that there is an increase in PCB contamination, then individuals may be willing to have some additional wetlands, improved water quality,

[a] CS is analogous to the compensating variation measure for price changes.
[b] A direct utility function is written in terms of the goods and services that the individual consumes to obtain utility; these goods and services are what directly provide utility. An indirect utility function is written in terms of prices and incomes because these only indirectly generate some level of utility in that they are the means by which an individual obtains goods and services.

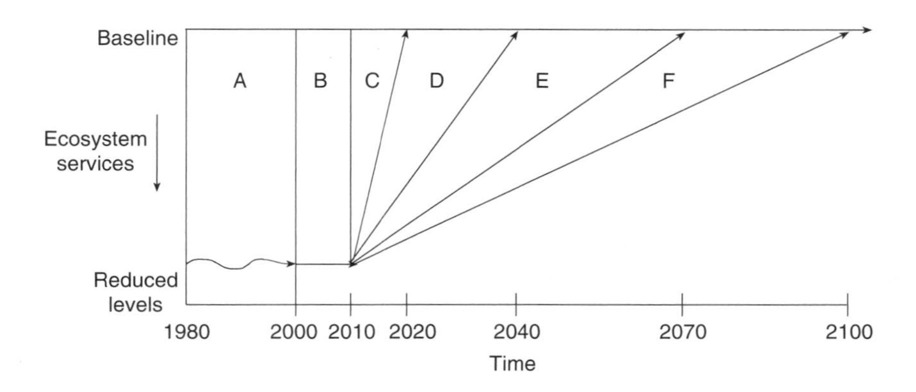

Figure 14.2 Interim losses under alternative time paths for a return to baseline.

or additional recreation facilities to compensate for PCB contamination in the environment. In this case, to maintain the same level of utility, u^0, individuals trade off wetlands, water quality, or recreation to compensate for increased PCB contamination and yet remain at the same level of utility. Thus, damages from PCB contamination (or benefits from cleanup) can be measured using either a dollar metric (WTP) or a resource service metric (e.g., ΔW, ΔQ, or ΔR).

Remediation Scenarios

To scale restoration to the level of PCB-induced losses, the TVE study determined what level of enhancements in the selected natural resource programs would have a value equivalent to the value of PCB-caused losses over various time periods for alternative remediation scenarios. The extent of PCB-induced losses depends on baseline conditions and on resource recoverability with or without PCB cleanup or remediation. Figure 14.2 illustrates how ongoing PCB-caused losses depend on the rate services increase under various cleanup scenarios. Area A represents losses experienced before initiation of cleanup begins at the site (assumed to be 2000); these losses are not addressed in the TVE study. Area B reflects an assumption of a 10-year period (2000–2009) for cleanup actions. During this time, limited (if any) recovery may occur. Areas C–F are ongoing losses after cleanup (if any), depending on the level of remediation. Several scenarios were considered:

1. *Intensive remediation.* This scenario assumes losses continue largely unabated during the cleanup period (Area B) and then linearly decline to baseline over another 10 years, so that Area C represents the losses after cleanup is completed. Hence, Areas B + C represent the total losses over the 20 years after cleanup is begun.
2. *Intermediate remediation.* This scenario assumes that losses continue largely unabated during a cleanup period (Area B) and then linearly decline to baseline over another 30 years (Areas C + D), for a total of 40 years on ongoing losses.
3. *Little or no additional remediation.* These scenarios consider limited remediation over 10 years (Area B), resulting in declining losses over an additional 60 years (Areas C + D + E) for a 70 year total, or resulting in declining losses over an additional 90 years (Areas C + D + E + F) for a 100 year total.

The TVE study design allows the calculation of the scale of restoration that provides services of equal value to the value of PCB-caused losses within any time period shown in Figure 14.2.

TVE Survey

The study was limited to a target population of residents from a 10-county area near Green Bay and the Lower Fox River and sampled from this population.[a] Each county is located almost entirely within 60 miles of Green Bay. Because of their proximity to the bay, individuals from these counties were expected to be more active users of, and more familiar with, the natural resources in the Green Bay area than individuals from outside the target population.

The TVE survey instrument was developed and pretested in a series of eight focus groups and three rounds of in-person pretest interviews conducted in northeastern Wisconsin with 182 subjects. Survey design involved the use of focus groups and in-person interviews to develop *common-language terminology* (see "Problem Formulation" in Chapter 9) for ecological concepts and trade-offs. To further ensure that the survey reflected professional standards, the survey instrument was peer reviewed at various stages. The survey was designed to be conducted by mail, with a telephone survey of nonrespondents.

The survey focused on the four types of natural resource restoration programs for the Green Bay area. The action levels for each program were selected to reflect relevant options and responses from respondents in survey focus groups and pretests and technical information on the feasible action levels. The survey described the four programs as follows:

1. *Restoration of wetlands* near the waters of Green Bay. Wetland restoration would provide increased spawning and nursery habitat and increased food for a wide variety of fish, birds, and other wildlife. Restoration levels ranged from taking no action up to a 20% increase in wetlands within five miles of Green Bay within Wisconsin.
2. *Reduction of runoff* that contributes to pollution of the waters of Green Bay. Controlling runoff would improve water quality by lessening algal growth and improving water clarity, especially in the southern part of the bay. This would improve aquatic vegetation and habitat for fish and some birds and would improve recreation. The runoff control levels considered range from no change in the amount of runoff up to a 50% reduction. The effects of runoff control were described in terms of water clarity and excess algae days.
3. *Enhancement of outdoor recreation* in counties surrounding Green Bay. Enhanced recreation would include increasing facilities at existing parks by adding picnic grounds, boat ramps, and biking and hiking trails, and by developing new parks. The levels of recreation enhancements considered range from no improvements up to a 10% increase in facilities at existing parks and a 10% increase in new park acreage.
4. *Remediation of PCBs* in the sediments of the assessment area. Removing PCBs would reduce the number of years until fish consumption advisories and injuries

[a] These counties are Brown, Calumet, Door, Kewaunee, Manitowoc, Marinette, Oconto, Outagamie, Shawano, and Winnebago.

to wildlife are eliminated. The levels of removal considered would affect the number of years until PCBs are at safe levels (i.e., a return to baseline conditions), ranging from 100 years (no additional removal) to 20 years with intensive remediation.

The survey asked a variety of questions to elicit preferences about the programs and the program levels. For each natural resource topic, the presentation began with information defining the resources in the topic area and other information found to be useful to respondents, such as historical trends and current status (i.e., baseline risks, as discussed in Chapter 9). The next questions identified the benefits associated with resource enhancements (or correspondingly, the impacts of current conditions) and asked how important it was to the respondent, if at all, to undertake resource-enhancement programs that would obtain these benefits. Each presentation was accompanied by diagrams or tables that provided supporting information and helped sustain respondent interest and attention.

Next, the survey included six conjoint-type, stated-preference choice questions, asking respondents to state their preferences by choosing which of two alternatives (A or B) they preferred, where each alternative had a specified level for each of the four restoration programs (see Appendix 5-A for more information on conjoint analysis).[17,18] To ensure that key features of the trade-off scenarios were clear, the choice section introduction reiterated key items of the natural resource programs, and the choice questions further identified how the proposed program levels compared to existing conditions. The first choice question also provided additional information on the baseline conditions and the specific differences between the two alternatives.

Figure 14.3 illustrates a choice question presented to respondents. In this example, respondents were asked to choose between a restoration alternative and the status quo: PCB removal resulting in a reduction to 40 years until PCBs are safe at a per household cost increase of $200 in Alternative A, or no additional resource enhancements and no additional household costs in Alternative B. By varying the program mixes and levels across questions and examining the choices made, mathematical methods (known as *random utility models*) were used to determine the willingness of respondents to trade off different levels of resource improvements against each other and against monetary costs. The alternatives were designed and combined into choice pairs to obtain sufficient independent variation in the attributes to statistically identify the separate influence of each natural resource attribute level on the choice of Alternative A or Alternative B. The survey included only six choice pairs to limit potential respondent fatigue from answering repetitive questions. Ten sets of choice pairs (i.e., 10 survey versions) were designed to obtain sufficient variation in choice pairs for statistical analysis, for a total of 60 alternatives (10 sets with six pairs each).

The mail survey of the stratified random sample of households in the 10 counties near Green Bay followed standard procedures for repeat-contact mail surveys,[20] except that an attempt to contact nonrespondents by telephone was added. Hence, the survey involved an initial mail survey package, a thank you and reminder postcard, and a combination telephone and mail follow-up. Of the 650 eligible respondent households, 470 responded, for a 72% response rate. An evaluation of the sampling plan and

14 **If you had to choose, would you prefer Alternative A or Alternative B?** *Check one box at the bottom.*

	Alternative A	Alternative B
Wetlands		
Acres	58,000 acres (current)	58,000 acres (current)
PCBs		
Years until safe for nearly all fish and wildlife	40 years until safe (60% faster)	100+ years until safe (current)
Outdoor Recreation		
Facilities at existing parks	0% more	0% more
Acres in new parks	0 acres (current)	0 acres (current)
Runoff		
Average water clarity in the southern Bay	20 inches (current)	20 inches (current)
Excess algae days in lower Bay	80 days or less (current)	80 days or less (current)
Added cost to your household		
Each year for 10 years	$200 more	$0 more
Check ( ) the box for the alternative you prefer	☐	☐

Figure 14.3 Sample choice question.

responses indicated that any potential sampling and response biases were likely to be small and thus had a minimal impact on the results.[10]

TVE Results: Awareness and Preferences

Respondents were asked how aware they were of each of the four natural resource topics presented (wetlands, PCBs, outdoor recreation, and runoff control) before receiving the survey. Respondents reported being moderately to highly aware of the topics, with over 80% reporting they were somewhat to very aware of each topic. The literature indicates that higher awareness can be expected to enhance the reliability of responses and to reduce the burden of communication in survey design.

Respondents were asked their preferences for actions that would improve or increase resource services. There was a stronger and more consistent preference for

Table 14.2 Compensatory Restoration Scaling: Illustrative Example Combinations

	Example Compensatory Restoration Combinations[a]		
PCB Remediation Scenarios	Wetland Acres[b]	Existing park Enhancement	Runoff Control[c]
Intensive: (0 to 20 years) (areas B and C in Figure 14.2)	3100	10%	14"/50%
	5500	8%	12"/45%
	11,000	0%	12"/45%
Intermediate: (0 to 40 years)[d] (areas B, C, and D in Figure 14.2)	24,100	10%	16"/55%
	16,000	20%	16"/55%

[a]Restoration is for PCB-related losses during the period indicated.
[b]Rounded to nearest 100 acres.
[c]Additional inches of water clarity/percentage decrease in number of excess algae days.
[d]Requires extrapolating beyond the range of actions considered for some or all programs.

PCB removal over other natural resource enhancement programs. Relative to PCB removal, runoff control and wetland enhancements attracted moderate interest. Limited interest was expressed in enhancing 120 regional parks, and almost no interest was expressed in adding new regional parks. The reported preferences varied by household characteristics. For example, households reported greater interest in doing more and spending more on the various programs if they had anglers active in fishing the waters of Green Bay, if they lived very close to Green Bay, and if they were previously very aware of the natural resource topic.

Econometric models were used to analyze the TVE responses.[a,10] The analysis demonstrated that respondents predominately answered in a manner consistent with expectations; they preferred more resource enhancements to fewer enhancements and lower costs to higher costs. Furthermore, the trade-offs respondents expressed seemed to be consistent with the awareness and preferences expressed in earlier questions. In addition, individuals revealed diminishing marginal utility for wetlands and runoff programs (i.e., more of such resources are viewed as better but with less additional utility added per unit the higher the level of wetlands or runoff control). They also revealed a preference for PCB cleanup sooner rather than later (i.e., a positive rate of time preference).

The resource trade-off questions were used to scale combinations of resource restoration programs that the public would consider to be equivalent in value (measured in utility) to eliminating the continuing PCB-caused losses. Table 14.2 provides examples of the scale of restoration projects that provide services with value equal to the ongoing PCB-caused losses for selected scenarios. Each line represents one possible mix of restoration projects. The listed examples are but a few of the large number of possible combinations that could be developed to provide services of equal value to the PCB-caused losses. The first three lines show example combinations for the scale of restoration providing services of value equal to the PCB-caused

[a] Statistical procedures were used to explore how well the econometric model explained the data. We examined the proportion of choices from choice pairs that are accurately predicted by the model. Based on standard procedures for using fitted choices, the model correctly predicts about 66% of the 2,784 choices in the data. Second, a pseudo-R^2 for the choice pairs is approximately 0.12, which is typical for cross-sectional data.

losses from 2000 until a return to baseline if an intensive level of remediation returns services to baseline by 2020:

A combination of 3,100 acres of wetlands restoration plus a 10% enhancement in existing park facilities plus an additional 14 inches of Green Bay water clarity from a runoff control program.

A combination of 5,500 acres of wetlands restoration plus an 8% increase in existing park facilities plus an additional 12 inches of Green Bay water clarity from a runoff control program.

A combination of 11,000 acres of wetlands restoration plus an additional 12 inches of Green Bay water clarity from a runoff control program.

Each combination of compensatory restoration would generate value equal to the ongoing losses if intensive remediation returned conditions to baseline by 2020. These are equal to the sum of the areas B and C in Figure 14.2.

The second block provides examples for the 40-year intermediate level of remediation or combinations of programs that generate values equal to the sum of the areas B, C, and D in Figure 14.2.

These illustrations do not include additional acres of new parks as a restoration approach because this approach was found to have a near-zero value in the 10-county area. While specific outdoor recreation enhancements would benefit some residents, most residents indicated limited interest in additional facilities and parks.

The TVE analysis presents important information for selecting and scaling restoration alternatives within the three identified project types (wetlands, runoff control, and outdoor recreational facilities). For instance, wetland (and likely other wildlife habitat) restoration programs and runoff control programs are preferred to, and more highly valued than, programs to enhance outdoor recreation in the area. Yet continued increases in the levels of wetland restoration programs increase benefits at a declining rate. As a result, increasing restoration well beyond the levels addressed in the study would most likely result in limited additional benefits to the public.

The TVE study also indicated that the value of PCB-caused losses is so much larger than the value of the restoration programs that it is difficult to generate benefits equivalent in value to the PCB-caused losses with improvements in just one program. Therefore, providing restoration of other resources with a value equal to the value of ongoing PCB-caused losses would most likely require a combination of several programs.

Furthermore, even the maximum combination of the wetlands, outdoor recreation, and runoff control programs considered were insufficient to offset PCB losses that last more than 40 years (i.e., areas E and F in Figure 14.2). To provide service flow benefits for PCB-caused losses beyond 40 years would require additional restorations, perhaps farther away from Green Bay.

INCORPORATING THE RESULTS OF THE TVE INTO A COMPREHENSIVE POSITION: THE RESTORATION AND COMPENSATION DETERMINATION PLAN

The USFWS synthesized previous work of the Green Bay NRDA in the restoration and compensation determination plan (RCDP).[10] The RCDP included all prior determinations, the "Recreational Fishing Damages from Fish Consumption Advisories in the Waters of Green Bay,"[21] the TVE results, and new work to help determine the costs of restoration scaled by the TVE. In the RCDP, the TVE study was used along with this new work to develop plans for restoration alternatives that would make the public whole by providing services other than PCB cleanup.

This involved first identifying the services that could be generated by actual restoration projects (i.e., different types and levels of wetland restoration or preservation or nonpoint-source pollution control that are feasible "on-the-ground" in the Green Bay area) and relating these to the services in the programs defined in the TVE study. Second, the costs of these potential restoration programs were calculated. Potential combinations of programs were then examined to determine which combination of programs could generate the required compensatory benefits at the lowest cost.

Also, as discussed in this section, estimated costs for restoration programs are compared with monetized damage estimates derived using WTP calculations from the TVE. This comparison shows that the values generated from compensatory restoration may be considerably more than the costs of providing those services, which should garner support from potentially responsible parties for restoration and lead to settlements in such cases.

Relating Realistic Restoration Projects to TVE-Scaled Restoration Programs

Scaling establishes the magnitude of restoration required to offset losses caused by PCBs. Scaling the three programs within the preferred restoration alternative was accomplished through value-to-value equivalency to determine the level of restoration required to compensate the public for the injuries to natural resources by determining the value to the public of the ecosystem services gained through restoration. A key component for determining restoration costs was relating realistic restoration opportunities in the field to the programs described in the TVE study.

Key uncertainties and omissions in the assessment were considered and assessed for the potential direction of bias. A conservative approach was taken at many steps in the analysis to not overstate potential benefits. Conservative approaches to benefit estimation in the TVE include the following: past damages were omitted; only about 15% of the Wisconsin households were considered; potential losses to Michigan household were not considered; potential losses to Tribal resources were not considered; population growth was not incorporated; and increasing environmental preferences were not considered. The effect of the sampling and nonresponse biases on the WTP values would most likely be an increase in the computed values, although the analysis suggests any such biases would be small, if they existed at all. It is expected that the likely overall effect of these omissions, biases, or uncertainties on the appropriate scale of compensatory restoration or WTP value measurements is an understatement of the true level or value.

The benefits provided by restored wetlands were expressed in the TVE as acres of wetlands restored. Different restored wetlands, though, can provide dramatically different levels of benefits such as floodwater retention, sediment and nutrient trapping, energy and carbon cycling, and plant and wildlife habitat. In addition, wetland restoration benefits increase from the onset of wetland restoration until the wetland

Table 14.3 General Approach Toward Combining Different Components of Wetlands Actions

Restoration Project Type	Approach for Developing Project Mixes	
Wetland restoration and preservation	3:1	ratio between acres of wetlands preserved and acres of wetlands restored
Wetland preservation	2:2:1	ratio between acres of coastal wetlands, acres of other high-quality wetlands, and acres of wetlands in more populated areas preserved
Coastal upland preservation (to help protect neighboring coastal wetlands)	9:1	ratio between acres of coastal wetlands and acres of coastal uplands preserved

becomes fully functional, while the benefits of wetland preservation do not begin until the wetland would have been lost or degraded without preservation. Differences in wetland services were evaluated to derive a relationship between different types of wetlands and the wetland services as described in the TVE. Table 14.3 is a general framework of how the different actions that constitute wetland preservation and restoration into an overall wetlands strategy were combined.

Nonpoint-source runoff delivers approximately 136,000 metric tons of sediment (TSS) per year and 643,000 kg of phosphorus per year to Green Bay.[10] TSS and phosphorus loads into Green Bay under current and alternative land management practices were modeled[10] to translate them into corresponding increases in water clarity and reductions in algal growth, the two parameters used in the TVE study to express the benefits of controlling nonpoint-source pollution. Since water clarity and excess algae days were perfectly correlated in the TVE study, it was unnecessary to model algae day reductions to estimate restoration benefits. Potential programs for conservation tillage and vegetated buffer strips and estimates of load reductions under altered land management practices (accounting for such factors as land cover type, soil characteristics, climate, topography, and current tillage practices) were translated to corresponding increases in water clarity using the relationship between phosphorus and water clarity that had been measured in Green Bay.

Estimating Costs of the Selected Restoration Alternative

The TVE study allowed for the calculation of the values of the losses as (1) WTP to shorten the time over which losses will occur or (2) the costs of enough restoration to offset the public's losses. The USFWS opted to use restoration costs as the metric for damages rather than values derived from WTP because the statute's purpose and the agency's preference is for restoration instead of money and because liability should be easier to resolve with the potentially responsible parties through settlement.

The different cost elements of restoration alternatives included direct costs such as land acquisition or easements, restoration actions, project maintenance, contingency, and monitoring and indirect costs (i.e., overhead). Land cost estimates were developed on a per acre basis for agricultural lands, with separate estimates for lands

Table 14.4 Examples of Estimated Overall Average Unit Costs

Restoration Action	Total Costs ($/acre)
Wetland restoration	$10,300
Wetland preservation[a]	$2600 to $3100
Conservation tillage	$140
Vegetated buffer strips	$2900

[a] Range covers coastal wetlands, coastal uplands, other high quality natural areas, and wetlands in more populated areas.

sold for continued use as agricultural land and for lands sold for diverting use; wetlands, including separate estimates for inland wetlands, bay and coastal wetlands, and inland wetlands along stream or river waterfronts; and bay coastal uplands and uplands along stream or river waterfronts. Active restoration costs included costs for restoring wetlands, converting farmland to conservation tillage, and installing vegetated buffer strips. Table 14.4 presents examples of estimated average unit costs for each component of the four types of restoration actions within the wetlands and runoff programs. The values shown represent reasonable estimates of the overall average costs for each of the elements.

The cost of improving existing park facilities was estimated based on an incremental increase in the current costs allocated to existing park facilities in the Green Bay area. The cost of improving these parks was estimated based on a percentage increase in the $9.4 million currently spent annually on county and state park facilities within the area. Indirect costs such as overhead costs were estimated using standard indirect rates for projects of a similar nature for a total overhead rate of 38.84%.

Combining Restoration Projects in the Preferred Alternative

Combinations of the three restoration programs that would offset PCB-caused losses were determined. Realistic opportunities for restoration within the assessment area, and declining marginal utility and increasing marginal cost for each program, limit the combinations that are realistic. The RCDP presented several examples of realistic combinations that were acceptable to the trustees based on factors considered in determining the relative amount of the restoration project types that constitute the preferred alternative:

Natural resource restoration is preferred over outdoor recreational facility improvements.
A mix of project types is preferred.
There are technical limitations on the maximum amount of each restoration type that is reasonably possible to implement.
The scale of restoration required affects the project mix.
A mix of actions is preferred to cost-effectively produce service flow benefits.

The columns of Table 14.5 list examples of mixes of restoration actions, including using different combinations of preserving and restoring wetlands, increasing water clarity through conservation tillage and buffer strips, and improving existing park facilities. These examples illustrate the types of combinations that could be considered

Table 14.5 Illustrative Examples of Project Mixes and Total Non-PCB Restoration Costs Under Different Time Periods of Injury

	For PCB Injuries from 0 to 20 Years into the Future (Intensive Remediation) (Area B+C from Fig.14.2)		For PCB Injuries from 0 to 40 Years into the Future (Intermediate Remediation) (Area B+C+D from Fig.14.2)	
Vegetated buffer strip (acres)	5500	23,500	12,000	23,500
Cropland converted to conservation tillage (acres)	106,000	477,000	254,000	852,000
Wetland acres preserved	8700	6900	9900	8700
Wetland acres restored	2900	2300	3300	2900
Percent improvement in park facilities	10	5	10	10
Total cost (millions)	$111	$191	$158	$268

under different possible remediation scenarios and timeframes of continuing injury. Estimated total costs for each of the combinations are also provided.

For instance, a program that would generate values equivalent to the losses from PCB contamination under intensive PCB remediation (areas B and C in Figure 14.2) could comprise 5500 acres of vegetated buffer strips, 106,000 acres of conservation tillage, 8700 acres of preserved wetlands, 2900 acres of restored wetlands, and a 10% improvement in park facilities. These actions combined would compensate for PCB losses if intensive PCB remediation returned the watershed resources to baseline in 20 years. The cost of these actions would be $111 million. An alternative approach, illustrated in the next column of Table 14.5, would generate the same benefits by employing more nonpoint-source runoff reduction and park improvement, but fewer wetlands, and would cost considerably more to implement. The table shows that if intensive remediation is conducted and baseline levels are achieved within 20 years, compensatory restoration costs are less than if intermediate remediation is conducted and PCB injuries continue for 40 years.

The combinations of wetland preservation, wetland restoration, reduced nonpoint-source runoff through improved tillage practices and riparian buffer strips, and improved park facilities would provide a broad array of environmental benefits, from improving habitat for birds, fish, and other biota to increasing bay water clarity to enhancing recreational opportunities. These actions would compensate for the resources and services lost because of PCB injuries by providing valuable environmental benefits within the Lower Fox River and Green Bay environment. By combining different levels of restoration activities, the trustees can take advantage of cost-effective settlement and restoration opportunities.

Comparing Restoration Costs to Willingness to Pay

The present values of WTP for PCB-caused losses between 2000 and a return to baseline were presented for alternative scenarios based on the population-weighted estimates of WTP. The aggregate values represent losses to the 346,700 households

Table 14.6 Present Value of Total WTP for Ongoing PCB-Caused Losses: Residents of 10 Wisconsin Counties (Millions of 1999 Dollars)

PCB remediation scenarios	Areas of Figure 14.2	Mean (range[a])
Intensive (20 years)	B + C	$254 ($143–$356)
Intermediate (40 years)	B + C + D	$362 ($204–$508)
Intermediate (70 years)	B + C + D + E	$515 ($349–$661)
Limited or none (100 years)	B + C + D + E + F	$610 ($445–$766)

[a]95% confidence interval.

in the 10 counties, assuming the population remains constant into the future. The aggregate values were computed using a 3% discount rate.[a] The total values for key remediation scenarios, and differences between remediation scenarios, are summarized in Table 14.6. They range from $254 million for ongoing losses with a 20-year return to baseline to $610 million for ongoing losses if there is little or no PCB removal. These estimates do not include past damages also covered by NRDA authorities.

The trustees used restoration costs instead of WTP to create the opportunity for the potentially responsible parties to settle at reasonable cost at the same time that the trustees fairly compensate the public. This strategy will be effective only if the potentially responsible parties are willing to take advantage of this opportunity, and the trustees are efficient in spending settlements on restorations consistent with the RCDP.

Incorporating Recreational Fishing Damages and Restoration Costs into a Comprehensive Position

Before the TVE was completed, the USFWS conducted a combined revealed preference and stated preference study of the compensable values of recreational fishing service-flow losses to the public (referred to as *recreational fishing damages*) as a result of releases of PCBs into the waters of Green Bay.[21] The USFWS determined that recreational fishing damages in Green Bay were worth between $106 million and $148 million, starting in 1981 and depending on cleanup scenarios for future timeframes. Next, the USFWS completed the TVE study, which was more comprehensive in its coverage of injuries and values, but able to project only future restoration program trades, and thus able to calculate only future costs and damages. Therefore, to avoid any potential double counting for future damages, the USFWS opted in the RCDP to combine past recreational fishing damages of $65 million with future restoration costs scaled by the TVE. This resulted in total damages of between $176 million and $333 million, depending on the PCB remediation scenario (i.e., 20 to 100 years) and the mix of the three restoration programs implemented.

[a] Discounting, when benefits and costs occur in different time periods, translates all values into present value (see "Cost–Benefit Analysis" in Chapter 5 or "Scale Preferred Restoration Alternatives" in Chapter 15).

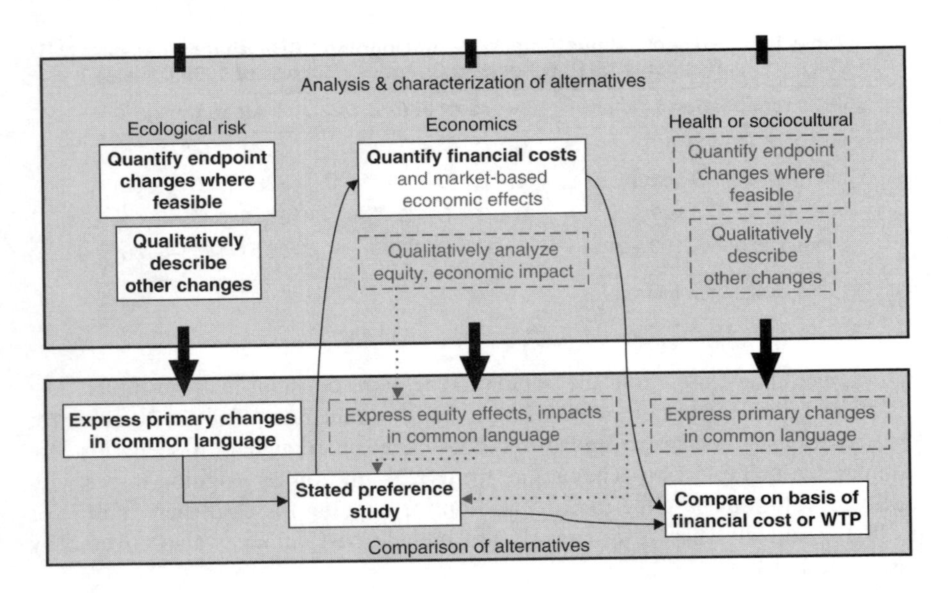

Figure 14.4 Techniques used for analysis, characterization, and comparison of management alternatives in the Lower Fox River and Green Bay in Wisconsin and Michigan, as compared to the example shown in Figure 9.3. White boxes and bold type show features included in this analysis.

DISCUSSION

This section compares the generalized conceptual approach for integrating economics and science in the TVE approach (see Figure 9.1). First, injury determination for the TVE (Step 1 of the Green Bay NRDA) corresponded to analysis and characterization of baseline risk in the conceptual approach. Next, restoration alternatives were formulated in two stages: the first, before analysis and characterization of alternatives and comparison of alternatives, involved specifying various PCB-remediation options and categorizing a long list of restoration projects that met minimal criteria. Rather than finalizing the alternatives for comparative analysis, as would be done for a USACE project, for example, the outcomes (i.e., ecological endpoints) were first compared using TVE. Following this analysis, risk analysis methods were used in a backward rather than forward fashion (to determine more definitively the required features of the restoration alternatives). These more-definitive alternatives were then the basis for determining total financial costs and WTP. These aspects are shown in Figure 14.4.

With respect to the other categories not highlighted in Figure 14.4, equity impacts were not specifically addressed in the TVE study, although subsistence fishing impacts were considered elsewhere as part of the NRDA process. The TVE evaluates trade-offs between different categories of restoration options rather than specific projects. As stated in the RCDP, "the degree to which a restoration project achieves environmental equity and justice is best evaluated at the individual project level rather than at the project category level."[10] Also with respect to Figure 14.4, health considerations were not considered separately from the TVE study because they are implicitly included in the total values elicited from the general public. Calculating values for health impacts separately might have led to double counting of the benefits from reduced exposure to PCBs.

The Lower Fox River and Green Bay have been the object of intensive investigation for many decades. The USFWS pursued a strategy for the Green Bay NRDA of distilling the large body of previous work, supplemented with focused new assessment work, into credible and relevant determinations that would benefit from the force of law. The USFWS invoked an administrative process that allowed for far more public participation than is required or typical, both to strengthen the legal standing of the NRDA and to gain acceptance by the scientific community and the general public.

The TVE study is one of the central elements used by the USFWS to ensure accurate reflection of public preferences about realistic restoration alternatives. The USFWS constrained restoration options under consideration by applying rigorous restoration criteria, rooted in NRDA and agency mandates, to a universe of realistic restoration opportunities at the site. However, the USFWS did not substitute its own values for the public's about the relative value of different acceptable restoration projects to offset PCB-caused losses. Instead, they commissioned and used the TVE study to scale acceptable restoration projects using public values and preferences.

The TVE study is of central importance for translating between the suite of subtle but long-lived injuries caused by PCBs in Green Bay and the realistic restoration opportunities that exist there. As such, the TVE represents the crux of integration between ecological risk assessment and economic benefits analysis at this large, complicated, and vitally important watershed.

There are many opportunities to use similar techniques at other sites. However, the willingness of agencies to incur the expense of such work, and the willingness of the scientific community and the public to accept it, may depend on the outcome of settlement discussions or litigation related to the Green Bay NRDA. Nevertheless, even if the trustees do not achieve all of their goals in Green Bay, we believe that the use of the TVE in the Green Bay NRDA was good public policy that deserves further consideration at other sites.

NOTES AND ACKNOWLEDGMENTS

P. David Allen II was the assessment manager for the Green Bay NRDA while working for the USFWS from 1992 until 2001. Many people contributed to the work described in this chapter, directly and indirectly. In addition to the authors of this chapter, the Green Bay NRDA determinations were authored by the staff of Stratus Consulting, including Michael Anderson, Mace Barron, Bill Breffle, David Cacela, Hector Galbraith, Russell Jones, Diana Lane, Josh Lipton, David Mills, and Sonya Wytinck. Funding was provided by the DOI Natural Resource Damage Assessment and Restoration Work Group and the USFWS. Staffing was provided by the USFWS Green Bay Ecological Services Field Office, the USFWS Green Bay Fisheries Resource Office, the USFWS Regional Office, the USGS Upper Midwest Environmental

Sciences Center, the USGS National Wildlife Health Center, the USGS Columbia Environmental Research Center, NOAA, the Oneida Tribe of Indians of Wisconsin, and the Menominee Indian Tribe of Wisconsin. Support and assistance was also provided by the states of Wisconsin and Michigan, and numerous local experts and concerned citizens. We would also like to thank Frank Horvath and William Hartwig for their leadership at USFWS.

REFERENCES

1. USFWS and Stratus Consulting, Lower Fox River/Green Bay NRDA: Injuries to avian resources, U.S. Fish and Wildlife Service, May 7, 1999. Accessed Oct. 24, 2003 at http://midwest.fws.gov/nrda/.
2. NOAA DARP, Habitat equivalency analysis: An overview, NOAA Damage Assessment and Restoration Program, 2000. Accessed Mar. 20, 2003 at http://www.darp.noaa.gov/pdf/heaoverv.pdf.
3. Mitchell, R.C. and Carson, R.T., *Using Surveys to Value Public Goods: The Contingent Valuation Method*, Resources for the Future, Washington, D.C., 1989.
4. Bjornstad, D.J. and Kahn, J.R., Eds., *The Contingent Valuation of Environmental Resources: Methodological Issues and Research Needs*, Edward Elgar, Cheltenham, UK., 1996.
5. USFWS and Stratus Consulting, Lower Fox River/Green Bay NRDA: Initial restoration and compensation determination plan, U.S. Fish and Wildlife Service, Sept. 14, 1998. Accessed Oct. 24, 2003 at http://midwest.fws.gov/nrda/.
6. USFWS and Stratus Consulting, Lower Fox River/Green Bay NRDA: Association between PCBs, liver lesions, and biomarker responses in adult walleye (*Stizostedium vitreum vitreum*) collected from Green Bay, Wisconsin, U.S. Fish and Wildlife Service, Apr. 13, 1999. Accessed Oct. 24, 2003 at http://midwest.fws.gov/nrda/.
7. USFWS and Stratus Consulting, Lower Fox River/Green Bay NRDA: PCB pathway determination, U.S. Fish and Wildlife Service, Aug. 30, 1999. Accessed Oct. 24, 2003 at http://midwest.fws.gov/nrda/.
8. USFWS and Stratus Consulting, Lower Fox River/Green Bay NRDA: Injuries to surface water resources, U.S. Fish and Wildlife Service, Nov. 8, 1999. Accessed Oct. 24, 2003 at http://midwest.fws.gov/nrda/.
9. USFWS and Stratus Consulting, Lower Fox River/Green Bay NRDA: Injuries to fishery resources, U.S. Fish and Wildlife Service, Oct. 8, 1999. Accessed Oct. 24, 2003 at http://midwest.fws.gov/nrda/.
10. USFWS and Stratus Consulting, Lower Fox River/Green Bay NRDA: Restoration and compensation determination plan, U.S. Fish and Wildlife Service, Oct. 25, 2000.
11. Patterson, D., Zhang, X., Jopke, P., and Steuer, J.J., The Green Bay Mass Balance Project, historical perspective and future projections, WDNR and the United States Geological Survey, Madison Wisconsin, 1994.
12. Fox River Global Meeting Participants, Fox River global meeting goal statement, Meeting at U.S. EPA Region 5 Offices, Chicago, IL, Oct. 27, 1997.
13. WDNR and Bureau of Watershed Management, Polychlorinated biphenyl (PCB) contaminated sediment in the Lower Fox River: Modeling analysis of selective sediment remediation, PUBL-WT-482-97, Wisconsin Department of Natural Resources, Madison, WI, Feb., 1997.
14. ThermoRetec Consulting and Natural Resource Technology, Draft remedial investigation: Lower Fox River, Wisconsin, Prepared by ThermoRetec Consulting Corporation, Seattle, WA, and Natural Resource Technology Inc., Pewaukee, WI, 1999.

15. ThermoRetec Consulting and Natural Resource Technology, Draft remediation investigation: Lower Fox River and Green Bay, Wisconsin, Prepared for WDNR by ThermoRetec Consulting corporation, Seattle, WA, and Natural Resource Technology Inc., Pewaukee, WI, 2000.
16. Freeman, A.M., *The Measurements of Environmental and Resource Values: Theory and Methods*, Johns Hopkins University Press, Baltimore, MD, 2003.
17. Ben-Akiva, M. and Lerman, S., *Discrete Choice Analysis: Theory and Application to Travel Demand*, The MIT Press, Cambridge, MA, 1985.
18. Louviere, J.J., Hensher, D.A., and Swait, J.D., *Stated Choice Methods: Analysis and Application*, Cambridge University Press, Cambridge, UK, 2000.
19. Breffle, W.S. and Rowe, R.D., Comparing choice question formats for evaluating natural resource tradeoffs, *Land Econ.*, 78, 298, 2002.
20. Dillman, D.A., *Mail and Internet Surveys: The Tailored Design Method*, John Wiley and Sons, New York, NY, 1999.
21. USFWS and Stratus Consulting, Lower Fox River/Green Bay NRDA: Recreational fishing damages from fish consumption advisories in the waters of Green Bay, U.S. Fish and Wildlife Service, Nov. 1, 1999.

CHAPTER 15

The Habitat-Based Replacement Cost Method: Building on Habitat Equivalency Analysis to Inform Regulatory or Permit Decisions under the Clean Water Act

P. David Allen II, Robert Raucher, Elizabeth Strange, David Mills, and Douglas Beltman

CONTENTS

1-56670-639-4/05/$0.00+$1.50

INTRODUCTION

This chapter discusses the habitat-based replacement cost (HRC) method for quantifying, monetizing, and scaling natural resource losses and presents a case study of its use. An HRC analysis compares losses of organisms caused by an environmental impact with the costs of sufficient habitat restoration to produce organisms to the level necessary to offset the losses. The HRC method is based on habitat equivalency analysis (HEA) and resource equivalency analysis (REA) (see Chapter 8). HEAs balance losses and gains of habitat or the ecological and human use services provided by the habitat; REAs balance gains and losses of organisms. In contrast, the HRC method balances losses of organisms with gains of habitat, and provides a monetary cost estimate associated with generating the necessary habitat gains. Therefore, the HRC method is a hybrid of the REA and HEA approaches. The HRC method was developed for the U.S. Environmental Protection Agency (USEPA) by Stratus Consulting in 2001-2002[1] and the method has been described by Strange et al.[2]

The HEA and REA analytical methods that the HRC approach is based on have often been conducted as part of litigation. However, in this chapter we describe an HRC analysis used for permitting purposes. First, we discuss the conceptual basis of the HRC method. Next, we present an overview of the method itself and provide a case study showing how the method has been used for permitting under the Clean Water Act (CWA). Finally, we discuss issues associated with the use and interpretation of the HRC method, as well as the relationship of the HRC method to the conceptual approach for integrating ecological risk assessments and economic benefits analyses, outlined in Chapter 9.

CONCEPTUAL BASIS AND RATIONALE FOR THE HRC METHOD

The need for more complete cost-benefit analyses (CBA) was the initial motivation for developing the HRC approach. A CBA is often required to evaluate proposed regulatory and permitting actions ("Cost-Benefit Analysis" in Chapter 5 describes the eight stages of CBA). However, the cost of regulatory actions is typically much easier to measure than the value of environmental benefits that regulations are designed to generate. Consequently, many CBAs include only a small subset of easily measured values, such as recreational and commercial fishing values, while omitting other benefits that may be associated with impacts of greater magnitude.

To prevent systematic bias that overestimates the cost per unit of value, a CBA should measure all values (i.e., the total value), including use values (benefits associated with actively enjoying, using, consuming, or observing environmental resources, e.g., hunting and fishing) and nonuse values (e.g., bequest values tied to enhanced environmental quality for use by others in the future and existence values that are not dependent on human use ever occurring).

While many use values can be quantified on the basis of marketed goods and services, nonuse values often involve nonmarket goods and services that are difficult to measure. Nonuse values are nonetheless important in the context of many environmental quality issues. Indeed, an extensive body of environmental economics literature demonstrates that the public often holds significant value for service flows from natural resources well beyond those associated with direct uses.[3–10] Studies have also documented public values for the ecological services provided by a variety of natural resources potentially affected by environmental impacts. Natural resources that have been studied in this way include fish and wildlife,[11,12] wetlands,[13] wilderness,[14] and critical habitat for threatened and endangered species.[15–17] Studied impacts include overuse of groundwater,[18] hurricane impacts on wetlands,[19] global climate change effects on forests,[20] bacterial impacts on coastal ponds,[21] oil impacts on surface water,[22] toxic substance impacts on wetlands,[23] shoreline quality,[24] and beaches, shorebirds, and marine mammals.[25]

In the context of CBA, if time and budgetary constraints or controversies about economic methods prevent direct, site-specific total value studies, then expedited approaches such as benefit transfer (see "Economic Value" in Chapter 5) and replacement costs may be used by permitting or regulatory agencies. The cost of a replacement action does not necessarily equate to its benefits to society. (Consider, for example, a technically elaborate project to convert an upland meadow, having well-drained soils and no water source, into a commonplace cattail marsh.) Therefore, agencies do not always accept replacement costs as a reliable estimate of benefits. For example, the USEPA's *Guidelines for Preparing Economic Analyses* state that use of the replacement cost method ". . . is justified only when individuals are proven willing to incur such replacement costs, through either their voluntary purchases or their support for public works projects. If so, the value of the service is at least as much as the replacement cost."[26] In other words, agencies prefer to have clear evidence of demand for the action, to substantiate its value.

However, the courts have upheld and commented on the advantages of using replacement costs based on natural production as a measure of damages, pursuant to the Comprehensive Environmental Response, Compensation, and Liability Act (CERCLA), the National Marine Sanctuaries Act,[a] and the oil spill-related provisions of the CWA.[b] CERCLA states that natural resource damages recovered shall be used only to "restore, replace, or acquire the equivalent of such [damaged] natural resources" and that the "measure of damages . . . shall not be limited by sums which can be used to restore or replace such resources."[c] One court explained that Congress intended that these two clauses be read together so that its guiding purpose of achieving restoration of damaged resources would not be interpreted to make restoration costs a "ceiling" on damages recovery, though they would typically provide the measure of damages in most cases (also noting that any recovery in excess of restoration costs would be directed to acquiring equivalent resources).[d] The court goes on to explain that:[e]

> [t]he fatal flaw of Interior's approach [which favored use values over restoration values], however, is that it assumes that natural resources are fungible goods, just like any other, and that the value to society generated by a particular resource can be

[a] United States v. Great Lakes Dredge and Dock Company, 259 F.3d 1300, 1304 (11th Cir. 2001).
[b] State of Ohio v. U.S. Department of Interior, 880 F2d 432, 444-46, 448, 450, 459 (D.C. Cir. 1989).
[c] State of Ohio, 880 F.2d at 444 (quoting 42 U.S.C. § 9607(f)(1)).
[d] State of Ohio, 880 F.2d at 444, n. 8, 445-46.
[e] State of Ohio, 880 F.2d at 456-57 (emphasis in the original), and 880 F.2d at 441, 445, 446, n. 13.

> accurately measured in every case — assumptions that Congress apparently rejected. As the foregoing examination of CERCLA's text, structure and legislative history illustrates, Congress saw restoration as the presumptively correct remedy for injury to natural resources. To say that Congress placed a thumb on the scale in favor of restoration is not to say that it foreswore the goal of efficiency. "Efficiency," standing alone, simply means that the chosen policy will dictate the result that achieves the greatest value to society. Whether a particular choice is efficient depends on how the various alternatives are valued. Our reading of CERCLA does not attribute to Congress an irrational dislike of "efficiency"; rather, it suggests that Congress was skeptical of the ability of human beings to measure the true "value" of a natural resource. . . . Congress' refusal to view use value and restoration cost as having equal presumptive legitimacy merely recognizes that natural resources have value that is not readily measured by traditional means.

The same court also noted that many scholars shared Congress' skepticism concerning our ability to adequately monetize the full value of natural resources. One of these scholars is quoted at some length by the court as follows:

> At first glance, restoration cost appears to be inferior, because it is a cost-based, supply-side measure, rather than a demand-side, value-based measure of natural resource value. For this reason, when natural resource economics advances far enough to provide an adequate demand-side measure, reliance on restoration cost will become inappropriate. At present, however, the economic tools for valuing natural resources are of questionable accuracy . . . [Using restoration costs as the measure of damages] acknowledges the current ignorance of economic valuation of resources by adopting a cautious, preservationist approach.[27] [Footnote in the original.]

The HRC method is a replacement cost method based on established methods such as HEA. HEA is used to scale the amount of restored or created habitat necessary to replace the environmental services lost from injuries or impacts to those habitats on an interim or permanent basis. Although estimating restoration costs can be difficult, generally they are more accurately and easily measured than the total value of an injured resource, and under certain circumstances restoration costs constitute a sensible alternative or supplement for use values or "demand-side" measures of the value of natural resources.

OVERVIEW OF THE HRC METHOD

HRC analysis has eight primary steps.[2] First, the injury is quantified by estimating total losses. Then, habitat requirements of habitat-limited species are identified, and the habitat restoration measures likely to be most effective at producing these species are identified. Next, estimates are made of the expected increases in the production of each species in the habitats of interest. Then, losses are divided by the habitat production estimates to determine the required scale of habitat restoration for each species. The total amount of all restoration alternatives required to replace all losses is determined next, while taking care to avoid "double-counting" problems. Finally, cost

Step 1: Quantify I&E losses by species
Step 2: Identify habitat requirements of I&E species
Step 3: Identify potential habitat restoration actions that could benefit I&E species
Step 4: Consolidate, categorize, and prioritize identified habitat restoration alternatives
Step 5: Quantify the benefits for the prioritized habitat restoration alternatives
Step 6: Scale the habitat restoration alternatives to offset I&E losses
Step 7: Estimate "unit costs" for the habitat restoration alternatives
Step 8: Develop total cost estimates for I&E losses

Figure 15.1 The eight steps of the HRC method.

estimates are made for implementing the restoration alternatives. The eight steps are shown in Figure 15.1 and detailed below.

Injury Quantification

Step 1 in an HRC analysis quantifies losses of habitat-limited species and life stages. This step is usually straightforward only if losses are acute, short-lived, easily measured, and involve a single species and life stage in a confined area. However, losses of organisms such as fish rarely meet these criteria. For example, losses often include many species with different life stages (eggs, larvae, juveniles, and adults), affected by acute and chronic impacts that may be difficult to monitor, particularly if immigration and emigration are taking place. Therefore, different kinds of data must be converted into some common metric. For example, total losses could be expressed in terms of a single equivalent life stage (e.g., age-1 equivalents) by estimating cumulative survival rates between life stages.

Identify Limiting Habitat of Species Lost

The second step in an HRC analysis identifies the habitat that is limiting for each impacted species. Literature searches and discussions with local resource managers, biologists, conservationists and restoration experts with specific knowledge of the species typically provide this information.

Identify Suitable Habitat Restoration Alternatives

The third step in an HRC analysis identifies actual habitat restoration alternatives that could potentially increase the local production of the organisms lost, emphasizing local habitat constraints and opportunities. Only biological understanding and engineering capability, not existing funding and administrative constraints, restrict the pool of alternatives. For example, local zoning or easements could restrict particular restoration options, but local experts may identify the engineering costs for feasible restorations that could be included in the analysis, perhaps including the administrative costs of altering zoning or purchasing easements.

Consolidate, Categorize, and Prioritize Restoration Alternatives

The fourth step in an HRC analysis consolidates, categorizes, and prioritizes restoration alternatives, providing a list of alternatives for each species of concern. Then, for each species, a single option is chosen as the preferred alternative.

Quantify Productive Capacity of Habitats

The fifth HRC step quantifies the expected increases in production of each species caused by the chosen habitat restoration actions. If possible, estimates of habitat productivity should be site-specific both before and after restoration activities take place. When such information is unavailable, catch-per-unit-effort data converted to densities, population or food web models, ratios of production (numbers of organisms) to biomass (weight) from the literature or the best professional judgment of knowledgeable biologists may be used to estimate productivity. The metrics used to estimate productivity must be comparable to the metrics used to estimate losses in the first step of the HRC.

Scale Preferred Restoration Alternatives

Step six of an HRC analysis scales the preferred habitat restoration alternatives so that increases in production would offset the losses for each species. In the simplest case, dividing the loss by the increase in production per area determines the number of habitat units, or scale, of restoration needed. For example, if 1 million age-1 equivalent winter flounder are lost per year, and restoring local salt marsh habitat is expected to provide an increase of 500 age-1 equivalent winter flounder per hectare per year, then offsetting the loss requires restoration of 2000 hectares of salt marsh (assuming that losses and restoration gains occur over the same time periods).

In most cases, the amount of habitat needed for different species sharing a preferred restoration alternative will vary, and more than one type of restoration will be required to account for all species lost. Thus, several estimates of the amount of each type of restoration can be calculated for a given alternative. To ensure that enough habitat is restored to offset losses of each species, the species requiring the greatest scale of implementation will determine the level of restoration. The increased production of one species will not necessarily offset the service losses associated with another species, since each species and life stage provides some unique services. In some cases, there may be no feasible or practical restorations available for a species or life stage. When this occurs, the service-to-service scaling used by HRC analysis, HEA and REA may not be appropriate, and alternative metrics such as value-to-value scaling may be necessary (i.e., services or natural resources of equal value to the public can be traded if identical services or resources cannot).

The expected increase in production may vary widely over the productive life of a habitat unit, especially in the early years. As a result, average annual increases should be estimated by specifying the expected increase over the productive life of the restored habitat and discounting the results to estimate an average present value equivalent. Use of a discount rate is based on the standard economic assumption that people place a greater value on having resources available in the present than on having availability delayed into the future, analogous to placing money in the bank at a given rate of interest (see "Cost-Benefit Analysis" in Chapter 5 for more information on discounting).

Develop Unit Cost Estimates

Seventh, an HRC analysis estimates unit costs for chosen habitat restoration alternatives. These estimates should include the costs of conceptualization, planning, design, implementation, administration, maintenance, and monitoring of each restoration action, plus contingency funding to account for unexpected circumstances that arise during implementation and monitoring. Unit costs and restoration actions should be presented in comparable habitat units. Cost estimates are made by dividing total project costs by the number of habitat units to be restored.

Develop Total Cost Estimate for Offsetting Total Losses

Finally, the total cost of implementing all of the necessary restorations at the appropriate scale to offset all losses is determined. This calculation is made by multiplying the scale of restoration by its associated unit cost and summing the costs of each alternative, being careful to avoid double counting.

Strengths, Limitations, and Uncertainties of the HRC Method

A major advantage of the HRC method is that it addresses the complete suite of ecological and human services associated with natural resources lost to human impacts by focusing on increasing the natural production of these resources in the holistic context of restored habitats. This advantage is significant because many of the ecological and indirect use service flows provided by the resources lost to impacts are often difficult to quantify and value and/or are poorly understood.

This advantage of the HRC method is clearly visible when compared with methods limited to measurements of active use losses such as commercial and recreational fishing impact valuations. For direct use benefits such as recreational angling, the predicted change in a recreational fishery typically affects recreational participation levels, and the value of an angling day is often associated with changes in the expected catch rate. However, impacts affect aquatic ecosystems and public use and enjoyment in many ways not addressed by typical commercial and recreational impact valuation methods.

Examples of ecological and public services not addressed by commercial and recreational fishing valuations include decreased numbers of ecological keystone, rare, sensitive, or special status (e.g., threatened or endangered) species; decreased numbers of popular species that are not fished, perhaps because the fishery is closed; increased numbers of exotic or disruptive species that compete well in the absence of species lost; disruption of ecological niches and ecological strategies used by aquatic species; disruption of organic carbon and nutrient transfer through the food web; disruption of energy transfer through the food web; decreased local biodiversity; disruption of predator-prey relationships; disruption of age-class structures of species; disruption of natural succession processes; disruption of public uses other than fishing, such as diving, boating, and nature viewing; and disruption of public satisfaction with a healthy ecosystem.[28–32] The HRC method can account for these ecological endpoints and, hence, services (use and nonuse). This eliminates the discrepancy that may arise when conventional benefits valuation methods cannot fully embody key biological injuries and service flows.

Another advantage of the HRC method is that it helps determine pragmatic and relevant restoration opportunities for potential mitigation in lieu of or in addition to technologies to reduce impacts. The HRC method incorporates the best available information from local resource experts and restoration programs, and has the advantage of being flexible. An HRC analysis can be updated to adapt to new data for any of the inputs, such as estimates of losses, increased species production or unit costs of restoration. The HRC method also partially alleviates gaps caused by over-reliance on targeted monitoring, because habitat restoration is likely to provide some increased production benefits to species lost but not monitored. In contrast, methods such as fishing impact valuations and hatchery and stocking cost estimates simply omit unmonitored losses.

The most significant limitation of the HRC method is that it calculates the cost of restoring the lost natural resources, rather than directly measuring the public's willingness to pay (WTP) to avoid or restore lost resources. As a result, comparison of an HRC estimate with costs for technologies for minimizing impacts is not a CBA comparison in strict economic terms. Nevertheless, HRC estimates have the same utility as other replacement costs, such as the hatchery-based replacement costs that are often used as a proxy for value in the evaluation of fish kills.[33,34]

Another weakness of the HRC method is that the quality and certainty of its results, and the consequences of those results for public confidence, are affected by the quality of the data used as inputs into the HRC analysis. This is not a methodological weakness, but can result from a failure to identify all species lost, uncertainties about species life histories and habitat needs, and abundance data that are poorly linked to productivity. However, by explicitly incorporating monitoring, HRC results can help improve the quality of future analyses by reducing data gaps and uncertainties.

CASE STUDY: HRC ANALYSIS APPLIED TO CLEAN WATER ACT PERMITTING

Permitting Context and Rationale for HRC Analysis

Cooling water intake structures are regulated by USEPA and the states, pursuant to section 316(b) of the CWA. Section 316(b) requires that adverse environmental impacts such as impingement and entrainment (I&E)[a] of aquatic organisms be minimized using the best technology available. Because the cost of installing, operating and maintaining the technology to minimize adverse environmental impacts, and the benefits of minimizing impacts, can be substantial, techniques are needed to help balance gains and losses at particular sites, whether to inform regulatory cost-benefit analyses and permit decisions, or to find compromises at disputed sites.

Unfortunately, it is typically much easier to obtain information about the costs of technologies than the benefits of minimizing I&E. Assessing benefits from I&E control using the benefits transfer approach focuses predominantly on readily measured recreational fishing benefits, such as the value tied to catch rates and commercial fishing benefits, estimated as producer and consumer surplus in commercial fishery markets. However, recreational and commercial fishery species are a very small portion of I&E losses.

Most I&E losses are of forage species for which methods of valuation are limited. Even converting forage fish into landed fish through food web modeling omits or appreciably undervalues many of the losses. Recent work by USEPA indicates that landed commercial and recreational fish constitute a small fraction of total I&E losses nationwide, but over 90% of the total estimates of monetized losses nationwide were associated with these losses alone.[1] Moreover, the ecological value of losses of eggs and larvae and other life stages up to age 1 are generally not considered. Therefore, the fact that aquatic food webs require orders of magnitude more organisms in the lower trophic levels to support harvested species and other top-level consumers is overlooked.

Many of these key understated benefits pertain to hard-to-quantify physical impacts of I&E, such as the full contribution that forage fish — and eggs, larvae, and juveniles (i.e., organisms younger than age 1) of all species — play in aquatic ecosystems. The scientific uncertainties and usual omissions in most I&E analyses compound the problem of undervaluing benefits provided by environmental technologies and projects

[a] *Impingement* generally refers to larger organisms or adults being trapped on intake structures, usually filtering screens. *Entrainment* generally refers to smaller organisms or life stages (e.g., eggs and larvae) that are lost while passing through intake structures and cooling systems.

required by the regulations and permits, apart from those values attributed to commercially and recreationally landed fish.

One approach to compensating for losses of forage fish is determining fish replacement costs based on the costs associated with producing and stocking hatchery fish.[35] The Socioeconomics Section of the American Fisheries Society (AFS) developed these estimates primarily as a way to determine restitution in court of damages from fish kills.[33,34] However, the forage species that make up most I&E losses are unlikely to be raised in hatcheries and are therefore not included in such estimates. In many cases, costs for even the limited subset of species considered are incomplete or poorly documented. For example, costs of post-release hatchery operations and of monitoring of stocking success are not included in the AFS estimates. Costs may also be underestimated because many public hatcheries are under-funded and do not perform sufficient maintenance and capital replacement to sustain activity over the long term.

Even though replacement costs of hatchery-based fish are used in many policy contexts, they miss important ecological services because hatchery fish are not biologically equivalent to wild fish. A recent extensive review identified some 54 problems with artificial propagation, including poor post-stocking performance and harm to wild fish populations.[36] A critical problem is that hatchery fish are often not self-sustaining, particularly when available habitats do not support all life stages, or are otherwise inadequate. In fact, artificial propagation and the stocking of hatchery fish have failed to restore fish populations where habitat destruction, over-fishing, or other human stressors persist, which is clear from the failure of decades of stocking to restore salmon and steelhead stocks in the Pacific Northwest.[37] Further, stocking young hatchery fish can be ineffective for increasing the abundance of older life stages.[36] Thus, many times more young hatchery fish must be released to achieve the number of adults to be replaced. Finally, because hatchery fish are reared in isolation from the natural ecosystem and are released only after attaining a certain size, the ecosystem does not benefit from the complex ecological interactions provided by early life stages. This includes the food that eggs and larvae provide for other parts of the food web.

In contrast to the hatchery-based approach, the HRC method develops replacement costs based on natural production. The natural production emphasis embodied in the HRC approach reflects the fact that artificially produced fish are not a substitute for losses of wild fish. The HRC approach also inherently recognizes that a full range of habitat services is required to produce the equivalent of organisms lost. In addition, hatchery and stocking costs must continue indefinitely to ensure a steady release of fish to the environment, whereas natural habitat produces fish indefinitely if properly restored and protected.

By using the HRC method, one can more easily and accurately quantify the cost of replacing all I&E losses. Moreover, HRC results provide costs for restoring lost fish in an ecologically sound manner. In addition, conventional techniques to value the benefits of technologies often omit important ecological and public services. In contrast, the HRC method addresses a broad range of ecological and human services affected by I&E losses that are either undervalued or ignored by conventional valuation approaches.

Applying the HRC method in this context is consistent with the use of replacement cost as a measure of damages in natural resource damages contexts. It is also conceptually consistent with the wetlands protection program under Section 404 of

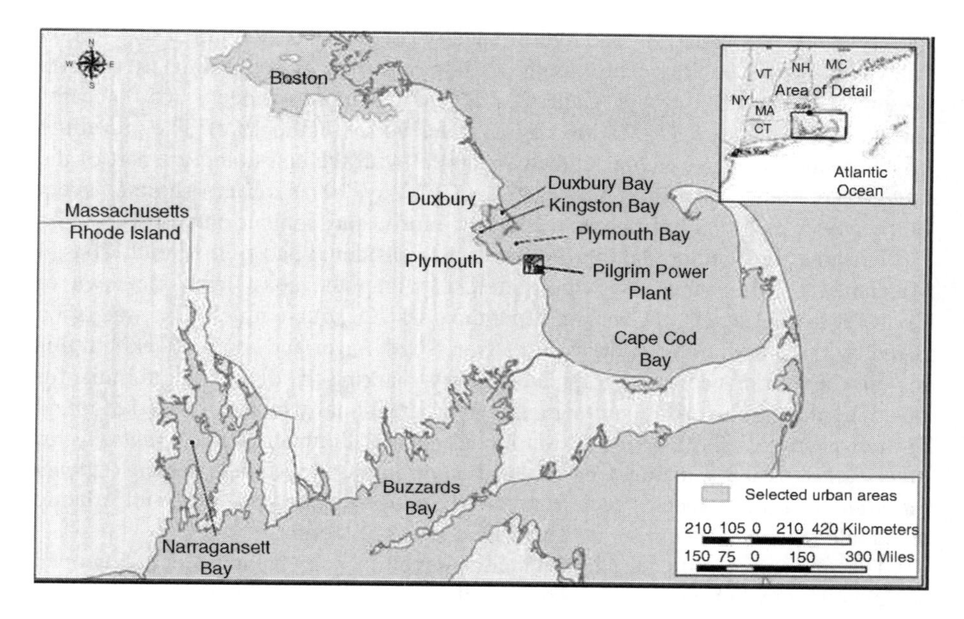

Figure 15.2 Location of the Pilgrim Nuclear Power Generating Station in Plymouth, Massachusetts.

the CWA, where it is accepted that ecosystem losses that cannot be avoided are to be offset with wetlands restoration or creation to replace the functions and values of the lost wetlands. Restoration of aquatic species in I&E-impacted waters clearly constitutes a significant public natural resource benefit and is an important public goal, as evidenced by the CWA goals of restoring the "biological integrity" of the Nation's waters and achieving water quality for the protection and propagation of fish.[a]

Location and Description of Case Study Facility

The HRC approach is being used in the permit development for the Pilgrim Nuclear Generation Facility (Pilgrim) in Massachusetts to (1) identify and scale a group of habitat restoration alternatives that, if implemented, would be expected to provide an increase in the natural production of aquatic organisms at a level that offsets I&E by the power plant; (2) estimate the total cost of implementing the scaled group of habitat restoration alternatives; and (3) provide an estimate of the cost required to restore fully the range of ecological and human use services eliminated by I&E. This analysis was completed to support the development of a new National Pollution Discharge Elimination System (NPDES) permit to be issued to Pilgrim by the USEPA, Region 1.

Pilgrim is a 670 MW nuclear power plant located in Plymouth, Massachusetts (Figure 15.2). Commercial operation of the Pilgrim station began in 1972.[38] Pilgrim uses water from Plymouth Bay as a coolant, and as water is drawn into the facility, aquatic organisms are entrained into the plant or are impinged on screens across the intake pipes.

[a] 33 U.S.C. § 1251(a)(2).

Plymouth Bay is part of, and hydrologically connected to, the larger water body known as Cape Cod Bay. The mouth of Plymouth Bay is approximately 4 miles northwest of the Pilgrim site. Cape Cod Bay covers approximately 365,000 acres and is approximately 23 miles long by 23 miles wide (Figure 15.2). The prevailing offshore currents move to the southeast, parallel with the coast, and are part of the large-scale, counterclockwise flow in Cape Cod Bay.[39] The western shore adjacent to the power plant is a mix of sand beaches, bluffs, and boulder outcrops.

The area surrounding the Pilgrim facility is considered part of the South Coastal Watershed in Massachusetts, which stretches along the coast from the town of Cohasset to the Cape Cod Canal and encompasses 220 square miles. This watershed is fed by three major rivers: the North River, South River and Jones River.[40] In this region, Cape Cod is a zoogeographic boundary, marking the distributional limits for many marine organisms.[41] Commercially and recreationally important species found in the waters near the Pilgrim station include winter flounder (*Pseudopleuronectes americanus*), Atlantic menhaden (*Brevoortia tyrannus*), and Atlantic herring (*Clupea harengus*).[41] Forage species such as cunner (*Tautogolabrus adspersus*) and Atlantic silverside (*Menidia menidia*) are also found near the facility.[42]

The area surrounding the Pilgrim facility supports many habitats, including open sandy and rocky bottoms, seagrass beds, salt marshes, tidal mud flats, sandy beaches and dunes, coastal ponds, and open water. Plymouth Bay supports a considerable amount of eelgrass (*Zostera marina*) habitat.[40] Eelgrass provides an important source of food and refuge for a number of fish species in the area, including Atlantic cod (*Gadus morhua*), pollock (*Pollachius virens*), and threespine stickleback (*Gasterosteus aculeatus*).

Use of the HRC Method in Permitting Process

As part of the permit review process for the Pilgrim facility, the injury to fishery resources was estimated using biological monitoring data reported by the facility.[43] The data consisted of records of impinged and entrained organisms sampled at intake structures and included organisms of all life stages, from newly laid eggs to mature adults.[44–53] These sampling counts were converted into standardized estimates of the annual numbers of fish impinged and entrained expressed as numbers of age-1 equivalents, i.e., the number of individuals of different ages impinged and entrained expressed as an equivalent number of age 1 fish.[54]

The HRC approach was used to estimate the costs of the type and amount of habitat restoration that would be required to offset I&E losses that would occur if engineering-based alternatives were not put in place. Application of the HRC method to the Pilgrim facility used published biological data wherever possible. Where published data were unavailable or insufficient to address HRC needs, unpublished data from knowledgeable resource experts were used.

For the Pilgrim HRC analysis, the total cost for preferred restoration alternatives was determined by multiplying the required scale of implementation for each restoration alternative by the unit cost for that alternative. For each restoration alternative, the scale of implementation was based on the amount of restoration required to offset losses of the single species requiring the greatest amount of restoration. The restoration

Table 15.1 Total HRC Estimates for Pilgrim I&E Losses

Preferred Restoration Alternative	Species Benefiting from the Restoration Alternative: Species	Species Benefiting from the Restoration Alternative: Average Annual I&E Loss of Age-1 Equivalents	Required Units of Restoration Implementation[a]	Units of Measure for Preferred Restoration Alternative	Total Annualized Unit Cost	Total Annualized Cost
Restore SAV	Northern pipefish	118	**4700**	m^2 of directly revegetated substrate	$12.34	$57,975
	Threespine stickleback	118	600			
	Atlantic cod	2439	Unknown			
	Pollack	525	Unknown			
Restore tidal wetland	Winter flounder	210,715	**2,429,812**	m^2 of restored tidal wetland	$1.95	$4,746,249
	Atlantic silverside	25,929	139,539			
	Striped killifish	90	527			
	American sand lance	4,116,285	Unknown			
	Grubby	879	Unknown			
	Striped bass	9	Unknown			
	Bluefish	2	Unknown			
Create artificial reefs	Cunner	993,911	**176,218**	m^2 of reef surface area	$24.85	$4,379,701
	Tautog	1076	36,699			
	Rock gunnel	4,862,872	Unknown			
	Radiated shanny	1,644,456	Unknown			
	Sculpin spp.	734,773	Unknown			
Install fish passageways	Alewife	4343	**0.49**	New fish passageway	$49,437.64	$49,438[b]
	Rainbow smelt	1,330,022	Unknown			
	Blueback herring	703	Unknown			
	White perch	73	Unknown			

(*Continued*)

Table 15.1 Total HRC Estimates for Pilgrim I&E Losses (*Continued*)

Preferred Restoration Alternative	Species Benefiting from the Restoration Alternative: Species	Species Benefiting from the Restoration Alternative: Average Annual I&E Loss of Age-1 Equivalents	Required Units of Restoration Implementation[a]	Units of Measure for Preferred Restoration Alternative	Total Annualized Unit Cost	Total Annualized Cost
Species not valued	Blue mussel	160,000,000,000	Unknown for all	Restoration measures unknown — survival and reproduction may be improved by other regional objectives such as improving water quality or reducing fishing pressure	N/A	N/A
	Fourbeard rockling	411,191				
	Atlantic herring	29,079				
	Windowpane	17,542				
	Atlantic menhaden	14,270				
	Atlantic mackerel	6662				
	Searobin	3767				
	Red hake	1774				
	Lumpfish	1297				
	Butterfish	399				
	American plaice	221				
	Scup	114				
	Little skate	78				
	Bay anchovy	18				
	Hogchoker	2				
		Total annualized HRC valuation				$9,233,362

[a]Numbers of units used to calculate costs for each restoration alternative are shown in bold and have been rounded to the nearest unit.

[b]Anadromous fish passageways must be implemented in whole units, and increased production data are lacking for most affected anadromous species. Therefore, one new passageway was assumed to be warranted.

needs of all species preferring that habitat were not summed because the habitat benefits each of the species simultaneously. The costs of each scaled restoration activity were then summed to determine the total cost necessary to offset all Pilgrim losses.

The total HRC estimates for the Pilgrim facility are provided in Table 15.1, along with the species requiring the greatest level of implementation for each restoration. The scale of implementation, the unit costs and the total costs in this table were rounded to two significant digits to avoid false precision. Resulting total costs also carry two significant digits. These costs can be converted to annualized values by specifying a time period and interest rate.

Results of the Pilgrim HRC analysis indicated that the present value cost of restoration to offset I&E would be at least $140 million. This is significantly greater than the $6–7 million of foregone recreational and commercial fishing value calculated in the Pilgrim case study for the USEPA's CWA Section 316(b) proposed rule for existing facilities.[1] Recreational and commercial fishing values are substantially lower primarily because they include only a small subset of the species, life stages and human use services associated with I&E losses. In this case, the HRC analysis might reasonably support the conclusion that more expensive technologies can be justified to prevent I&E than when fishing alone is considered in the analysis of the benefits of species protection. However, the proper use and context for HRC in CWA Section 316(b) rulemaking and permit issuance is still being debated by the USEPA, the Office of Management and Budget, the Department of Energy, states, environmental groups, and the regulated community. As of this writing, HRC analyses are in the administrative records for the national rulemaking and specific NPDES permits. However, both the rulemaking and specific permits are likely to undergo additional review, and possibly litigation. At a minimum, the HRC analyses are likely to liven the debate about whether and how various costs and benefits should be compared when issuing regulations and permits under Section 316(b). Hopefully, HRC analyses will also point to additional data and analyses that can make future comparisons of costs and benefits more useful, more complete, and more equitable.

In addition to broadening the species, life stages, and services addressed relative to the conventional valuation of I&E losses through recreational and commercial fishing impacts, the Pilgrim HRC analysis provides a roadmap for mitigating I&E losses and for closing critical data gaps through effective monitoring. Effective restorations with reliable biological data can broaden analyses of public losses and increase the production of fish per restoration dollar. As a result, the public can benefit from more effective restoration and additional technology implementation justified by a more comprehensive evaluation of the benefits of reducing I&E.

DISCUSSION

Relation of the HRC Method to the Conceptual Approach

Figure 15.3 shows how the HRC approach follows the conceptual approach in Figure 9.1. First, Step 1 of the HRC method (quantifying losses by species) is similar to analysis and characterization of baseline risk. Next, formulation of

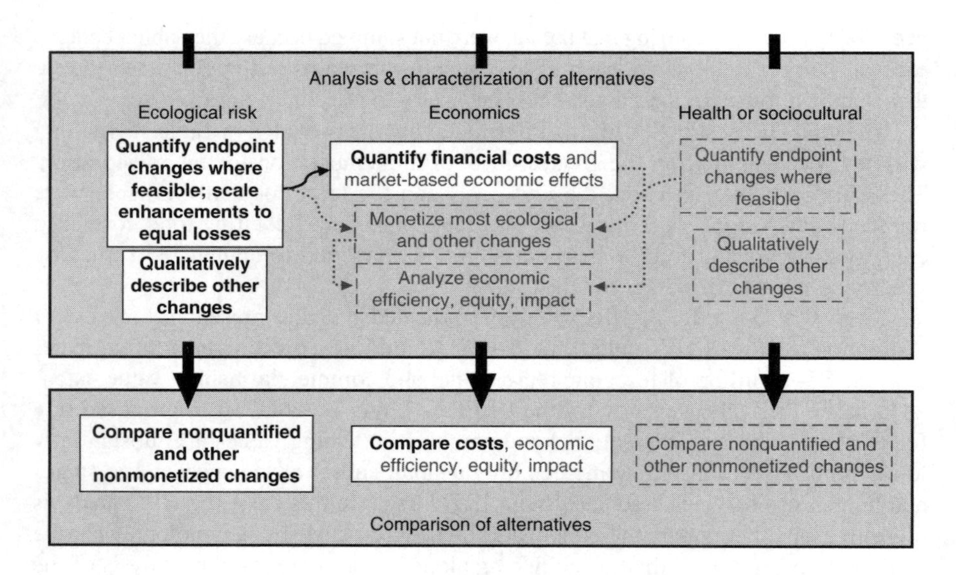

Figure 15.3 Techniques used for analysis, characterization, and comparison of management alternatives for Pilgrim I&E losses, as compared to the example shown in Figure 9.2. White boxes and bold type show features included in this analysis.

alternatives occurs in three stages. The first (HRC Step 2, identify habitat requirements of impacted species) involves determining the habitat constraints on the organisms. HRC Step 3 then identifies all relevant options that would increase the production of the organisms. To offset losses, a scaling step is required to match the production increase. Uncertainty in the biology and engineering aspects constrain this step. HRC Step 4 (consolidating, categorizing, and prioritizing alternatives) completes the formulation of alternatives. Once the preferred alternative for each species is chosen, unit costs are developed to estimate a total cost for offsetting losses. These are described as financial costs. These restoration costs can be compared to the costs of technologies for reducing losses to determine the most cost-effective option.

Efficacy of the HRC Method

As illustrated by the Pilgrim case study, the HRC method can be applied in several contexts under the CWA and other statutes. It enables analysts to do the following:

1. Identify specific suites of restoration options that provide the types of ecological service outputs that correspond to the regulatory or permit-specific impacts of concern (e.g., coastal wetland and eelgrass programs to support nurseries for species of relevance).
2. Define the scale and scope of the restoration program elements so that they match the level of environmental damage (or benefit) of relevance (e.g., how many acres

of coastal wetlands are required to promote fishery yields comparable to the regulatory target).

3. Estimate the cost of the restoration options, and compare those costs to estimates from alternative regulatory compliance strategies (e.g., technological controls) to identify which suite of actions is most cost-effective at generating the desired types and levels of ecological outputs.

These three outputs of the HRC process can be used (individually or collectively) for various regulatory and permitting applications. If one needs to identify restoration options that might offset an environmental impact from an accidental release or permit violation, the HRC approach provides an ecologically scaled answer. Or, an analyst can define the type and scale of restoration needed to offset impacts that would occur without various technology-based pollution control approaches. Cost estimates then allow comparisons of technology-based options versus suitably scaled and targeted restoration alternatives, which are useful for both the regulatory agency and the permitee. Even when a full assessment of the value (or public benefit) of the resources that could be saved via regulatory or permit-driven improvements is required, a restoration cost-based analysis can still provide useful supplemental information about relative cost-effectiveness.

Costs versus Values

Finally, in evaluating the HRC method and its uses, we note that it is important to recognize that fundamental distinctions exist between the costs of supplying a resource or service and the values that are derived from those resources and related services. Values are based on the demand for a good or service and the opportunity costs associated with obtaining the good or service. The cost of providing such services may exceed the combined use and nonuse values of the services or may be less than these values. In recent discussion of the costs of replacing wetland function, natural resource economists argued that "assessment of these costs can serve as an indicator of a lower bound on the economic value of the associated wetland function."[55]

Unfortunately, few studies address the relationship between costs and values. However, information exists that suggests that the cost of habitat restoration may be less than the value people place on the natural resources that have been damaged and are to be restored, and that people are aware of the ecological significance and value of habitat restoration, at least for some sites. In such cases, restoration cost can serve as a proxy for value in the absence of reliable data from stated preference surveys or other valuation studies.

Two examples of studies that estimated both resource values and restoration costs are natural resource damage assessments for the *Nestucca* oil spill off the Washington coast and for PCB contamination in Green Bay, Wisconsin and Michigan (see Chapter 14). Results suggest that in each case restoration costs are significantly less than public values for the restored natural resources. The *Nestucca* oil spill released 231,000 gallons of fuel oil in 1988, killing 41,000 seabirds, and oiling 200 miles of coastline.[56]

A contingent valuation survey estimated the value of lost resources and services from the spill.[a] Total losses were estimated to be worth $162 million (1994 U.S. dollars). This compares with total costs for restoration, including increasing seabird populations by restricting the use of gill-nets, of $10 million to $30 million (1994 U.S. dollars).

The Green Bay Natural Resource Damage Assessment[57] included estimates of the value of PCB-caused service losses within the Lower Fox River and Green Bay, as well as the cost for restoration of wetland habitat and improved water clarity in Green Bay, scaled to be equivalent in value to losses, using a stated preference survey.[58] Survey results show that sufficient restoration would cost $111 million to $268 million, whereas the WTP for this restoration was $254 million to $362 million.[b]

Voluntary habitat restoration to improve the production of aquatic organisms is another indication that habitat replacement may be worth its cost. A voluntary commitment of resources suggests economic efficiency and positive net benefits (as opposed to mandated actions that do not necessarily reveal values of those required to pay). In addition, long-standing legislation to preserve or restore aquatic habitats is a broad indication that habitat restoration is widely perceived as being worth its cost to society.[c] Site-specific information about WTP does not exist for many sites and natural resource losses. However, a number of studies indicate a WTP for habitat restoration, and survey data could be developed to test the value of habitat restoration for section 316(b) and other permitting situations.

ACKNOWLEDGMENTS

Funding for this work was provided in part by the New England Interstate Water Pollution Control Commission and the USEPA in support of the USEPA's national rulemaking and NPDES permitting. Useful comments, guidance, and insights and previous versions of portions of this paper were provided by Phil Colarusso and Mark Stein (USEPA, Region 1) and Lynne Tudor (USEPA, Office of Science and Technology). Any errors and omissions should be attributed to the authors. The use of USEPA funding does not imply that USEPA necessarily accepts or endorses this paper or its findings.

[a] The contingent valuation method and other stated preference techniques such as conjoint analyses address ecological services and nonuse values, and have been upheld in federal court (State of Ohio v. U.S. Department of Interior, 880 F.2d 432, 444–446, 448, 450, 459 [D.C. Cir. 1989]) and supported by a National Oceanic Atmospheric Administration panel co-chaired by two Nobel Laureate economists, although the methods remain controversial.[59]

[b] The total value equivalency study, described more completely in Chapter 14, found WTP as high as $610 million to avoid the "no action" cleanup option corresponding to 100 years of continuing PCB-caused injuries. However, there was not enough realistic habitat and nonpoint source restoration in and near Green Bay to offset the value of 100 years of PCB-caused injuries. Therefore, we compare only the costs and values of 20-year and 40-year scenarios here.

[c] As national examples, the Chesapeake Bay Program and the National Estuaries Program, authorized and funded since 1983 and 1987, respectively, identify, restore, and protect nationally significant estuaries and habitats within the United States. Local boards direct habitat restoration (and other) activities within designated estuaries, further suggesting that habitat restoration often is highly valued. As a further example, Save the Bay (a nonprofit organization dedicated to restoring and protecting the environmental quality of Narragansett Bay and its watershed) voluntarily restores submerged aquatic vegetation because it explicitly recognizes the value of this habitat to sustaining and increasing fish populations. Similar examples of voluntary habitat restorations are likely to exist near many § 316(b) facilities where HRC analyses could be conducted.

REFERENCES

1. USEPA, National pollutant discharge elimination system – Proposed regulations to establish requirements for cooling water intake structures at Phase II existing facilities, *Fed. Reg.*, 67, 17122, 2002.
2. Strange, E.M., Allen, P.D., Beltman, D., Lipton, J., and Mills, D., The habitat-based replacement cost method for assessing monetary damages for fish resource injuries, *Fisheries*, 29, 17, 2004.
3. Fisher, A. and Raucher, R., Intrinsic benefits of improved water quality: Conceptual and empirical perspectives, *Adv. Appl. Micro-Econ.*, 3, 37, 1984.
4. Boyd, J., King, D., and Wainger, L.A., Compensation for lost ecosystem services: The need for benefit-based transfer ratios and restoration criteria, *Stanford Environ. Law J.*, 20, 393, 2001.
5. Fischman, R.L., The EPA's NEPA Agency duties and ecosystem services, *Stanford Environ. Law J.*, 20, 497, 2001.
6. Heal, G., Daily, G.C., Ehrlich, P.R. et al., Protecting natural capital through ecosystem service districts, *Stanford Environ. Law J.*, 20, 333, 2001.
7. Herman, J.S., Culver, D.C., and Salzman, J., Groundwater ecosystems and the service of water, *Stanford Environ. Law J.*, 20, 479, 2001.
8. Ruhl, J.B. and Gregg, R.J., Integrating ecosystem services into environmental law: A case study of wetlands mitigation banking, *Stanford Environ. Law J.*, 20, 365, 2001.
9. Salzman, J., Thompson, B.H., Jr., and Daily, G.C., Protecting ecosystem services: Science, economics, and law, *Stanford Environ. Law J.*, 20, 309, 2001.
10. Wainger, L.A., King, D., Salzman, J., and Boyd, J., Wetland value indicators for scoring mitigation trades, *Stanford Environ. Law J.*, 20, 413, 2001.
11. Stevens, T.H., Echeverria, J., Glass, R., Hager, T., and More, T.A., Measuring the existence value of wildlife: What do CVM estimates really show?, *Land Econ.*, 67, 390, 1991.
12. Loomis, J., Kent, P., Strange, L., Fausch, K., and Covich, A., Measuring the total economic value of restoring ecosystem services in an impaired river basin: Results from a contingent valuation survey, *Ecol. Econ.*, 33, 103, 2000.
13. Woodward, R.T., Kent, P., Strange, L., Fausch, K., and Covich, A., The economic value of wetland services: A meta-analysis, *Ecol. Econ.*, 37, 257, 2001.
14. Walsh, R.G., Loomis, J.B., and Gillman, R.A., Valuing option, existence, and bequest demands for wilderness, *Land Econ.*, 60, 14, 1984.
15. Whitehead, J.C. and Blomquist, G.C., Measuring contingent values for wetlands: Effects of information about related environmental goods, *Wat. Resour. Res.*, 27, 2523, 2001.
16. Hagen, D., Vincent, J., and Welle, P., Benefits of preserving old growth forests and the spotted owl, *Contemp. Policy Issues*, 10, 13, 1992.
17. Loomis, J. and Ekstrand, E., Economic benefits of critical habitat for the Mexican spotted owl: A scope test using a multiple bounded contingent valuation survey, *J. Agric. Resour. Econ.*, 22, 356, 1997.
18. Feinerman, E. and Knapp, K., Benefits from groundwater management: Magnitude, sensitivity, and distribution, *Am. J. Agric. Econ.*, 65, 703, 1983.
19. Farber, S., The value of coastal wetlands for protection of property against hurricane wind damage, *J. Environ. Econ. Manage.*, 14, 143, 1987.
20. Layton, D. and Brown, G., Heterogeneous preferences regarding global climate change, *Estuaries*, 7, 478, 1998.
21. Kaoru, Y., Differentiating use and nonuse values for coastal pond water quality improvements, *Environ. Resour. Econ.*, 3, 487, 1993.

22. Cohen, M.A., The costs and benefits of oil spill prevention and enforcement, *J. Environ. Econ. Manage.*, 13, 167, 1986.
23. Hanemann, M., Loomis, J., and Kanninen, B., Statistical efficiency of double-bounded dichotomous choice contingent valuation, *Am. J. Agric. Econ.*, 73, 1255, 1991.
24. Grigalunas, T.A., Opaluch, J.J., French, D., and Reed, M., Measuring damages to marine natural resources from pollution incidents under CERCLA: Applications of an integrated ocean systems/economic model, *Mar. Resour. Econ.*, 5, 1, 1988.
25. Rowe, R.D., Shaw, W.D., and Schulze, W., Nestucca oil spill, in *Natural Resource Damages*, Ward, K. and Duffield, J., Eds., Wiley and Sons, New York, 1992, 527.
26. USEPA, Guidelines for preparing economic analyses, EPA/240/R-00/003, U.S. Environmental Protection Agency, Washington, D.C., 2000.
27. Cross, F.B., Natural resource damage valuation, *Vand. L. Rev.*, 42, 269, 1989.
28. Summers, J.K., Simulating the indirect effects of power plant entrainment losses on an estuarine ecosystem, *Ecol. Model.*, 49, 31, 1989.
29. Peterson, C.H. and Lubchenco, J., Marine exosystm services, in *Nature's Services, Societal Dependence on Natural Ecosystems*, Daily, G.C., Ed., Island Press, Washington, DC, 1997, 177.
30. Postel, S. and Carpenter, S., Freshwater ecosystem services, in *Nature's Services, Societal Dependence on Natural Ecosystems*, Daily, G.C., Ed., Island Press, Washington, DC, 1997, 195.
31. Holmlund, C.M. and Hammer, M., Ecosystem services generated by fish populations, *Ecol. Econ.*, 29, 253, 1999.
32. Strange, E.M., Fausch, K.D., and Covich, A.P., Sustaining ecosystem services in human-dominated watershed: Biohydrology and ecosystem processes in the South Platte River Basin, *Environ. Manage.*, 24, 39, 1999.
33. AFS (American Fisheries Society), *Sourcebook for Investigation and Valuation of Fish Kills*, American Fisheries Society, Bethesda, MD, 1993.
34. AFS (American Fisheries Society), *Investigation and Monetary Values of Fish and Freshwater Mussel Kills*, American Fisheries Society, Bethesda, MD, 2003.
35. State of Maryland, Monetary Value of Tidal Water and Non-Tidal Water Aquatic Animals, Title 08 Department of Natural Resources: Subtitle 02 Tidewater Administration: Chapter 09 Aquatic Animals, 4923 COMAR 08.02.09.01, Available at http://www.dsd.state.md.us/comar/08/08.02.09.01.htm. Accessed March 30, 2004.
36. White, R.J., Karr, J.R., and Nehlsen, W., Hatchery reform in the northwest: Issues, opportunities, and recommendations, Report prepared for Oregon Trout, Portland, OR, 1997.
37. Hilborn, R.W., Hatcheries and the future of salmon in the Northwest, *Fisheries*, 17, 5, 1992.
38. ENSR, Redacted Version 316 demonstration report — Pilgrim Nuclear Power Station, Prepared for Entergy Nuclear Generation Company, Westford, MA, 2000.
39. USEPA, Determination regarding issuance of proposed NPDES Permit No. MA0025135 (Pilgrim Power Plant Unit II), Mar. 11, 1977.
40. Manomet Center for Conservation Sciences, Southeastern Massachusetts Natural Resource Atlas, Available at http://www.manoment.org/semass/atlas.html. Accessed November 29, 2001.
41. Kelly, B., Lawton, R., Malkoski, V., Correia, S., and Borgatti, M., Pilgrim Nuclear Power Station marine environmental monitoring program report Series No. 6. Final report on Haul-Seine Survey and impact assessment of Pilgrim Station on Shore-Zone Fishes, 1981-1991, Prepared by the Department of Fisheries, Wildlife, and Environmental Law Enforcement, Massachusetts Division of Marine Fisheries for Boston Edison Company, Sandwich, MA, 1992.

42. Entergy Nuclear Generation Company, Marine ecology studies related to operation of Pilgrim Station, Semi-Annual Report Number 55, January 1999-December 1999, 2000.
43. Stratus Consulting, Habitat-based replacement costs: An ecological valuation of the benefits of minimizing impingement and entrainment at the cooling water intake structure of the pilgrim nuclear power generating station in Plymouth, Massachusetts, Prepared for The New England Interstate Water Pollution Control Commission and The U.S. Environmental Protection Agency, Region 1, February 5, 2002.
44. Boston Edison Company, Marine ecology studies related to operation of Pilgrim Station, Semi-Annual Report Number 37, January 1990-December 1990, April 30, 1991.
45. Boston Edison Company, Marine ecology studies related to operation of Pilgrim Station, Semi-Annual Report Number 39, January 1991-December 1991, April 30, 1992.
46. Boston Edison Company, Marine ecology studies related to operation of Pilgrim Station, Semi-Annual Report Number 41, January 1992-December 1992, April 30, 1993.
47. Boston Edison Company, Marine ecology studies related to operation of Pilgrim Station, Semi-Annual Report Number 43, January 1993-December 1993, April 30, 1994.
48. Boston Edison Company, Marine ecology studies related to operation of Pilgrim Station, Semi-Annual Report Number 45, January 1994-December 1994, April 30, 1995.
49. Boston Edison Company, Marine ecology studies related to operation of Pilgrim Station, Semi-Annual Report Number 46, January 1995-June 1995, October, 1995.
50. Boston Edison Company, Marine ecology studies related to operation of Pilgrim Station, Semi-Annual Report Number 47, January 1995-December 1995, April 30, 1996.
51. Boston Edison Company, Marine ecology studies related to operation of Pilgrim Station, Semi-Annual Report Number 49, January 1996-December 1996, April 30, 1997.
52. Boston Edison Company, Marine ecology studies related to operation of Pilgrim Station, Semi-Annual Report Number 51, January 1997-December 1997, April 30, 1998.
53. Boston Edison Company, Marine ecology studies related to operation of Pilgrim Station, Semi-Annual Report Number 53, January 1998-December 1998, April 30, 1999.
54. Goodyear, C.P., Entrainment impact estimates using the equivalent adult approach, FWS/OBS-78/65, U.S. Fish and Wildlife Service, 1978.
55. Florax, R., Nijkamp, P., Willis, K., Eds., *Comparative Environmental Economic Assessment*, Edward Elgar Publishing Inc., Northampton, MA, 2002, 249.
56. Rowe, R.D., Schulze, W.D., Shaw, W.D., Schenk, D.S., and Chestnut, L.G., Contingent valuation of natural resource damage due to the Nestucca Oil Spill. Final Report, Prepared for Department of Wildlife, State of Washington, Olympia, WA by the British Columbia Ministry of Environment, Victoria, BC, Environment Canada, Vancouver, BC, 1991.
57. USFWS and Stratus Consulting, Restoration and Compensation Determination Plan (RCDP). Lower Fox River/Green Bay Natural Resource Damage Assessment, Prepared for U.S. Fish and Wildlife Service, U.S. Department of the Interior, U.S. Department of Justice, Oneida Tribe of Indians of Wisconsin, National Oceanic and Atmospheric Administration, Little Traverse Bay Bands of Odawa Indians, Michigan Attorney General, October 25, 2000. http://MIDWEST.FWS.GOV/NRDA/RCDP-1.pdf. Accessed June 1, 2004.
58. Breffle, W.S. and Rowe, R.D., Comparing choice question formats for evaluating natural resoure tradeoffs, *Land Econ.*, 78, 298, 2002.
59. Arrow, K.J., Solow, R., Portney, P.R. et al., Report of the NOAA panel on contingent valuation, *Fed. Reg.*, 58, 4601, 1993.

CHAPTER 16

CONCLUSIONS

Randall J.F. Bruins and Matthew T. Heberling[a]

CONTENTS

[a] The views expressed in this chapter are those of the authors and do not necessarily reflect the views or policies of the U.S. Environmental Protection Agency.

1-56670-639-4/05/$0.00+$1.50

Part I of this text introduced fundamental concepts and methods in ecological risk assessment (ERA) and economic analysis of environmental problems as applied to watersheds, and it developed a conceptual approach for their integration for watershed management (see Chapter 9, and especially Figure 9.1). Part II described and evaluated case studies from six U.S. watershed settings. Three of these (Chapters 10–12) were U.S. Environmental Protection Agency (USEPA)–funded research studies in which watershed-ERA (W-ERA) was conducted initially, followed several years later by economic analysis. These studies were carried out in the Big Darby Creek watershed in Ohio, the Clinch Valley of Virginia and Tennessee, and the Platte River in Colorado, Wyoming, and Nebraska. A study funded by the Hudson River Foundation (Chapter 13) addressed regional planning needs in two watersheds comprising Dutchess County, New York, and the U.S. Fish and Wildlife Service (USFWS), acting as natural resource cotrustee, funded an economic study as part of natural resource damage assessment for the Fox River and Green Bay, in Wisconsin and Michigan (Chapter 14). Finally, the USEPA funded a study in Cape Cod Bay, Massachusetts, as part of permit development for a cooling water intake. In each case, elements of ERA and economic analysis were used, and the extent of their integration has been critically evaluated. This closing chapter offers a brief overview of the case studies, highlighting their diversity and similarities, and it provides concluding observations with respect to the integration process.

OVERVIEW OF CASE STUDIES

Ecological Threats

Species that were foci of concern in these cases included fish, shellfish, and aquatic-dependent bird populations (Table 16.1). Four of the cases involved species listed as threatened or endangered, although the authorities of the Endangered Species Act (ESA) were invoked in only one case. Recreationally or commercially caught fish species were often important as well. In two locations, freshwater mussels were studied owing to their extreme vulnerability; a commercially important marine mussel was among organisms impacted in a third study. In addition to particular species, ecosystem condition was frequently a concern; multimetric indices determined the integrity of the fish communities, bottom-dwelling communities, and stream habitat quality, both in settings of outstanding conservation importance such as Big Darby Creek and the Clinch Valley and in less distinctive settings such as the Hudson River tributaries of Dutchess County.

In three studies, threats were due to point sources, albeit involving very different kinds of stressors. Fox River and Green Bay had experienced a long-time point discharge of polychlorinated biphenyls (PCBs), although by the time of the study, contaminated sediments had become a widely distributed source of these toxic compounds. The Cape Cod Bay case studied the destruction of aquatic organisms by an intake structure for power plant cooling water. At several key points along the Platte River, water is withdrawn for various uses, and the resultant flow reduction is a primary stressor in critical downstream reaches. The other three studies were

Table 16.1 Summary and Comparison of Selected Features of Six Watershed Case Studies[a]

Chapter and Case Study Location	Ecological Entities or Services of Concern	Source/Stressors	Management Action or Policy Evaluated	Means of Public Input	Evaluation Approach or Criterion	Decision Context
10. Big Darby Creek, OH	Endangered and rare fish and mussels, biodiversity, recreation	Suburban development/run off, degraded riparian habitat	Limit housing density or preserve agriculture	CVM survey of watershed residents, near-residents, and other OH residents	Maximize mean individual utility (WTP)	Not specified
11. Clinch Valley, VA & TN	Endangered and rare fish and mussels, biodiversity, recreation, tourism	Agriculture/runoff, degraded riparian habitat	Establish agriculture-free zone along river and tributaries	Conjoint survey of watershed residents	Maximize mean individual utility (WTP)	Not specified
12. Platte River, CO, WY, & NE	Endangered and other distinctive migratory birds, tourism	Water withdrawal/ degraded channel and floodplain habitat	Increase instream flow and restore wet meadow areas	Preference survey of residents of three states	Maximize sum or product or minimize difference in utility of interest groups	Three-state negotiation to meet ESA requirements
13. Hudson River (Dutchess County), NY	Aquatic ecosystem biotic integrity, quality of life	Suburban development/ runoff, degraded riparian habitat	Improve regional development and land-use planning	Stakeholder meetings using multicriteria decision aid (proposed)	Multicriteria evaluation of economic, social, and environmental indicators	Goal-setting and planning activities of an intergovernmental watershed council
14. Fox River & Green Bay, WI & MI	Endangered birds, sport fish, recreational fishing	Historic PCB releases/ contaminated sediments	Reduce agricultural runoff, enhance wetlands and regional parks, remove contaminated sediments	Conjoint survey of residents near Green Bay	Mean individual utility of remedy compensates for harm	NRDA process and settlement negotiations under CERCLA
15. Cape Cod Bay, MA	Fish and shellfish populations, recreational and commercial fishing	Power plant cooling water intake/impingement and entrainment	Increase habitat of affected species' populations	None (resource replacement is assumed to make the public whole)	Species-by-species replacement of number of individuals killed	Determination of appropriate costs of permit requirements under CWA

[a] Each category lists the primary subject (source, stressor, endpoint, etc.) addressed by the ecological and economic case study, which is not necessarily the only or the most important issue in this watershed. See respective chapters for more details; see front matter for definition of acronyms.

primarily concerned with land-use impacts, including nonpoint-source pollution and the degradation or loss of riparian or floodplain habitat due to housing development or agriculture.

Proposed Management Strategies

The early history of watershed resources management in the United States was characterized by inadequate appreciation for the importance of the land–water connection; and early aquatic environmental protection strategies, which focused on point sources, ignored the land–water connection as well (see Chapter 2). By contrast, the management actions or policies evaluated in these six case studies were vitally concerned with uses of land and were only indirectly concerned with pollution control in the conventional sense (Table 16.1). Improved planning of suburban development and restoration of riparian zones and wetlands were studied most often. Interestingly, this was true even where impacts were due to point sources. Since the remediation of PCB-contaminated sediments in the Fox River and Green Bay was feasible only to a limited degree, other avenues for restoring ecosystem services were the focus of study, and these included wetland creation and restoration, measures to reduce agricultural runoff, and enhancement of regional parks. Similarly, because of the difficulty of eliminating harm to fish and shellfish populations from cooling water intake, the restoration of various kinds of wetland habitats favorable to the species impacted was evaluated as an alternative. In the Platte River case, restoration of flow remained the priority for satisfaction of ESA provisions, but the restoration of thousands of acres of wet meadows was also a formal requirement.

Measurement of Preferences

In four of the six cases, approaches for determining social preferences or values associated with a proposed action or policy included surveying members of the public (Table 16.1). Three of these studies used stated-preference methods to estimate mean individual utility (see Appendix 5-A); the contingent valuation method (CVM) was used in the Big Darby Creek case, and conjoint analysis was used for both the Clinch Valley and Fox River and Green Bay. In these three cases, utility was determined as willingness to pay (WTP), although utility itself, independent of a monetary measure, was used in the Fox River and Green Bay case to balance harm and proposed remedy. The fourth survey approach, used in the Platte River case, also elicited respondents' preferences but did not determine WTP; instead, a utility measure was used in the evaluation of various bargaining model solutions. The latter approach treated interest groups, rather than individuals, as the basis for modeling, and individual preferences were aggregated according to group membership. Another unique aspect of this case study was its hypothesis that among-group differences in preference could be reduced through education, changing the bargaining model outcomes.

Two cases did not employ surveys. The Hudson River case study proposed to explore preferences or values through an analytical and deliberative process involving stakeholder interactions and the use of a multicriteria decision aid (see also

Appendix 9-B). A stated goal of this approach was to make explicit, rather than numerically boil down, the complex trade-offs involved in regional planning decisions. The researchers in this case assumed that a greater awareness of consequences would have a salutary effect on the planning process. The absence of a survey or other preference measurement approach in the Cape Cod Bay case illustrates a distinctive feature of restoration-scaling approaches that are based on replacement of equivalent ecological resources or habitats. When resources or habitats proposed to be restored provide services different in kind from those lost, as was true in the Fox River and Green Bay case, a survey may be needed to determine value equivalence. In the Cape Cod Bay case, however, fish species to be restored were judged equivalent in kind to those lost, obviating a separate measure of value. Once such a judgment is made, economic benefit need not be measured, and the subsequent role of the economist is limited to that of determining restoration costs. Indeed, the replacement cost method does not actually determine the monetary value of the loss or the net benefit of the restoration action; it only ensures that the action is at least compensatory.

Prospects for Success

The management approaches employed by this small sample of projects would seem to reflect the oft-repeated claim that future improvements in aquatic ecosystem quality will depend more on the implementation of watershed-based strategies — including the improved management or restoration of uplands, floodplains, riparian zones, and wetlands — than on further refinement of point-source control methods. The extent to which the particular strategies applied in these studies will be effective remains unclear, however. In the Big Darby Creek and Clinch Valley case studies, decision mechanisms that might employ the study findings have not been clearly identified. In the Platte River and the Fox River and Green Bay cases, the decision processes involved are clearly defined (Table 16.1), but negotiations have been extremely protracted, and outcomes are uncertain. The habitat replacement cost method applied in Cape Cod Bay has not yet been fully accepted by regulatory agencies, and the use of integrated modeling and stakeholder-engaged planning processes in the Hudson River watersheds of Dutchess County is still getting underway.

But even if unbridled enthusiasm about the value of these integration techniques is unwarranted, and even though most watershed applications still have experimental features, the underlying methods are in most cases well researched and scientifically accepted. Arguably, the main tasks remaining have more to do with improving interdisciplinary communication, coordination, and methodological integration than with theoretical advances. The following section offers specific observations with respect to the improvements needed.

OBSERVATIONS

The following observations do not constitute a comprehensive list of recommendations for integrating ERA and economic analysis. (The conceptual approach for integration presented in Chapter 9 is more complete in that regard.) Rather, they are a set of important statements drawn from an analysis of the six case studies. For the most part, they leave aside issues that are particular either to ERA itself or to economic analysis and focus on the problem of their integration. These conclusions provide further insight on certain topics raised by the conceptual approach, but additional studies are still needed to test that approach more fully.

Achieving Ecological–Economic Integration Requires a Coherent Strategy

The central conclusion arising from evaluation of the case studies is that watershed problems should be approached with a coherent strategy for assessment and management. If decision-makers need to consider both ecological risks and economic factors (and perhaps health and sociocultural factors), a strategy that guides their integration is necessary. The conceptual approach described in Chapter 9 provides one such strategy. The approach is based on the USEPA *Framework for Ecological Risk Assessment*,[1,2] and it modifies or augments that framework as needed to accommodate economic analysis and to address a broader management context. Its elements are similar to those of other frameworks that have been used in environmental management (see Table 9.4 and Appendix 9-A). Although this book presents the conceptual approach before the case studies, to serve as a guide to their evaluation, it was developed following the completion of the three USEPA-sponsored research studies (Chapters 10–12; see also "Goal and Genesis of This Book" in Chapter 1) and should be considered as integral to these concluding observations.

The case studies help illustrate the need for the conceptual approach. The W-ERA components of the three USEPA research studies were not undertaken with economic integration as a goal. The economic components, initiated years later, did have such a goal, but used only a limited set of guiding principles; that is, each economic analysis was to address the same system, problems, and ecological assessment endpoints analyzed by the W-ERA, and it was to be relevant to decision-making. The approaches used were novel, and the results are potentially useful, but in each case their usefulness could have been improved by a more comprehensive approach, as is detailed in the following sections. For example, the lack of an interdisciplinary assessment planning and problem formulation process contributed in one case to divergent views of goals and endpoints. In two cases (Clinch and Platte), management alternatives were formulated for economic analysis, but the likely ecological effects of those alternatives were not quantitatively assessed, limiting the scope of the conclusions. Also, in two cases (Darby and Clinch) the economic analysis tools chosen were not clearly aligned to the relevant decision context; in other words, it was not shown that they were developed with a set of decisions and decision-makers in mind. Use of the conceptual approach for integration theoretically could have helped avoid these limitations.

Restoration-scaling approaches, introduced in Chapter 8 and applied in Chapters 14 and 15, differ from typical project-planning approaches, such as the six-step approach described in Chapter 4 and the more ad-hoc approaches applied in Chapters 10–13. Like other project-planning approaches, restoration-scaling approaches first identify a problem and then conduct planning and analysis to select the best available remedy. Their unique aspect, however, is that once a set of restoration alternatives has been identified, analyzed, and determined to be potentially suitable, the final mix and scale of projects is determined by back-calculation, so as to provide neither more nor less than a compensatory level of ecosystem services or economic value.

This difference may seem to suggest that scaling approaches are not described well by the conceptual approach presented in Chapter 9. However, some degree of scaling can result from any outcome-focused negotiation, where parties do not want to pay any more than necessary to achieve a particular regulatory standard or environmental condition. Scaling may also occur later as part of adaptive implementation, after an initial management step has proven either more or less effective than required.

Integration Requires Assessment Planning and Problem Formulation to be Interdisciplinary

Our conceptual approach emphasizes the need for ecologists and economists (and other specialists, as required) to participate together in the steps of assessment planning and problem formulation. In the three USEPA-sponsored research studies, the fact that the W-ERA and economic analysis were done sequentially, rather than in a more integrated fashion, limited their value for management. In the W-ERA efforts, planning and problem formulation were systematic and painstaking, but economists were not involved. When the economic studies were later initiated, informational meetings were held with members of the W-ERA teams, but these focused on assessment and did not reopen fundamental questions about the nature of the management problems, so views about management goals and objectives were not necessarily the same.

Of these three studies, the lack of a common view was most pronounced in the Platte River case. The W-ERA team viewed the vegetative diversity and dynamic character of the braided-river-channel landscape mosaic as an endpoint in itself, as well as the diversity of fauna using its various habitats. The economic team focused more narrowly on current efforts among the three Platte River states and the federal government to reach agreement on provisions to meet the needs of three endangered species, the interior least tern (*Sterna antillarum athalassos*), the piping plover (*Charadrius melodus*), and the whooping crane (*Grus americana*). The W-ERA analyzed conditions affecting the use of river segments by sandhill cranes (*Grus canadensis*), whose needs overlap substantially with the endangered species', but they also analyzed the effects of landscape patch size on grassland and woodland breeding birds, whose needs are less relevant to, and in some cases conflict with, those of the endangered species. In the Darby and Clinch studies, there was not a significant divergence of views, but the lack of a joint assessment planning exercise may have contributed to the failure to identify a decision context for the economic assessment, as well as certain other weaknesses discussed later in this chapter.

Research is Needed on the Development and Use of Integrated Conceptual Models

A conceptual model is a graphical depiction, typically a box-and-arrow diagram, of the hypothesized relationships between human activities, ecological stressors, and ecological assessment endpoints (refer to "Problem Formulation" in Chapter 3 for an explanation, and Figure 11.2 for an example). According to the conceptual approach for integration, an interdisciplinary problem formulation process should include the development of extended conceptual models (see "Problem Formulation" in Chapter 9). In extended models, risk hypotheses show how sources and stressors affect economic endpoints, or *services*, as well as ecological assessment endpoints. An extended model also includes risk management hypotheses, which we have defined as explanations of how management alternatives are expected to affect sources, exposures, effects, and services. Their development should involve environmental program managers, if the management actions are in the form of programs or policies, and environmental engineers or restoration specialists, if the actions involve structural changes to ecosystems. Their development also requires the involvement of landowners and other stakeholders whose active cooperation may be instrumental in solving the environmental problem. Extended models were not developed in these case studies, and at present we are aware of very few examples of the use of these extended models in a risk assessment context. The National Center for Environmental Assessment of the USEPA's Office of Research and Development is presently conducting work to gain experience with their development and use.[a]

Clearly Formulated Management Alternatives Facilitate Integrated Analysis

Describing management alternatives is an important way to frame the integration problem. Any given alternative will entail a unique bundle of ecological, economic, and other changes. Some of those changes may be judged beneficial and some detrimental — some to a greater or lesser degree — and resulting gains or losses in human well-being may accrue unevenly across groups or locations. The heart of the integration problem usually includes somehow evaluating the signs, magnitudes, and distributional effects of all these changes collectively, to determine whether one alternative can be clearly preferred over another. However, ERA is sometimes conducted only for the purpose of understanding threats and their causes so that potential management approaches can be identified. Assessors may stop short of helping to develop and analyze specific approaches. Without assessments of both ecological and economic outcomes of specific alternatives, the potential for integration is reduced, as was true in some of these studies.

In the Big Darby watershed, three possible land-use scenarios (low-density ranchettes, low-density clusters, and preservation of agriculture) and a most-likely base case (high-density residential) were described in some detail, and their respective ecological, economic, and quality-of-life impacts were rated by the researchers. Using the CVM, respondents were able to jointly value the different impacts.

[a] Contact the authors of this chapter for additional information.

Although each respondent was posed only one of the three possible choices, mean WTP served as a kind of referendum on these three alternatives.

In the Platte study, the problem of determining a preferred policy was viewed not as one of determining mean WTP but rather as one of determining what policy alternatives were likely to be mutually acceptable to competing stakeholder groups. Like the first two, it elicited responses to preference questions that combined ecological and economic dimensions, but unlike them it used this information to model a negotiation process. Using principles from game theory (the study of interacting decision-makers),[3] the model analyzed 125 hypothetical policies for meeting endangered species needs, where a given policy described the method of meeting those needs, its cost, and who would pay. As in the Clinch case study, the expected ecological outcomes of the policies were not analyzed.

In the Clinch Valley study, two hypothetical policies for establishing voluntary agriculture-free riparian zones (i.e., a narrow zone and a wider zone), compensated by property or income tax revenues, were employed in a conjoint survey. The choice sets included in the survey were generated as random combinations of these policies and other attributes describing potential ecological outcomes and individual payments, and therefore the choices did not correspond to specific policy scenarios. However, the resulting choice model could be used to generate a mean WTP for obtaining any policy scenario that could be described from those attributes (as compared to the status quo), and such a value would have an interpretation similar to the Big Darby result. As mentioned above, however, the expected ecological outcomes of such a policy were not estimated.

The Hudson River watershed study described an economic and ecological modeling system being set up to assess impacts of regional development scenarios. The focal scenario was a planned expansion of the regional semiconductor industry; regional land-use policies were to be examined in conjunction. The modeling system is intended to compute values for economic, social, and environmental indicators corresponding to scenarios.

The Fox River and Green Bay study identified various kinds of project alternatives for restoring diminished ecosystem services in the region because complete removal of PCB-contaminated sediments was infeasible. Once these project categories were identified, a survey was designed to evaluate their comparative value for area residents, and ecological analysis quantified the level of services a given project would provide.

In habitat replacement cost studies, such as the one conducted in Cape Cod Bay, and in habitat equivalency assessments more generally,[a] habitat restoration is used to restore lost ecosystem services. In such cases, the selection of management alternatives — in this case the identification of specific types and locations of habitat for restoration — is a key point for ecological–economic collaboration. Analysts must determine that the resources to be restored are both ecologically and economically equivalent to those lost (see "Quantifying Ecosystem Services" in Chapter 8). Since management alternatives are important for economic analysis and for decision-making, their formulation should receive careful attention from all parties involved in an assessment, and their ecological outcomes should be estimated.

[a] Habitat replacement cost is a hybrid of the resource equivalency and habitat equivalency methods described in Chapter 8.

Careful Effort is Required to Relate Ecological Endpoints to Economic Value

An important step in the problem-formulation phase of ERA is the selection of ecological assessment endpoints. Assessment endpoints are chosen that are considered ecologically relevant, susceptible to the stressors of concern and relevant to the environmental management objectives (see "Problem Formulation" in Chapter 3). The likelihood of adverse effects on these endpoints is described in the risk-characterization phase ("Risk Characterization" in Chapter 3). The challenge of ERA–economic integration includes determining economic value (see Chapter 5) associated with those changes as well as characterizing other linkages between the ecological system, management actions, and economic value (see "Problem Formulation" in Chapter 9).

In these case studies, endpoints chosen for ERA because of their ecological importance sometimes posed a challenge for economic analysis. Whereas the freshwater mussel faunas of the Big Darby and Clinch systems are considered ecologically significant, members of the general public who are unaware of their diversity and threatened status may be unconcerned about their survival. To counter this problem, in the Clinch study the survey text mentioned mussels ten times in its brief introductory paragraphs, explaining their unusual degree of diversity in the Clinch Valley, their usefulness as an indicator of water quality, and their sensitivity to pollutants and susceptibility to crushing by the hooves of cattle, before posing the choice sets. In the central reach of the Platte River, where management concerns have centered on endangered or conspicuous migratory waterfowl, the landscape–ecological viewpoint employed in the W-ERA treated landscape diversity, and several less conspicuous bird species, as additional endpoints. These endpoints did not factor in the economic (i.e., game theoretic) analysis, but if efforts had been made to value these endpoints, similar problems would have been faced.

Another complication occurred when the measurement methods that were used to express the ecological endpoints, or were a surrogate for the endpoints, were not readily understandable to the public. For example, even if the public considers a diverse stream fauna to be important, they may have difficulty determining what they would be willing to give up obtain, for example, a 3-point or 10-point improvement in a multimetric ecological index. Since these indices may be composites of ten or so individual measures, it is impossible to make a scientifically precise statement about the meaning of any such change. Yet because these indices are becoming widely relied upon to indicate the presence or absence of biological impairment ("Water Quality Standards and Ecological Risk Assessment" in Chapter 6 and Appendix 6-A), they are likely to be a critical part of the available knowledge base about ecological risk in a watershed. Ways must be found to adequately communicate their meaning if individuals are to comprehend how such changes affect their welfare.

In the Big Darby CVM study, the index of biotic integrity (IBI, a fish assemblage indicator) and invertebrate community index (ICI, a stream-bottom community indicator) were used as risk assessment endpoints. CVM survey respondents were shown a table (see Table 10.1a) in which each of the four land use scenarios was rated from "low" to "high" for each of four stressors (nutrients, sediments, toxins, and flow pattern) and were told that a "high" level posed a "risk to stream integrity." In the Clinch Valley study, where the study of risks relied heavily on IBI, respondents were presented with choice sets in which one of six attributes of the choice was "aquatic life," and the possible levels were "full recovery," "partial recovery," or "continued decline" (see Tables 11.3 and 11.4). The supporting text (see Appendix 11-B) explained that "partial recovery" meant "some improvement" in the Clinch River but not in its tributaries, whereas "full recovery" meant "improvement" in both the Clinch River and its tributaries. Both surveys avoided direct presentation of the indices, using instead qualitative description. While the descriptors for the Darby study could be related back to the results of scenario impact analysis, those for the Clinch were not as easily related to a given physical change.

In the Fox River and Green Bay study, care was taken to use survey language that, although very approximate, was meaningful to both respondents and ecologists (Figure 14.3). Different levels of PCB cleanup were described in units of "years until safe for nearly all fish and wildlife." Wetland habitat supporting fish and wildlife populations was expressed in acres. Eutrophication outcomes were described in units of inches of water clarity, or days with excess amounts of algae, in a specific area of Green Bay.

Where environmental management may have particular objectives — for example, the protection of water quality and stream biological integrity — the results of management actions can affect additional endpoints as well. Therefore, the ecological information set needed for economic analysis may be broader than that envisioned in the ERA (if problem formulation for the ERA did not include consideration of management alternatives). In the Clinch Valley, for example, management actions examined in the economic study included hypothetical policies to compensate farmers for voluntarily restricting agriculture from a riparian buffer area. Besides improvements in the diversity of native mussels and fish, which were the ERA endpoints, the economists expected that such policies would improve sport fishing and enhance the presence of songbirds, which were not included in the ERA. Consideration of songbirds turned out to be unimportant in this case, since respondents did not appear to value them significantly (see Table 11.8), but sport fishing was important. Full analysis of the economic benefits of these policies therefore would have required analysis of sport fish response. Other potential economic benefits of riparian zone restoration may result from enhanced nonavian wildlife habitat, reduced nutrient export, increased sequestration of carbon, and improved value of river-corridor recreation such as canoeing. Had an attempt been made to capture these values as well, additional ecological and economic endpoints only tangentially related to the original management goal would have been required.

Estimating the benefits of a given change in the ecological condition of water bodies requires better procedures. In 1986, Mitchell and Carson[4] reported on a national survey of U.S. households to quantify water quality-related WTP. They used a *water quality ladder* that established progressively increasing use levels (i.e., boating, game fishing, swimming, drinking) for surface waters. These levels equated to points on a cardinal scale determined as a combined index of five conventional water quality parameters: fecal coliforms, dissolved oxygen, biological oxygen demand, turbidity, and acidity (pH). Thus, the benefit of any change in those parameters could be associated with WTP, but the index reflected only a narrow set of pollutants and did not include any direct measurements of stream biological communities.

Since then, substantial progress has been made in the development of state programs for biological monitoring and the use of indices such as IBI and ICI in water quality standards (WQS). These programs have not required a detailed understanding on the public's part of the measures that underlie the indices or a feel for their numerical scales. Work must be done to expand the scientific basis of the water quality ladder to include a broader set of ecological measures, or in some cases to replace exposure (pollutant) measures with response (biological) measures. In addition, the uses, or rungs, that were originally examined need to be expanded to better reflect the full variety of uses that have been designated in state WQS programs[a] as well as other levels of quality to which the public may attribute value. For example, respondents in the Big Darby and Clinch Valley studies probably recognized freshwater mussels as valued components of those aquatic communities, yet a level of water quality sufficient to support game fish (the highest rung of the ladder) may not be sufficient to promote "full recovery" of mussels.

Thus, informed decisions (i.e., ones where decision-makers understand the inherent trade-offs) require techniques that link the kinds of indices ecologists currently measure to values held by the public. Part of the challenge, therefore, is translating indicators into common language.[5] By the same token, ecological measures may require adjustment. For example, if the public values the response of the instream biological community to stream corridor restoration, then ecological measures of the efficacy of such projects should not be limited to modeled changes in water quality parameters. Similarly, if sport fishing and bird watching are among the values the public places on such projects, measuring aquatic community integrity alone is not sufficient. Further, since ecological measurements are highly variable, and model predictions are highly uncertain, methods are needed that enable the public to understand and account for ecological uncertainty, such as when responding to stated preference survey questions.

The Appropriate Tools for Analysis and Comparison of Alternatives Depend on the Decision Context

To weigh management alternatives, analysts should select comparison methods that fit the decision context. If decision criteria are constrained by statute or regulation, the comparison procedure must include any required information and be capable of segregating any precluded information. For example, regulatory impact analyses conducted by the USEPA (see "Water Quality Standards and Economic Analysis" in Chapter 6) may require an analysis in which all costs and benefits are monetized to the greatest extent feasible. By contrast, U.S. Army Corps of Engineers project evaluation procedures maintain separate accounts for changes in the national output of goods and services (expressed in monetary units) and changes in the net quantity and quality of desired ecosystem resources (expressed in physical units).[6] These decision contexts imply particular comparison procedures, whereas in other contexts

[a] Personal communication with John Powers, USEPA, Office of Water (2003).

procedures can be subjective or ad hoc. Other important differences in context may be as follows:

- *One entity* has clear authority to decide vs. *many parties* will negotiate.
- *One decision* will be made affecting a large area vs. *many small decisions* will be made, each affecting only one land parcel, stream segment, or political jurisdiction.
- Decision-makers expect to reach a *decision point* once analysts have presented all information vs. decision-makers expect to examine data, construct alternatives, and engage in an active *decision process*.

To ensure successful integration researchers need to categorize environmental management situations based on the decision context and evaluate the full complement of comparison procedures available, to identify compatibilities between context and procedure. The contexts of these six case studies varied considerably (Table 16.1), and the compatibility of the analytical methods chosen was variable and in many ways is still uncertain (see "Prospects for Success" earlier in this chapter). Flexibility of the analytic tool appears to be a key trait. Flexibility enables restoration scaling and other modifications required by negotiation or adaptive implementation. Conjoint-style models (used in the Clinch, Platte, and Fox River and Green Bay studies) and input–output models (as in the Hudson River study) may be more useful than the less-articulated models typically derived from CVM.

Research is Needed on Transferring the Value of Ecological Endpoint Changes

Environmental management problems tend to be highly unique, complicating the direct transfer of economic findings from one watershed setting to another. The novel methods developed in each of these case studies undoubtedly could be adapted for use in other systems. Given the expense and time of conducting surveys, however, analysts need to understand whether there are dimensions of value that are less variable across systems. The Big Darby results suggest that one might be able to improve the comparability of WTP estimates by using the numerical IBI change and area affected (or perhaps stream miles affected) as normalizing factors. Work in that case is still ongoing to determine whether ecological value can be estimated as a fraction of WTP. The Clinch Valley case study decomposed WTP according to a set of attributes, determining part-worths for each. We might hypothesize that such a partial value, if normalized for magnitude and the extent of stream improvement, would be less variable across situations than would a more bundled estimate. These assumptions require validation, however.

The Role of Ecological Risk Information in the Measurement of Preferences Requires Further Research

When outcomes are uncertain, individuals' preferences reflect their expectations, and expectations depend on beliefs.[7] When individuals know little about an environmental management problem, the information provided in a survey will have an important influence on the construction of beliefs and the statement of preferences.[8]

The purpose of ERA is to develop accurate information about the nature, magnitude, and certainty of adverse effects to ecological resources, given present circumstances and sometimes under different prospective management regimes. The challenge of integration therefore goes beyond determining how to associate preferences with risk outcomes and includes determining the appropriate use of risk information to inform (or even construct) preferences.

The treatment of information and belief in the Clinch Valley and the Fox River and Green Bay studies was the most conventional, in that the mailed questionnaire included some introductory and explanatory text to help respondents understand the conjoint-style choice questions. In addition, it asked questions about respondent age, income, education, and environmental preferences to help characterize the respondent population and determine the factors that underlie preference. The Platte River study similarly employed a mailed survey with an informative preamble and demographic questions, but it took the additional step of asking respondents' agreement or disagreement with a series of statements about the environmental management problem. These were intended to determine not only attitudes but also knowledge, since the statements were considered to have known, correct answers,[a] and responses were used to score respondents' knowledge level (see "Utility of Policy Attributes" in Chapter 12). This information was then used to speculate about the potential effects of better information on negotiation outcomes. By contrast, the Big Darby survey used an in-person presentation approach, with a detailed script and a computer-based slide show including many photographs, to clearly illustrate each of the development scenarios and their anticipated outcomes. Risk assessors have often recognized risk communication as an important field for research and development of practical techniques. The differing approaches used in these surveys highlight the importance of defining best practices and exploring novel techniques for risk communication in survey design and in other stages of decision-making.

The stakeholder-group processes to be used in Dutchess County, which are to combine expert information, group interaction, and the use of a multicriteria preference instrument, may prove to be a useful way to bring together education and the elicitation of preferences. Other work on the ranking of health and ecological hazards found that the use of a process combining individual and group surveys, and holistic and multicriteria ranking methods, tended to inform the participants, yield consistent results, move group members toward agreement, and produce high participant satisfaction ratings.[9]

FINAL WORD

Because watershed boundaries often encompass areas that are ecologically and socially complex, assessment and management of watershed problems can be complex as well. Processes to support watershed decision-making need to be flexible and adaptable to a given context, and multidisciplinary analyses are often required. Differences in methodology between the disciplines, especially between the natural and social

[a] In reality there was some ambiguity about this distinction, since not all of the answers could be clearly established by documentation.

sciences, can complicate the decision-making task, but as these case studies have shown, they also provide fertile ground for the development of unique approaches. The conceptual approach for integration of ERA and economic analysis presented in this book offers a set of principles and procedures that can help ensure that analyses are constructively focused and mutually supportive. It also offers a coherent framework within which other novel, analytical approaches should be explored.

REFERENCES

1. USEPA, Guidelines for ecological risk assessment, EPA/630/R-95/002F, Risk Assessment Forum, U.S. Environmental Protection Agency, Washington, D.C., 1998.
2. USEPA, Framework for ecological risk assessment, EPA/630/R-92/001, Risk Assessment Forum, U. S. Environmental Protection Agency, Washington, D.C., 1992.
3. Varian, H., *Microeconomic Analysis*, W.W. Norton and Company, New York, 1992.
4. Mitchell, R.C. and Carson, R.T., The use of contingent valuation data for benefit/cost analysis in water pollution control, Final Report, EPA Assistance Agreement # CR 810224-02, Resources for the Future, Washington, D.C., 1986.
5. Schiller, A., Hunsaker, C.T., Kane, M.A., et al., Communicating ecological indicators to decision-makers and the public, *Conserv. Ecol.*, 5, 19 [online], 2001. http://www.ecologyandsociety.org/vol5/iss1/art19/index.html. Accessed Oct. 06, 2004.
6. USACE, Planning guidance notebook, ER 1105-2-100, U.S. Army Corps of Engineers, Washington, D.C., 2000.
7. Diamond, P.A. and Hausman, J.A., Contingent value: Is some number better than no number?, *J. Econ. Perspect*, 8, 45, 1994.
8. Gregory, R., Lichtenstein, S., and Slovic, P., Valuing environmental resources: A constructive approach, *J. Risk Uncertain.*, 7, 177, 1993.
9. Willis, H.H., DeKay, M.L., Morgan, M.G., Florig, H.K., and Fischbeck, P.S., Ecological risk ranking: Development and evaluation of a method for improving public participation, *Risk Anal.*, 24, 363, 2004.

Index

A

B

C

D

E

I

K

L

M

N

O

P

V

W